Lecture Notes in Mathematics

Edited by A. Dold and B. Eckmann

642

Theory and Applications of Graphs

Proceedings, Michigan May 11–15, 1976

Edited by
Y. Alavi and D. R. Lick

Springer-Verlag
Berlin Heidelberg New York 1978

Editors
Yousef Alavi
Don R. Lick
Department of Mathematics
Western Michigan University
Kalamazoo, MI 49008/USA

AMS Subject Classifications (1970): 05CXX, 94A20

ISBN 3-540-08666-8 Springer-Verlag Berlin Heidelberg New York
ISBN 0-387-08666-8 Springer-Verlag New York Heidelberg Berlin

Printing and binding: Beltz Offsetdruck, Hemsbach/Bergstr.
2141/3140-543210

DEDICATED TO THE
MEMORY OF

D. R. Fulkerson
Fernando Escalante

PREFACE

This volume constitutes the proceedings of the International Conference on the Theory and Applications of Graphs, held at Western Michigan University in Kalamazoo, May 11-15, in the year of the Bicentennial of the American Independence, Nineteen Hundred Seventy-six. Conference participants included research mathematicians from Colleges, Universities, and Industry, as well as graduate and undergraduate students. In all 27 states and eight countries were represented. The contributions to this Volume* include a great variety of topics in current research in both the theory and applications of graphs, as well as a special Bicentennial histori cal sketch, as the lead article, which is based on Professor Robin Wilson's Banquet presentation.

*J. Bosák was unable to present his paper at the Conference, but his contribution is included as part of the published record.

ACKNOWLEDGEMENTS

For their generous contributions to the success of the International Conference and the preparation of these proceedings, the editors gratefully acknowledge

The financial support of Western Michigan University;

The financial support of U. S. Army Research Office-Durham;

The overall support of the Department of Mathematics and Professor A. Bruce Clarke, Chairman;

The assistance of the WMU Office of Research Services;

The outstanding contributions of our Co-Directors of the Conference, Professors:

Gary Chartrand
S. F. Kapoor
Mark Jungerman
Arthur T. White

The fine administrative assistance of Mrs. Darlene Lard;

The excellent work of our Conference Assistant Directors:

James M. Benedict	Brian Garman
John F. Fink	Ronald Gould
Annette Friedhoff	Joseph Straight

The dedicated work of our team of Conference Assistants:

Vithal N. Bhave	Marie Smith
Nancy Boynton	Susan Steffens
David Burns	Sui-Lung Tang;

The general and Secretarial assistance of Stacy George, Darlene Lard, and Margo Johnson.

And, most definitely

the outstanding work of Mrs. Margo Johnson for her skill ful typing of the manuscripts as well as her fine assistance in numerous chores.

Y.A.
D.R.L.

LIST OF THE PARTICIPANTS IN

THE INTERNATIONAL CONFERENCE

ON THE

THEORY AND APPLICATIONS OF GRAPHS

Yousef Alavi *
Michael Albertson
David K. Baxter
Pierre Bayle
Thomas Beasley
Mehdi Behzad
Lowell Beineke
James Benedict **
Claude Berge
David M. Berman
Arthur Bernhart
Frank R. Bernhart
Vithal Bhave ***
N. L. Biggs
Gary S. Bloom
Paul T. Boggs
Béla Bollobás
Juraj Bosák
André Bouchet
Nancy Boynton ***
David Burns ***
Michael Capobianco
Paul A. Catlin
Gary Chartrand *
Chang-Yun Chao
Phyllis Chinn
V. Chvátal
E. J. Cockayne
L. J. Cummings
Anastasia Czerniakiewicz
William H. E. Day
Richard T. Denman
John K. Doyle
Richard A. Duke
Roger B. Eggleton
Thomas Eitner
Vance Faber
Ralph Faudree
John Fink **
J. Sutherland Frame
Annette Friedhoff **
Marianne L. Gardner
Brian L. Garman **
Cyril W. L. Garner
Henry Glover
Donald Goldsmith
Martin Golumbic
Ronald Gould **
R. L. Graham
Donald L. Greenwell
Charles M. Grinstead
Jonathan L. Gross

Ram P. Gupta

Gary Haggard

S. Louis Hakimi

Frank Harary

Jehuda Hartman

Clare Heidema

Arthur M. Hobbs

Derek A. Holton

Charlotte Huang

Joan P. Hutchinson

Derbiau Hsu

Judy Johnson

Mark Jungerman *

Paul Kainen

S. F. Kapoor *

John W. Kennedy

Hudson Kronk

John LeFever

Rochelle Leibowitz

Linda M. Lesniak-Foster

Hank Levinson

Roy B. Levow

Don R. Lick *

Judith Q. Longyear

Paul J. McCarthy

Joseph Malkevitch

Ben Manvel

David Matula

John S. Maybee

Zevi Miller

John Mitchem

John C. Molluzzo

J. W. Moon

V. Kumar Murty

Ladislav Nebeský

Edward A. Nordhaus

Phillip Ostrand

Edgar M. Palmer

Torrence D. Parsons

Raymond E. Pippert

Albert D. Polimeni

Geert Prins

Louis V. Quintas

K. Brooks Reid

Bruce Richmond

Richard D. Ringeisen

Gerhard Ringel

Fred Roberts

John Roberts

Neil Robertson

Michael Rolle

David Roselle

Richard H. Schelp

Seymour Schuster

Allen J. Schwenk

Herbert Shank

Gustavus J. Simmons

William Simpson

Marie Smith ***

Saul Stahl

Susan Steffens ***

M. James Stewart

Paul K. Stockmeyer

Joseph Straight **

David Sumner

Siu-Lung Tang ***

J. Dalton Tarwater

Carsten Thomassen

William T. Trotter, Jr.

Alan Tucker

William T. Tutte

Donald W. Vanderjagt

Paul Vincent

Curtiss E. Wall

Mark E. Watkins

Arthur T. White *

E. G. Whitehead, Jr.

Susan Whitesides

Eva Williams

James Williamson

Robin J. Wilson

D. R. Woodall

Daniel Younger

Conference - Co-Director *
Conference - Assistant Director **
Conference - Assistant ***

TABLE OF CONTENTS*

* On joint papers the speaker's name is listed first.

200 YEARS OF GRAPH THEORY - A GUIDED TOUR*

Robin J. Wilson
The Open University
London, ENGLAND

In this paper we shall survey some of the most significant developments in graph theory over the period 1736-1936, from the first paper on the subject (Euler's treatment of the Königsberg bridge problem) to the first book (König's Theorie der endliche und unendliche Graphen [17]). In particular, we shall attempt to set the record straight on various myths which have grown up over the years. Our treatment will of necessity be very brief - for fuller details on all these topics, the reader is referred to the recent book of Biggs, Lloyd and Wilson [2].

1. Euler and the Königsberg bridges.

It is generally agreed that the first paper to incorporate graph-theoretical ideas was Euler's paper on the Königsberg bridge problem [10]. In this paper, Euler proved the impossibility of traversing the seven bridges of Königsberg exactly once, and he described a method for determining whether the corresponding problem could be solved for a general arrangement of regions and bridges. Although his method is correct, he did not prove that his necessary condition for a walk to be possible (namely, that there should be exactly zero, or two, regions with an odd number of bridges emanating from them) is also sufficient - he simply outlined a general method for obtaining a proof, saying that the details could be filled in 'after a little thought'.

* An illustrated version of this talk was presented at the Conference Banquet.

A correct proof did not appear for over 130 years, when Hierholzer, who was unaware of Euler's work, supplied the missing details in a posthumous paper in 1873 [14].

In fact, there is little evidence to suggest that Euler's paper was well known. It had been mentioned two or three times in the early nineteenth century, but was unfamiliar to most mathematicians. This situation was remedied in 1851 when Coupy [8] published a French translation of Euler's paper, and applied Euler's methods to the corresponding problem of the bridges over the River Seine. In 1876, Saalschütz [27] published a short note announcing that an extra bridge had been built, thereby making the problem possible, and he enumerated all the possible solutions.

A closely-related type of problem was studied by Poinsot, Clausen and Listing, namely the problem of how many separate pen-strokes are needed to draw a given diagram, without repeating any line or lifting one's pen from the paper. Clausen's paper [7] was written in Latin and concerned a problem in astronomy, but ended with two paragraphs (in German!) asserting that the diagram in Figure 1 needs at least four separate pen-strokes.

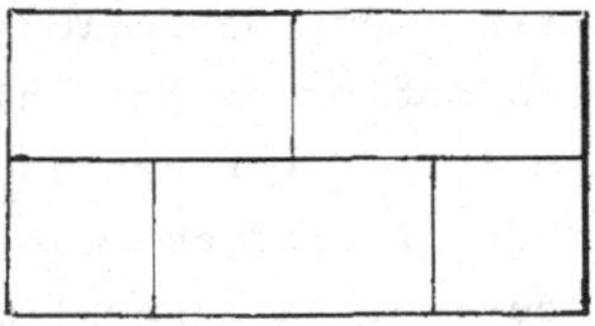

Figure 1

This diagram was also discussed by Listing [21] in his classic paper 'Vorstudien zur Topologie' - incidentally, the first time the word 'topology' had ever been used. Listing's paper dealt with those geometrical problems which depended only on positional relationships rather than metrical ones, and he included sections on screws, knots, links, and diagram-tracing puzzles of the kind just described. It is unfortunate that Listing's paper is so little known.

There were, of course, other types of problems which are closely related to the ones just described. Such problems included the domino problem and the tracing of mazes and labyrinths, but there is no space to describe them here. A fuller treatment of these problems may be found in the recreational books of Rouse Ball [26], Lucas [23] and Ahrens [1], or in [2].

2. '<u>Hamiltonian</u>' <u>Graphs</u>.

One of the earliest examples of a Hamiltonian-type problem was the knight's-tour problem - can a knight visit all the squares of an 8 x 8 chessboard exactly once, and return to where it started from? This problem was solved by Euler [12] in 1759, and the corresponding problem for a 5 x 5 chessboard was proved impossible. Sixteen years later, Vandermonde wrote a beautiful paper [30] in which he described a systematic method for obtaining a knight's tour; his paper also included a mathematical discussion (with diagrams) of plaits and stocking-stitch.

'Hamiltonian' graphs were first considered in general by the Rev. T. P. Kirkman, an enthusiastic amateur mathematician who was later elected a Fellow of the Royal Society. Kirkman [16] considered the problem of when one could find a circuit passing through all the vertices of a polyhedron, and he gave a necessary and sufficient condition (incorrect!) for such a circuit to exist. He also observed that 'if we cut in two the cell of the bee' we obtain a graph which contains no such circuit (see Figure 2). Kirkman's papers are difficult to read due to the fact that he invented his own terminology and notation; for example, a polyhedron with p faces and q vertices would be called a 'p-edral q-acron'!

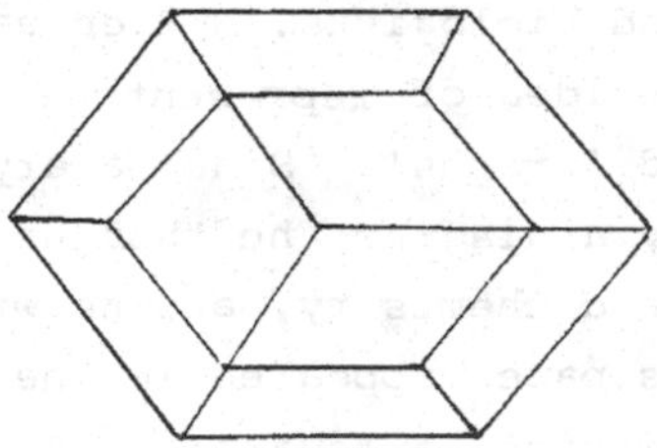

A few months later, as a consequence of his work on quaternions (or the 'Icosian calculus' as he called it), Hamilton became interested in the existence of circuits passing through all the vertices of a dodecahedron. He later marketed a game, called the 'Icosian Game' or

'A Voyage round the World' which consisted of a flat or solid dodecahedron with twenty pegs - the idea was to attach the pegs to the vertices of the dodecahedron in such a way that one traces out Hamiltonian circuits satisfying certain given conditions. Fuller details of this game, together with Hamilton's instructions for playing it, may be found in [2].

Although priority for the study of Hamiltonian circuits should properly be given to Kirkman, there is little doubt that it was due to Hamilton's influence, and the general interest in his Icosian Game, that these circuits came to be known as 'Hamiltonian'.

3. Trees, and the Origin of the term 'Graph'.

In 1857 Cayley wrote an important paper [4] in which he introduced the term 'tree', and established a formula which could be used to enumerate rooted trees with a given number of branches. It was not until several years later that he was able to solve the more difficult problem of enumerating unrooted trees. He presented this latter paper [5] to the British Association in 1875, and applied his result to the enumeration of chemical molecules, such as the paraffins C_nH_{2n+2}. Cayley's interest in chemistry had been aroused by this enumeration problem, and he wrote several papers on the subject.

Cayley's friend Sylvester also became interested in chemistry, but for a different reason. Sylvester was interested in invariant theory and the study of binary quantics, and was trying to find a way of conveying the flavor of the subject to 'a mixed society consisting of physicists, chemists and biologists.' Eventually (at 3 o'clock one morning) he hit on the idea of representing these invariants by diagrams - which he called 'graphs'. Since every chemical molecule can also be represented by a diagram, he thought he saw a close link between invariant theory and chemistry, and he wrote a lengthy paper [28] on the subject. This paper appeared in the first volume of the American Journal of Mathematics, a journal founded by Sylvester shortly after his arrival as first mathematics professor at Johns Hopkins University. Incidentally, it was in this paper, and in the abstract of it which appeared in Nature, that the term 'graph' was used for the first time. It is unclear whether the choice of the word 'graph' was due to Sylvester, or to his friend W. K. Clifford, who was working along similar lines, but the term has certainly far outlived the 'chemico-algebraic theory' which gave rise to it.

4. Euler's Polyhedral Formula.

Euler's polyhedral formula $(v + f = e + 2)$ was first mentioned in a letter from Euler to Goldbach in 1750. Euler was interested in the various relationships which exist between the numbers of vertices, edges, faces and angles of a polyhedron, and he stated his formula, confessing that it was too difficult for him to prove. Two years later he published a proof [11] using a dissection method; although ingenious, this proof later turned out to be deficient.

It has sometimes been asserted that Descartes discovered a proof of the polyhedral formula around 1640, but there is little evidence to support this claim. In one of his papers Descartes found an expression for the sum of the angles of all the faces of a polyhedron, from which the desired result can easily be deduced, but Descartes never actually made the deduction. Many years later, in 1794, Legendre [19] used an approach similar to Descartes', and successfully completed the proof; however unlike Descartes, he had the advantage of knowing what he wanted to prove.

A few years later, in 1811-1813, significant contributions to the theory of polyhedra were made by Cauchy and Lhuilier. Cauchy [3] showed how one could obtain a planar graph by projecting the polyhedron onto a plane, and he obtained the corresponding formula relating the numbers of vertices, edges and faces of a planar graph. Lhuilier [20] gave various examples of polyhedra for which the formula fails to hold; these included polyhedra containing a hole, and polyhedra not all of whose faces are simply-connected. He also obtained the corresponding formula $(v + f = e + 2 - 2g)$ for a closed surface of genus g . This last formula was the starting point for an extensive study of higher surfaces by Listing [22]. Listing extended the polyhedron formula to 'abstract complexes', and thereby set the scene for the work of Poincaré and others on the foundations of algebraic topology. It is of interest to note that Listing's paper includes a description of the Möbius strip; it was rediscovered by Möbius a short time later.

5. Non-Planar Graphs.

Around the year 1840, Möbius included the following problem in one of his lectures: 'There was once a king with five sons. In his will he stipulated that after his death his kingdom was to be divided into five regions in such a way that each of the sons' regions had a boundary line in common with the boundaries of each of the other regions. Is such a division of land possible? This problem was later elaborated by Tietze who stated that the sons were also required to build a palace in each region, and that these palaces were to be connected by roads which did not intersect. In either case, the impossibility of finding a solution follows from the fact that the complete graph K_5 is non-planar.

It is clear that if five mutually neighbouring regions had existed, then the four-color conjecture would have been false, but the two problems are not equivalent; one cannot deduce the truth of the four-color conjecture from the fact that K_5 is non-planar. These two problems were later confused by Isabel Maddison [24] who consequently attributed the four-color problem to Möbius. (The true origins of the four-color problem will be described in the next section.) This error was repeated by Rouse Ball [26] and others, and quickly became part of the folk-lore of the subject. It was not until 1959 that the record was put straight, by Coxeter [9].

The origins of the corresponding problem for the complete bipartite graph $K_{3,3}$ are unknown. Sam Loyd Jr. states that his father 'brought out' the puzzle around 1900, but its origins are certainly older than that. The problem became known as the 'Utilities Problem' after Dudeney included it in his book Amusements in Mathematics under the title 'Water, gas and electricity'.

The importance of the two graphs K_5 and $K_{3,3}$ arises from the fact that every non-planar graph contains a subgraph homeomorphic to one of them, as was proved by Kuratowski [18]. It is less well known that essentially the same proof was found a few months later by Frink and Smith, but by the time their paper came to be refereed Kuratowski's proof had just been published. Their paper was consequently rejected.

6. The Four-Color Problem.

The four-color problem can be traced back to a letter sent by De Morgan to Hamilton in October 1852. In this letter he described how a student of his had asked him whether four colors were sufficient for coloring any map; the student, Frederick Guthire, had learned the problem from his brother Francis who had discovered it while coloring a map of England. So Francis Guthire was the originator of the four-color problem. Hamilton replied to De Morgan that he was 'not interested in your quaternion of colors'.

The next reference to the four-color problem occurs in an unpublished manuscript of C. S. Peirce at Harvard. Peirce claims to have proved the conjecture and to have presented his proof at a Harvard seminar in the 1860's. Peirce states that he learned about the problem from a note by De Morgan in the Athenaeum around 1860, but we have been unable to locate the exact reference.

The problem obtained prominence when Cayley asked at a London Mathematical Society meeting in June 1878 whether the problem had ever been solved. Shortly afterwards, Cayley wrote a short note [6] explaining where the difficulty lay, reducing the problem to trivalent maps, and saying that he was unable to solve it. Kempe's paper [15] containing his famous (fallacious) proof appeared soon afterwards, and he followed this with two simplified (fallacious) proofs. Tait [29] also entered the fray and gave two or three fallacious proofs of his own, as well as rephrasing the four-color conjecture in terms of edge-colorings. There is no reason to doubt that any of these proofs were completely accepted by mathematicians of the day. The problem also became known in recreational circles, and was mentioned by Lewis Carroll, and solved (incorrectly!) by the Bishop of London, Frederick Temple.

The error in Kempe's paper was eventually noticed by P. J. Heawood, who wrote an important paper [13] in which he managed to salvage enough of Kempe's argument to prove the five-color theorem. But Heawood also made an error. In trying to generalize coloring problems to higher surfaces, he gave a formula for the number of colors needed for a surface of genus g. But his proof yielded the formula only as an upper bound, and he did not prove that this bound is always attained; the gap in the proof was eventually filled, after a long and difficult struggle, by Ringel and Youngs in 1967 (see [25]).

REFERENCES

Papers indicated with an asterisk have been translated (where necessary) and are included as extracts in the book by Biggs, Lloyd and Wilson [2].

1. Ahrens, W. Mathematische Unterhaltungen und Spiele, Leipzig, 1901.
2. Biggs, N. L., Lloyd, E. K. and Wilson, R. J., Graph Theory 1736-1936, Oxford Univ. Press, 1976.

3*. Cauchy, A. - L., Recherches sur les polyèdres, J. Ecole Polytech. 9 (Cah. 16) (1813), 68-86 = Oeuvres (2), Vol. 1, 7-25.

4*. Cayley, A., On the theory of the analytical forms called trees, Phil. Mag. (4) 13 (1857), 172-176 = Math. Papers. Vol 3, 242-246.

5. Cayley, A., On the analytical forms called trees, with application to the theory of chemical combinations, Rep. Brit. Assoc. Advance. Sci. 45 (1875), 257-305 = Math. Papers, Vol. 9, 427-460.

6*. Cayley, A., On the colouring of maps, Proc. Roy. Geog. Soc. 1 (1879), 259-261 = Math. Papers, Vol. 11, 7-8.

7. Clausen, T., De linearum tertii ordinis proprietatibus, Astron. Nachr. 21 (1844), col. 209-216.

8. Coupy, E., Solution d'un problème appartenant a la géométrie de situation, par Euler, Nouv. Ann. Math. 10 (1851), 106-119.

9. Coxeter, H. S. M., The four-color map problem, Math. Teacher 52 (1959), 283-289.

10*. Euler, L. Solutio problematis ad geometriam situs pertinentis. Comm. Acad. Sci. Imp. Petropol. 8 (1736), 128-140 = Opera Omnia (1), Vol. 7, 1-10.

11. Euler, L. Demonstratio nonnullarum insignium proprietatum quibus solida hedris planis inclusa sunt praedita, Novi. Comm. Acad. Sci. Imp. Petropol. 4 (1752-3), 140-160 = Opera Omnia (1), Vol. 26, 94-108.

12. Euler, L., Solution d'une question curieuse qui ne paroit soumise à aucune analyse, Mém. Acad. Sci. Berlin 15 (1759), 310-337 = Opera Omnia (1), Vol. 7, 26-56.

13*. Heawood, P. J., Map-colour theorem, Quart. J. Pure. Appl. Math. 24 (1890), 332-338.

14*. Hierholzer, C., Ueber die Möglichkeit, einen Linienzug ohne Wiederholung und ohne Unterbrechnung zu umfahren, Math. Ann. 6 (1873), 30-32.

15*. Kempe, A. B., On the geographical problem of the four colours, Amer. J. Math. 2 (1879), 193-200.

16*. Kirkman, T. P., On the representation of polyedra, Phil. Trans. Roy. Soc. London 146 (1856), 413-418.

17. König, D., Theorie der endlichen und unendlichen Graphen, Akademische VerlagsgeseUschaft Leipzig, 1936. Reprinted by Chelsea, New York, 1950.

18*. Kuratowski, K. Sur le problème des courbes gauches en topologie, Fund. Math. 15 (1930), 271-283.

19. Legendre, A. M., Eléments de géométrie, Firmin Didot, Paris, 1794.

20*. Lhuilier S., (abridged by J. D. Gergonne). Mémoire sur la polyédrométrie, Ann. Math. 3 (1812-13), 169-189.

21*. Listing, J. B., Vorstudien zur Topologie, Göttinger Studien 1 (1847), 811-875.

22. Listing, J. B., Der Census räumlicher Complexe oder Verallgemeinerung des Euler'schen Satzes von den Polyëdern, Abh. K. Ges. Wiss Göttingen Math. Cl. 10 (1861-2), 97-182.

23. Lucas, E., Récréations mathématiques, 4 vols. Gauthier-Villars, Paris, 1882-94.

24. Maddison, I., Note on the history of the map-coloring problem, Bull. Amer. Math. Soc. 3 (1896-7), 257.

25. Ringel, G., Map color theorem, Springer-Verlag, Berlin, 1974.

26. Rouse Ball, W. W., Mathematical recreations and essays (11th edn.), Macmillan, London, 1939.

27. Saalschütz, L., [Untitled], Schr. Phys. - Ökon. Ges. Königsberg Prussia 16 (1876), 23-24.

28. Sylvester, J. J., On an application of the new atomic theory to the graphical representation of the invariants and covariants of binary quantics, Amer. J. Math. 1 (1878), 64-125 = Math. Papers, Vol. 3, 148-206.

29*. Tait, P. G., On the colouring of maps. Proc. Roy. Soc. Edinburgh 10 (1878-80), 501-503, 729.

30*. Vandermonde, A.-T., Remarques sur les problèmes de situation, Hist. Acad. Sci. (Paris) (1771), 566-674.

CHROMATIC NUMBER AND SUBGRAPHS OF CAYLEY GRAPHS

László Babai
Eötvös L. University
Budapest, HUNGARY

Abstract

Results, problems and conjectures concerning the chromatic number and subgraphs of finite and infinite Cayley graphs are presented. We prove that every group has a Cayley graph of chromatic number $\leq \omega$; for solvable groups the minimum chromatic number is ≤ 3. The Cayley graph of an irreducible group presentation contains neither $K_{3,5}$ nor K_4^- as a subgraph. Every group has a Cayley graph containing neither K_4^- nor $K_{5,17}$. We mention that every graph is a induced subgraph of some Cayley graph of any sufficiently large group.

1. Introduction. The Cayley graph $X = X(G, H)$ of the group G with respect to the generating set H is defined by

$$V(X) = G,$$
$$E(X) = \{[g, hg] : g \in G, h \in H\}.$$

It is almost a century since Cayley graphs have been used to visualize group structure, and Burnside and Maschke have already investigated their imbeddability on surfaces. By now, a powerful theory of imbeddings of Cayley graphs is developed with several applications (see COXETER, MOSER [7] and A. T. WHITE [23] for details). Some other problems, such as that of the connectivity and the 1-factors, are settled generally for graphs with transitive automorphism groups (the connectivity by WATKINS [21], MADER [17, 18] and JUNG

[13]; the 1-factors by LITTLE, GRANT and HOLTON [15]; LOVÁSZ has shown that vertex-transitive graphs with an even number of vertices are even <u>bicritical</u> (personal communication). For both problems, see Chapter 12 of the excellent book [16] by L. LOVÁSZ.)

Almost nothing seems to be known, however, about the relationship of Cayley graphs to a number of classical graph theoretic notions, such as chromatic number, Hamilton paths, containment of certain subgraphs, etc. The present note is by no means intended to provide a systematic treatment of the subject. It is the author's hope to stimulate research in this area by presenting a few results and mentioning numerous problems and conjectures.

Concerning Hamilton circuits, we must rest satisfied with conjecturing that each Cayley graph has a Hamilton circuit, and each connected graph with a transitive automorphism group has a Hamilton path. The former can be easily proven for abelian groups, as remarked by J. PELIKÁN.

In this paper, we investigate the minimum chromatic number of Cayley graphs, and the problem as to whether a subgraph necessarily occurs in any Cayley graph of a group.

Acknowledgement: the author is indebted to P. Frankl for his valuable suggestions.

2. Preliminaries.

We deal with undirected graphs without loops or multiple edges. In order that the Cayley graph $X(G, H)$ be loopless, we always suppose $1 \notin H$. (1 denotes the unit of G.) For basic information on Cayley graphs see [19, 22]. We let $\operatorname{ch} X$ stand for the chromatic number of the graph X; K_m denotes the complete m-graph and $K_{m,n}$ the complete (m, n) bipartite graph.

Throughout, G denotes a (finite or infinite) group. For $H \subseteq G$, H is an <u>irreducible generating set</u>, if $\langle H \setminus \{h\} \rangle \neq G$ for any $h \in H$.

H is a *quasi-irreducible generating set*, if $\langle H \rangle = G$ and H can be well-ordered such that no member of H is contained in the subgroup generated by its predecessors.

THEOREM 2.1. *If H is a quasi-irreducible generating set of G then* $\mathrm{ch}\, X(G, H)$ *is at most countable.*

The proof follows in Section 3.

CONJECTURE 2.2. *There is a finite m such that for H a quasi-irreducible (irreducible, resp.) generating set of G,* $\mathrm{ch}\, X(G, H) \geq m$.

We shall give a stronger version of this conjecture in purely graph theoretic terms (3.5).

Note that:

PROPOSITION 2.3. *Every group has a quasi-irreducible generating set.*

Proof. Let "<" denote a well-order relation on G and set

$$H = \{g\colon g \notin \langle h; h < g \rangle\} .$$

Obviously, H is a quasi-irreducible generating set.

In section 4, we shall prove

THEOREM 2.4. *Every (finite or infinite) solvable group has a Cayley graph whose chromatic number is at most 3.**

Let us remark that Conjecture 2.2 seems to be unsolved even for finite cyclic groups. It holds trivially, with $m = 3$, for elementary abelian groups. N. Biggs has asked whether (contrary to 2.2), for such complicated groups G as the (non-cyclic) simple ones,

---

* Independently, for finite groups, this has been proven by Saul Stahl (personal communication by A. T. White.)

ch $X(G, H)$ is necessarily large if H is a sufficiently large irreducible generating set. The answer is no: for n odd, A_{3n} has an irreducible set of $n + 1$ generators such that the resulting Cayley graph is at most 9-chromatic. (The generators are the cycles $(1, 2, 3), (4, 5, 6), \ldots, (3n - 2, 3n - 1, 3n)$ and $(3, 6, 9, \ldots, 3n)$.)

3. The proof of Theorem 2.1 and a graph theoretic conjecture.

DEFINITION 3.1. A graph X is called a _no pied circuit_ graph if its edges can be colored such that

(i) each vertex is incident with at most two edges of any color;

(ii) each circuit contains at least two edges of the same color.

If we replace (ii) by

(iii) for any edge of any circuit there is another edge of the same color in that circuit, we call X a _no lonely color_ graph.

PROPOSITION 3.2. _For_ H _a quasi irreducible generating set of_ G, $X(G, H)$ _is a_ no pied circuit _graph_. _If_ H _is irreducible then_ $X(G, H)$ _is a_ no lonely color _graph_.

Proof. We may assume that if $h, h^{-1} \in H$ then $h^2 = 1$. Let us take H as the set of colors.

We assign color h to the edge $[g, hg]$ $(g \in G, h \in H)$. The edges of color h incident with g are $[g, hg]$ and $[g, h^{-1}g]$ (they coincide iff $h^2 = 1$), hence (i) is satisfied. A circuit having colors $h_1, h_2, \ldots, h_k$ (in this order) means $h_1^{\pm 1} h_2^{\pm 1} \cdots h_k^{\pm 1} = 1$. Hence any _lonely_ color can be expressed by the rest of the colors in this circuit. Thus if H is irreducible (iii) is satisfied. If H is quasi-irreducible and h_j is the greatest among $h_1, \ldots, h_k$ in the corresponding well-ordering of

H, then, by definition, h_j cannot be expressed by other colors of this circuit; hence $h_i = h_j$ for some $i \neq j$, $1 \leq i \leq k$, and (ii) is satisfied.

LEMMA 3.3. <u>A</u> no pied circuit <u>graph does not contain any</u> $K_{5,17}$ <u>subgraph</u>. <u>A</u> no lonely color <u>graph does not contain any</u> $K_{3,5}$.

Proof. Assume, to the contrary, that $K_{5,17}$ is a <u>no pied circuit</u> graph. Let $V(K_{5,17}) = A \cup B$, $|A| = 5$, $|B| = 17$, all edges going between A and B. Let $b \in B$. By 3.1 (i), there are at least 3 members a_1, a_2, a_3 of A such that the $[b, a_i]$ $(i = 1, 2, 3)$ have different colors. Let γ_i denote the color of $[b, a_i]$. By 3.1 (i) there are at most 5 edges joining a_i to $B \setminus \{b\}$ whose color belongs to $\{\gamma_1, \gamma_2, \gamma_3\}$; hence there is a vertex $b' \in B \setminus \{b\}$ such that none of the edges $[b, a_i]$ has any of the colors $\gamma_1, \gamma_2, \gamma_3$. On the other hand, there are at least two different ones among the colors of $[b', a_i]$, say $[b', a_1]$ and $[b', a_2]$ have different colors. But now, the circuit $b - a_1 - b' - a_2$ is <u>pied</u> (has 4 different colors), a contradiction.

The proof of the other statement of the lemma goes similarly; we omit it.

Observe that we used condition 3.1 (ii) (and (iii), resp.) for 4-circuits only.

Theorem 2.1 follows directly from Proposition 3.2, Lemma 3.3 and the following theorem of ERDŐS and HAJNAL:

THEOREM 3.1 (P ERDŐS and A. HAJNAL [8, Cor.5.6, p. 72]). <u>If the chromatic number of a graph</u> X <u>is at least</u> ω_1 <u>then</u> X <u>contains a</u> K_{m,ω_1} <u>for any finite</u> m.

Now we formulate a stronger version of Conjecture 2.2 which has the advantage for graph theorists that no groups occur in it.

CONJECTURE 3.5. There is a finite m such that the chromatic number of any no pied circuit graph (no lonely color graph, resp.) does not exceed m.

Of course, this would not be true if we assume 3.1 (ii) ((iii), resp.) for circuits of bounded lengths only, as there are finite graphs of arbitrarily large girth and chromatic number (P. ERDŐS, see [9, p. 54]).

4. Minimum chromatic number of Cayley graphs.

It is natural to define graph theoretic characteristics of a group as the extremum of the corresponding numbers for its Cayley-graphs (cf. [23, p. 69] for the definition of the genus of a group).

For a group G, let

$$\operatorname{ch} G = \min \{\operatorname{ch} X(G, H) : H \text{ generates } G\} .$$

Thus Theorem 2.4 asserts:

THEOREM 4.1. For G a solvable group, $\operatorname{ch} G \leq 3$.

LEMMA 4.2. For $N \lhd G$, $N \neq G$ we have

$$\operatorname{ch} G \leq \operatorname{ch}(G/N) .$$

Proof. Let $\varphi: G \to G/N$ denote the natural epimorphism and let $X = X(G/N, H)$ be a Cayley-graph $(1 \notin H)$ with $\operatorname{ch} X = \operatorname{ch}(G/N)$. Let $Y = X(G, \varphi^{-1}(H))$. Clearly, $\varphi^{-1}(H)$ generates G. We assert that $\operatorname{ch} Y \leq \operatorname{ch} X$. For this observe that $[x, y] \in E(Y)$ iff $[\varphi(x), \varphi(y)] \in E(X)$; thus the coloring of X can be lifted to an appropriate coloring of Y.

Since the n-cycle is a Cayley-graph of the cyclic group Z_n, Lemma 4.2 makes Theorem 4.1 trivial for finite groups.

CONJECTURE 4.3. *For any finite group,* $\mathrm{ch}\, G \le 4$. (Probably, even $\mathrm{ch}\, G \le 3$ holds.)

Lemma 4.2 implies that it suffices to prove the conjecture for finite simple groups.

PROPOSITION 4.4 *If every finite simple group can be generated by* 2 *of its members* (as conjectured by group theorists, cf. [10]) *then* 4.3 *is true. If one of the generators can always be chosen to be an involution than even* $\mathrm{ch}\, G \le 3$ *holds for any finite group* G .

Proof. By 4.2 it suffices to prove that $\mathrm{ch}\, G \le 4$ ($\mathrm{ch}\, G \le 3$, resp.) for G simple. If $H = \{\alpha , \beta\}$ then $X(G , H)$ has degree 4(3, if α is an involution). Hence the proposition follows from Brooks' theorem, since, as H is irreducible, $X(G , H)$ does not contain K_4 (nor K_4^- , by 5.3 and 5.4), and a triangle in the second case would mean $\beta^3 = 1$ (5.3, 5.4). In this latter case ($\alpha^2 = \beta^3 = 1$) let $N = (\alpha , \beta^{-1}\alpha\beta , \beta^{-2}\alpha\beta^2)$.

Clearly, $N \lhd G$, hence (G being simple) $N = G$. Thus G is generated by 3 involutions. The Cayley graph $X(G , \{\alpha , \beta^{-1}\alpha\beta , \beta^{-2}\alpha\beta^2\})$ has again degree 3 and it does not contain any triangle.

PROPOSITION 4.5. $\mathrm{ch}\, S_n = 2$ $(n \ge 2)$, *and* $\mathrm{ch}\, A_n = 3$ $(n \ge 3)$.

Proof. $\mathrm{ch}\, A_n \le 3$ follows, as in the above proof, from the observation that A_n can be generated by an involution and another permutation (for $n \ge 4$; $A_3 = Z_3$) . $\mathrm{ch}\, S_n = 2$ and $\mathrm{ch}\, A_n \ne 2$ both follow from the next observation:

PROPOSITION 4.6. *For a* (finite or infinite) *group* G , $\mathrm{ch}\, G = 2$ *iff* G *has a subgroup of index* 2 .

Proof. Let $|G:N| = 2$; then by 4.2 we have

$$\text{ch } G \leq \text{ch } (G/N) = \text{ch } Z_2 = 2 .$$

Conversely, if $X(G, H)$ is a connected bipartite graph then the members having the same color as the unit form a subgroup of index 2.

We mention a corollary of 4.2 and 4.5.

COROLLARY 4.7. _If_ G _is the automorphism group of a finite planar graph, then_ $\text{ch } G \leq 3$.

Proof. By [3, Theorem 9.3, p. 72], the factors of the composition series of G are cyclic and alternating groups. Now we may apply 4.2 and 4.5.

The proof of 4.1 will be based on the following:

LEMMA 4.8. $\text{ch}(Z_{p\infty}) = 3$.

($Z_{p\infty}$ denotes the additive group of the fractions a/p^k mod 1 ($a \in Z$, $k \in Z$, $k \geq 0$) for p a prime. Any proper subgroup of $Z_{p\infty}$ is finite. This implies by 4.6 that $\text{ch}(Z_{p\infty}) \geq 3$.)

Proof. Let $q = 4$ for $p = 2$ and $q = p$ otherwise. Let $H = \{1/q^k\colon\ k = 1, 2, \ldots\}$. We assert that $\text{ch } X(Z_{p\infty}, H) = 3$. The colors will be called $\underline{0}$, $\underline{1}$ and $\underline{2}$.

Let us fix a 3-coloring $a_0, a_1, \ldots, a_{q-1}$ of a $(q-1)$ - path such that $a_0 = a_{q-1} = \underline{0}$ (and $a_i \neq a_{i+1}$, of course). Let $H_n = \{1/q^k\colon\ k = 1, \ldots, n\}$, $G_n = \langle H_n \rangle$. We shall successively 3-color the graphs $X_n = X(G_n, H_n)$; the color of x will be denoted by $c(x)$.

X_1 is a q-circuit; we 3-color it arbitrarily.

Let X_{n-1} be 3-colored. Let $x \in G_n \setminus G_{n-1}$; thus

$$x = y + s/q^k \text{ for some } s, \quad 1 \leq s \leq q-1 \text{ and } y \in G_{n-1} .$$

(s and y are unique.)

Set

$$c(x) \equiv c(y) + a_s \qquad (\text{mod } 3) .$$

Assuming that c is a good coloring of X_{n-1}, $a_0 = a_{q-1} = \underline{0}$ and $a_i \neq a_{i+1}$ guarantee that $c(x) \neq c(x')$ if $x \equiv x' + 1/q^n \pmod 1$ $(x, x' \in V(X_n))$. If $x \equiv x' + 1/q^k \pmod 1$ for some k, $1 \le k < n$, then $c(x) - c(x') \equiv c(y) = c(y') \pmod 3$; hence this is $\not\equiv 0$ as y and y' are neighbors in X_{n-1}. Thus, by induction, c is a good 3-coloring of $X(Z_{p^\infty}, H) = \bigcup_{n=1}^{\infty} X_n$.

Now we can turn to the proof of Theorem 4.1. As G is solvable, $G/G' = G_1$ is an abelian group, $|G_1| \ge 2$. Then G_1/pG_1 is an elementary abelian p-group for any prime p; thus it has Z_p as a factor group except if $G_1 = pG_1$. Thus either G has Z_p as a factor group for some p and $\operatorname{ch} G \le 3$ follows from 4.2, or G_1 is a divisible group, and hence the direct sum of copies of Z_{p^∞} $(p = 2, 3, 5, \ldots)$ and Q (the additive group of the rationals) (cf. FUCHS [11, Thm. 23.1]). Now $Q/Z \cong \sum_p Z_{p^\infty}$; hence in any of these cases, G_1 (and G) has some Z_{p^∞} as a factor group, proving $\operatorname{ch} G \le 3$ by 4.8 and 4.2.

5. Subgraphs of Cayley graphs

We start this section with proposing

PROBLEM 5.1. Which graphs Y are avoidable in the sense that every group has a Cayley graph not containing Y as a subgraph?

By 3.2 and 3.3, $K_{5,17}$ is avoidable (and, for finite groups, even $K_{3,5}$ is avoidable).

Recall a simple lemma [4, Proposition 5.6].

LEMMA 5.2 *Let H be a generating set of G. Then H contains a subset K, generating G, such that for any $x_1, x_2, x_3 \in K$ and $\epsilon_1, \epsilon_2, \epsilon_3 = \pm 1$,*

$$x_1^{\epsilon_1} x_2^{\epsilon_2} x_3^{\epsilon_3} = 1$$

implies $x_1 = x_2 = x_3$, $x_1^3 = 1$.

Let us call such a generating set K 3-_product-free_.

5.3. A generating set K is _reduced_ if it is quasi-irreducible and 2-product-free. By 2.3 and 5.2, _every group has a reduced generating set_. Note that an irreducible generating set is necessarily reduced.

PROPOSITION 5.4. _For_ H _a reduced generating set of_ G, $X(G, H)$ _does not contain_ K_4^- (K_4 minus an edge). _If_ $X(G, H)$ _contains a triangle then one of the generators has order_ 3.

Proof. Trivial.

COROLLARY 5.5. K_4^- _is avoidable_.

We now show that the n-cycles are simultaneously unavoidable.

PROPOSITION 5.6. _The Cayley graphs of_ Z_3^n _are pancyclic_ (they contain cycles of all possible lengths $3, 4, \ldots, 3^n$).

Proof. It is readily checked that the Cartesian product (cf. [20]) of a 3-cycle by a path of length $k \geq 2$ is pancyclic. Now it follows by induction that the Cartesian product C_3^n of n copies of C_3 is pancyclic. But clearly, $X(Z_3^n, H) \cong C_3^n$ for any irreducible generating set H of Z_3^n.

As our proof of 5.5 shows, Problem 5.1 leads to questions concerning relations among generators of a group. One of the simplest problems of this kind is the following:

PROBLEM 5.7. _Given an infinite group_ G, _does it contain a subset_ H _of power_ $|G|$ _such that for any_ $x, y, z \in H$,

$$x^{-1} y = y^{-1} z \text{ implies } x = z?$$

Partial results concerning 5.7 have been applied in [4] to prove that, _given an infinite group_ G, _there is a digraph whose automorphism group is transitive, and isomorphic to_ G.

Concerning 5.7, see also [12].

Let us close our remarks on problem 5.1 with:

CONJECTURE 5.8. _A graph is avoidable (if and) only if it has an avoidable finite subgraph._

5.9. Finding the possible induced subgraphs of Cayley graphs is much easier than problem 5.1. H.W. LEVINSON has shown that _any connected graph on_ n _vertices may be embedded (as a induced subgraph) in some Cayley graph of a free group of rank_ $n - 1$ [14].

This result can be improved:

THEOREM 5.11. _There is a mapping_ f _of the class of cardinals into itself such that, given a graph_ X _of order_ n, _any group of order_ $\geq f(n)$ _has a Cayley graph containing_ X _as a induced subgraph. For_ n _finite,_ $f(n)$ _is finite;_ $f(\omega) = \omega$ _(where_ ω _is the cardinal number of the integers) and_ $f(\lambda) = \lambda$ _for any weakly compact cardinal_ λ; _and_ $f(\varkappa) \leq (\sup_{\mu<\varkappa} 2^{\mu})^{+}$ _for any infinite_ $\varkappa$. _Under the generalized continuum hypothesis,_ $f(\varkappa) \leq \varkappa^{+}$. ($\varkappa^{+}$ is the successor of cardinal $\varkappa$.)

The proof of 5.11 (see [6]) is based on the partial answer to 5.7 given in [4, Lemma 3.2].

PROBLEM 5.12. _Does_ 5.11 _not hold with_ $f(\omega_1) = \omega_1$?

PROBLEM 5.13. Which graphs may be embedded (as induced subgraphs) in Cayley graphs of _irreducible_ group presentations?

6. Remark. The investigation of _contractions_ of Cayley graphs (and of graphs with prescribed automorphism groups) seems to be worth

mentioning here. For lack of space we have to content outselves with mentioning one of the results, which is a corollary to [1, Lemma 3] (cf. also [2]).

LEMMA 6.1. *For* G *a subgroup of the group* H, *any Cayley graph of* H *is contractible onto some Cayley graph of* G.

This lemma has a number of applications. It in turn implies the Nielsen-Schreier theorem saying that any subgroup of a free group is free, A.T. White's conjecture [23, problem 7-7] stating that the genus of a subgroup never exceeds that of the group, etc. For details see [5].

REFERENCES

1. L. BABAI, Groups of graphs on given surfaces, *Acta Math. Acad. Sci. Hung.* 24(1973), 215-221.
2. L. BABAI, Automorphism groups of graphs and edge-contraction, *Discrete Math.* 8(1974), 13-20.
3. L. BABAI, Automorphism groups of planar graphs II, *Infinite and Finite Sets, Keszthely 1973* (A. Hajnal et al. eds.), North-Holland, Amsterdam 1975, 29-84.
4. L. BABAI, Infinite digraphs with given regular automorphism groups, *J. Comb. Theory B*, to appear.
5. L. BABAI, Some applications of graph contractions, to appear.
6. L. BABAI, Embedding graphs in Cayley graphs, to appear.
7. H. S. M. COXETER and W. O. J. MOSER, *Generators and relations for discrete groups*, Springer, Göttingen 1957.
8. P. ERDŐS and A. HAJNAL, On chromatic number of graphs and set-systems, *Acta Math. Acad. Sci. Hung.* 17(1966), 61-99.
9. P. ERDŐS and J. SPENCER, *Probabilistic methods in combinatorics*, Akadémiai Kiadó and Academic Press, Budapest-New York, 1974.
10. W. FEIT, The current situation in the theory of finite simple groups, *Actes du Congrés Internat. des Math. I*, Gauthier-Villars 1971, 55-93.
11. L FUCHS, *Infinite abelian groups*, I, Academic Press, N. Y. 1970.
12. J. L. HICKMAN and B. H NEUMANN, A question of Babai on groups, *Bull. Austral. Math. Soc.* 13(1975), 355-368.

13. H. A. JUNG, Ober den Zusammenhang von Graphen, mit Anwendungen auf symmetrische Graphen Math. Ann. 202(1973), 307-320.

14. H. W. LEVINSON, On the genera of graphs of group presentations III, J. Comb. Theory 13(1972), 209-302.

15. C. H C. LITTLE, D. D. GRANT and D. A. HOLTON, On defect-d matchings in graphs, Discrete Math. 13(1976), 41-54.

16. L. LOVASZ, Combinatorial Problems and Excercises, Akadémiai Kiadó-North-Holland, book to appear.

17. W. MADER, Über den Zusammenhang symmetrischer Graphen, Archiv der Math., 21(1970), 331-336.

18. W. MADER, Eine Eigenschaft der Atome endlicher Graphen, Archiv der Math. 22(1971), 333-336.

19. G. SABIDUSSI, On a class of fixed-point-free graphs, Proc. Amer. Math. Soc. 9(1958), 800-804.

20. G. SABIDUSSI, Graphs with given automorphism group and given graph theoretical properties, Canad. J. Math. 9(1957), 515-525.

21. M. E. WATKINS, Connectivity of transitive graphs, J. Comb. Theory 8(1970), 23-29.

22. M. E. WATKINS, On the action of non-abelian groups on graphs, J. Comb. Theory 11(1971), 95-104.

23. A. T. WHITE, Graphs, groups and surfaces, North-Holland, Amsterdam 1973.

THE MULTICOLORINGS OF GRAPHS AND HYPERGRAPHS

Claude Berge
University of Paris
Paris, FRANCE

Abstract

Let $H = (X, \xi)$ be a hypergraph with vertex-set $X = \{x_1, x_2, \dots, x_n\}$. We have a <u>multicoloring</u> of H with λ colors if we can assign to each $x \in X$ one or several of these colors so that in each edge $E \in \xi$ every color occurs exactly once.

Clearly, if there exists a multicoloring of H, one color defines a set of vertices which is both strongly stable and transversal; no general condition for the existence of such a set exists in the literature. The purpose of this paper is to provide existence conditions not only for a strongly stable transversal set, but also for a multicoloring.

1. Multicolorings for balanced hypergraphs.

In this section, we investigate the existence of a multicoloring for the vertices of a balanced hypergraph. A hypergraph H is said to be <u>balanced</u> (cf. [1]) if every odd cycle $(x_1, E_1, x_2, E_2, \dots, x_{2k+1}, E_{2k+1}, x_1)$ contains an edge E_i which contains three of the x_i's.

Let $H = (X, \xi)$ be a hypergraph on X; consider a function $\varphi: X \to R$ such that

$$\varphi(x) > 0 , \qquad (x \in X)$$

$$\sum_{x \in E_i} \varphi(x) = 1 , \qquad (i \in I) .$$

Following Csima [4], such a function will be called a _positive stochastic function_ of H . The following result generalizes Theorem 10 given in chapter 20, of [3].

THEOREM 1. _If H is multicolorable, then H has a positive stochastic function; furthermore, when H is balanced, the converse is also true._

Proof. If H has a multicoloring with λ colors, let $\lambda(x)$ be the number of colors assigned to the vertex x . Clearly, $\lambda^{-1}\lambda(x)$ is a positive stochastic function for H .

Conversely, let H be a balanced hypergraph with a stochastic function $\varphi(x)$. Since the vector $(\varphi(x_1), \varphi(x_2), \ldots, \varphi(x_n))$ is a solution of a system of linear inequalities with integer coefficients, we may assume that all the $\varphi(x_i)$'s are rational numbers, and we write for all i :

$$\varphi(x_i) = \frac{p_i}{p} ,$$

where p_i and p are positive integers. Let $H' = (X', (E_i'))$ be the hypergraph obtained from H by duplicating p_i times the vertex x_i, that is:

$$X' = X_1 \cup X_2 \cup \cdots \cup X_n ,$$

$$|X_i| = p_i ,$$

$$X_i \cap X_j = \emptyset , \qquad (i \neq j)$$

and

$$E_j' = \bigcup \{X_i \mid x_i \in E_j\} .$$

Clearly, the hypergraph H' is also balanced. Furthermore, H' is p-uniform, because for all j ,

$$|E_j'| = \sum_{x_i \in E_j} |X_i| = \sum_{x_i \in E_j} p_i = p \sum_{x \in E_j} \varphi(x) = p .$$

Therefore, by Theorem 4, Chapter 20 of Berge [3], it is possible to color the vertices of H' with p colors such that each edge E_i has all its vertices with different colors. This coloring defines also a multicoloring of H with p colors, which completes the proof.

Remark. If H is not balanced, it is not necessarily true that the existence of a positive stochastic function yields the existence of a multicoloring. The graph K_3 has a stochastic function (with value 1/2 at each vertex), but no multicoloring.

We can easily see that Theorem 1 is a generalization of the Birkhoff-Von Neumann Theorem on bistochastic matrices: Every bistochastic matrix is a barycentre of permutation matrices. Let A be an n x n bistochastic matrix. We shall show that $A = \Sigma \lambda_i P_i$, where the P_i's are permutation matrices, $\lambda_i > 0$ and $\Sigma \lambda_i = 1$. Let H be a hypergraph whose vertices are the positions of the non-zero entries of A , an edge of H being defined by all the positions located on a same column or on a same row. This hypergraph is balanced (because H has no odd cycles) and possesses a stochastic function (defined by the entries of the matrix A). By Theorem 1, H has a multicoloring. Let λ_1 be the least positive coefficient of A , and let x be the corresponding vertex of H . Let α_1 be the color assigned to x . The (0 , 1)-matrix P_1 , with a 1 in each position corresponding to a vertex of H with color α_1, is a permutation matrix; $A' = (A - \lambda_1 P_1)(1 - \lambda)^{-1}$ is a bistochastic matrix having more zero entries than A . If we repeat the same reduction with A' , as many times as needed, we get the required decomposition.

THEOREM 2. (Csima) Let $H = (X, \xi)$ be a hypergraph on X; for $E_i \in \xi$, let $\varphi_i(x)$ be the characteristic function of E_i. There exist integers $p_1, p_2, \dots, p_m \in Z$ such that:

$$\sum_{i=1}^{m} p_i = 0,$$

$$(1) \qquad \sum_{i=1}^{m} p_i \, \varphi_i(x) \geq 0, \qquad (x \in X)$$

$$\sum_{i=1}^{m} p_i \, \varphi_i(x) \neq 0, \qquad \text{for some } x,$$

if and only if H does not possess a stochastic function.

From Theorem 1, we obtain: The existence of p_i satisfying (1) is a necessary and sifficient condition that a balanced hypergraph H not possess a multicoloring.

In some cases, like the bipartite graphs, this condition can be simplified. A better statement can be obtained for the interval hypergraphs, whose vertices are points of the line, and whose edges are intervals.

THEOREM 3. An interval hypergraph has no multicoloring if and only if there exist two edges A and B such that $A \subset B$ and $A \neq B$. (In other words, an interval hypergraph is multicolorable if and only if it is a "Sperner hypergraph").

1. If there exist two edges A and B such that $A \subset\subset B$, then a multicoloring with λ colors is impossible, because A would contain λ colors, and therefore B would contain more than λ colors, which is a contradiction.

2. Let $H = \{A_i \mid i \in M\}$ be an interval hypergraph satisfying

$$A_i \subset A_j \Rightarrow i = j .$$

We shall show that there exists a multicoloring of H. It suffices to show that every vertex x_o is contained in a set S which is both strongly stable and transversal.

Let $H(x_o)$ be defined by the intervals which contain x_o. Let $H^+(x_o)$ be defined by the intervals having their initial endpoint at the right side of x_o. Let $H^-(x_o)$ be defined by the intervals having their terminal endpoint at the left side of x_o. Thus

$$H(x_o) + H^+(x_o) + H^-(x_o) = H$$

Let A_o be the (unique) interval of $H(x_o)$ with a terminal endpoint as far as possible from x_o; the set

$$\bigcap (A_i - A_o) ,$$

with $A_i \in H^+(x_o)$ and $A_i \cap A_o \neq \emptyset$, is a non-empty interval (because H is a Sperner hypergraph). Let x_1 be the terminal endpoint of this intersection, and let A_1 be the interval of $H(x_1)$ with a terminal endpoint as far as possible from x_1. We can similarly define a vertex x_2, etc. The set $S^+ = \{x_o, x_1, x_2, \ldots\}$ is both transversal and strongly stable for the union $H(x_o) + (H^+(x_o)$. By the same method, we can define a set $S^- = \{x_o, x_{-1}, x_{-2}, \ldots\}$ which is transversal and strongly stable for $H(x_o) + H^-(x_o)$. The union $S^+ \cup S^-$ is transversal and strongly stable for H, which completes the proof.

2. <u>The multicoloring index of a graph</u>.

Let $G = (X, E)$ be a simple graph, with vertex set X, with order $n = |X|$, and with maximum degree

$$\Delta(G) = \max_{x \in X} d_G(x) \quad .$$

For two disjoint sets of vertices A and B, we denote by $m_G(A, B)$ the number of edges in G having one extremity in A and the other extremity in B; all notations used here are the same as in [3]. To generalize the classes of graphs having the edge-coloring property, we consider the following problem: Is it possible to cover the edge set E of G by perfect matchings? If the answer is yes, what is the least number of perfect matchings needed to cover E?

It is sometimes easier to think of this problem in terms of multicolorings: we have a multicoloring of G with k colors if we assign to each edge of G one or several of the colors 1, 2, 3, ..., k so that in each vertex every color occurs exactly once. In this section, we shall give necessary and sufficient conditions for the existence of a multicoloring. If G has a multicoloring, we can ask for the least number of colors needed for a multicoloring of G; such a number is called the multicoloring index of G. It is well known that the Petersen graph P_{10} is not 3-edge-colorable; nevertheless, P_{10} has a multicoloring with 5 colors, as shown in Figure 1, and Laskar [1] has checked that the multicoloring index of P_{10} is exactly 5.

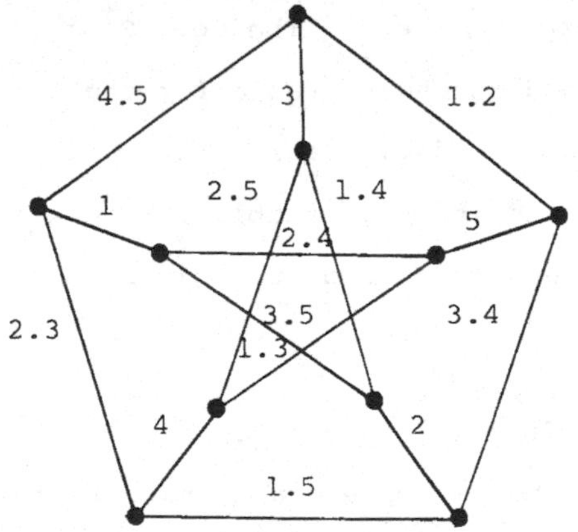

Fig. 1. Multicoloring of the edges P_{10} with colors 1, 2, 3, 4, 5.

Let G be a simple graph, and H be a multigraph obtained from G by replacing one edge $[x, y]$ by several parallel edges between x and y. We shall say that H is obtained by one edge-multiplication.

THEOREM 4. *A simple graph* $G = (X, E)$ *has a multicoloring if and only if some regular multigraph* $H = (X, F)$ *obtained from* G *by edge multiplications satisfies*:

$$m_H(S, X - S) \geq \Delta(H) \qquad (1)$$

for every subset $S \subset X$ *with odd cardinality.*

Proof. Assuming that the edges of G are multicolorable with r colors, let H be the multigraph obtained from G by replacing an edge of G that received q colors by q parallel edges. Thus H is r-regular and its chromatic index is equal to the maximum degree $\Delta(H) = r$.

If S is an odd subset of X, each color defines a perfect matching of H which leaves at least one unsaturated vertex in the subgraph H_S induced by S. Hence at least one edge between S and X - S receives the color i. The condition (1) follows.

Conversely, let H be a multigraph obtained from G by duplicating some of the edges so that H is regular and satisfies (1). We shall show that each edge of H belongs to some perfect matching (which implies easily that the edges of G are multicolorable).

First, we shall show that H has a perfect matching. Without loss of generality we may assume that H is connected. Let $S \subset X$ and let $p_i(S)$ denote the number of odd connected components $C_1, C_2, \ldots, C_k, \ldots$ of the subgraph of H induced by X - S. We have

$$\Delta(H)\,|S| = \sum_{x \in S} d_H(x) = 2m_H(S, S) + m_G(S, X - S) \geq m_H(S, X - S) \geq$$

$$\sum_k m_H(S, C_k) = \sum_k m_H(C_k, X - C_k) \geq p_i(S)\,\Delta(H)$$

Hence,

$$|S| \geq p_i\ (S)\,, \qquad (S \subseteq X)\ .$$

By Tutte's Theorem, this shows that H has a perfect matching (and therefore G has also a perfect matching).

Let $F_o \subset F$ be a perfect matching of H. To prove the existence of a multicoloring, it suffices to show that every edge $[y, z] \notin F_o$ belongs to some perfect matching, or equivalently, is on an even cycle with edges alternately in F_o and in $F - F_o$.

Assume that this is false (proof by contradiction). Consider the multigraph R obtained from G by adding a new vertex x_o and by replacing the edge $[y, z]$ by $[y, x_o]$ and $[x_o, z]$. These two edges and the edges in $F - F_o$ are called <u>light</u>, and the edges in F_o are called <u>dark</u>; if there exists no alternating chain (with no repeated edge) from x_o to x, this vertex x is called <u>inaccessible</u>, and we write $x \in I$.

If all the alternating chains to x terminate with a dark edge, x is called a <u>dark</u> vertex, and we write $x \in D$. If all the alternating chains to x terminate with a light edge, x is called a <u>light</u> vertex, and we write $x \in L$. If some alternating chains to x terminate with a light edge, and other terminate with a dark edge, x is called a <u>mixed</u> vertex, and we write $x \in M$. Furthermore, we orient toward x all the edges leading to x on such an alternating chain.

We have $x_o \notin D \cup L \cup M \cup I$, and by the hypothesis, $y, z \in L$.

We shall use a theorem of Gallai (Theorems 9 and 10, Chapter 8 in [3]) which states: If E_o is a maximum matching of R, then each connected component C of the subgraph R_M is odd, and the only edges incident to M are: one dark edge (oriented toward C) and some light edges (directed out of C). Thus, by (1), the number of light edges incident to C is at least $h - 1$.

Now, let $\overline{R}$ be the graph obtained from R by contracting each connected component C of R_M into one single vertex. Then E_o induces in $\overline{R}$ a matching $\overline{E_o}$ having only one unsaturated vertex, namely x_o, and there are no more mixed vertices. The vertex set of $\overline{R}$ can be partitioned as before by $\overline{D}$, $\overline{L}$, $\overline{I}$ and $\{x_o\}$. If $x_o \in \overline{D}$, the number of light edges directed out from x is at least $h - 1$, and we write

$$d^+(x) \geq h - 1, \qquad (x \in \overline{D}) .$$

Let $x \in \overline{L}$; then the number of light edges incident to x and directed toward x is at most $h - 1$, and we write

$$d^-(x) \leq h - 1, \qquad (x \in \overline{L}) .$$

Since the directed dark edges define a bijection from $\overline{D}$ to $\overline{L}$, we have $|\overline{D}| = |\overline{L}|$. Hence

$$(h-1)|\overline{L}| = (h - 1)|\overline{D}| \leq \sum_{x \in \overline{D}} d^+(x)$$

$$= (\text{number of directed light edges}) - 2$$

$$\leq \sum_{x \in \overline{L}} d^-(x) - 2 \leq (h - 1)|L| - 2 .$$

The contradiction follows.

Note that Theorem 1 shows immediately that there exists a multicoloring for the Petersen graph (Figure 1). The proof given above is in fact similar to the one of a generalization of Petersen's

Theorem to r-regular graphs which appeared for the first time, as far as we know, in [2] (Theorem 13, Chapter 8 in [3]).

Remark. Theorem 1 is a partial answer to a question of Nash-Williams [5] which asked "if for some $h > 0$, the condition

$$m_H(S, X - S) \geq \Delta(H), \qquad (S \subset X, \quad |S| \text{ odd})$$

could imply that an h-regular graph H has the edge coloring property."

To apply Theorem 1, the main difficulty is to know if a regular multigraph H can be obtained from G by edge-multiplications. The following result can be used.

THEOREM 5. Let G be a simple graph with vertex set

$$X = \{x_1, x_2, \ldots, x_n\} .$$

It is possible to obtain from G a regular multigraph by edge multiplications if and only if there is no integral vector $(p_1, p_2, \ldots, p_n) \in Z^n$ such that

(1) $\sum_{i=1}^{n} p_i = 0$,

(2) $[x_i, x_j] \in E \Rightarrow p_i + p_j \geq 0$, and

(3) $p_i + p_j > 0$ for some edge $[x_i, x_j]$.

Proof. Let the p_i's be positive, negative or zero integers satisfying (1), (2), (3), and assume that a regular multigraph H can be obtained from G by multipling by λ_j each edge e_j. Let $((a_j^i))$ be the incidence matrix of G. We have

$$\sum_{i=1}^{n} a_j^i \, p_i \geq 0, \qquad (j \leq m)$$

and this inequality is strict for some j . Hence,

$$\sum_j \lambda_j \sum_i a_j^i p_i > 0 .$$

Therefore,

$$\sum_i p_i \sum_j \lambda_j a_j^i = \sum_i p_i \Delta(H) > 0 .$$

Since we have (1), the contradiction follows.

The "only if" part can be deduced from the Farkas Lemma. The proof will be omitted here, since it is a special case of Csim 's Theorem [4].

The existence or the non-existence of a multicoloring is easier to show for a bipartite graph.

THEOREM 6. _Let_ $G = (X, Y, \Gamma)$ _be a bipartite graph. For a vertex_ v, _we shall denote by_ $\Gamma(v)$ _the set of the neighbors of_ v _and by_ $E(v)$ _the set of edges in_ G _which are incident to_ v . _For_ $A \subset X$, _we put_

$$\Gamma A = \bigcup_{x \in A} \Gamma(x), \qquad \text{if } A \neq \emptyset,$$

$$\Gamma A = \emptyset, \qquad \text{if } A = \emptyset .$$

The following conditions are equivalent:

(1) G _has a multicoloring;_

(2) _a regular multigraph can be obtained from_ G _by edge multiplication;_

(3) _two sets_ $A \subset X$ _and_ $B \subset Y$ _with_ $|A| = |B|$ _will not satisfy_

(3a) $\bigcup_{x \in A} E(x) \supsetneqq \bigcup_{y \in B} E(y)$, nor

(3b) $\bigcup_{x \in A} E(X) \subsetneqq \bigcup_{y \in B} E(y)$;

(4) <u>for every</u> $A \subset X$ <u>and every</u> $B \subset Y$,

(4a) $|\Gamma A| \geq |A|$,

(4b) $|\Gamma B| \geq |B|$,

(4c) $|\Gamma A| = |A| \Rightarrow \Gamma(\Gamma A) = A$, and

(4d) $|\Gamma B| = |B| \Rightarrow \Gamma(\Gamma B) = B$.

<u>Proof</u>. (1) $\Rightarrow$ (2). Assume that there exists a multicoloring of the edges of G with λ colors, and that the edge e_i receives λ_i colors. By multipling each edge e_i by λ_i, we obtain a λ-regular multigraph, and (2) follows.

(2) $\Rightarrow$ (3). Let H be a regular multigraph obtained from G. Assume first the existence of two sets $A \subset X$ and $B \subset Y$ with $|A| = |B|$ satisfying (3a). We shall show that this leads to a contradiction. Put

$$p_i = \begin{cases} +1, & \text{if } x_i \in A, \\ -1, & \text{if } x_i \in B, \\ 0, & \text{if } x_i \in (X - A) \cup (X - B). \end{cases}$$

The p_i's satisfy the conditions (1), (2), (3) of Theorem 2 and therefore no regular multigraph H can be obtained. The contradiction follows.

(3) $\Rightarrow$ (4). Let G satisfy (3). Clearly, G has no isolated vertex, and if $|X| < |Y|$, the set $A = X$ and a subset $B \subset Y$ with cardinality $|B| = |A|$ satisfy (3a), which is a contradiction. Hence, $|X| = |Y|$.

If a set $A \subset X$ satisfies

$$|\Gamma A| < |A|$$

then there exists a set $B \subset Y$ such that $|B| = |A|$ and $B \supset \Gamma A$. So the two sets A and B satisfy (3b), and the contradiction follows.

If a set $A \subset X$ satisfies

$$|\Gamma A| = |A| , \qquad \Gamma(\Gamma A) \neq A ,$$

then the sets A and $B = \Gamma A$ satisfy (3b), and the contradiction follows.

So, we have (4a) and (4c), and by the same argument, we get (4b) and (4d).

(4) ⇒ (1). Let G be a graph satisfying (4a), (4b), (4c), (4d). Let $[x_o , y_o]$ be an edge of G with $x_o \in X , y_o \in Y$. It suffices to show that every connected component $G_1 = (X_1 , Y_1 , \ldots , \Gamma_1)$ of the subgraph of G induced by $(X - \{x_o\}) \cup (Y - \{y_o\})$ has a perfect matching. By the König-Hall Theorem, this is equivalent to:

$$|\Gamma_1 A| \geq |A| , \qquad (A \subset X_1) ,$$

$$|\Gamma_1 B| \geq |B| , \qquad (B \subset Y_1) .$$

Let $A \subset X_1$. We shall only show that $|\Gamma_1 A| \geq |A|$.

CASE 1: We have $y_o \notin \Gamma A$. Then

$$|\Gamma_1 A| = |\Gamma A| \geq |A|$$

CASE 2: We have $y_o \in \Gamma A$. Then

$$|\Gamma_1 A| = |\Gamma A| - 1 \geq |A| - 1 .$$

We cannot have $|\Gamma_1 A| = |A| - 1$, because this would imply $|\Gamma A| = |A|$; hence, from (4c), $\Gamma(\Gamma A) = A$ and $x_o \in \Gamma(\Gamma A) = A$. Since $x_o \notin X_1$ and $A \subset X_1$, the contradiction follows. Thus,

$$|\Gamma_1 A| \geq |A|$$

By the same argument, $|\Gamma_1 B| \geq |B|$ for every $B \subset Y_1$, and the proof is achieved.

REFERENCES

1. C. Berge, Sur certains hypergraphs généralisant les graphes bipartis, _Combinatorial Theory and its Applications_, (Erdös, Renyi, Sós editors), North Holland, Amsterdam-London, 1970, 119-133.

2. C. Berge, _Théorie des graphes et ses applications_, Dunod, Paris, 1958.

3. C. Berge, _Graphs and Hypergraphs_, North-Holland, New York, 1973.

4. J. Csima, Stochastic functions on Hypergraphs, _Combinatorial Theory and its Applications_, (Erdös, Rényi, Sós, editors), North Holland, Amsterdam-London, 1970, pp. 247-355.

5. J. C. Nash-Williams, Unexplored and semi-explored territories in graph theory, _New Directions in the Theory of Graphs_, Academic Press, New York, 1973, pp. 149-186.

IRREDUCIBLE CONFIGURATIONS AND THE FOUR COLOR CONJECTURE

Frank R. Bernhart
Shippensburg State College
Shippensburg, PA 17257

Abstract

The classical four color reduction process takes on a new appearance in the light of the recently begun theory of open sets of colorings. In this paper we show that specific configurations and clusters can be simply classified as either reducible or irreducible, without appealing to the truth or falsity of the Four Color Conjecture (4CC). By treating irreducibility and reducibility together, we hope to round out the theory and gain a better understanding of why clusters do or do not reduce.

The new methods are illustrated by giving a condensed proof of irreducibility for all reasonable candidates on the order of the six-ring or less. The principle tools are the union and splicing properties of open sets, and the rotation of "antiset" pairs (Lemma 1). Complete details and extension to higher rings will come in a later paper or papers.

At the conclusion it is shown that the 4CC is equivalent to the set of irreducible clusters (in our definition) being infinite!

1. Basic Concepts of Reduction

Naive induction is a natural approach to the Four Color Conjecture (4CC). This approach finds difficulties. The more sophisticated process of reduction attempts to meet these difficulties, at least in part.

In a map, the typical target for reduction is a cluster (Figure 1a), or an explicit arrangement of a small number of faces inside a circuit. It is customary to show the half-edges leaving the cluster also. We may then reckon contacts between faces in the cluster and the cyclic sequence of faces outside (the bounding ring). The ring is a proper ring if its faces form an annular region without interior vertices (in the case of two faces, any multiply connected region), as in Figure 1b.

FIGURE 1

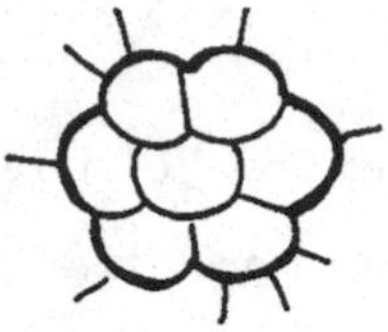
a.

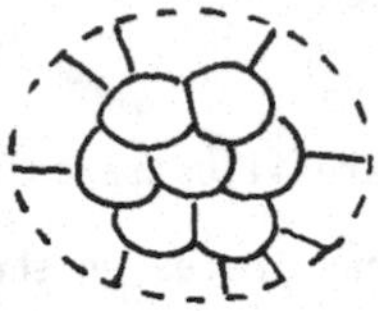
b.

We will be using the dual approach, in which maps and clusters are replaced by their dual graphs. Figure 2 translates Figure 1, and shows the simple and the augmented form of the cluster. An occurrence of a cluster in a plane graph is an intuitive concept, which depends on a type of inclusion mapping. The interior (simple form) corresponds to a subgraph, but the image of the ring need not even be a true circuit. A proper ring must correspond to a true circuit which has no diagonal chords (edges). Cyclic order of edges at interior vertices is preserved.

FIGURE 2

a.

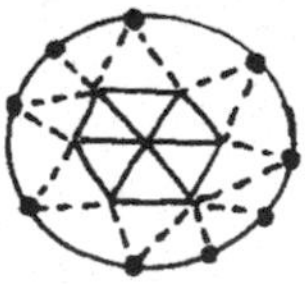
b.

The augmented diagram carries more information, but the simple diagram is less clumsy. A suitable compromise is found by coding the degrees of vertices of the simple diagram after the manner of W. Stromquist or H. Heesch (Figure 3). We will adopt the coding of Heesch, which uses these symbols for degrees five to nine:

•	✳	○	◘	Δ
5	6	7	8	9 .

FIGURE 3

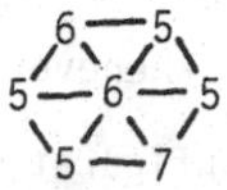

a.

b.

The process of reduction replaces a given cluster with something smaller, and usually highly degenerate. Figure 4 shows an example for a map cluster (the dotted line is where the circuit used to be) and the graph equivalent. Here we use an ordinary diagonal to mean a contact of two nonconsecutive ring elements, and we use a doubled diagonal (essentially an elongated equal sign) to indicate a merger of two ring elements. Single and double diagonals may also be treated as negative and positive edges of signed graphs, where two vertices joined by a negative (positive) edge have to have different (the same) colors in a coloring.

FIGURE 4

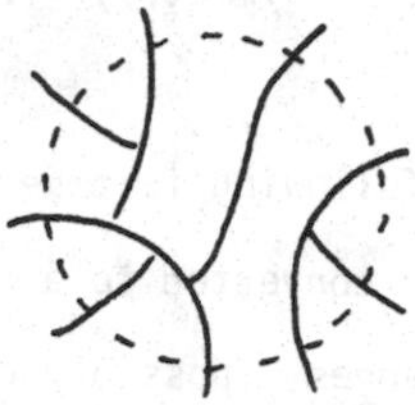

a.

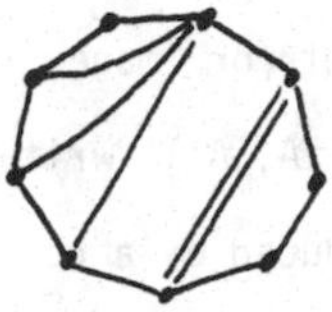

b.

A reduction is often obtained merely by contracting edges of the cluster (erasing boundaries of the map) but the limitation is not necessary in the theory. We will not admit graphs with loops (trivially noncolorable), and so must take care in reduction not to form loops. When the bounding ring is proper, no such problem can arise.

From now on, the only plane graphs considered are the members of T^6, the "almost 6-connected" triangulations of the plane (or sphere). These have at least five vertices, no proper rings of length four or less (and hence no vertices of degree less than five), and no <u>five-rings</u> (length exactly five) except the bounding rings of 5-vertices (vertices of degree five).

It is possible to construct a 4-coloring for any planar graph from colorings of appropriate members of T^6. Moreover, all minimal graphs belong to T^6, where a <u>minimal graph</u> is a graph without loops or multiple edges, not colorable (in four colors), and such that all graphs with fewer vertices are colorable. The pioneer paper of Birkhoff is still highly readable, and is recommended [ ].

2. Reducibility versus Irreducibility

A circuit or ring of length n will be denoted by Q_n. Whenever Q_n is a proper ring in a plane graph G, we decompose G into two configurations or clusters, each bounded by Q_n. We write $G = A + Q_n + B$ where A,B are the two configurations. The problem of coloring G is now identical with finding a coloring for A and a coloring for B which establish the same color pattern on ring Q_n.

If K is a cluster of m interior vertices bounded by ring Q_n, then the formal unreduced fraction m/n is the index of K, and $|K|$ is the total number of vertices of K. If the n elements of the ring Q_n are actually distinct, then $|K| = m + n$, otherwise $|K| < m + n$. We assume that Q_n is labeled in some way, so that K is distinguished (in the absence of symmetry) from its rotations, and its mirror image.

Given clusters A, A', write $A' \rightarrow A$ if the following is true for each coloring of Q_n induced by a coloring of A' : it can be converted to a coloring compatible with A by a sequence of Kempe Color Interchanges (possibly empty). We do not make the KCI concept more precise, since it is scheduled to be replaced shortly by a variant terminology, the "open sets" of Whitney and Tutte [13]. In their terminology, $A' \rightarrow A$ means that the A'-extensible colorings (of Q_n) are "immersible" in the set of A-extensible colorings. In effect, a coloring of $G' = A' + Q_n + B$ can be converted into a coloring of $G = A + Q_n + B$ by steps which are all independent of the structure of B.

If $A' \rightarrow A$ and also $|A'| < |A|$, say that A' reduces A. In that case a 4-coloring of a graph G' of form $A' + Q_n + B$ implies (by construction) the existence of a 4-coloring for the larger graph $G = A + Q_n + B$, and A cannot occur in any minimal graph.

Moreover, for fixed A, the number of colorings of Q_n is finite, and the number of A' with $|A'| < |A|$ is finite. For each A' a constructive finite test decides whether $A' \rightarrow A$. Independently of the truth or falsity of the 4CC therefore, A is classifiable as reducible (for some A', A' reduces A), or else irreducible.

The classical reduction theory was invented by Kempe, and expanded and given clear mathematical form by Birkhoff. It has persisted for sixty years, almost without significant change, except the use of computers to handle lengthy and tedious calculation. No reduction of a cluster in a sense outside our definition has been found, which gives the definition a degree of permanence. Unfortunately, it was not usual to make an exhaustive search of possible reducers, and perhaps this is the reason that the term "irreducible" in the literature was not treated on a par, logically, with "reducible." It remained vague, or was associated exclusively with the structure (and existence) of a minimal graph. It even seemed tnat the 4CC would imply every configuration was reducible. We prove a result at the end of the paper which strongly conflicts with that expectation.

Our arguments are based solely on the above definition, and a slight strengthening of the conditions using equations, on the grounds that this encompasses all known reducibility results. A treatment which does not require an assumption of a minimal map is desirable for other reasons. Whitney and Tutte described the collapse of a proof of the 4CC [13], where the error (it occurred in a flawed computer program) at first escaped notice. Doubt first arose when it was shown that the assumption of a minimal map for purposes of indirect proof was superfluous.

An earlier example is the famed proof by Kempe, refuted by Heawood [8]. What Heawood in effect showed was that a configuration Kempe believed reducible was in fact irreducible.

3. Realizable and Open Sets of Color Schemes

It is desirable to regard as equivalent any two ways of coloring a circuit Q_n which produce the same partition of vertices into color classes. In other words, the colorings differ only by permutation of colors. A color scheme is a partition of the vertices of Q_n into four or fewer color classes, and is represented indifferently by any coloring which yields that partition. By orthodox scheme we mean the representation of a color scheme by a sequence 1,2,... made up of colors 1,2,3,4, with the first occurrence of 4 not preceding the first occurrence of 3. If the ring Q_n has ρ_n orthodox schemes, we find

n	3	4	5	6	7	8	9	10	...	n
ρ_n	1	4	10	31	91	274	820	2461	...	$(3^{n-1} + (-1)^n + 2)/8$.

Let J_n denote the set of schemes for Q_n . In lexicographical order the ten members of J_5 are #1= 12123, #2= 12132, #3= 12134, #4= 12312, #5= 12313, #6= 12314, #7= 12323, #8= 12324, #9= 12342, #10= 12343 .

If A is a cluster, write J(A) for the set of schemes induced by colorings of A (colorings of Q_n extendible to A, schemes compatible with A, etc.). Sometimes we will find it convenient to use a single symbol for a specific cluster A and for its associated scheme set J(A), to be distinguished by context. For example, each degree n vertex (denote by V_n) is the single interior vertex of a Q_n cluster called a <u>wheel</u>, and here denoted W_n. Cluster W_5 has scheme set W_5 consisting of all members of J_5 above in which color 4 is absent, namely #1, #2, #4, #5, #7, and similarly for W_n in general. The complement of set W_n, denoted W_n' is the set of "true 4-colorings" of Q_n.

The subsets U of J_n which are of form J(A) for some A are <u>realizable</u>. The 4CC is true just in case the set W_5' is <u>not</u> realizable. It is just not practical to test given U for realizability, since the potential clusters A form an infinite set. What can be tested is certain necessary conditions, derived from the KCI process (or certain equations). Whitney and Tutte [13] define a subset U of J_n as <u>open</u> if it is consistent with the KCI conditions. Most of the open sets significant for the theory, including all those treated here, also satisfy a set of equations in a certain sense; these are developed at length in the author's dissertation [5], and are derivable also from certain equations in the theory constrained chromatic polynomials, specialized to four colors . The equations were used for reductions by the author's father A. Bernhart in the forties, following correspondence with D.C. Lewis. An independent development of the same structure is found in the thesis of D. Cohen [7].

Equation consistent open sets will be called <u>absolutely</u> open, since no other effective limitation has been found. Each realizable set is absolutely open. This fact, plus the algebraic properties to be introduced, form the basis of the results shortly to be presented. This basis is simply assumed for reasons of space -- a large part of the fundamentals can be pieced together from the papers cited in the previous paragraph.

The most powerful property required is splicing. It was explicitly introduced by W. Stromquist [12], although prefigured in Shimamoto as described in [13]. We refer to the general <u>splice diagram</u> of Figure 5a, consisting of a plane graph in which three paths join two vertices. Occasionally a degeneracy of one path is considered (Figure 5b). The plane is divided into three regions, each with a circuit for boundary; <u>left</u>, <u>right</u>, and <u>outside</u>.

FIGURE 5

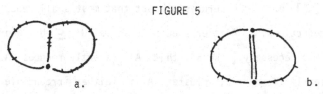

a. b.

Each way of 4-coloring the splice diagram generates a set of three color schemes, say s_L on the left circuit, s_R on the right circuit, and s on the outside circuit. These form a triple (s_L, s_R, s). Now assign to each of the left and right circuits a specific open set (or more generally, any set), say U_L and U_R. We collect all triples where s_L is in U_L and s_R is in U_R. The set of third members s is $U = U_L * U_R$, the <u>splice</u> of U_L and U_R. The definition abstracts from the splice of configurations A_L (in the left region) and A_R (in the right region) to form a combined configuration A, since in this case $J(A) = J(A_L)*J(A_R)$.

The following properties are crucial.

#A. On a fixed circuit, the union of (absolutely) open sets is (absolutely) open.

#B. Given a splice diagram, the splice of (absolutely) open sets is again (absolutely) open.

#C. Splice distributes over union: $(U_1 \cup U_2)*V = U_1*V \cup U_2*V$.

The essentials of the classic reduction process follow almost immediately from #A. For, if S is any subset of J_n, #A guarantees a unique largest (absolutely) open set disjoint from S. This is the <u>conjugate</u> $\tilde{S}$ (<u>absolute conjugate</u> $\hat{S}$). The complement of $\tilde{S}$ contains S and is the <u>closure</u> $\overline{S}$.

The reduction property $A' \to A$ presented earlier on an intuitive basis is now effectively defined by $U' \subseteq \overline{U}$, or equivalently $U' \subseteq \tilde{\tilde{U}}$, where U', U are the realizable sets $J(A')$, $J(A)$. By hand, the process consisted of showing that each scheme in U' was a member of $\overline{U}$. For computer implementation, the set $\overline{U}$ is completed first, since the search for U' is less easy. An excellent discussion of these and similar points is given by Allaire and Swart [1], who have made the catalog of reducible configurations for $n \leq 10$ nearly complete.

The reducer A' is unnecessary if $\tilde{U}$ is empty ($\overline{U} = J_n$). We say that A in this case is <u>freely reducible</u> (Allaire and Swart: colorable, Heesch: D-reducible). Haken [2] notes the surprising fact that most small reductions are possible as free reductions. Another special case is $U' \subseteq U$, where no computation of $\overline{U}$ is necessary. We say that A' directly reduces A (Heesch: A-reduction). Clearly if $\tilde{\tilde{U}} = U$, cluster A is either irreducible or directly reducible. The latter has no known instances, so that Allaire and Swart [1] can conjecture A is irreducible when $\tilde{\tilde{U}} = U$.

In the above the conjugate can be replaced by the absolute conjugate, but it is not known if this obtains the reduction of any cluster not previously reducible. Computational difficulties are eased by computing $\hat{U}$ only after $\tilde{U}$.

The relation $A' \to A$ is transitive, by #A and #B we also have A reducible if subconfiguration B is reducible (given $B' \to B$, form A' by putting B' for B in A, then $A' \to A$). And, A or A' is reducible if it is not a cluster found in T^6 graphs (because of "holes"). So the most likely reducer A' is a Q_n plus single and double diagonals (<u>totally degenerate</u>).

4. <u>General Technique for Irreducibility Proofs</u>

Between the forties and the sixties little was published on reducibility theory. We can logically speak of an early period, and a modern period. Aids such as the computer, and a new interest in abstract methods also set off the modern period. The most outstanding feature of the early period is ingenuity in the calculation of reductions, with a combination of intuition and luck in choosing clusters A and their reducers A' helping to shorten the time. There were few shortcuts, and no good predictions for reducibility.

The situation for irreducibility was worse, there being no shortcut against testing all feasible reducers A' and calculating the entire set $\overline{U}$ (for U = J(A) as before).

Principally due to Heesch [9], enormous progress has been made in attempting to build a theory of reducibility. A rule-of-thumb (see Appel and Haken [2]) for reducibility has been found, and seems to be necessary but not quite sufficient. However the tedious calculation of $\overline{U}$ remains.

It seems that satisfactory understanding of either reducibility or irreducibility requires knowledge of both. Although we cannot go farther in attempting to prove that the rule-of-thumb actually works, we can make a few steps in the direction of proving irreducibility by theoretical tools. The reader will after reading the below condensed proofs of irreducibility appreciate both the short cuts which new theory provides, and the seemingly great difficulties in obtaining a complete generalization.

Our strategy is to detour the calculation of the conjugate $\tilde{U}$ and instead construct by means of splice and union an open set V disjoint from U (of course that implies $V \subseteq \tilde{U}$). Two open sets will be called <u>antisets</u> when they are disjoint. After affirming that (U, V) are indeed a pair of antisets, we must prove that (U', V) are not antisets for U' = J(A'), and A' any feasible reducer of A. Then we conclude that A is irreducible.

Our main lemma now follows. We refer to the splice diagram of Figure 5 and assign three open sets to the three circuits, say U, V, W. Which circuit is left or right or outside is immaterial because of the symmetry. In any case U*V = V*U is an open set on the same circuit as W, and similarly for U*W = W*U and V, or V*W = W*V and U. Then we have the following associativity property, which we prefer to call the <u>rotation of antisets</u>.

LEMMA 1: <u>If one of the pairs</u> (U*V, W), (U*W, V), (V*W, U) <u>is a pair of antisets then all three are pairs of antisets</u>.

Proof: The condition for any pair to be a pair of antisets reduces to the claim that all triples of schemes (see above) fail to be compatible with at least one of U, V, W.

Next we consider annexing . A cluster B_1 with boundary Q_n becomes a cluster B_2 in several ways such that the interior of B_2 is the interior of B_1 plus one element of ring Q_n. Say that the new vertex v has power p if there are p - 2 more edges meeting in the augmented diagram for B_2 than in the augmented diagram for B_1. The annexing is then effected merely by assigning a power $p \geq +1$ to some element of Q_n. The bounding ring is seen to be Q_m with $m = n + p - 3$, and the case $p = +1$ is a degenerate splice not often used.

We consider the construction of clusters by starting with an n-wheel W_n for $n \geq 5$. A cluster is normal if it is possible to construct it with nondecreasing ring size, i.e. with powers assigned all positive, and good (for reduction) if the power of all interior vertices is at most three.

All known reducible clusters are good, and satisfy two other properties as well (see Appel and Haken [2]). Furthermore, if a cluster is not normal, it must either have a reducible subconfiguration or a large number of interior vertices -- and hence (if this be assumed) we need only try to reduce normal clusters. For reasons not elaborated here, we limit the search to nondegenerate normal clusters.

The potential reducers are usually highly degenerate. Call cluster A' or set U' = J(A') totally degenerate if A' has no interior vertices. This will be abbreviated to td-cluster and td-set. A few facts about such clusters are needed. Proof follows easily from inspection or induction on the number of degeneracies (diagonals of Q_n, single or double).

LEMMA 2: Let A' be any td-cluster on Q_n, for $n \geq 5$. Then

(a) The set U' = J(A') contains a 3-color scheme and a true 4-color scheme.

(b) The complement of U' is absolutely open, more particularly, it is the union of J(B) where B ranges over the td-clusters which have a single diagonal where A has a double diagonal, or vice versa.

An open set W is strong if (U,W) is not an antiset pair for all td-sets U. Lemma 2a is equivalent to the statement that W_n and its complement W_n' are strong.

5. Condensed Irreducibility Proofs

Consider first Q_n with $n \leq 4$; there are of course no nondegenerate normal clusters. An interesting case arises if $G = A + Q_4 + B$ is a minimal graph (not 4-colorable). Suppose that A =  is degenerate, and B is not. Both A. Bernhart [3] and Birkhoff and Lewis [6] found that "twisting" the diagonal or replacing G =  + Q_4 + B by G' =  + Q_4 + B gave a colorable graph. In a sense of reduction more general than defined here,  or  is reducer for the union 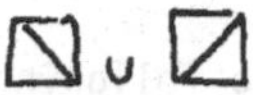 . Birkhoff and Lewis found a similar result for G =  + Q_5 + B when  was replaced by any of  ,  ,  , or  . This last has been rediscovered more than once since then. We now consider Q_5 and Q_6 in turn.

5.1 Irreducible Five-rings

The only nondegenerate normal cluster is W_5 or  , and the only feasible reducers are the ten td-clusters. The claim that (W_5, W_5') are antisets reduces to showing that W_5' (which we symbolize as ) is open. That claim is in essence Heawood's refutation [8] of the alleged proof [10] of Kempe. We will shortly show that W_n' is open. Since W_n' is strong by Lemma 2a, we conclude

THEOREM 1: Cluster  is the only irreducible five-ring cluster, which is normal.

Birkhoff's result [6] is slightly stronger: at most  and  are irreducible. Here  stands for a hypothetical realization of W_5'.

The antiset pair  , 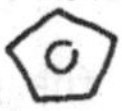 furnishes us with other pairs by rotation (i.e. by Lemma 1). These are  ,  and  ,  and  ,  . By Lemma 2b we have containments 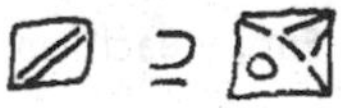 and  and 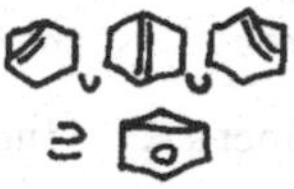 . A simple check on sizes is enough to improve the first two to set equalities 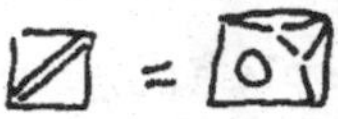 and  .

Here the reader should understand the diagram conventions. The same symbol is used for a cluster and its associated open set, and results of complex splices are drawn in a natural manner analogous to splices of hypothetical realizations.

5.2 Irreducible Six-Rings

Normal six-ring clusters are formed by annexing vertices to W_5 or W_6. The three smallest are in Figure 6, in simple diagrams, not augmented.

FIGURE 6

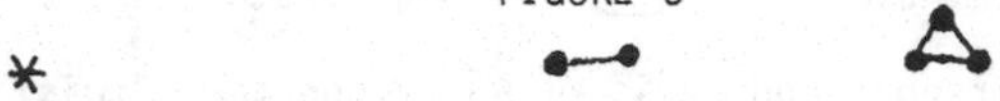

The first is the 6-vertex (simple) or six-wheel (augmented). First we show that the pair (W_6, W_6') or  , 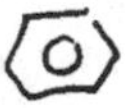 is really a pair of antisets. That is, we show that  is open. Consider the following easily checked fact: every true 4-coloring in J_n, $n \geq 6$, becomes a true four-coloring of order n-1 when a suitably chosen digit in the orthodox representation is struck out. For example, scheme #31 = 123434 loses the final digit: #10 = 12343 in J_5. The equation  immediately follows, and we conclude  is (absolutely) open, if  is.

Moreover, the equation remains true if the terms of the union on the right are pruned to three consecutive. This follows from 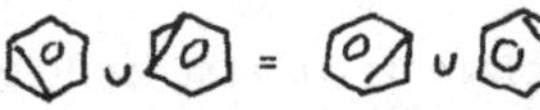 which follows from 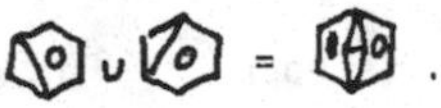 . The truth of this last set equation follows from open set property #C on substituting 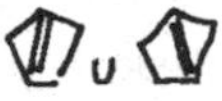 for  in the set  .

No td-cluster can reduce W_6, because W_6' is strong. By the rotational symmetry of  , we are left to consider the reducer possibility  . To prove that ,  are not antisets, replace the left member of the pair by proper subset  and rotate to the pair  , . Again we "depress" the left member and finally get  ,  , clearly not a pair of antisets. For, each nondegenerate cluster A with $|A| = k$, k less than the size of a minimal graph, must have $U = J(A)$ strong. Otherwise $A + Q_m + B$ with B totally degenerate gives a minimal graph (no loops). Since a minimal graph must have at least 40 vertices (Ore-Stemple [11]), small clusters allow the conclusion.

The next cluster to consider is  , the 55-edge, which augments to  . Our first step is to produce an antiset. But we can show that

 ,  is a pair of antisets. Rotate the pair to  ,  and then break down the right member by substitution of  =  to get  ,  . This is a special case of S-construction [4], which depends on work of Shimamoto, as found in Whitney and Tutte [13].

We next claim that pairs  ,  and  ,  and  ,  are not antisets, where B is any totally degenerate set, of appropriate order (order 6,5, 4). Proof is similar in all three cases, so we give only the first, which is longest. Rotate the pair to get  ,  and now show the left member is strong, breaking it down by substitutions:  . Further simplification uses the easily verified set equation 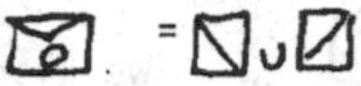 , we finally get 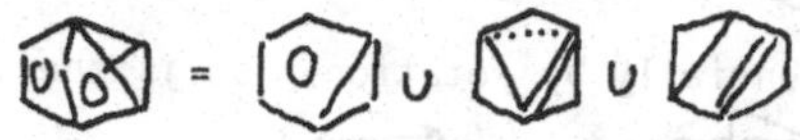 . Proof that this union is a strong open set is left to the reader, and is easy by considering what diagonals the td-cluster B can have.

By reduction of the argument to these three cases, it is not hard to see that  meets all sets which are rotations of  , or  , or totally degenerate, except for possibly  and  . But the pair  ,  rotates to  , and the remaining pair can be checked by hand. Thus the 55-edge is irreducible.

The next cluster is  (simple) or  (augmented). This we call the 555-triad, or just <u>triad</u>. The antiset was first specified by A. Bernhart [3] and ranks as a new primitive building block beside  . Ore and Stemple [11] comment that knowledge of neighborhoods containing the triad is the single factor most inhibiting their theorem being improved (it is an approximation to the 4CC). We write T for the triad, and T' for the antiset, also symbolizing these two by  ,  . Note that a 3-fold symmetric cluster like T has a 3-fold symmetric conjugate by property #A. The symbol for T' (due to Stromquist) has 3-fold rotational symmetry.

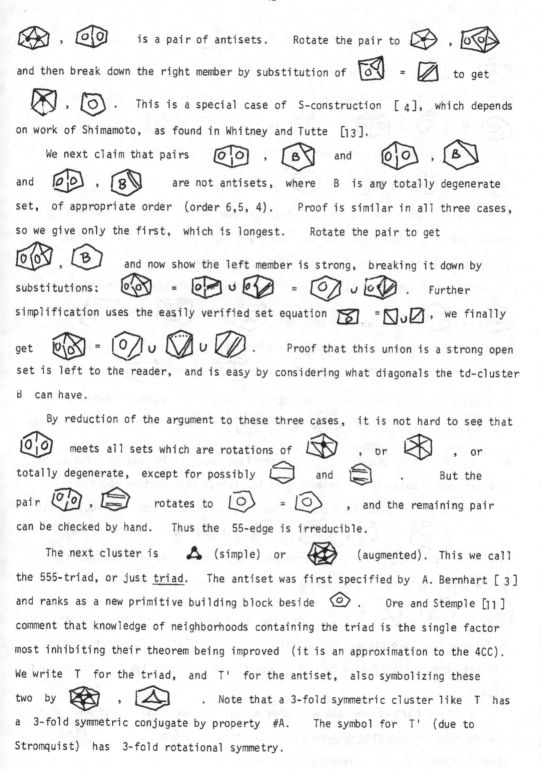

Rotate the pair T, T' to get  ,  . We can infer that 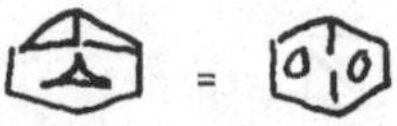 if we assume the fact (not here proved, see [3]) that the 55-edge has a unique nonempty antiset. From this we can get the containment  by recalling that 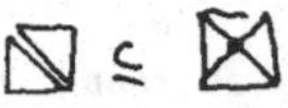 . From the rotational symmetry we get a containment that is actually a set equation:  . The fact that T' exists and is defined be this equation (a poor definition) can be reduced to the labor of showing that the intersection is (a) open and (b) disjoint from T. This is then clearly the maximum antiset, or the conjugate.

Rotation of the pair T, T' gives us an important equation, namely 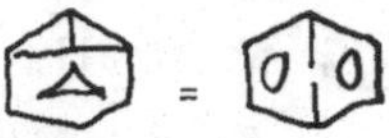 ; we also compute  . The first of these sets is strong, and inferentially by rotation, so is the second. Thus T' meets any set of form  or  where B is any totally degenerate set. That includes (because of 3-fold symmetry), the set  , rotations of  and  , and all td-sets except these two:  and  . For the last two, it is easy to check by hand that each has a color scheme belonging to the triple intersection above.

Attempt to avoid any hand checking at all; we find we need new equations, such as 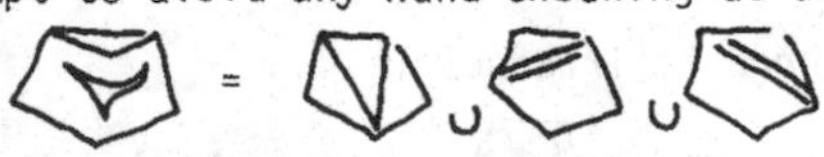 not provided by previous theory. This is symptomatic, because a systematic approach to splicing equations turns up new primitive open sets occasionally, and surprises like the following:  . Our conclusion is this: that open sets are inadequate as a theoretical approximation to realizability, or perhaps the intricate complexity of an actual 4CC counterexample underlies the odd relationships.

It is not hard to see that any other normal cluster construction must give a cluster with the <u>six-ring diamond</u> as sub-cluster:  or  . Since Birkhoff found this reducible, we conclude:

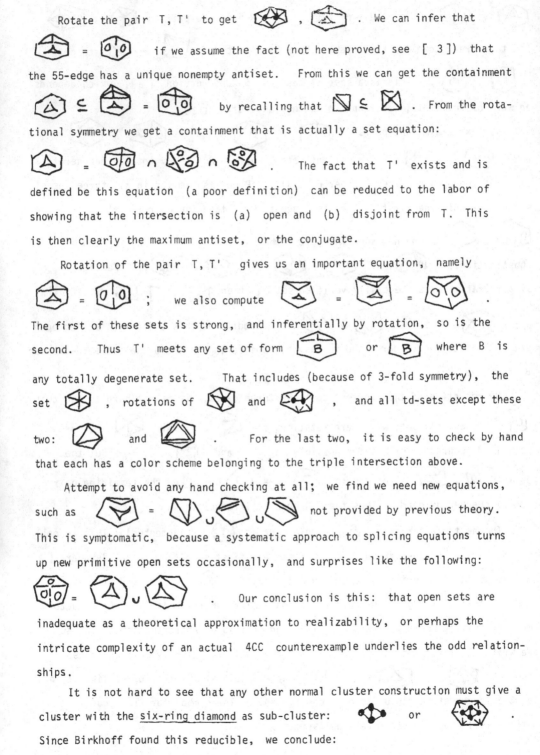

THEOREM 2: The only nondegenerate normal irreducible six-ring clusters are the six-vertex, the 55-edge, and the triad.

A. Bernhart [3], building on Birkhoff [ ], found that besides these three with their antisets, there were to within rotational symmetry only three other pairs of strong antisets. He stated strong geometrical conditions on the realizability of any of them, actually obtained (private conversation) by the splicing of hypothetical configurations -- which ranks as a precursor of open set splicing.

6. General Irreducibility

The irreducibility proofs preceding are defective on two counts, the lack of details, and the fact that the results are essentially known. The primary interest is in terms of method. We feel that a rigorous development of the material needed for irreducibility and extension to higher ring orders should go together. Because the prospect is long and tedious, a condensed account aimed at motivation was considered worthwhile.

In the author's dissertation [5], a systematic search for reducible clusters resulted in nine best possible clusters, seven already published. In the course of the research, absolutely open antisets for non-reduced clusters were identified, and two examples were given. We hope to extend this work to a proof of irreducibility of about three dozen nine ring clusters: the irreducibility of all subclusters follows, and this will extend the current frontier to include all clusters of the seven, eight, and nine rings.

Birkhoff realized that since the number of realizable sets was finite, there must be a finite number of irreducible clusters for any given Q_n. We prove that the 4CC is equivalent to an infinity of irreducibles.

THEOREM 3: The 4CC is equivalent to the conjecture that the set of irreducible clusters (definition of this paper) is infinite.

Proof: First assume the 4CC, then any realizable open set is nonempty. We claim that no wheel W_n is reducible. Already we know W_n' (the complement) is open, so that $\overline{W_n} = W_n$. By generalizing Lemma 2a slightly, we find that for

A any cluster with a degeneracy, the set $J(A)$ must contain a true 4-color scheme. Thus $J(A)$ meets the antiset W_n' and A does not reduce W_n. This leaves as a possible reducer only a nondegenerate A, but then $|A| \geq |W_n|$. (Note we assume here that $n \geq 5$).

To prove the other direction, we need only to show that for large enough n there is a Q_n cluster A with at most n vertices, such that $J(A)$ is empty. Then A is a universal reducer for that ring.

If the 4CC is false, say for a graph of N vertices (if the graph is minimal, then N by definition is the Birkhoff number), we obtain a Q_3 configuration which is a nondegenerate cluster with $N-3$ interior vertices, and the associated open set is empty. If $n \geq 2N - 3$, we can construct a cluster with at least $N-3$ double diagonals and containing the Q_3 configuration as its only nondegenerate part. This finishes the proof.

References

1. F. Allaire and E.R. Swart, A systematic approach to the determination of reducible configurations in the four-color conjecture, to appear in Journ. Comb. Theory (B).
2. K. Appel and W. Haken, An unavoidable set of configurations in planar triangulations, to appear in Journ. Comb. Theory (B).
3. A. Bernhart, Six-rings in minimal five-color maps, Amer. Journ. Math. 64 (1947), 391-412.
4. F. Bernhart, Splicing and the four color conjecture, to appear.
5. F. Bernhart, Topics in Graph Theory Related to the Five Color Conjecture, dissertation, Kansas State University, 1974.
6. G.D. Birkhoff, The reducibility of maps, Amer. Journal Math. 35 (1913), 114- 128.
7. D. Cohen, Small Rings in Critical Maps, thesis, Harvard, 1975.
8. P J. Heawood, Map-colour theorem, Quart. Journ. Math. 24 (1890), 322-338.
9. H. Heesch, Untersuchungen zum Vierfarbenproblem, Bibliog. Instit. AG, Mannheim, 1969.
10. A.B. Kempe, On the geographical problem of the four colours, Amer. Journ. Math. 2 (1879), 193-204.
11. O. Ore and J. Stemple, Numerical calculations on the four-color problem, Journ. Comb. Theory(B) 8 (1970), 65-78.
12. W. Stromquist, Some Aspects of the Four Color Problem, thesis, Harvard, 1975.
13. H. Whitney and W.T. Tutte, Kempe chains and the four color conjecture, Utilitas Math. 2 (1972), 241-281.

NUMBERED COMPLETE GRAPHS, UNUSUAL RULERS, AND ASSORTED APPLICATIONS

Gary S. Bloom
California State University, Fullerton
Fullerton, CA 92634

and

Solomon W. Golomb
University of Southern California
Los Angeles, CA 90007

Abstract

A variety of physical processes can be modelled by assigning integer values to the points and edges of complete graphs. A survey was made of three such numberings, their relation to "ruler models," and their applications to x-ray crystallography, to codes for radar, missile guidance, and angular synchronization, to convolutional codes, to addressing in communications networks, and to an integral voltage generator.

INTRODUCTION

A graph Γ with vertex set $V(\Gamma)$ and edge set $E(\Gamma)$ is numbered if each vertex $x \epsilon V(G)$ is assigned a value $\psi(x)$ (usually an integer), and each edge $(x,y) \epsilon E(\Gamma)$ is assigned a value determined by the values given to its endpoints. That is,

$$\psi(x,y) = |\psi(x) - \psi(y)| .$$

Since every graph may be numbered in infinitely many ways, one is interested in classes of numberings which are determined by imposing additional constraints of either mathematical or physical significance. Graphs with graceful numberings have been of interest in recent years.

1. Graceful Graphs

A graph is considered graceful (or to have a β-valuation) if (1) all e edges of Γ are distinctly labeled with the integers from 1 to e, (2) the n vertices are labeled with distinct non-negative integers, among which (3) the largest value, $\max_{x \epsilon V(\Gamma)} \psi(x)$, equals n. There are many graceful graphs. Figure 1 shows the graphs K_1, K_2, K_3, and K_4 with graceful numberings. It has been previously shown by Golomb [6] that for $n \geq 5$, K_n is not graceful. That is, for $n \geq 5$, K_n cannot be gracefully numbered. An easily understood proof of that property utilizes a ruler

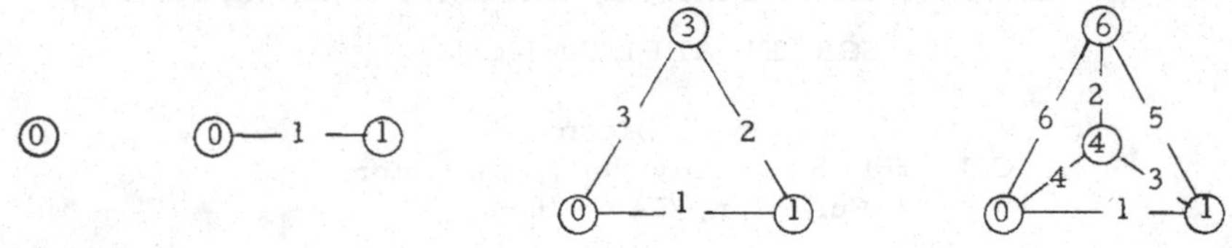

Figure 1. Graceful numberings of K_1, K_2, K_3, K_4.

model of numbered complete graphs.

2. Ruler Models

A ruler model may be constructed from any numbered K_n with distinct vertex numbers in the following way: For each vertex x of K_n, place a mark at $\psi(x)$ on the ruler. By this process the $\binom{n}{2}$ distances which the ruler can measure are numerically equal to the edge numbers of K_n. Figure 2 shows the rulers corresponding to the previous numberings of K_2, K_3, and K_4. Additional, but equivalent, graceful numberings of K_3 and K_4 are obtained by numbering from the right end of the rulers shown.

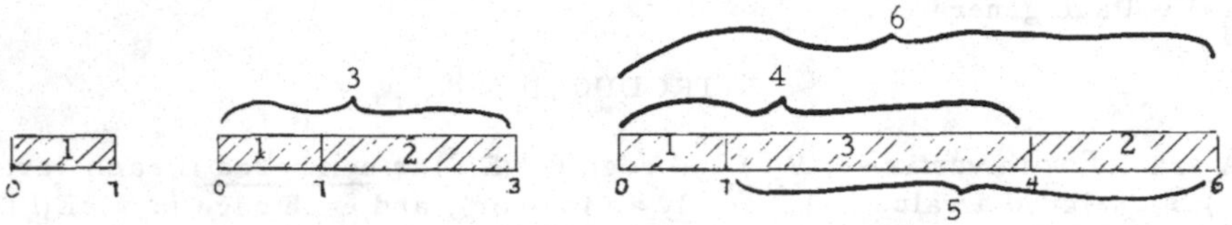

Figure 2. Ruler Models of the Graceful Numberings of K_2, K_3, and K_4.

With this ruler model Simmons [12] gave a simple proof of the lack of grace of K_n, $n \geq 4$. His argument is worth repeating for two reasons: (1) It emphasizes the theorem, whose truth was our motivation for seeking alternative numberings which are "as graceful as possible." (2) It emphasizes the intimate connection between ruler models and numbered complete graphs.

Proof. A graceful n-mark, (n-1)-section ruler must have length $\binom{n}{2} = \frac{1}{2}(n^2-n)$, the largest edge number of the graceful graph. This length partitions into n-1 distinct integers uniquely. That is, the n-1 integers in the set $\{1, 2, \ldots, n-1\}$ must comprise the n-1 sections of the ruler. The problem is to assemble these ruler sections in a graceful order. Because a graceful numbering is non-redundant, no measurement can be duplicated on the ruler. Because the single ruler sections measure all lengths up to n-1, the section of length 1 can be adjacent only to the section of length n-1. Since the individual sections and the adjacent pair [1 | n-1] measure all lengths up to n, the length 2 section can be adjacent only to the section of length n-1, also. The resulting new measurements are n+1 and n+2: [1 | n-1 | 2] . Assembly of a graceful ruler can go no farther. Because of the adjacency restrictions on the length 1 and 2 ruler sections, there can be no additional sections appended to the illustrated ruler. Hence, the length of the middle ruler section is 3 and the maximum value of n is 4. We conclude that no ruler with more than three pieces (or four marks) can be graceful.

It follows that rulers corresponding to numberings of K_n, in general, must either be longer than $\binom{n}{2}$ (if no measurements are repeated) or must repeat measurements (if all consecutive integers must be present); or they will do both, if these special restrictions are not made. We will report on applications in each of these three classes, starting with the non-redundant numberings that repeat no edge number.

NON-REDUNDANT NUMBERINGS

Applications in X-ray diffraction crystallography and in a variety of coding problems are encompassed by numbering K_n in this manner.

1. Ambiguities in X-Ray Crystallography

Ruler models have long been utilized in X-ray crystallography. In that science positions of atoms in crystal structures are determined by measurements made on X-ray diffraction patterns. These measurements indicate the set of inter-atom distances in the crystal lattice, but, in general, do not necessarily, unambiguously specify the absolute positions of atoms. Mathematically, one finite set of integers $R = \{0 = a_1 < a_2 < \ldots < a_n\}$ (corresponding to atom positions) and another set $S = \{0 = b_1 < b_2 < \ldots < b_n = a_n\}$ (corresponding to other atom positions) may have exactly the same set of differences $D(R) = D(S) = \{a_i - a_j \mid i > j\}$. Since the diffraction pattern determines the set of differences, D(R), it is impossible to determine which of the homometric sets R or S produced it, and consequently which crystal lattice gave rise to the diffraction pattern. Less than 2 years ago, Franklin [4] displayed pairs of homometric sets and indicated how to generate certain families of these sets.

For our purposes, the homometric set problem may be viewed as a determination of non-equivalent rulers which make identical sets of measurements. The sets R and S designate the positions of the n marks on two rulers and D(R) and D(S) are their respective sets of $\binom{n}{2}$ measurements.

For many years one class of diffraction patterns was believed to be unambiguous. This class corresponds to a set of differences which has no repeated elements, that is, to a non-redundant set.

In 1939 S. Piccard [10, p. 31] stated that no two non-equivalent rulers could measure a set of all distinct distances. Last August Bloom [2] determined the falsity of this "theorem" with the pair of rulers shown in Figure 3. Since that time we have proved that this is the smallest such counterexample. That is, there are no non-redundant rulers with fewer than 6 marks or of length less than 17. In addition, we have shown that these 2 rulers are members of the following

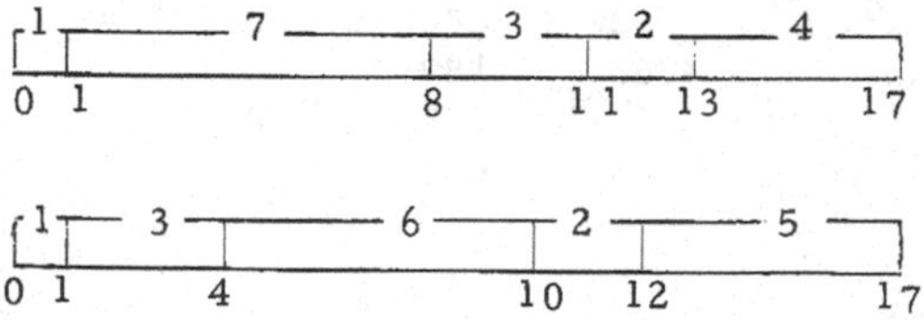

Measurements made by these rulers:

1, 2, 3, 4, 5, 6, 7, 8, 9, 10, 11, 12, 13, 16, 17

Figure 3. The shortest, non-redundant, homometric pair of rulers and the $\binom{5}{2} = 10$ intervals which they measure.

2-parameter, family of infinitely many homometric ruler pairs.

$$R = \{0, u, u+v, 4u+2v, 6u+2v, 8u+3v\}$$

$$S = \{0, u, 5u+v, 5u+2v, 7u+2v, 8u+3v\} .$$

The original smallest example occurs when $u = 1$, $v = 3$. A second two-parameter family of non-redundant homometric pairs is given by

$$R' = \{0, y, y+w, 4y+2w, 6y+4w, 8y+5w\}$$

$$S' = \{0, y, 3y+w, 3y+2w, 7y+4w, 8y+5w\} .$$

We have developed a variety of algorithms and theorems which determine and describe these sets. Among these are algorithms which allow us to generate the general embedding family of solutions given any <u>one</u> particular counter-example in that family. We also have formulated a very efficient, computerized search algorithm that finds all homometric pairs among all rulers with n marks and length L. Table 1 gives current "census" data for rulers with up to length 55 with 10 marks. An intriguing problem which currently exists is to explain why

marks	length	rulers	homometric pairs
6	17	4	1
	18	4	1
	19	20	1
	20	35	1
	21	86	1
	22	101	0
	23	203	2
	24	282	0
	25	419	1
	26	520	1
	27	861	1
	28	1007	1
	29	1442	2
7	25	5	0
	26	7	0
	27	20	0
	28	60	0
	29	147	0
	30	190	0
	31	429	0
	32	655	0
	33	1048	0
8	34	1	0
	35	9	0
	36	14	0
	37	48	0
	38	91	0
	39	192	0
	40	379	0
	41	763	0
	42	1102	0

marks	length	rulers	homometric pairs
9	44	1	0
	45	1	0
	46	4	0
	47	21	0
	48	40	0
	49	96	0
	50	167	0
	51	398	0
	52	699	0
	53	1425	0
10	55	1	0
	56	0	0
	57	0	0
	58	1	0
	59	2	0
	60	14	0
	61	31	0
	62	74	0
	63	189	0
	64	357	0
	65	660	0

TABLE 1. Current results of computer search for homometric pairs of rulers.

the only homometric ruler pairs found in this range are members of the two families just described. What makes these six-mark rulers so special, and what other sets, if any, have such special properties?

2. Radar Pulse Codes

The original pair of non-redundant, homometric rulers shown in Figure 3 has an additional property which makes them useful to designers of several classes of codes. No shorter non-redundant rulers with 6 marks exist (although two others are just as short). Gardner [5] called these n-mark, non-redundant, minimum-length,semi-gracefully numbered measuring devices Golomb Rulers. To date all such rulers with fewer than 12 marks have been found (see [1, p.27-30]) and are shown in Table 2. The discovery of Golomb Rulers with more marks as well as a method for generating this class remain open questions.

TABLE 2. Golomb Rulers for n ≤ 1.

n Nodes	Length f(n)	Divisions	Marks at
2	1	1	0.1
3	3	1,2	0.1.3
4	6	1,3,2	0.1.4.6
5	11	1,3,5,2	0.1.4.9.11
		2,5,1,3	0.2.7.8.11
6	17	1,3,6,2,5	0.1.4.10.12.17
		1,3,6,5,2	0.1.4.10.15.17
		1,7,3,2,4	0.1.8.11.13.17
		1,7,4,2,3	0.1.8.12.14.17
7	25	1,3,6,8,5,2	0.1.4.10.18.23.25
		1,6,4,9,3,2	0.1.7.11.20.23.25
		1,10,5,3,4,2	0.1.11.16.19.23.25
		2,1,7,6,5,4	0.2.3.10.16.21.25
		2,5,6,8,1,3	0.2.7.13.21.22.25
8	34	1,3,5,6,7,10,2	0.1.4.9.15.22.32.34
9	44	1,4,7,13,2,8,6,3	0.1.5.12.25.27.35.41.44
10	55	1,5,4,13,3,8,7,12,2	0.1.6.10.23.26.34.41.53.55
11	72	1,3,9,15,5,14,7,10,6,2	0.1.4.13.28.33.47.54.64.70.72
		1,8,10,5,7,21,4,2,11,3	0.1.9.19.24.31.52.56.58.69.72

Radar distance ranging is accomplished by transmitting a pulse or train of pulses with the speed of light and waiting for its return. During its time of flight, the pulse or pulse-train reaches its target, bounces off, and returns. Because of the dispersion of energy occuring both during transmission of the signal and its scattering during reflection, only a small fraction of the transmitted energy ever returns to the detector. Greater energy in a transmitted signal effectively extends the range of a radar system, but there is an offsetting factor to simply increasing power.

The exactitude of the measurement of signal's time-of-flight determines the accuracy in measuring target distance. It is thus desirable to have a very narrow transmitted radar pulse whose instant of return can be accurately determined. Since generally a larger energy pulse is "wider" as well as "higher" than a low energy pulse,resolution is lost in time-interval measurements for large pulse systems. A solution to this problem is to design the signal as a pulse-train with the energy evenly divided among a set of low-amplitude, narrow-width pulses. The total energy in the pulse-train unit can be increased by increasing the number of pulses in the train.

If a series of n radar "pulses" are transmitted at times corresponding to marks on a non-redundant ruler, it is easy to determine precisely when the pulse-train returns. A signal of relative amplitude n will be generated when the returning signals precisely align with an array of detectors distributed like a template of the transmitted pulse-train. At any other time, no more than one pulse can excite any detector in the template. Moreover, if the temporal positions of the pulses occur at mark positions on a Golomb Ruler, the overall duration of the train will be minimized. Figure 4 schematically shows a returning pulse-train and the associated detector array, as well as the autocorrelation function of the pulse-train. This illustrated pulse-train uses the marks shown on the first of the rulers in Figure 3. The resulting autocorrelation is six units high when the incom-

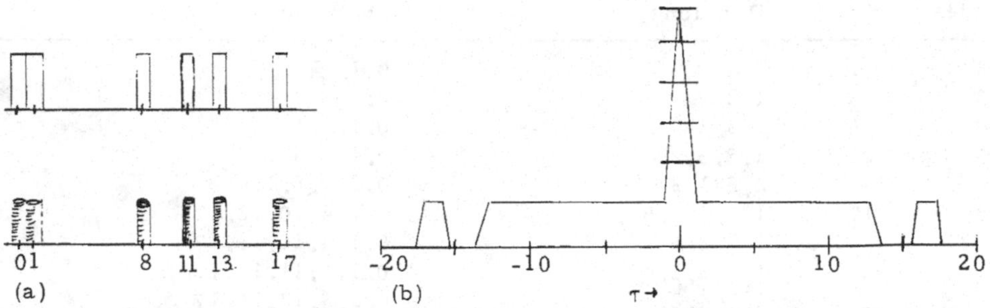

Figure 4. a. Returning pulse train in perfect alignment with detector array at $\tau = 0$.

b. Autocorrelation of pulse train indicating relative signal amplitudes as a function of time after perfect alignment.

ing pulse string aligns perfectly with the detector. At all out-of-synchronization positions corresponding to possible ruler measurements, a single slit aligns and a "noise" signal of amplitude 1 is generated. For alignment times corresponding to non-measurable lengths on the ruler, no slits align and the autocorrelation is zero.

3. Missile Guidance Codes

Eckler [3] investigated the related problem of designing missile guidance codes. In an airborne missile, an incoming signal train is passed down a delay line which taps the signal at times corresponding to the intervals between incoming pulses. When the sum of the tapped signals exceeds a preset threshold, a control action is initiated. A variety of controls is possible if each action is associated with its own set of taps on the control line awaiting the proper triggering signal train.

The design of a set of triggering pulse trains is a problem similar to that of the single train, e.g. the trains should be of minimum overall duration, and should have minimum out-of-synch autocorrelation amplitude. An obvious additional constraint requires minimizing the chance that any two control commands be confused--even in the face of possible jamming efforts or electrical storms. Thus, the cross-correlation between the different pulse trains should be minimal, e.g. no more than 1 unit at all times. In terms of ruler models, designing a set of n commands requires determining n different rulers on which no measurement between any pair of marks is repeated. All rulers are to have n marks. This makes n the relative magnitude of the out-of-synch autocorrelation for each control action, which in turn sets the triggering level for the control actions. Figure 4 shows two mutually non-redundant, 4-mark rulers of minimum length. The $2\binom{4}{2} = 12$ measurements made by this <u>set of rulers</u> omits only length 10. The comparable numbered complete graphs are also shown in the figure. Determination of optimal non-redundant ruler sets remains an open problem.

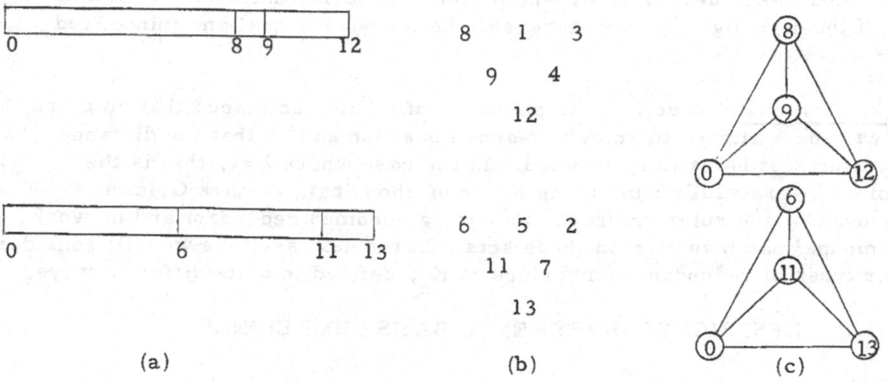

Figure 5. a. A set of mutually non-redundant rulers used to generate a 2 message, 4 pulse missile code of minimum length.

b. The sets of measurements for each ruler.

c. The comparable numbered graph with two K_4 components.

4. <u>Convolutional Codes</u>

<u>Convolutional codes</u> are a broad class of algebraic codes with error recovery properties in which source messages are partitioned into short blocks whose encoding depends not only on the messages in that block but upon some fixed number of previous blocks as well. <u>Self-orthogonal codes</u> are a class of convolutional codes designed with sets of mutually non-redundant rulers as the underlying model. These codes which were initially studied in 1967 by Robinson and Bernstein [1] have the property that d-1 marks on each of the underlying ruler models separate all code words by distance d; that is, each of transmitted binary code words differs from every other code word in exactly d bit positions. The number of rulers needed is equal to the number of "information bits" in each code block.

The transmission rate of self-orthogonal codes is maximized by making the longest ruler in any set as short as possible. Consequently, solutions to the best missile guidance codes are also solutions here. In this light, the rulers in Figure 5 give parameters needed to construct an optimal self-orthogonal code with 2 information bits per block in which code words differ from one another by 4 bits.

5. Synch Set Codes

In 1974 Simmons [12] designed these codes in order to synchronize the relative angular position of a photodetector on one side of a rotating disk with a stationary, coaxial target light source on the other side. If "noise" were no problem in this system, synchronization could be provided by lining up a single slit in an otherwise opaque annular region on the disk with an identically sized slit in front of the detector. Since the amount of light so passed is insufficiently low, "slit codes" were developed by which identical performation of the detector mask and of the disk significantly increased the in-synch signal and minimized passage of out-of-synch light.

An $S(k, \lambda)$ synch set specifies the positions of k holes, so placed that no more than λ holes can be aligned in an out-of-synch position and so that the distance between the furthest holes is minimized. In the case where $\lambda = 1$, this is the problem of semi-gracefully numbering K_k or of choosing a k-mark Golomb Ruler. When $\lambda > 1$ the ruler or graph numbering becomes redundant and no work beyond Simmons' has investigated these sets. In the next sections we will consider some other types of redundant numberings of K_n, defined in quite different ways.

RESTRICTED DIFFERENCE BASIS NUMBERINGS

1. General

When $n > 4$ all of the requirements for gracefully numbering K_n cannot be met. In the previous section, we discarded the requirement that all integer values be present as edge numbers from 1 to $\binom{n}{2}$, the number of edges. However, we retained, both the requirements that there be no redundancy, that is, no repeat of edge numbers, and that the maximum edge value, max $\psi(u, v)$, equal $G(K_n)$, the largest node value. In this section, we relax the redundancy requirement, so that more than one edge of K_n may be assigned the same number. The largest edge value must still equal $G(K_n)$. Moreover, in these cases gaps must not occur in the edge number sequence $1, \ldots, G(K_n)$. Also, in these numberings, n must be as large as possible for each $G(K_n)$ chosen. The vertex numbers of such a set were called a restricted difference basis (RDB) by Leech [8].

It is natural once again to view these numberings of K_n in ruler fashion; indeed, such a ruler will measure (with repeats) every integral distance from 1 unit to its length, $G(K_n)$. For each n, there is a maximum length ruler that will satisfy the constraints. Examples of maximum length restricted difference bases are shown in Table 3. Families of these bases have been discovered and investigated in the last 20 years. Miller [9], in fact, has exhaustively calculated all RDB's for every N, $1 \leq N \leq 68$ as well as for many other values of N.

2. Applications

Utilization of this model may appear in a variety of contexts, in which a need for an uninterrupted sequence of integer values exists. We will briefly describe two such contexts.

n	Restricted difference basis	Number of Edges
8	{0, 1, 4, 10, 16, 18, 21, 23}	23
9	{0, 1, 2, 14, 18, 21, 24, 27, 29}	29
10	{0, 1, 3, 6, 13, 20, 27, 31, 35, 36}	36

3. Examples of restricted difference bases of maximum length for 8, 9, and 10 marks.

Integral valued voltages are needed, let us say, for testing. One would like to construct a box with the least number of terminals between whose terminal pairs all integral voltages up to the value of the maximum voltage could be tapped. That is, at least one pair of terminals has each chosen voltage up to the maximum value as its difference. Naturally, for a given number of terminals, n, one would like to have the sequence of consecutive integral voltages extend as far as possible. Assigning the values of the elements of an n member restricted difference basis to the terminals realizes all of these objectives.

A second application is the design of an efficient addressing system for the terminals and links in a communication network. For a network of e links between terminals, one can assign the identifying numbers 1 through e in a manner determined automatically by the terminals' own numerical designations. This is accomplished by setting each links' identification number to the absolute difference of the identification numbers of the two connected terminals. To avoid ambiguity, no two links may join pairs of terminals with a common difference between their identification numbers.

The requirements above cannot be realized for links between all pairs in an n-terminal network, for this would give a graceful numbering of K_n, which we have shown was impossible for $n > 4$. Rarely, however, are all terminals directly connected. This situation leads naturally to asking "How large a subgraph of K_n can be gracefully numbered?"

3. The Largest Graceful Subgraph of K_n

This question was posed by Golomb in 1974 [7] and subsequently studied until Leech observed its solution by restricted difference bases. If one numbers the nodes of K_n with an n-element RDB, all consecutive integers from 1 to the maximum node value, $G(K_n)$, are assigned to edges. A duplication of numbers to edges occurs on $\binom{n}{2} - G(K_n)$ edges. Some set E_1 of edges with duplicate numbers must be removed. The resulting graph, $K_n - E_1$, is a maximal graceful subgraph of K_n.

One RDB gives rise to more than one graceful subgraph, contingent upon which of the edges of K_n with duplicate numbers are eliminated. Figure 6 shows all four possible graceful numberings resulting from eliminating edges with duplicate numbers from K_6, when the RDB {0, 1, 2, 6, 10, 12} is used.

There are some interesting properties of these subgraphs of K_n which time precludes discussing in length here. For example, from Figure 6 one can observe that a particular RDB will give rise to as many distinct graceful numberings of maximal graceful subgraphs of K_n as the product of the cardinalities of elements in the set of differences resulting from that basis. Symbolically, let B be a RDB, and let c(i) be the number of repetitions of i in the set of differences D(B) for RDB B. Let M(B) be the number of graceful numberings derived from basis B. Then $M(B) = \prod_{i \in D(B)} c(i)$.

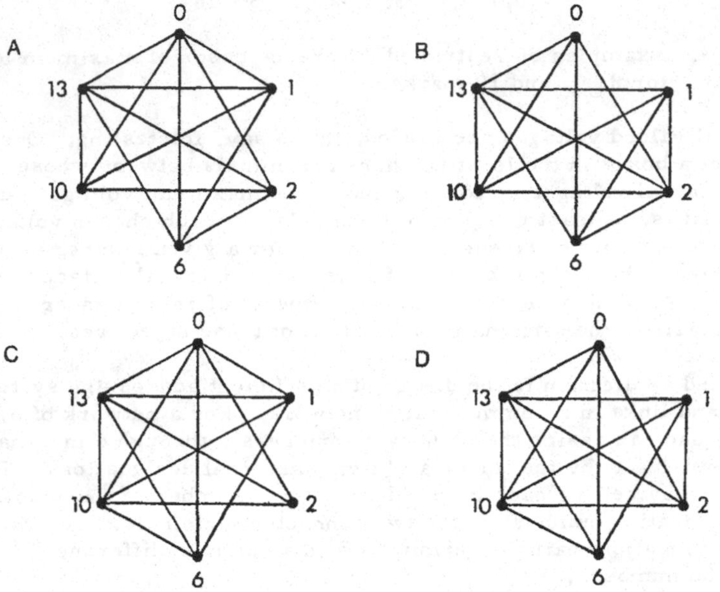

Figure 6. The graceful numberings of the largest graceful subgraph of K_6.

From the example one can see that the numberings associated with an RDB B can be partitioned into equivalence classes by isomorphisms of the subgraphs of K_n that they number. In Figure 6 A, B, and D illustrate the numberings of isomorphic graphs. Comparing their degree sequence with that of the graph C shows that the latter is not isomorphic to the others.

One of the most important properties of the graphs numbered this way to the communication network problem is the limiting number of sequentially labelled links that the graph with n nodes can contain. By inverting a bound of Leech's (subsequently improved by Wichmann [13] and by Miller [9]), we have shown that when n is large, approximately 1/3 of the edges of K_n must be graceful. This, then, is the limit for the number of communication links that our n-terminal network can contain with this addressing method.

If one more constraint of the restricted difference basis numbering is dropped, additional sequentially labeled edges can be included in the numbered K_n subgraph.

UNRESTRICTED DIFFERENCE BASES NUMBERINGS

On the ruler models for numberings of K_n in the previous section, one could measure all integral lengths from 1 to the ruler's length. If the constraint is removed that a ruler can be no longer than the longest consecutive measurement it makes, then the range of the consecutively made integral measurements can be extended. That is, if L_R is the length of an m-mark unrestricted difference set,

L_u that of an unrestricted difference basis, then $L_R \leq L_c \leq L_u$, where L_c is the longest integral interval measured consecutively (starting from 1 unit). The sets of integers constituting the mark positions in these sets are called unrestricted difference bases (UDB) and were also studied by Leech and Miller.

By using a UDB to label the vertices of K_n, and then, first, discarding edges with duplicated numbers and, second, edges numbered above the consecutive sequence, a larger subgraph of K_n can be retained than by labeling with a RDB. The resulting graph numbering is not graceful since the largest vertex number does not equal the largest edge number. However, since the graph has a non-redundant numbering of edges with all consecutive integral lengths up to the largest, we term the graph and its numbering quasi-graceful. Figure 7 gives a comparison of a maximal graceful graph on 8 vertices (generated by a RDB) and a maximal quasi-graceful graph on 8 vertices (generated by a UDB).

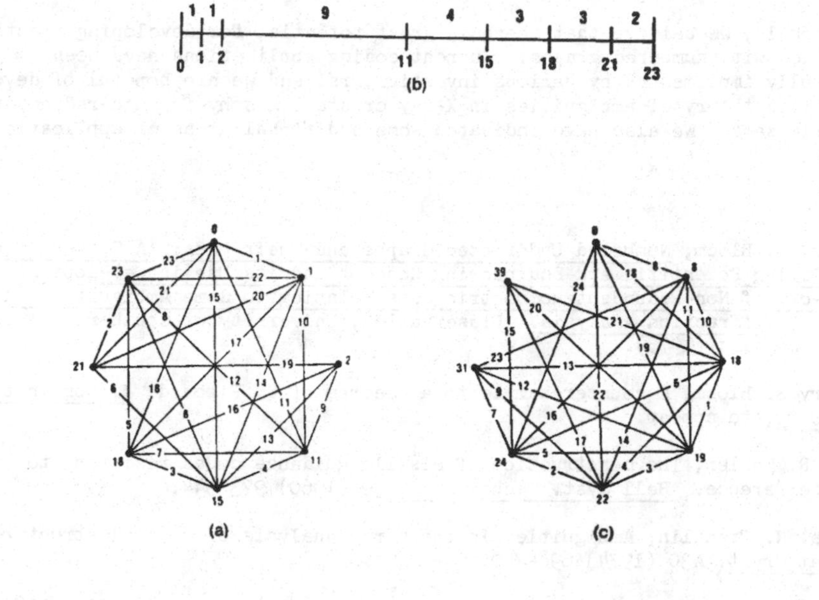

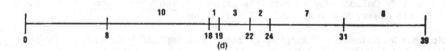

Figure 7. A comparison of maximal graceful and quadi-graceful graphs on 8 vertices.
(a) A maximal graceful graph on 8 vertices. 5 chords not used: 0-2, 1-2, 2-23, 11-21, 18-21.
(b) The "ruler" generating (a).
(c) A maximal quasi-graceful graph on 8 vertices. 4 chords not used: 0-31, 0-39, 8-39, 31-39.
(d) The "ruler" generating (c).

In the addressing plan for the communication networks terminals and links, the UDB numbering allows more links. Nevertheless, the asymptotic properties of these numberings have not been studies and whether in the limit of many nodes fewer than 1/3 of the links would need to be discareded is unknown. Usiang a UDB for the voltage generation box allows one to extend the box's range. To so do may require using some substantially higher node voltages than are covered in the sequential range, as we showed in the previous example.

CONCLUSIONS

In this brief survey we have attmepted to leave two principal impressions. First of all, mathematically, we have tried to indicate that interesting work can be done in numbered graphs. In this talk, we have emphasized the equivalence of numbered complete graphs and ruler models in several contexts. We have cited some of the recent results in investigating these models, but hopefully both by our expression of remaining problems and by our omission of citing any all-embracing theory, we have left the impression that many questions are still unanswered. Of course, many results were not cited because of time constraints, but we have been finding new results in each class of models we have investigated.

Secondly, we believe that there is great potential for developing practical mathematics with numbered graphs. Current coding applications have been successfully implemented by various investigators, and we are hopeful of developing a definitive theory of ambiguities in X-ray crystallography for non-redundant homometric sets. We also have indicated some additional areas of applications.

REFERENCES

1. Gary s. Bloom, Numbered Undirected Graphs and Their Uses: A Survey of a Unifying Scientific and Engineering Concept and its Use in Developing a Theory of Non-Redundant Homometric Sets Relating to Some Ambiguities in X-Ray Diffraction Analysis. Dissertation, University of Couthern California, Los Angeles (1975).

2. Gary S. Bloom, A counterexample to a theorem of S. Piccard. J. Combinatorial Theory, to appear.

3. A. R. Eckler, The construction of missile guidance codes resistant to random interference. Bell Syst. Technical J. 39 (1960) 973-944.

4. Joel N. Franklin, Ambiguities in the s-ray analysis of crystal structures. Acat Cryst. A30 (1974) 698-702.

5. Martin Gardner, Mathematical Games. Scientific American 226 (1972) 118-121.

6. Solomon W. Golobm, How to number a graph. Graph Theory and Computing. Academic Press, New York (1972) 23-37.

7. Solomon W. Golomb, The largest graceful subgraph of the complete graph. Amer. Math. Monthly 81 (1974) 499-501.

8. John Leech, On the representation of 1,2,...,n by differences. J. London Math. Soc. 31 (1956) 160-169.

9. J. C. P. Miller, Difference bases, three problems in additive number theory. Computers in Number Theory (eds. A. D. L. Atkin and B. J. Birch). Academic Press, London (1971) 299-322.

10. S. Piccard, Sur les Ensembles de Distance des Ensembles d'un Espace Euclidean. Librarie Gauthier-Villars & Cie, Paris (1939).

11. John R. Robinson and Arthur J. Bernstein, A class of binary recurrent codes with limited error propagation. IEEE Trans. on Information Theory IT-13 (1967) 106-113.

12. G. J. Simmons, Synch sets: A variant of difference sets. Proceedings of the Fifth Southeastern Conference on Combinatorics, Graph Theory, and Computing. Utilitas Mathematica Publishing, Winnipeg (1974) 625-645.

13. B. Wichmann, A note on restricted difference bases. J. London Math. Soc. 38 (1963) 465-466.

CYCLES AND SEMI-TOPOLOGICAL CONFIGURATIONS

Béla Bollobás
University of Cambridge
ENGLAND

and

University of Calgary
CANADA

Given a graph G we denote by $|G|$ the order of G (i.e. the number of vertices) and by $e(G)$ the size of G (i.e., the number of edges). Recently I obtained the following result [2], proving a conjecture of Burr and Erdös [6].

Theorem 1. Let k be an odd natural number and let h be an arbitrary integer. Then there exists a constant $c = c(h, k)$ such that if G is any graph satisfying

$$e(G) \geq c|G|$$

then G contains a cycle of length h modulo k.

The existence of this constant had been proved earlier for two values of h. Erdös proved it for $h = 2$ and Robertson proved it for $h = 0$ (cf. [6]).

It is easily seen that Theorem 1 can not be extended to even values of k since for every constant $c > 0$ there is a bipartite graph G satisfying $e(G) > c|G|$.

In fact Theorem 1 is a consequence of the following result.

Theorem 2. Given natural numbers s and t there is a constant $c = c_1(s, t)$ such that if

$$e(G) \ge c|G|$$

then G contains a path P, a vertex x not on P and t paths of length s, say $P_1, P_2, \ldots, P_t$, such that P_i joins x to a vertex x_i on P. This vertex x_i is the only common vertex of P and P_i and any two paths P_i and P_j have only the vertex x in common.

To see that Theorem 2 implies Theorem 1 choose s, $1 \le s \le k$, such that

$$2s \equiv h (\mathrm{mod}\ k)$$

and put $t = k + 1$. Let $c = c(h, k) = c_1(s, t)$ and suppose $e(G) \ge c|G|$. Let $\{P, x, P_1, P_2, \ldots, P_t\}$ be the system guaranteed by Theorem 2. Then there are at least two endvertices of the paths $P_1, P_2, \ldots, P_t$, say x_i and x_j, such that the distance of x_i from x_j on P, say $d'(x_i, x_j)$, is divisible by k. Then the cycle formed by P_i, P_j and the segment of P joining x_i to x_j has length

$$2s + d'(x_i, x_j) \equiv h (\mathrm{mod}\ k).$$

Theorem 2 happens to be one of a number of results of similar nature. Pósa [11] proposed the following exercise in a Hungarian journal: Prove that if $e(G) \ge 2|G| - 3 \ge 5$ then G contains a cycle and one of its diagonals. The solution of Czipszer [11] of this exercise shows that if $e(G) \ge (k+1)|G|$ then G contains a cycle and k diagonals incident with a vertex. Erdős and this writer [4] proved that if $|G| = n \ge 4$ and $e(G) > \left\lfloor \frac{3}{2}(n-1) \right\rfloor$ then G contains a cycle C, a vertex x_0 not on the cycle and two edges joining x_0 to vertices on C. Mader [9] proved that if $e(G) \ge 3 \cdot 2^{p-3}|G|$ and $p \ge 4$ then G contains a subdivision of K_p, the complete graph of order p, i.e., G contains a TK_p, a <u>topological</u> complete graph of order p .

In [3] this writer defined the following generalization of all these configurations. Let F be a multigraph and let F_o be a spanning subgraph of F . A <u>semi-topological graph</u> F <u>with kernel</u> F_o is a graph obtained from F by subdividing some of the edges <u>not</u> belonging to F_o . Also in [3] this writer proved the following theorem which, in a certain sense, extends all the results mentioned so far (with the exception of Theorem 1 which, as we have seen, is a consequence of Theorem 2).

<u>Theorem 3</u>. Given a pair (F_1F_o) as above, there is a constant $c = c(F_1F_o)$ such that whenever

$$e(G) \geq c|G|$$

the graph G contains an $ST(F_1F_o)$ if and only if F_o is a forest. If F_o is a forest of order p then $c(F_1F_o) = 2^e(p-1)$ is an appropriate constant, where $e = e(F) - e(F_o)$.

Note that the "only if" part of this result is trivial so the theorem states that the obvious necessary condition for the existence of a constant is also sufficient.

Let me state a conjecture that would also imply Theorem 2.

<u>Conjecture</u>. Given a constant $\alpha > 0$ and a natural number k there is a constant $c = c(\alpha, k)$ with the following property. If $e(G) \geq c|G|$ then there is a graph H with $e(H) \geq \alpha|H|$ such that the graph obtained from H by subdividing each edge of H into exactly k edges is a subgraph of G .

Given a graph G and vertices x and y in G, the vertex connectivity of G between x and y, denoted by $\kappa(x, y)$, is the maximal number of vertex disjoint paths joining x to y . Let $\kappa(G)$ be the maximum of $\overline{\kappa}(x, y)$ as x and y run over the vertices of G . Denote by $e(n; \overline{\kappa} \leq k)$ the maximal number of edges in a graph G of order n provided $\overline{\kappa}(G) \le k$. As a problem in a competition for undergraduates in Hungary, Erdős [5] asked the value of $e(N; \overline{\kappa} \le 2)$. It turned out that $e(n; \overline{\kappa} \le 2) = \left[\frac{3}{2}(n-1)\right]$. In [1] I showed $e(n; \overline{\kappa} \le 3) = 2n - 2$ $(n \ge 4)$ and Sørensen and Thomassen [12] proved that $e(n; \overline{\kappa} \le 4) = \left[\frac{8}{3}n\right] - 4$ $(n \ge 6, n \ne 7$ and $n \ne 12)$. It is easily seen that

$$e(n; \overline{\kappa} \le k) \le \left[\frac{(k+1)(n-1)}{2}\right]$$

and Erdős and this writer conjectured that if n is sufficiently large compared to k then we have equality here. This was disproved by Leonard [7] and Mader [8]. One can check that $\lim_{n\to\infty} n^{-1} e(n; \overline{\kappa} \le k)$ exists. Putting

$$\lim_{n\to\infty} n^{-1} e(n; \overline{\kappa} \le k) = \frac{k+1}{2} + \alpha_k$$

Mader's result implies $\alpha_k \ge \frac{c}{k}$ $(k_i \ge 4)$. Sørensen and Thomassen [12] proved that $\alpha_k \ge \frac{k-3}{4k-2}$ if $k \ge 4$. Our next result implies that, in fact, $\alpha_k \to \infty$.

<u>Theorem 4</u>. Suppose p is a prime power and $p^2 + p + 1 \le k$. Then

$$e(n; \overline{\kappa} \le k) > \frac{k(k+1) - 2p^3}{2(k+1-p)} (n - (p^2+p+1)k) .$$

<u>Proof</u>. The proof is based on the following lemma whose straightforward verification we leave to the reader.

<u>Lemma</u>. Let G_i, $1 \le i \le p^2 + p + 1$, be disjoint graphs with $\overline{\kappa}(G_i) \le k$. Let $\{x_0^i, x_1^i, \dots, x_p^i\}$ be the vertex set of a K_{p+1} in G_i such that

$$\overline{\kappa}(x_j^i, x_h^i) \le k - p^2, \quad 1 \le i \le p^2 + p + 1, \; 0 \le j < h \le p .$$

Suppose $x_{p+1}^1 \ne x_0^1$ is a vertex of G_1 joined to each

x_i^1, $1 \le i \le p$, and

$$\kappa(x_{p+1}, x_i^1) \le k - p^2, \quad 1 \le i \le p .$$

Let G be the graph obtained from the vertex disjoint union of the graphs $G_1, G_2, \ldots, G_{p^2+p+1}$ by identifying some of the vertices x_j^i with each other in such a way that the sets

$$\{x_o^i, x_1^i, \ldots, x_p^i\}, \quad 1 \le i \le p^2 + p + 1,$$

become the lines of a projective plane of order $p^2 + p + 1$. Then

$$\overline{\kappa}(G) \le k \quad \text{and} \quad \kappa(x_j^1, x_h^1) \le k - p^2, \; 1 \le j < h \le p + 1 .$$

Armed with this lemma we proceed to the proof of Theorem 4. We may assume without loss of generality that $n \ge k + 1$. Let r be fixed, $0 \le r \le (p^2 + p)k - p^3$, and let $m \ge 0$. We shall prove by induction on m that there exists a graph H_m of order $n = k + 1 + r + ((p^2 + p)k - p^3)m$ such that

$$e(H_m) > (m-1)(p^2 + p)\left(\binom{k+1}{2} - p^3\right),$$

$$\overline{k}(H_m) \le k$$

and H_m contains a k_{p+1} with vertex set $\{z_o^m, z_1^m, \ldots, z_p^m\}$ such that

$$k(z_j^m, z_h^m) \le k - p^2, \quad 0 \le j < h \le p .$$

Note that the second and third conditions are exactly the ones imposed on G_i, $i \ge 2$, in the Lemma. Let H_o consist of a k_{p+1} and isolated vertices. Suppose we have constructed $H^{m-1} (m \ge 1)$. Let G_i, $1 \le i \le p^2 + p$, be obtained from a K^{k+1} by selecting a k_{p+2} in it with vertex set $\{x_o^i, x_1^i, \ldots, x_{p+1}^i\}$ and omitting p^3 of the edges such that each of the vertices $x_1^i, x_2^i, \ldots, x_p^i$ has degree $k - p^2$. Construct

H_m from $G_1, G_2, \ldots, G_{p^2+p}, G_{p^2+p+1} = H^{m-1}$ as described in the Lemma with $x_j^{p^2+p+1} = z_j^{m-1}$, $0 \le j \le p$. Then

$$|H_m| = |H_{m-1}| + (p^2 + p)(k - p) + p^2 = |H_{m-1}| + (p^2 + p)k - p^3$$

and

$$e(H_m) = e(H_{m-1}) + (p^2 + p)\left(\binom{k+1}{2} - p^3\right).$$

Then the Lemma implies that H_m has the desired properties with $z_i^m = x_{i+1}^1$, $1 \le i \le p + 1$. This completes the proof of the theorem.

To conclude this note we give a simple proof of a theorem of Tutte [13] and Nash-Williams [10] on the existence of k edge disjoint spanning trees. It happens to be more natural to formulate the problem for multigraphs. Let G be a multigraph with vertex set V and let P be a partition of V into p disjoint non-empty sets $V_1, V_2, \ldots, V_p$. We write $|P| = p$ and denote by $H = G|P$ the multigraph of order p whose vertices are $V_1, V_2, \ldots, V_p$ and in which the number of edges joining a vertex V_i to a vertex V_j $(1 \le i < j \le p)$ is equal to the number of $V_i - V_j$ edges of G. If G contains k edge disjoint spanning trees then clearly $e(G|P) \ge k(p-1) = k(|P|-1)$. Tutte and Nash-Williams proved that this trivial necessary condition is also sufficient.

<u>Theorem 5</u>. Suppose G is a multigraph such that

$$e(G|P) \ge k(|P| - 1) \qquad (1)$$

for every partition of the vertex set V. Then G contains k edge disjoint forests.

Proof. By applying induction on $n = |G|$, it is easily seen, (cf. [10]) that it suffices to prove the result under the following additional hypothesis:

$$e(G) = k(n-1) . \qquad (2)$$

Consider systems of k edge disjoint forests in G. Call a system $\mathfrak{F} = (F_i)_1^k$ maximal if $\sum_1^k e(F_i)$ is maximal. We have to show that a maximal system consists of trees. Suppose this is not so. Then if $\mathfrak{F} = \{F_i\}_1^k$ is a maximal system, there is an edge α of G that does not belong to any of the forests F_i.

Call $R = (\alpha, \mathfrak{F})$ a maximal pair. As $F_i + \alpha$ is not a forest, F_i contains a path $a_0 a_1 \cdots a_h$ joining the two endvertices of α. Put $\alpha' = a_j a_{j+1}$ for some j, $0 \le j < h$. Put furthermore $F_i' = F_i + \alpha - a_j a_{j+1}$,

$$\mathfrak{F}' = \{F_1, F_2, \ldots, F_{i-1}, F_i', F_{i+1}, \ldots, F_h\} \text{ and}$$

$$R' = \{\alpha', \mathfrak{F}'\} .$$

Clearly R' is also a maximal pair. We call it a simple shift of R. Call a maximal pair R^* a shift of R if it can be obtained from R by applying a sequence of simple shifts.

Fix a maximal pair $R_0 = (\alpha_0, \mathfrak{F}_0)$, $\mathfrak{F}_0 = (F_1, \ldots, F_k\}$. Call an edge superfluous if it is the first element of a shift of R_0. Denote by S the set of superfluous edges and put

$$V_0 = \{x \in V : xy \in S \text{ for some } y \in V\} .$$

As usual, (W, T) denotes the graph with vertex set W and edge set T. If H is a graph then we put $H = (V(H), E(H))$ and if $W \subset V(H)$ then $H[W]$ denotes the subgraph of H apanned by the vertices in W.

By the definition of a simple shift any two superfluous edges are contained in a path containing only superfluous edges so $G_0 = (V_0, S)$ is connected.

Our next aim is to show that $F_i[V_0]$ is connected for each i, $1 \le i \le k$. Suppose this is not so, say $V_0 = U_0 \cup W_0$, $U_0 \cap W_0 = \emptyset$, $U_0, W_0 \neq \emptyset$, and no edge of F_1 joins U_0 to W_0. The maximality of $\mathfrak{F}_0$ implies that α_0 is not a $U_0 - W_0$ edge. Let $R' = (\alpha_0', \mathfrak{F}_0')$ and suppose α_0' is a $U_0 - W_0$ edge. Then $\alpha_0' \notin E(F_1)$ so $F_1' = F_1$. The path in $F_1' = F_1$ joining the two endvertices of α_0' must leave V_0. As each edge in this path is superfluous, there is at least one superfluous edge with an endvertex not in V_0, contradicting the definition of V_0. Consequently α_0' does not join U_0 to W_0 and no edge of F_1' joins U_0 to W_0, i.e. our initial assumptions are satisfied to R' as well. Therefore no essential edge joins U_0 to W_0 and this is impossible.

Since $F_i[V_0]$ is connected for each i, $1 \le i \le k$, $V_0 \neq V$. Putting $q = |V_0|$, we have

$$e(G[V_0]) \ge \sum_1^k e(F_i[V_0]) + 1 = k(q-1) + 1, \qquad (3)$$

since $\alpha \in E(G[V_0]) - \bigcup_1^k E(F_i)$. Let P be the partition of V into V_0 and $n-q$ 1-element sets. Then $p = |P| = n - q + 1$ and (2) and (3) give

$$e(G|P) = e(G) - e(G[V_0]) \le k(n-1) - k(q-1) - 1 = k(p-1) - 1.$$

This contradicts (1), completing the proof.

References

1. B. Bollobás, On graphs with at most three independent paths connecting any two vertices, Studia Sci. Math. Hung. 1(1966), 137-140.

2. B. Bollobás, Cycles modulo k , to appear.

3. B. Bollobás, Semi-topological configurations, to appear.

4. B. Bollobás, and P. Erdös, On extremal problems in graph theory (in Hungarian), Mat. Lapok 13(1962) 143-152.

5. P. Erdös, Problem in 3 of Schweitzer competition, Soltn by P. Bartfai, Mat. Lapok 11 (1960) 175-176.

6. P. Erdös, Some recent problems and results in graph theory, Proc. Seventh Southeastern Comb. Conf., to appear.

7. J. L. Leonard, On a conjecture of Bollobás and Erdös, Per. Math. Hung. 3(1973) 281-284.

8. W. Mader, Ein Extremalproblem des Zusammenhaugs von Graphen, Math. Z. 131 (1973) 223-231.

9. W. Mader. Hinreichende Bedingungen für die Existenz von Teilgraphen, die zu einem vollständigen Graphen homeomorph sind, Math. Nachr. 53(1972) 145-150.

10. C. St. J. A. Nash-Williams, Edge-disjoint spanning trees in finite graphs, J. Lond. Math. Soc. 36 (1961) 445-450.

11. L. Pósa, Problem No. 127 (in Hungarian), Mat. Lapok 12 (1961) 254, Soltn by J. Czipszer, 14(1963) 373.

12. B. A. Sørensen and C. Thomassen, On K-rails in graphs, J. Comb. Theory 17(1974) 143-159.

13. W. T. Tutte, On the problem of decomposing the graph into n connected factors, J. Lond. Math. Soc. 36 (1961) 221-230.

GRAPHS WITH UNIQUE WALKS, TRAILS OR PATHS OF GIVEN LENGTHS

Juraj Bosák
Mathematical Institue of the Slovak Academy of Sciences
Bratislava, CZECHOSLOVAKIA

Abstract

In this paper we consider directed, undirected, or mixed graphs G. We also assume that for any two vertices u and v of G there exists exactly one walk [trail, path] from u to v whose length is in a given interval. Mixed Moore graphs (as special kinds of such graphs) are also studied. It is shown that there exist infinitely many mixed Moore graphs of diameter two. Some of the proofs are indicated and some will be published elsewhere.

We consider here graphs with loops, directed links and undirected links (every loop joins a vertex with itself; every link joins two different vertices - cf. [10]). A graph is said to be directed [undirected] if all its links are directed [undirected]. A graph is said to be mixed if it contains at least one directed link and at least one undirected link. Infinite graphs and multiple edges are also admissible. In all other aspects we follow the terminology and notation of [6]. As a rule, we do not distinguish between isomorphic graphs. The directed [undirected] graph induced by a cycle of length n will be denoted by Z_n $[C_n]$.

Let a and b be integers with $0 \le a \le b$. A graph G is said to be a W_a^b - graph if G has at least one vertex and for any two vertices (distinct or not) of G there exists in G exactly one walk from u to v whose length c satisfies the inequality $a \le c \le b$. We shall refer to this as the W_a^b condition.

No restrictions are imposed concerning the walks whose lengths are outside the interval $\langle a, b \rangle$. An example of a W_1^2-graph is given in Fig. 1, and a W_3^3-graph is shown in Fig. 2.

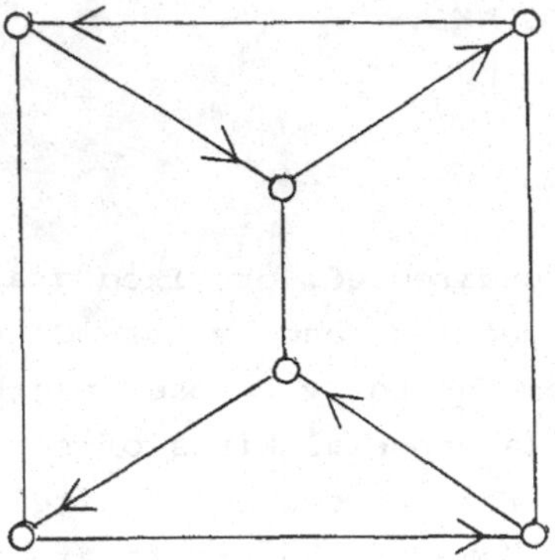

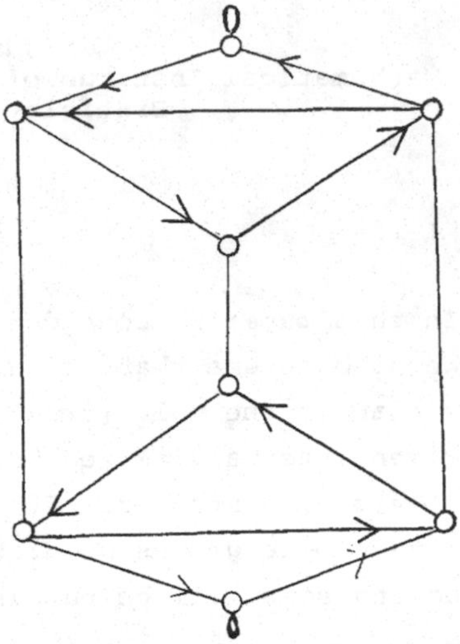

Fig. 1. A W_1^2-graph. Fig. 2. A W_3^3-graph.

If in the definition of the W_a^b condition the requirement concerning the upper bound b is omitted, we obtain a new condition that will be denoted by W_a^∞. If $a = 0$ and G has finite diameter b [or infinite diamter ∞], we write W instead of W_a^b [W_a^∞].

Similarly, we have the conditions T_a^b, T_a^∞, T and P_a^b, P_a^∞, P when "walk" is replaced by "trail" and "path" respectively.

This paper is concerned with the following problem: when there exists a graph with a given number of vertices satisfying some of the above nine conditions, if possible, describe all such graphs.

T-graphs were studied also under the name "strongly geodetic graphs" [3, 4, 9] - sometimes no existence but only the uniqueness of the corresponding trails has been demanded; in this case the notation "T-graphs" can be used.

Note that the following implications obviously hold:

$$\begin{array}{ccccc} W_0^\infty & \Longrightarrow & T_0^\infty & \Longrightarrow & P_0^\infty \\ \Downarrow & & \Downarrow & & \Downarrow \\ W & \Longrightarrow & T & \Longrightarrow & P \end{array}$$

Moreover, for a directed graph G we have $W_0^\infty \Leftrightarrow T_0^\infty$ and $W \Leftrightarrow T$, for a loopless undirected graph G we have $T_0^\infty \Leftrightarrow P_0^\infty$ and $T \Leftrightarrow P$.

Let G be a (directed, undirected or mixed) graph. Denote by G^o the loopless graph obtained from G by deleting all the loops of G. Denote by G^* the directed graph obtained from G by replacing each undirected link by two oppositely directed links. We need the following trivial but useful results:

<u>Lemma 1</u>. (i) A graph G satisfies the conditon W_a^b [W_a^∞, W, P_a^b, P_a^∞, P] if and only if the directed graph G^* satisfies the same condition.

(ii) A graph G satisfies the condition P_a^b [P_a^∞, P] if and only if the loopless graph G^o satisfies the same condition.

Lemma 1 enables us to restrict ourselves to directed (and in the last three cases even loopless) W_a^b-, W_a^∞-, W-, P_a^b-, P_a^∞- and P-graphs. However, an analogous reduction cannot be done for conditions T_a^b, T_a^∞ or T. For example, the graph G of Fig. 1 satisfies T and G^* satisfies T_1^2, but not conversely.

<u>Theorem 1</u>. Let a, b and p be integers such that $0 \le a \le b$, and $p \ge 1$. A directed W_a^b-graph G with p vertices exists if and only if one of the following cases occurs:

(i) $b = a$, $p = d^a$, where d is a positive integer.

(ii) $b = a + 1$, $a \ge 1$, $p = d^a + d^b$, d is a positive integer.

(iii) $b = a + 1$, $a = 0$, p is a positive integer (G is a complete digraph).

(iv) $b \ge a + 2$, $p = b - a + 1$ (G is Z_p).

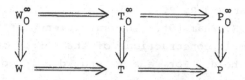

<u>Proof</u>. (outline). That the conditions (i) - (iv) are necessary, can be proved using the method of eigenvalues of adjacency matrices due to A. J. Hoffman and R. R. Singleton [7] in the form described by J. Plesník and S. Znám [9]. The sufficiency of (i) - (iv) can be demonstrated by simple constructions of the corresponding graphs. For example, let (ii) hold. For $a \geq 1$ and $d \geq 1$ construct a directed graph $B(d, a)$ as follows. The vertex set of $B(d, a)$ is $\{1, 2, 3, \dots, d^a + d^{a+1}\}$. From a vertex α a directed edge goes to all vertices β such that

$$\alpha = sd + i,$$
$$\beta = j(d^a + d^{a-1}) - s,$$

where s, i, j are integers satisfying the inequalities

$$0 \leq s \leq d^a + d^{a-1} - 1,$$
$$1 \leq i \leq d,$$
$$1 \leq j \leq d.$$

It is easy to check that $B(d, a)$ is a W_a^{a+1}- graph. ∎

A graph G is said to be regular of degree d if

$$id(v) = od(v) = d$$

for every vertex v of G. Here the symbol $id\ v\ [od(v)]$ denotes the number of edges incident with v that are not directed links going from [to] v.

The most important results concerning directed W-graphs proved in [9, Theorems 1-3] can be summarized as follows:

<u>Theorem 2</u>. Any directed W-graph is a regular graph of a finite diameter. Given cardinal numbers d and k (k finite), a regular directed W-graph G of degree d and diamter k exists if and only if at least one of the following cases occurs:

(i) d is arbitrary, $k = 0$ (G is the graph with one vertex and d loops).

(ii) $d = 1$, k arbitrary (G is Z_{k+1}).

(iii) $d \geq 1$, $k = 1$ (G is the complete digraph on $d + 1$ vertices).

(iv) d is infinite, k arbitrary.

Some of the considered nine classes of graphs are not interesting, which follows from the following results:

Theorem 3.

(i) W_a^∞ -graphs do not exist except the case $a = 0$ and $G = K_1$.

(ii) A graph G is a T_0^∞-graph if and only if G is an undirected tree.

(iii) A graph G is a T_1^∞-graph if and only if G is $Z_p (p \geq 1)$.

(iv) T_a^∞-graphs and P_a^b-graphs do not exist for $a \geq 1$.

The proofs are simple.

We can describe all P_0^∞-graphs (cf. Lemma 1):

Theorem 4. A loopless directed graph G is a P_0^∞-graph if and only if G is a connected graph whose every block is a $Z_n (n \geq 2)$.

Proof. Evidently, each graph of the described kind is a P_0^∞-graph. The converse needs more detailed considerations. ∎

The study of T_0^b-graphs (P_0^b-graphs) can be reduced to that of T-graphs (P-graphs, respectively):

Theorem 5. (i) A graph G is a T_0^b-graph if and only if G is a T-graph of diamteter b or an undirected tree of diameter less than b.

(ii) A graph G is a P_0^b-graph if and only if G is a P-graph of diamter $\leq b$.

Proof (outline). Evidently, a graph G is a T_0^b-graph [P_0^b-graph] if and only if G is a T-graph [P-graph, respectively] of diamter $\leq b$. Thus it is sufficient to show that every T_0^b-graph of diamter less than b is an undirected tree. But this is not difficult to prove.

Our knowledge concerning T_a^b-graphs for $a \geq 1$ is very incomplete. We mention (without proof) only one theorem and its simple corollary.

Theorem 6. Let $1 \le a \le b$. Then any T_a^b-graph contains exactly one quadratic factor F induced by disjoint cycles whose lengths are in the interval $\langle a, b \rangle$. Any edge not belonging to F has at least one end in a cycle of length b.

Corollary. The number of vertices of any finite T_b^b-graph $(b \ge 1)$ is a multiple of b.

The smallest T_2^2-graphs are shown in Fig. 3.

Fig. 3. Two T_2^2-graphs.

It remains to study T-graphs and P-graphs.

Theorem 7. Every T-graph is either an undirected tree or a regular graph with a finite diameter.

This theorem generalizes results known for undirected [4] and directed [9] graphs.

A regular T-graph of degree d with a finite diameter k will be called a tied graph (cf. [3, 4]) of type (d, k). Finite tied graphs are called Moore graphs [1, 4, 5, 7, 8].

Evidently, tied graphs of type (d, 0) are just the graphs with one vertex and d loops and tied graphs of type (d, 1) are the complete graphs, more precisely, the graphs G such that $G^* = (K_p)^*$ for some cardinal number $p \ge 2$. Therefore we may suppose $k \ge 2$. We start with $k \ge 3$.

Conjecture. Let d and k be integers, $d \ge 0$, $k \ge 3$. A finite graph G is a tied graph of type (d, k) if and only if either $d = 1$ and G is Z_{k+1} or $d = 2$ and G is C_{2k+1}.

Evidently, Z_{k+1} and C_{2k+1} are tied (Moore) graphs. The converse has been proved for undirected graphs in [1] and [5]. For directed graphs the conjecture follows from [9] or Theorem 2 as for directed graphs the notion of a W-graph coincides with that of a a T-graph.

According to Theorem 7 every T-graph of diameter 2 is either a tree of diameter two (that is, $K_{1,n}$ for $n \geq 2$) or a tied (i.e. Moore) graphs of type $(d, 2)$.

A regular graph G is said to be totally regular with directed degree z and undirected degree r if for every vertex v of G exactly z directed links going from [to] v and v is incident with exactly r undirected links.

Theorem 8. Let G be a finite tied graph of type $(d, 2)$. Then G is a loopless totally regular graph. If z is the directed degree of G and r is the undirected degree of $G (d = z + r)$, then G has

(1) $$p = (z + r)^2 + z + 1 \geq 3$$

vertices and just one of the following cases occurs:

A. $z = 1$, $r = 0$, G is Z_3.

B. $z = 0$, $r = 2$, G is C_5.

C. There exists a nonnegative integer c such that

(2) $$c \text{ divides } (4z + 5)(4z - 3)$$

and

(3) $$r = \frac{1}{4}(c^2 + 3) .$$

If G is a graph, denote by G^+ the graph with no pair of oppositely directed links such that $(G^+)^* = G^*$. For example, the graph $(B(2, 1))^+$ is shown in Fig. 1. (The graph $B(d, a)$ has been defined in the proof of Theorem 1.)

Theorem 9. For every integer $d \geq 2$ there exists a Moore (i.e., a finite tied graph) of type $(d, 2)$ with directed degree $z = d - 1$, undirected degree $r = 1$, and $d^2 + d$ vertices.

Proof. The graph $(B(d, 1))^+$ has all the required properties. ∎

In the case of undirected graphs (where the directed degree $z = 0$) as well-known [7] there exist Moore graphs of diameter two (they have $d^2 + 1$ vertices) only for degrees $d = 2$ (the pentagon), $d = 3$ (the Petersen graph [6, 7]), $d = 7$ (the Hoffman-

Singleton graph [2, 7]) and possibly d = 57. R. Damerell [5, p. 227] remarks: "Thus Moore graphs are interesting and it is a pity there should be so few of them." However, if we admit mixed Moore graphs of diameter two:

p	c	z	r	example
6	1	1	1	$B^+(2, 1)$
10	3	0	3	The Petersen graph
12	1	2	1	$B^+(3, 1)$
18	3	1	3	M
20	1	3	1	$B^+(4, 1)$
30	1	4	1	$B^+(5, 1)$
40	3	3	3	unknown
42	1	5	1	$B^+(6, 1)$
50	5	0	7	The Hoffman-Singleton graph
54	3	4	3	unknown

Here M denotes the graph with 18 vertices whose "toroidal" representation (in the sense that the opposite sides of the great rectangle are to be identified) is given in Fig. 4.

<u>Conjecture</u>. Let p, c, z and r be nonnegative integers such that (1), (2) and (3) hold. Then there exists a tied (Moore) graph of type (z + r, 2) with directed degree z, undirected degree r and p vertices.

From Theorem 9 it follows that the conjecture holds for c = 1 (r = 1).

Examples of P-graphs can be constructed using the following theorem.

<u>Theorem 10</u>. Let G be a graph such that either

(i) G is a P_0^∞-graph,

or

(ii) there exists a T-graph H with $H^* = (G^*)^{\cdot}$. Then G is a P-graph.

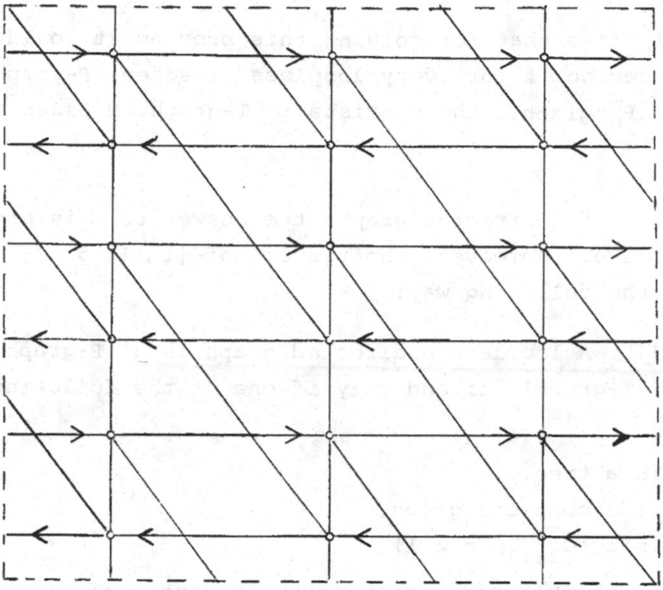

Fig. 4. A tied graph M of type (1 + 3 , 2) with 18 vertices.

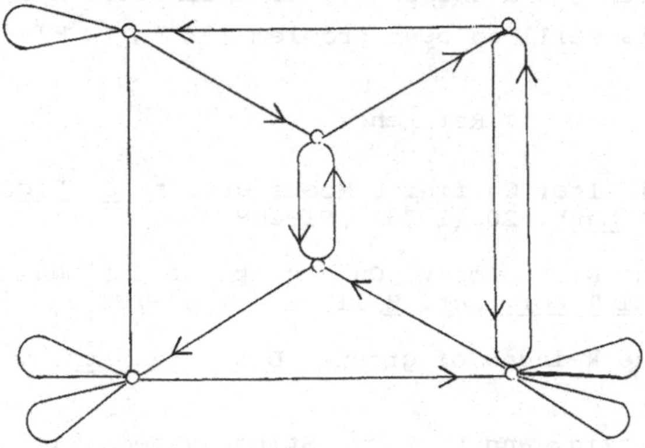

Fig. 5. A P-graph.

For example, the P-graph G of Fig. 5 can be obtained thus from the T-graph H of Fig. 1.

Problem. Is the converse to Theorem 10 true?

Lemma 1 implies that for solving this problem it would be sufficient to decide if for every loopless directed P-graph G that is not a P_0^∞-graph, there exists a T-graph H such that $H^* = G$.

In the case of undirected graphs the answer to this problem is obviously positive. Moreover, the results of [1 , 4 , 5 , 7] can be summarized in the following way:

Theorem 11. A loopless undirected graph is a P-graph (or, equivalently, T-graph) if and only if one of the following cases occurs:

(i) G is a tree.

(ii) G is a complete graph.

(iii) G is a C_{2n+1} $(n \geq 1)$.

(iv) G is the Petersen graph (with 10 vertices).

(v) G is the Hoffman-Singleton graph (with 50 vertices).

(vi) G is a Moore graph of type (57, 2)(with 3250 vertices).

(vii) G is an infinite tied graph [3 , 4] .

The P-graphs with loops can be obtained using Lemma 1, part (ii). Note that the existence and uniqueness of undirected Moore graphs of type (57, 2) is still an open problem [1 , 5 , 7 , 8].

References

1. E. Bannai and T. Ito, On finite Moore graphs. J. Fac. Sci. Univ. Tokyo Sect. I A Math. 20 (1973) 191-208.

2. C. T. Benson and N. E. Losey, On a graph of Hoffman and Singleton. J. Combinatorial Theory Ser. B 11 (1971) 67-79.

3. J. Bosák, On the k-index of graphs. Discrete Math. 1 (1971) 133-146.

4. J. Bosák, A. Kotzig, and S. Znám, Strongly geodetic graphs, J. Combinatorial Theory 5 (1968) 170-176.

5. R. Damerell, On Moore graphs, Proc. Cambridge Philos. Soc. 74 (1973) 227-236.

6. F. Harary, Graph Theory, Addison-Wesley, Reading, Mass. (1969).

7. A. J. Hoffman and R. R. Singleton, On Moore graphs with diameters 2 and 3. IBM J. Res. Develop. 4 (1960) 497-504.

8. J. Plesník, One method for proving the impossibility of certain Moore graphs. Discrete Math. 8 (1974) 363-376.

9. J. Plesník and S. Znám, Strongly geodetic directed graphs. Acta Fac. Rerum Natur. Univ. Comenian. Math. 29 (1974) 29-34.

10. W. T. Tutte, Connectivity in Graphs. Toronto Univ. Press, Tronto (1966).

TRIANGULAR IMBEDDINGS INTO SURFACES OF A JOIN OF EQUICARDINAL INDEPENDENT SETS FOLLOWING AN EULERIAN GRAPH

André Bouchet
Centre Universitaire du Mans
Le Mans, FRANCE

Abstract

If G is a graph and $m \geq 1$ an integer, $G_{(m)}$ is the graph obtained from G by replacing each vertex v by m vertices $(v,1), (v,2), \ldots, (v,m)$ and joining two vertices (v,i) and (w,j) iff v and w are joined in G. $G_{(m)}$ is a particular case of a join in the sense of J. L. Jolivet [5]. Suppose that G has a triangular imbedding into a surface S (orientable or not); can we find a triangular imbedding of $\overline{G} = G_{(m)}$ into a surface $\overline{S}$ which has the same orientability characteristic as S? We give some sufficient conditions to have an affirmative answer to this question.

0. Introduction

We consider graphs without loops and/or multiple edges. A graph is said to be eulerian if it is connected and if its degrees are all even.

Let G be a connected graph and $m \geq 2$ an integer. $G_{(m)}$ is the graph constructed from G by replacing each vertex v by m vertices $(v,1), (v,2), \ldots, (v,m)$ and joining two vertices (v,i) and (w,j) by an edge of $G_{(m)}$ if and only if v and w are joined by an edge of G. $G_{(m)}$ is a particular case of a join in the sense of J. L. Jolivet [5]; $G_{(m)}$ is also equal to the strong tensor product $K_m \underline{\otimes} G$ introduced by B. L. Garman, R. D. Ringeisen and A. T. White [4]. If $G = K_n$, then $G_{(m)} = K_{n(m)}$ (the regular complete n-partite graph of order mn.)

Question: If a connected graph G has a triangular imbedding into a surface S (orientable or not), can we find a triangular imbedding of $\overline{G} = G_{(m)}$ into a surface $\overline{S}$ which has the same orientability characteristic as S ?

The answer may be negative. For example although K_4 has a triangular imbedding into the sphere, M. Jungerman [6] has verified that no triangular imbedding of $K_{4(3)}$ can be constructed into an orientable surface.

The purpose of this paper is to state sufficient conditions which imply an affirmative answer when G is an eulerian graph (so K_4 is not considered). For example the answer will be affirmative for the three following cases: m is an odd integer (Section 4), G is triangularly imbedded into an orientable surface with a bichromatic dual (Section 5), G is a complete multipartite graph (Section 6). Two tables in Section 7 indicate the cases where we can construct a triangular imbedding into a surface of a regular complete multipartite graph.

1. Definitions and elementary properties.

Our terminology about graphs and their imbeddings into surfaces agrees with [10] and [15]. The vertex-set and the edge-set of a graph G will be denoted respectively $V(G)$ and $E(G)$. An edge with end points x and y will be denoted $[x, y]$.

An abstract simplicial complex K of dimension 2 is defined by a graph K^1 called the 1-skeleton of K and a set $T(K)$ of cycles of K^1 with length 3, called the triangles of K. The vertices and edges of K^1 are also called the vertices and edges of K; we put $V(K) = V(K^1)$ and $E(K) = E(K^1)$. A triangle of K defined by three vertices x, y and z will be denoted $[x, y, z]$.

K will be said to be connected if K^1 is a connected graph, and strongly connected if for every pair of triangles t and t' there exists a sequence of triangles beginning with t and ending with t' such that two successive triangles always have a common edge.

The link of a vertex x, denoted $L(x, K)$, is the graph spanned by the vertices y such that $[x, y] \in E(K)$ and the edges $[y, z]$ such that $[x, y, z] \in T(K)$.

We define a triangulation as a connected abstract simplicial complex K of dimension 2 where each link is a cycle. It is easy to verify that a triangulation K is also strongly connected and that each edge belongs to exactly two triangles. The usual methods of identification provide a surface S(K) and a triangular imbedding of the graph K^1 into S(K). So we shall often define a triangular imbedding of a connected graph G into a surface by a triangulation K such that $K^1 = G$; the Euler characteristic of S(K) is then equal to the integer $|V(K)| - |E(K)| + |T(K)| = |V(G)| - |E(G)|/3$.

Let x be a vertex of a triangulation K. A local orientation around x is an orientation O_x of the cycle L(x, K). It will be convenient to consider that O_x is a cyclic permutation on the set of vertices of L(x, K).

Consider a family $O = (O_x)x \in V(K)$ of local orientations around the vertices of K and an edge [x, y]; the vertex $O_x(y)$ is either equal to $O_y^{-1}(x)$ or $O_y(x)$ (Figures 1a and 1b). In the first case we say that [x, y] is an edge coherent with O; in the second case it is an edge incoherent with O. The family of local orientations O is said to be coherent if each edge of K is coherent with O. The surface S(K) is orientable if there exists a coherent family of local orientations for K; otherwise S(K) is non orientable. The orientability characteristic of K and S(K) is + 1 if S(K) is an orientable surface; otherwise the orientablity characteristic is -1.

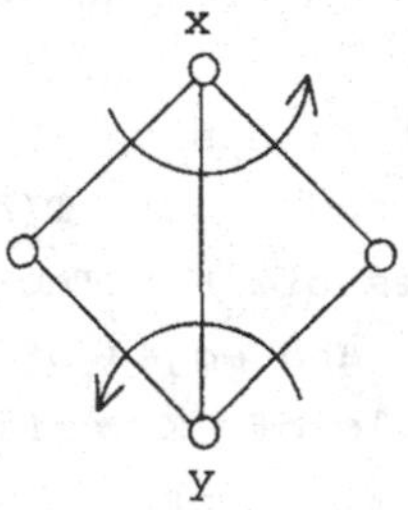

Figure 1a.

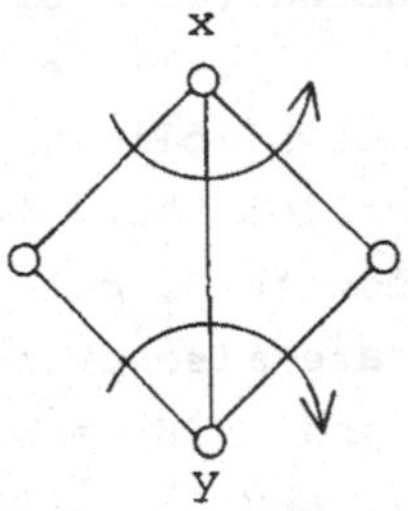

Figure 1b.

ORIENTABILITY LEMMA. <u>Let O be a family of local orientations around the vertices of a triangulation K. The surface S(K) is orientable if and only if the number of edges incoherent with O contained in a cycle of K^1 is always even.</u>

<u>Proof</u>: Consider a subset of vertices $V' \subset V(K)$. Change the local orientation O_x into O_x^{-1} for each $x \in V'$, obtaining a new family of local orientations denoted O(V'). The coherence of an edge is distinct in O and O(V') if and only if this edge belongs

to the cut-set of K^1 generated by V'. Since any family of local orientations is equal to some $O(V')$, there exists a coherent family of local orientations if and only if the set of edges which are coherent with O is a cut-set of K^1. We know that a set of edges is a cut-set if and only if its intersection with each cycle has an even cardinality, so the lemma is proved.

2. Generative m-valuation of an even triangulation.

Consider a triangulation K, an integer $m \geq 2$, and a map $\varphi: T(K) \to Z/m\,Z$. φ will be called an m-valuation of K. The expansion of K by φ is the simplical complex $\overline{K}$ of dimension 2 defined by:

$$V(\overline{K}) = V(K) \times (Z/m\,Z) ;$$

$$[(x , i) , (y , j)] \in E(\overline{K}) \Leftrightarrow [x , y] \in E(K) ;$$

$$[(x , i) , (y , j) , (z , k)] \in T(\overline{K}) \Leftrightarrow$$

$$t = [x , y , z] \in T(K) \quad \text{and} \quad i + j + k = \varphi(t) .$$

It is clear that the 1-skeleton $\overline{K}^1$ is equal to $K^1_{(m)}$, so $\overline{K}$ is connected. It is easy to verify that each edge of $\overline{K}$ belongs to two triangles exactly; therefore if we consider a vertex $\overline{x} \in V(\overline{K})$, each connected component of the link $L(\overline{x} , \overline{K})$ is a cycle.

Suppose now that K^1 is an eulerian graph. We shall say in that case that K is an even triangulation. Consider $x \in V(K)$, with the degree of x in K^1 equal to 2d. Denote by $t_1 , t_2 , \dots , t_{2d}$ the triangles of K incident to x ordered following a local orientation O_x (Figure 2). Put

$$\dot{\varphi}(x) = \sum_{j=1}^{2d} (-1)^j \varphi(t_j) \quad .$$

$\dot{\varphi}(x)$ is an element of $Z/m\,Z$ which depends on the local orientation O_x and on the first triangle t_1. However if we change O_x and /or t_1 the single other value which can be obtained is $-\dot{\varphi}(x)$. We shall say that φ is a generative m-valuation of K if each element $\dot{\varphi}(x)$ generates the cyclic group $Z/m\,Z$.

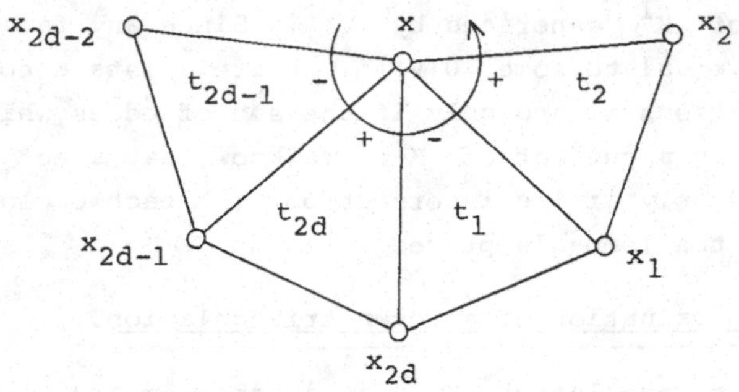

Figure 2.

THEOREM 1. <u>Let</u> K <u>be an even triangulation and</u> φ <u>a generative m-valuation of</u> K . <u>The expansion</u> $\overline{K}$ <u>of</u> K <u>by</u> φ <u>is a triangulation which has the same orientability characteristic as</u> K.

Let us make a preliminary computation before the proof of this theorem. Let $d \geq 1$ be an integer, and $a_1, a_2, \ldots, a_{2d}$ be elements of Z/m Z with the indices in Z/2d Z . Denote by $\hat{j}$ the residue class modulo 2d of an integer j . For every integer $\alpha \geq 1$, put

$$\psi_\alpha = \sum_{j=1}^{\alpha} (-1)^j a_{\hat{j}} .$$

It is easy to verify that the following equality holds for every integer $\alpha \geq 1$ and $k \geq 0$:

(E1) $$\psi_{\alpha+2kd} = \psi_\alpha + k\,\psi_{2d} .$$

<u>Proof of the Theorem 1</u>:

Let $\overline{x}$ be a vertex of $\overline{K}$ with degree equal to D . We shall construct a sequence of vertices in $L(\overline{x}, \overline{K})$, $S_{\overline{x}} = (\overline{x_1}, \overline{x_2}, \ldots, \overline{x_n}, \ldots)$, which has the three following properties for every integer $\alpha \geq 1$ and $\beta \geq 0$:

(i) $\overline{x}_{\alpha+\beta} = \overline{x}_\alpha$ if β is a multiple of D ;

(ii) $\overline{x}_{\alpha+\beta} \neq \overline{x}_{\alpha}$ if β is not a multiple of D ;

(iii) $[\overline{x}_{\alpha}, \overline{x}_{\alpha+1}]$ is an edge of $L(\overline{x}, \overline{K})$.

The properties (i) and (ii) signify that $S_{\overline{x}}$ is periodic and that its period is equal to D . Every subsequence of D successive elements contains then all the vertices of $L(\overline{x}, \overline{K})$. Since each connected component of $L(\overline{x}, \overline{K})$ is a cycle, the property (iii) implies that $L(\overline{x}, \overline{K})$ is a single cycle. So it will be proved that $\overline{K}$ is a triangulation.

Put $\overline{x} = (x, i)$, where x is a vertex of K and i is an element of $Z/m\,Z$. Put the degree of x equal to 2d, so the degree of $\overline{x}$ is $D = 2dm$. Choose an orientation 0_x of the cycle $L(x, K)$, and denote by $x_1, x_2, \ldots, x_{2d}$ the successive vertices of that cycle with its orientation, with the indices in $Z/2d\,Z$. For each j in $Z/2d\,Z$, put $t_{j+1} = [x, x_j, x_{j+1}]$ (see Figure 2). Denote by $\hat{j}$ the residue class modulo $2d$ of an integer j .

Let $\alpha \geq 1$ be an integer, and put:

$$\theta_{\alpha} = \sum_{j=1}^{\alpha} (-1)^{j} \varphi(t_{\hat{j}}) \quad \text{if } \alpha \text{ is even;}$$

$$\theta_{\alpha} = -\left(i + \sum_{j=1}^{\alpha} (-1)^{\alpha} \varphi(t_{\hat{j}})\right) \quad \text{if } \alpha \text{ is odd;}$$

$$\overline{x}_{\alpha} = (x_{\hat{\alpha}}, \theta_{\alpha}) \quad .$$

We study now the sequence $S_{\overline{x}} = (\overline{x}_{\alpha})_{\alpha \geq 1}$ where $\overline{x}_{\alpha}$ is defined by the preceding equalities. It is easy to verify that

(E2) $i + \theta_{\alpha} + \theta_{\alpha+1} = \varphi(t_{\widehat{\alpha+1}})$.

(E2) implies that $[\overline{x}, \overline{x}_{\alpha}, \overline{x}_{\alpha+1}]$ is a triangle of $\overline{K}$; therefore $[\overline{x}_{\alpha}, \overline{x}_{\alpha+1}]$ is an edge of $L(\overline{x}, \overline{K})$, so (iii) is verified.

Replace $a_{\hat{j}}$ by $\varphi(t_j)$ in the preliminary computation. We have then $\Phi(x) = \psi_{2d}$.

Let us consider an integer $k \geq 0$. If α is even, (E1) implies

$$\theta_{\alpha+2kd} = \psi_{\alpha+2kd} = \psi_{\alpha} + k\,\psi_{2d} = \theta_{\alpha} + k\,\dot{\varphi}(x)\ .$$

If α is odd, (E1) implies

$$\theta_{\alpha+2kd} = -i - \psi_{\alpha+2kd} = -i - \psi_{\alpha} - k\,\psi_{2d} = \theta_{\alpha} - k\,\dot{\varphi}(x)\ .$$

The two preceding equalities can be grouped in

(E3) $$\theta_{\alpha+2kd} = \theta_{\alpha} + k(-1)^{\alpha}\,\dot{\varphi}(x)\ .$$

Consider an integer $\beta \geq 0$ which is a multiple of $D = 2dm$. If we put $\beta = 2\mu md$, (E3) implies

$$\theta_{\alpha+\beta} = \theta_{\alpha} + m\mu(-1)^{\alpha}\,\dot{\varphi}(x)\ .$$

The element $m\mu\,\dot{\varphi}(x)$ is equal to 0 in $Z/m\,Z$, so we have $\theta_{\alpha+\beta} = \theta_{\alpha}$. We have also $x_{\widehat{\alpha+\beta}} = x_{\hat{\alpha}}$, for $\hat{\beta}$ is equal to 0 in $Z/2d\,Z$. Therefore we have $\overline{x}_{\alpha+\beta} = \overline{x}_{\alpha}$ and (i) is verified.

Suppose now that β is not a multiple of D. We can put $\beta = 2\mu\,md + r$ with $0 < r < 2md$. If r is not a multiple of $2d$, $\hat{r}$ is not equal to 0 in $Z/2d\,Z$, and therefore we have $x_{\widehat{\alpha+\beta}} = x_{\hat{\alpha}+\hat{r}} \neq x_{\hat{\alpha}}$, so $\overline{x}_{\alpha+\beta}$ is not equal to $\overline{x}_{\hat{\alpha}}$. If r is a multiple of $2d$, consider the integer $q = r/2d$. We have then $\beta = 2\mu\,md + 2qd$ with $0 < q < m$. The equality (E3) implies

$$\theta_{\alpha+\beta} = \theta_{\alpha} + m\mu(-1)^{\alpha}\,\dot{\varphi}(x) + q(-1)^{\alpha}\,\dot{\varphi}(x) = \theta_{\alpha} + q(-1)^{\alpha}\,\dot{\varphi}(x)\ .$$

The element $q\,\dot{\varphi}(x)$ is not equal to 0 in $Z/m\,Z$ because $0 < q < m$ and $\dot{\varphi}(x)$ is a generator of $Z/m\,Z$. Therefore $\theta_{\alpha+\beta}$ and θ_{α} are distinct and $\overline{x}_{\alpha}$ is also distinct from $\overline{x}_{\alpha+\beta}$, so (ii) is verified.

We have now proved that $\overline{K}$ is a triangulation. Two preliminary properties are necessary to study the orientability of $\overline{K}$.

(2.1) PROPERTY. There exists a local orientation $\overline{O}_{\overline{x}}$ around the vertex $\overline{x}$ such that the two following equalities are true for every vertex $\overline{y} = (y\,,\,j)$ of the link $L(\overline{x}\,,\,\overline{K})$:

(E4) $$\overline{O}_{\overline{x}}\ (\overline{y}) = \Big(O_{x}(y)\,,\,\varphi\big([x\,,\,y\,,\,O_{x}(y)]\big) - i - j\Big)\ ;$$

(E5) $$\overline{O}_{\overline{x}}^{-1}\ (\overline{y}) = \Big(O_{x}^{-1}(y)\,,\,\varphi\big([x\,,\,y\,,\,O_{x}^{-1}(y)]\big) - i - j\Big)\ .$$

<u>Proof</u>: The correspondence $\overline{O}_{\overline{x}}: \overline{x}_\alpha \to \overline{x}_{\alpha+1}$ is a cyclic permutation of the vertices of $L(\overline{x}, \overline{K})$ because D is the period of the sequence $S_{\overline{x}}$. There exists an integer $\alpha \geq 2$ such that $\overline{y} = \overline{x}_\alpha$. We have then $y = x\hat{_\alpha}$, $j = \theta_\alpha$, $\overline{O}_{\overline{x}}(\overline{y}) = \overline{x}_{\alpha+1}$, $\overline{O}_{\overline{x}}^{-1}(\overline{y}) = \overline{x}_{\alpha-1}$. In order to verify that the local orientation $\overline{O}_{\overline{x}}$ satisfies (E4) and (E5), we have to verify:

$$O_x(x\hat{_\alpha}) = x\hat{_{\alpha+1}} \; ; \quad O_x^{-1}(x\hat{_\alpha}) = x\hat{_{\alpha-1}} \; ;$$

$$\varphi\big([x , x\hat{_\alpha} , O_x(x\hat{_\alpha})]\big) - i - \theta_\alpha = \theta_{\alpha+1} \; ;$$

$$\varphi\big([x , x\hat{_\alpha} , O_x^{-1}(x\hat{_\alpha})]\big) - i - \theta_\alpha = \theta_{\alpha-1} \; .$$

The definition of O_x gives the first two equalities. We have then $[x , x\hat{_\alpha} , O_x(x\hat{_\alpha})] = t\hat{_{\alpha+1}}$ and $[x , x\hat{_\alpha} , O_x^{-1}(x\hat{_\alpha})] = t\hat{_\alpha}$, so the two last equalities are equivalent to (E2) applied with the values α and $\alpha - 1$.

(2.2) PROPERTY. To each family $O = (O_x)x \in V(K)$ of local orientations around the vertices of K, there corresponds a family $\overline{O} = (\overline{O}_{\overline{x}})\overline{x} \in V(\overline{K})$ of local orientations around the vertices of $\overline{K}$ satisfying the following property: an edge $[(x , i) , (y , j)] \in E(\overline{K})$ is coherent with $\overline{O}$ if and only if the edge $[x , y] \in E(K)$ is coherent with O.

<u>Proof</u>: To each vertex $\overline{x} = (x , i)$ of the triangulation $\overline{K}$ associate the local orientation $\overline{O}_{\overline{x}}$ satisfying the equalities (E4) and (E5) of the Property 2.1. We prove that the family of local orientations $\overline{O} = (\overline{O}_{\overline{x}})\overline{x} \in \overline{K}$ satisfies the Property 2.2. Put $\overline{x} = (x , i)$ and $\overline{y} = (y , j)$. If $[x , y]$ is coherent with O we have the equality $O_x(y) = O_y^{-1}(x)$; therefore the triangles $[x , y , O_x(y)]$ and $[x , y , O_y^{-1}(x)]$ are equal. Then the equalities (E4) and (E5) imply $\overline{O}_{\overline{x}}(\overline{y}) = \overline{O}_{\overline{y}}^{-1}(\overline{x})$, so the edge $[\overline{x} , \overline{y}]$ is coherent with $\overline{O}$. Similarly we see that $[\overline{x} , \overline{y}]$ is not coherent with $\overline{O}$ if $[x , y]$ is not coherent with O.

<u>Orientability of $\overline{K}$</u>: If K is orientable, we choose a coherent family of local orientations O around the vertices of K

and the corresponding family of local orientations $\overline{0}$ around the vertices of $\overline{K}$. Every edge of $\overline{K}$ is coherent with $\overline{0}$, so $\overline{0}$ is coherent.

If K is nonorientable, we choose an arbitrary family of local orientations 0 around the vertices of K and the corresponding family of local orientations $\overline{0}$ around the vertices of $\overline{K}$. The orientability lemma (see Section 1) implies the existence in the graph K^1 of a cycle $(x_1, x_2, \ldots, x_n)$ with an odd number of edges incoherent with 0. $\big((x_1, 0), (x_2, 0), \ldots, (x_n, 0)\big)$ is a cycle of $\overline{K}^1$ for $\overline{K}^1 = K^1_{(m)}$; the incoherent edges of this cycle are in 1 - 1 correspondence with the incoherent edges of the first cycle, so their number is odd and $\overline{K}$ is nonorientable.

We now have to study the existence of a generative m-valuation. This will be completely solved in Section 4. However it is interesting to state here an elementary corollary of Theorem 1 which implies the existence of triangular imbeddings for some regular complete multipartite graphs.

(2.3) COROLLARY. <u>Let G be an eulerian graph which can be triangularly imbedded into a surface S and $m \geq 2$ an integer. Then $G_{(m)}$ can be triangularly imbedded into a surface $\overline{S}$ with the same orientability characteristic as S if the following two conditions are verified:</u>

(i) <u>the number of vertices of G is an integer equal to</u> 3 (<u>mod</u> 4)

(ii) <u>there exists a vertex of G which is joined to all the other vertices of G</u>.

<u>Proof</u>: Consider a triangulation K such that $K^1 = G$. Let the number of vertices of G be equal to $4k + 3$, and consider a vertex $x \in V(G)$ which is joined to the other vertices of G. The degree of x is equal to $4k + 2$. Let $t_1, t_2, \ldots, t_{4k+2}$ be an ordering of the triangles of K incident to x and suppose that this ordering is compatible with a local orientation 0_x. Let $\mathfrak{F}$ be the set of these triangles with an even index; the number of triangles in $\mathfrak{F}$ is equal to $2k + 1$. Consider a subset $\mathfrak{F}^+ \subset \mathfrak{F}$ with a number of triangles equal to $k + 1$; let $\mathfrak{F}^-$ be the set of the k remaining triangles of $\mathfrak{F}$. Define the following m-valuation:

$$\varphi(t) = 0 \quad \text{if} \quad t \notin \mathfrak{T} ;$$

$$\varphi(t) = 1 \quad \text{if} \quad t \in \mathfrak{T}^+ ;$$

$$\varphi(t) = -1 \quad \text{if} \quad t \in \mathfrak{T}^- .$$

It is easy to verify that $\dot{\varphi}(x) = \pm 1$. A vertex $y \neq x$ belongs to one and only one triangle in $\mathfrak{T}$; therefore $\dot{\varphi}(y) = \pm 1$. So φ is a generative m-valuation and we find a triangular imbedding of $G_{(m)}$ defined by the triangulation $\overline{K}$ equal to the expansion of K by φ.

(2.4) COROLLARY. <u>Let</u> $s \geq 0$ <u>and</u> $m \geq 1$ <u>be integers</u>. $K_{12s+3(m)}$ <u>and</u> $K_{12s+7(m)}$ <u>are triangularly imbeddable into orientable surfaces</u>. $K_{12s+15(m)}$, $K_{12s+19(m)}$, $K_{12s+11(m)}$ <u>are triangularly imbeddable into non-orientable surfaces</u>.

<u>Proof</u>: The complete graphs K_{12s+3}, K_{12s+7}, K_{12s+15}, K_{12s+19} and K_{12s+11} satisfy the conditions of the preceding corollary. We know [8] that K_{12s+3} and K_{12s+7} are triangularly imbeddable into orientable surfaces; K_{12s+15}, K_{12s+19} and K_{12s+11} are triangularly imbeddable into non-orientable surfaces.

3. <u>Diagonal components of a triangulation</u>

Let K be a triangulation which is not necessarily even. The diagonal graph of K, denoted D_K, is defined in the following way:

$V(D_K) = V(K)$;

$[v, w] \in E(D_K) \Leftrightarrow$

$\exists [x, y] \in E(K) : [x, y, v]$ and $[x, y, w] \in T(K)$.

An edge of D_K will be called a diagonal edge and, a connected component of D_K will be called a diagonal component. A diagonal component will be said to be even (resp. odd) if its number of vertices is even (resp. odd). Diagonal components and their parities will be used in Section 4 to study the existence of a generative m-valuation.

(3.1) PROPERTY. <u>A triangulation</u> K <u>has at most three diagonal components</u>. <u>The number of vertices of a triangle</u> t <u>which belong to a diagonal component</u> D <u>does not depend on</u> t .

<u>Proof</u>. For every $x \in V(K)$ denote by $D(x)$ the diagonal component which contains x . For every $t \in T(K)$ denote by $V(t)$ the set of the three vertices of t . The property (3.1) will be true if there exists for each pair of triangles (t, t') a 1-1 map $i_{t't}: V(t) \to V(t')$ such that $D(x) = D(i_{t't}(x))$ when $x \in V(t)$. Suppose first that t and t' have a common edge $[x, y]$, and let z be the third vertex of t and z' the third vertex in t'. $D(z)$ is equal to $D(z')$, for $[z, z']$ is a diagonal edge, so we can put $i_{t't}(x) = x$, $i_{t't}(y) = y$ and $i_{t't}(z) = z'$. Consider now an arbitrary pair of triangles (t, t'). Since K is strongly connected (see Section 1) there exists a sequence of triangles $t_o, t_1, \dots, t_n$ where two successive triangles have always a common edge and such that $t_o = t$ and $t_n = t'$. We can put $i_{t't} = i_{t_k t_{k-1}} \circ \cdots \circ i_{t_2 t_1} \circ i_{t_1 t_o}$.

<u>Definition</u>: A diagonal component will be said to be simple (resp. double, triple) if it contains one (resp. two, three) vertices of each triangle. If K has three diagonal components, they are simple. If K has two diagonal components, one of them is simple, the other one is double. If K has a single diagonal component, it is triple.

(3.2) PROPERTY. <u>Let</u> x <u>be a vertex of a triangulation</u> K. <u>The vertices of the cycle</u> $L(x, K)$ <u>belong either to the same diagonal component or they belong alternatively to two diagonal components</u>. <u>In the second case the vertex</u> x <u>has an even degree</u>.

(3.3) PROPERTY. <u>A triangulation</u> K <u>has three diagonal components if and only if the chromatic number of the graph</u> K^1 <u>is equal to</u> 3. <u>In this case the diagonal components of</u> K <u>are equal to the chromatic classes of</u> K^1.

The two preceding properties are easy to verify. We study now the structure of a triangulation which has two or three diagonal components. Denote by K^1_s the subgraph of K^1 spanned by the vertices of a simple diagonal component D_s (this component is unique if K has two diagonal components, otherwise we choose for D_s an arbitrary diagonal component of K). Denote by K^1_d the subgraph of K^1

spanned by the vertices which do not belong to D_s (the vertices of K_d^1 are the vertices of the double diagonal component if K has two diagonal components, otherwise they are the vertices of the simple diagonal components distinct from D_s).

(3.4) PROPERTY. K_s^1 *is an independent set*. K_d^1 *is equal to the union of the links of* K *determined by the vertices of* K_s^1.

Proof: There cannot exist an edge $[x, y]$ in K_s^1, for a triangle $[x, y, z]$ containing this edge would have two vertices in the single diagonal component D_s. If x is a vertex of K_s^1 and $[y, z]$ is an edge of $L(x, K)$, the triangle $[x, y, z]$ has a single vertex in D_s, and thus y and z are two vertices of K_d^1. Conversely if $[y, z]$ is an edge of K_d^1 and $t = [x, y, z]$ is a triangle which contains this edge, then the triangle t contains a vertex in D_s and this vertex is necessarily equal to x because y and z do not belong to D_s. So K_d^1 is the union of the links of K determined by the vertices of K_s^1.

(3.5) PROPERTY. K_d^1 *is a bipartite graph if and only if* K *has three diagonal components*.

This property is an immediate consequence of (3.2) and (3.3). The meaning of the Property (3.4) appears if we consider the triangular imbedding of the graph K^1 into the surface $S(K)$.

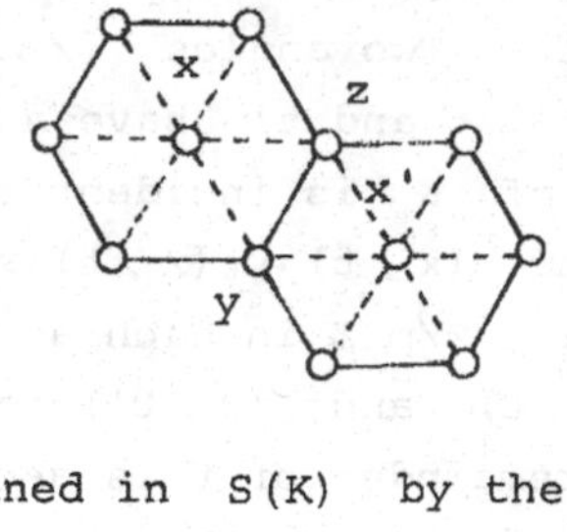

Figure 3

Each region determined in $S(K)$ by the graph K_d^1 is a 2-cell, for it is equal to the union of the triangles incident to some vertex of K_s^1. Two of these regions which have the same edge $[y, z] \in E(K_d^1)$ in their boundaries are distinct, for the triangles $[x, y, z]$ and $[x', y, z]$ which contain this edge are distinct. Therefore the graph K_d^1 defines a map in the surface $S(K)$ (Figure 3). Conversely consider a map defined by a 2-cell imbedding of a graph G into a surface S . Put a central vertex inside each country and joint this central vertex to the vertices of G which lie on the boundary of that country. We thus define a triangulation K . The central vertices form a single diagonal component of K . If G is a bipartite graph, the two chromatic classes of G form two other simple diagonal components of K ; otherwise the vertices of G form a double diagonal component of K .

4. Conditions for the existence of a generative m-valuation

We consider an even triangulation K and an integer $m \geq 2$. We shall prove the following two theorems:

THEOREM 2. *There exists a generative m-valuation of K if m is an odd integer.*

THEOREM 3. *There exists a generative m-valuation of K for an even integer m if and only if one of the following three properties holds:*

(i) K *has a single diagonal component;*

(ii) K *has two diagonal components and the double component is even;*

(iii) K *has three diagonal components and these components have the same parity.*

(4.1) *Angles of K*: We have seen in Section 2 that the values $\dot{\varphi}(x)$ associated with an m-valuation φ are not uniquely defined. We describe here a way to define them uniquely.

An angle of K is an ordered pair (x , t) with a vertex x and a triangle t such that $x \in t$. Two angles (x , t) and (x' , t') are adjacent if $x = x'$, $t \neq t'$, t and t' have a common edge (Figure 4). Since each vertex of K is incident to an even number of triangles, we can find a mapping $(x , t) \to \varepsilon(x , t)$ which associates to each angle the value +1 or -1 in $Z/m\,Z$ in such a way that $\varepsilon(x , t) + \varepsilon(x , t') = 0$ if (x , t) and (x , t') are adjacent angles. We fix such a mapping for the remainder of this section. If φ is an m-valuation of K, it is easy to verify that

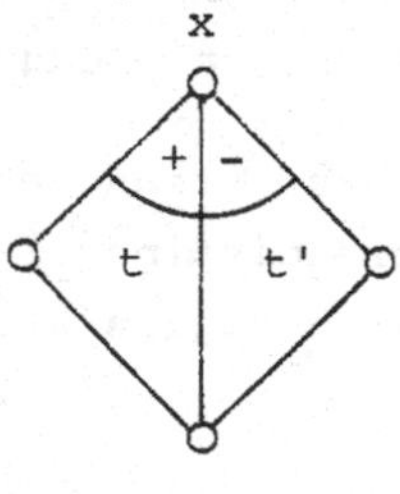

Figure 4

(E1) $$\forall x \in V(K) \; : \; \dot{\varphi}(x) = \sum_{\substack{x \in t \\ t \in T(K)}} \varepsilon(x, t) \cdot \varphi(t)$$

So the mapping $\dot{\varphi}: V(K) \to Z/m\,Z$ is now defined uniquely.

(4.2) Chains and chain-groups: We recall now the elementary definitions about chains and chain-groups introduced by W. T. Tutte [14]. Let E be a finite set and R be a commutative ring. A chain on E over R is a mapping $f: E \to R$. If x is an element of E, the value $f(x)$ is called the coefficient of x in f. If f and g are chains on E over R and if λ is an element of R, the sum $f + g$ is the chain on E over R satisfying

$$(f + g)(x) = f(x) + g(x)$$

for each $x \in E$, and the product λf is the chain on E over R satisfying

$$(\lambda f)(x) = \lambda\, f(x)$$

for each $x \in E$.

A chain-group on E over R is a class of chains on E over R that is closed under the operations of addition and multiplication by an element of R.

(4.3) The triangle chain-group $\dot{T}$: Consider again the even triangulations K and the integer $m \geq 2$. We shall be working with chains on $V(K)$ over the ring $Z/m\,Z$. In order to abbreviate the terminology they will simply be called chains and the chain group on $V(K)$ over $Z/m\,Z$ will be called simply a chain-group. To each vertex x we associate the chain $\tilde{x}$ satisfying $\tilde{x}(x) = 1$ and $\tilde{x}(y) = 0$ if $y \neq x$; so if f is a chain we have the equality

$$f = \sum_{x \in V(K)} c_x \; \tilde{x} \quad ,$$

where each c_x is the coefficient of x in f. f will be called a generative chain if each coefficient generates the additive group $Z/m\,Z$.

To each triangle t we associate the chain t satisfying

(E2) $$\dot{t} = \sum_{x \in t} \varepsilon(x, t) \cdot \tilde{x} .$$

The chain-group generated by the linear combinations of the chains $\dot{t}$ when t ranges over $T(K)$ will be denoted $\dot{T}$ and called the triangle chain-group. If φ is an m-valuation of K, it is easy to verify that the equalities (E1) and (E2) imply

(E3) $$\sum_{t \in T(K)} \varphi(t) \cdot \dot{t} = \sum_{x \in V(K)} \dot{\varphi}(x) \cdot \tilde{x}$$

So we obtain the following property:

(4.4) PROPERTY. <u>There exists a generative m-valuation of</u> K <u>if and only if there exists a generative chain in</u> $\dot{T}$.

(4.5) LEMMA <u>Let</u> (v, w) <u>be an ordered pair of distinct vertices of</u> K <u>which belong to the same diagonal component. There exists a chain</u> $\dot{\alpha}(v, w) \in \dot{T}$ <u>such that</u> $\dot{\alpha}(v, w) = \tilde{v} + \varepsilon \tilde{w}$ <u>with</u> $\varepsilon = \pm 1$.

<u>Proof</u>: We prove the result by induction on the distance d between v and w in the diagonal graph D_K .

If $d = 1$ then $[v, w]$ is an edge of D_K . There exists a triangle $t = [x, y, v]$ and a triangle $t' = [x, y, w]$. Consider the chain

$$\dot{t} + \dot{t}' = \varepsilon(v, t)\,\tilde{v} + \varepsilon(w, t')\,\tilde{w} + \Big(\varepsilon(x, t) + \varepsilon(x, t')\Big)\,\tilde{x} + \Big(\varepsilon(y, t) + \varepsilon(y, t')\Big)\,\tilde{y} .$$

$\varepsilon(x, t) + \varepsilon(x, t')$ is equal to 0 since the angles (x, t) and (x, t') are adjacent; similarly $\varepsilon(y, t) + \varepsilon(y, t')$ is equal to 0. So the lemma is proved with the chain $\dot{\alpha}(v, w) = \varepsilon(v, t) \cdot (\dot{t} + \dot{t}')$.

Suppose that the lemma is proved for every pair of vertices with a distance $d - 1$ in D_K and consider a pair of vertices (v, w) with a distance d in D_K . Let μ be a vertex of D_K with a distance to v equal to $d - 1$ and a distance to w equal to 1 . There exist ε and ε' equal to ± 1 such that $\tilde{v} + \varepsilon\tilde{\mu}$ and $\tilde{\mu} + \varepsilon'\tilde{w}$ are chains in $\dot{T}$. We consider then the chain

$\dot{\alpha}(v, w) = \tilde{v} + \epsilon\tilde{\mu} - \epsilon(\tilde{\mu} + \epsilon'\tilde{w})$.

(4.6) Proof of Theorem 2 and of the sufficient conditon of Theorem 3:

We consider the different cases depending on the number of diagonal components and their parities; for each case we determine a generative chain $\dot{\varphi}$ in $\dot{T}$.

(4.6.1) Each diagonal component of K is even. We partition these components into pairs (v_i, w_i) and we put $\phi = \sum \dot{\alpha}(v_i, w_i)$.

(4.6.2) K has a single diagonal component D, and this component is odd. We consider a triangle $t = [x, y, z]$; we partition $D - \{x, y, z\}$ into pairs (v_i, w_i), and we put $\dot{\varphi} = \dot{t} + \sum \dot{\alpha}(v_i, w_i)$.

(4.6.3) K has a double diagonal component D_d and a single diagonal component D_s . We consider a triangle $t = [x, y, z]$ and suppose that $x \in D_s$, y and $z \in D_d$. In the case where D_d is odd, we consider a vertex $y' \in D_d$ which is distinct from y and z, and we suppose that $\dot{\alpha}(y, y') = \tilde{y} + \epsilon_1 \tilde{y}'$. In the case where D_s is even, we consider a vertex $x' \in D_s$ which is distinct from x, and we suppose that $\dot{\alpha}(x, x') = \tilde{x} + \epsilon_2\tilde{x}'$.

(4.6.3.1) D_d is even, D_s is odd. We partition $D_d - \{y, z\}$ and $D_s - \{x\}$ into pairs (v_i, w_i), and we put $\dot{\varphi} = \dot{t} + \sum \dot{\alpha}(v_i, w_i)$.

(4.6.3.2) D_d is odd, D_s is odd. We partition $D_d - \{y', y, z\}$ and $D_s - \{x\}$ into pairs (v_i, w_i) and we put

$$\begin{aligned}\dot{\varphi} &= \dot{t} + \epsilon(y, t) \cdot \dot{\alpha}(y, y') + \sum \dot{\alpha}(v_i, w_i) \\ &= 2\,\epsilon(y, t) \cdot \tilde{y} + \epsilon(x, t) \cdot \tilde{x} + \epsilon(z, t) \cdot \tilde{z} + \epsilon_1 \cdot \epsilon(y, t) \cdot \tilde{y}' \\ &\quad + \sum \dot{\alpha}(v_i, w_i) \ .\end{aligned}$$

$\dot{\varphi}$ is a generative chain if m is odd.

(4.6.3.3) D_d is odd, D_s is even. We partition $D_d - \{y', y, z\}$ and $D_s - \{x, x'\}$ into pairs (v_i, w_i), and we put

$$\dot{\varphi} = \dot{t} + \varepsilon(x, t) \cdot \dot{\alpha}(x, x') + \varepsilon(y, t) \cdot \dot{\alpha}(y, y') + \sum \dot{\alpha}(v_i, w_i)$$

$$= 2\varepsilon((x, t) \cdot \tilde{x} + 2\varepsilon(y, t) \cdot \tilde{y} + \varepsilon(z, t) \cdot \tilde{z}$$

$$+ \varepsilon_1 \cdot \varepsilon(y, t) \cdot \tilde{y}' + \varepsilon_2 \cdot \varepsilon(x, t) \tilde{x}' + \sum \dot{\alpha}(v_i, w_i) .$$

$\dot{\varphi}$ is a generative chain if m is odd.

(4.6.4) K has three diagonal components D, D' and D''. These components are simple. The case where D, D' and D'' are even has been considered in(4.6.1). It remains to consider the cases where $0, 1$ or 2 of these components are even. We consider a triangle $t = [x, x', x'']$ and we suppose that x(resp. x', x'') belongs to D(resp. D', D''). In the case where D(resp. D') is even, we consider a vertex y(resp. y') in D(resp. D') distinct from x(resp. x'). We suppose that $\alpha(x, y) = \tilde{x} + \varepsilon_1 y$ (resp. $\dot{\alpha}(x', y') = \tilde{x}' + \varepsilon_2 \tilde{y}'$) .

(4.6.4.1) D, D' and D'' are odd. We partition $D - \{x\}$, $D' - \{x'\}$ and $D'' - \{x''\}$ into disjoint pairs (v_i, w_i), and we put $\dot{\varphi} = \dot{t} + \sum \dot{\alpha}(v_i, w_i)$.

(4.6.4.2) D is even, D' and D'' are odd. We partition $D - \{x, y\}$, $D' - \{x'\}$ and $D'' - \{x''\}$ into disjoint pairs (v_i, w_i), and we put

$$\dot{\varphi} = \dot{t} + \varepsilon(x, t) \cdot \dot{\alpha}(x, y) + \sum \dot{\alpha}(v_i, w_i)$$

$$= 2\varepsilon(x, t) \cdot \tilde{x} + \varepsilon(x', t) \cdot \tilde{x}' + \varepsilon(x'', t) \cdot \tilde{x}'' + \varepsilon_1 \cdot \varepsilon(x, t) \cdot \tilde{y}$$

$$+ \sum \dot{\alpha}(v_i, w_i) .$$

$\dot{\varphi}$ is a generative chain if m is odd.

(4.6.4.3) D and D' are even, D'' is odd. We partiton $D - \{x, y\}$, $D' - \{x', y'\}$ and $D'' - \{x''\}$ into disjoint pairs (v_i, w_i), and we put

$$\dot{\varphi} = \dot{t} + \varepsilon(x, t) \cdot \dot{\alpha}(x, y) + \varepsilon(x', t) \cdot \dot{\alpha}(x', y') + \sum \dot{\alpha}(v_i, w_i)$$

$$= 2\varepsilon(x, t) \cdot \tilde{x} + 2\varepsilon(x', t')\tilde{x} + \varepsilon(x'', t)\tilde{x}''$$

$$+ \varepsilon_1 \cdot \varepsilon(x, t) \cdot \tilde{y} + \varepsilon_2 \cdot \varepsilon(x', t) \cdot \tilde{y}' + \sum \dot{\alpha}(v_i, w_i) .$$

φ is a generative chain if m is odd.

(4.7) LEMMA. *Let $\mu \geq 2$ be an integer and suppose that m is a multiple of μ. If there exists a generative m-valuation of K, then there exists a generative μ-valuation of K.*

Proof: Consider the canonical morphism H of the ring $Z/m\,Z$ onto the ring $Z/\mu\,Z$. If g is a generator of the additive group $Z/m\,Z$, then $H(g)$ is a generator of the additive group $Z/\mu\,Z$.

Consider an m-valuation φ of the triangulation K and put $\psi = H \circ \varphi$. Then ψ is a μ-valuation of K. The equality (E_1) (cf. (4.1)) implies

$$(E'1) \qquad \forall x \in V(K): \quad H(\dot{\varphi}(x)) = \sum_{\substack{x \in t \\ t \in T(K)}} H(\varepsilon(x, t)) \cdot H(\varphi(t))$$

For each angle (x, t), $H(\varepsilon(x, t))$ is equal to ± 1 in $Z/\mu\,Z$. The equality $\varepsilon(x, t) + \varepsilon(x, t') = 0$ for each pair of adjacent angles implies $H(\varepsilon(x, t)) + H(\varepsilon(x, t')) = 0$. Therefore we have

$$(E''1) \qquad \forall x \in V(K): \quad \dot{\psi}(x) = \sum_{\substack{x \in t \\ t \in T(K)}} H(\varepsilon(x, t)) \cdot \psi(t)$$

(E'1) and (E"1) imply $\dot{\psi}(x) = H(\dot{\varphi}(x))$ for each $x \in V(K)$, so that if φ is generative, ψ is also generative.

(4.8) Proof of the necessary conditions of the Theorem 3: Following the preceding lemma the necessary condition must only be proved for $m = 2$. So we have now to show that the existence of a generative 2-valuation implies:

(A) if K has a double diagonal component, then that component is even;

(B) if K has three diagonal components, then these components have the same pairity.

Since $m = 2$, the chains on $V(K)$ over $Z/m\,Z$ are the

characteristic mappings of the subsets of $V(K)$. If $P \subset V(K)$, we put

$$\widetilde{P} = \sum_{x \in P} \widetilde{x} \ ;$$

$$\overline{w}_P = 0 \quad \text{if} \quad P \quad \text{has even cardinality;}$$

$$\overline{w}_P = 1 \quad \text{if} \quad P \quad \text{has odd cardinality.}$$

For each angle (x, t) we now have $\varepsilon(x, t) = 1$, since $+1 = -1$ in $Z/2\,Z$. Therefore for each triangle t, we have

$$\dot{t} = \sum_{x \in t} \widetilde{x} = \widetilde{t} \ .$$

If φ is a generative 2-valuation, the equality (E3) in (4.3) implies:

(E'3)
$$\sum_{t \in T(K)} \varphi(t) \cdot \widetilde{t} = \widetilde{V(K)} \ .$$

For each pair (f, g) of chains on $V(K)$ over $Z/2\,Z$ with coefficient $(f_x)_{x \in V(K)}$ and $(g_x)_{x \in V(K)}$ respectively, we put

$$\langle f, g \rangle = \sum_{x \in V(K)} f_x \cdot g_x \ .$$

The mapping $(f, g) \to \langle f, g \rangle$ is a symmetric bilinear form.
Let D be a diagonal component of K. We have

$$\langle \widetilde{D}, \widetilde{V(K)} \rangle = \overline{w}_D \ .$$

Therefore the equality (E'3) implies

(E"3)
$$\sum_{t \in T(K)} \varphi(t) \cdot \langle \widetilde{D}, \widetilde{t} \rangle = \overline{w}_D \ .$$

<u>Case A</u>: D is a double diagonal component of K. Each triangle t has exactly two vertices in D, so we have $\langle \widetilde{D}, \widetilde{t} \rangle = 0$ for each $t \in T(K)$. The equality (E"3) implies then that $\overline{w}_D = 0$, D must be even.

Case B: D is a simple diagonal component of K . Each triangle t has exactly one vertex in D , so we have $\langle \tilde{D}, \tilde{t} \rangle = 1$ for each $t \in T(K)$. The equality (E"3) implies then

$$\sum_{t \in T(K)} \varphi(t) = \bar{w}_D .$$

Therefore if we have three diagonal components D , D' and D" , we must have $\bar{w}_D = \bar{w}_{D'} = \bar{w}_{D''}$.

5. Some applications of the Theorems 1 and 2.

An immediate corollary of Theorems 1 and 2 is:

(5.1) COROLLARY. If an eulerian graph G has a triangular imbedding into a surface S , then for every odd integer $m \geq 1$, $G_{(m)}$ has a triangular imbedding into a surface $\bar{S}$ with the same orientability characteristic as S.

If K is an even triangulation and $m \geq 2$ is an even integer, there is no generative m-valuation of K in the two following cases (Theorem 3):

(A) K has a diagonal component which is odd and double;

(B) K has three diagonal components with different parities.

The case A is illustrated by Figure 5 which shows a triangulation K' of the torus; the case B is illustrated by Figure 6 which shows a triangulation K" of the sphere.

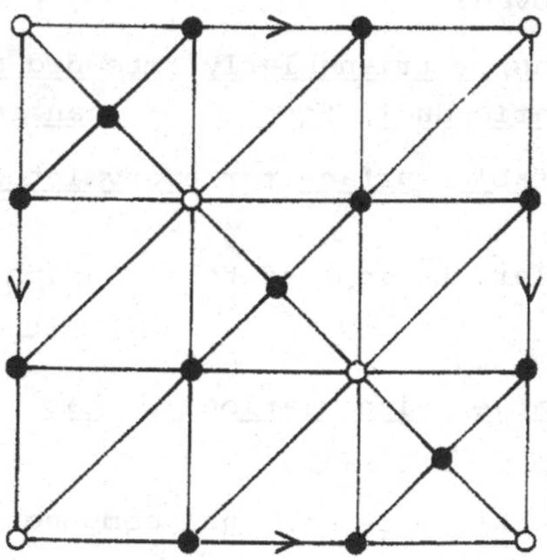

Figure 5.

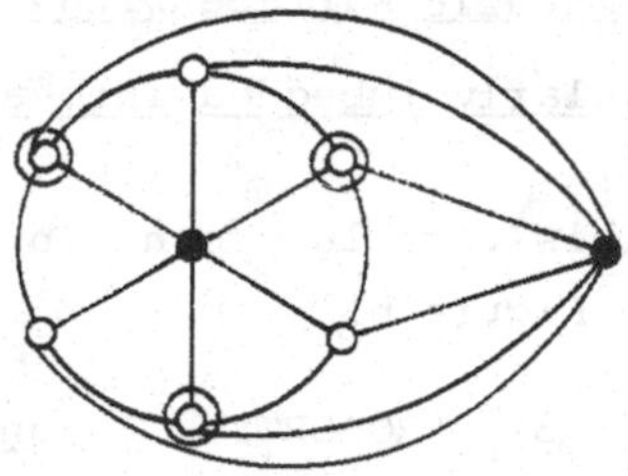

Figure 6.

Denote by G' and G" respectively the 1-skeletons of K' and K" . There does not exist a generative 2-valuation of K' , and we have verified that there does not exist a triangular imbedding of $G'_{(2)}$ into an orientable surface. However, we shall see that there exists a triangular imbedding of $G''_{(m)}$ into an orientable surface for every integer $m \geq 1$.

For a graph G which is 2-cell imbedded into a surface S , denote by G* the dual pseudograph for this imbedding. Say that the imbedding of G has bichromatic dual if $\chi(G^*) = 2$. In the particular case where the imbedding of G is triangular, consider the triangulation K associated with this imbedding. Define a 2-coloring of T(K) as an assignment of the signs + and - to the triangles of K in such a way that any two distinct triangles sharing a common edge have opposite signs. It is clear that the triangular imbedding of G has bichromatic dual if and only if there exists a 2-coloring of T(K). Note also that the existence of a 2-coloring of T(K) implies that K is an even triangulation.

B. L. Garman, R. D. Ringeisen and A. T. White [4] have proved that if a graph G has a triangular imbedding into an orientable surface with a bichromatic dual, then for every integer $\alpha \geq 0$, the graph $G_{(2^\alpha)}$ also has a triangular imbedding into an orientable surface with a biochromatic dual. More generally, consider an integer $m \geq 2$, the greatest integer α such that 2^α divides m, and the odd integer $\mu = m/2^\alpha$. Put $G' = G_{(2^\alpha)}$. Then Corollary (5.1) implies that $G'_{(\mu)} = G_{(m)}$ has a triangular imbedding into an orientable surface. So we have proved:

(5.2) COROLLARY. <u>If a graph G can be triangularly imbedded into an orientable surface with a bichromatic dual, then $G_{(m)}$ can be triangularly imbedded into an orientable surface for every integer $m \geq 2$</u> .

This corollary can, in particular, be applied to the graph G" of the Figure 6.

(5.3) PROPERTY. <u>If an orientable triangulation K has three diagonal components, then T(K) has a 2-coloring</u>.

<u>Proof</u>: Denote by D_1, D_2, D_3 the three diagonal components of K . Assign the color i = 1 , 2 , or 3 to each vertex x which belongs to D_i . Choose an orientation of S(K). Put the sign +

(resp. -) on a triangle t if the orientation of S(K) induces the ordering 1 , 2 , 3 (resp. 3 , 2 , 1) on the colors of its vertices. It is easy to verify that we thus obtain a 2-coloring of T(K).

(5.4) COROLLARY. Let G be an eulerian graph triangularly imbedded into an orientable surface. If the triangulation K determined by the imbedding of G has no double diagonal component with odd cardinality, then $G_{(m)}$ can be triangularly imbedded into an orientable surface for every integer $m \geq 1$.

This corollary is obtained by a combination of Theorem 3, Corollary (5.2) and Property (5.3).

(5.5) Remark: Consider a triangulation K with a 2-coloring of T(K) and a generative m-valuation φ of K, and let $\overline{K}$ be the expansion of K by φ. Assign the same sign to the triangle $\left[(x, i), (y, j), (z, k)\right] \in T(\overline{K})$ as is assigned to the corresponding triangle $[x, y, z] \in T(K)$. It is easy to verify that we thus obtain a 2-coloring of $T(\overline{K})$. Therefore if G has a triangular imbedding into a surface with bichromatic dual, then any triangular imbedding of $G_{(m)}$ constructed by an expansion of the triangulation associated with the imbedding of G also has a bichromatic dual. This remark is also true for Corollaries (5.2) and (5.4).

6. An application to complete multipartite graphs.

The purpose of this Section is to prove the following theorem:

THEOREM 4. Let G be a complete multipartite graph which is eulerian and $m \geq 2$ an integer. If there exists a triangular imbedding of G into a surface S, then there exists a triangular imbedding of $G_{(m)}$ into a surface $\overline{S}$ which has the same orientability characteristic as S.

Before proving the theorem, we give a necessary condition to have a triangular imbedding of $K_{m_1,m_2,\ldots,m_n}$ into a surface. A geometric interpretation of that necessary condition allows us to define particular types of complete multipartite graphs, the limit graphs which are also involved in the discussion of the number of diagonal components of a triangulation K whose 1-skeleton is a complete multipartite graph.

(6.1) PROPOSITION. In order to have a triangular imbedding of the complete n-partite graph $K_{m_1,m_2,\ldots,m_n}$ into a surface it is

<u>necessary that</u> $n \geq 3$ <u>and</u> $2m_i + m_j \leq m_1 + m_2 + \cdots + m_n$ <u>for every ordered pair of distinct indices</u> (i, j) <u>such that</u> $1 \leq i, j \leq n$.

<u>Proof</u>: We must have $n \geq 3$, for otherwise the graph is bipartite or an independent set. Let $S_1, S_2, \ldots, S_n$ be the maximal independent sets with $|S_i| = m_i$ for each i such that $1 \leq i \leq n$. Suppose that there exists a triangulation K such that $K^1 = K_{m_1, m_2, \ldots, m_n}$. Let (i, j) be an ordered pair of distinct indicies with $1 \leq i, j \leq n$. Consider a vertex x_j in S_j. The set of vertices belonging to the link $L(x_j, K)$ is equal to $V(K) - S_j$. Two vertices of S_i cannot be consecutive on the cycle $L(x_j, K)$ for S_i is an independent set; therefore the cardinality of $V(K) - S_j - S_i$ must be at least the cardinality of S_i. Thus we obtain

$$m_1 + m_2 + \cdots + m_n - m_j - m_i \geq m_i .$$

(6.2) COROLLARY. <u>In order to have a triangular imbedding of</u> K_{m_1, m_2, m_3} <u>into a surface it is necessary that</u> $m_1 = m_2 = m_3$.

<u>Proof</u>: Let i and j be two distinct indices such that $1 \leq i, j \leq 3$, and let k be the third index. The inequality $2m_i + m_j \leq m_1 + m_2 + m_3$ is equivalent to $m_i \leq m_k$. So we must have $m_1 \leq m_2 \leq m_3 \leq m_1$.

S. Stahl and A. T. White [13] have already proved the preceeding property, and they have also shown that it is sufficient for the orientable case (a solution is also given in [11]).

(6.3) <u>Geometric interpretation</u>: Associate with the complete n-partite graph $K_{m_1, m_2, \ldots, m_n}$ the values $x_i = m_i/(m_1 + m_2 + \ldots + m_n)$ We shall say that the point $x = (x_1, x_2, \ldots, x_n)$ of the n-dimensional affine space R^n is the projection of $K_{m_1, m_2, \ldots, m_n}$.

It is clear that a point $x = (x_1, x_2, \ldots, x_n)$ of R^n is the projection of a complete n-partite graph if and only if it is a rational point (i.e. its coordinates are rational numbers) with strictly positive coordinates. In that case there exists one and only one graph

$G^x = K_{\mu_1, \mu_2, \ldots, \mu_n}$ with a projection equal to x such that the integers $\mu_1, \mu_2, \ldots, \mu_n$ are relatively prime; the other complete n-partite graphs with a projection equal to x are of the form $G^x_{(m)}$ where $m \geq 1$ is an arbitrary integer.

Suppose $n \geq 3$. The inequalities of Proposition (6.1) imply

(E_{ij}) $\quad 2x_i + x_j \leq 1$ for every pair of distinct indices

(i, j) such that $1 \leq i, j \leq n$.

We also have the additional constraints:

(c) $x_1 + x_2 + \cdots + x_n = 1$;

(ci) $0 \leq x_i$ for every integer i such that $1 \leq i \leq n$.

The constraints (E_{ij}), (c), (ci) define a convex polyhedron P_n in the space F^n. Since these constraints have rational coefficients, the vertic. of P_n are rational points. The vertices of P_n whose coordinates are all distinct from 0 are particularly interesting for they correspond to limit cases where the necessary conditions to have a triangular imbedding of a complete n-partite graph into a surface are satisfied. Therefore a complete n-partite graph whose projection is a vertex of P_n will be called a limit n-partite graph.

(6.4) PROPOSITION. <u>A point</u> $\ell = (\ell_1, \ell_2, \ldots, \ell_n)$ <u>with coordinates different from</u> 0 <u>is a vertex of</u> P_n <u>if and only if</u> $n - 1$ <u>of the coordinates are equal to</u> $1/(2n-3)$, <u>the</u> n^{th} <u>coordinate being equal to</u> $(n-2)/(2n-3)$.

The proof of this proposition is not within the scope of this paper. In order to characterize the limit n-partite graphs, we can suppose without loss of generality that $\ell_1 = (n-2)/(2n-3)$, and $\ell_2 = \ell_3 = \cdots = \ell_n = 1/(2n-3)$, so we see that a limit n-partite graph is of the form $K\ell(n, \alpha) = K_{(n-2)\alpha, \alpha, \alpha, \ldots, \alpha}$ where $\alpha \geq 1$ is an arbitrary integer.

(6.5.1) <u>Remark</u>: The limit 3-partite graph $K\ell_{(3, \alpha)}$ is equal to the regular complete 3-partite graph $K_{\alpha,\alpha,\alpha}$.

(6.5.2) Remark: A vertex of $K\ell(n, \alpha)$ which belongs to the independent set of cardinality $(n-2)\alpha$ has degree equal to $(n-1)\alpha$; any other vertex has degree equal to $2(n-2)\alpha$. Therefore $K\ell(n, \alpha)$ is an eulerian graph if and only if $(n-1)\alpha$ is an even integer.

(6.6) PROPOSITION. Let $n \geq 3$ be an integer and K a triangulation whose 1-skeleton K^1 is a complete n-partite graph, then:

(i) if $K^1 = K\ell(3, \alpha)$ then K has three diagonal components with the same cardinality α;

(ii) if $K^1 = K\ell(n, \alpha)$ with $n \geq 4$, then K has two diagonal components and the cardinality of the double component is equal to $(n-1)\alpha$;

(iii) if K^1 is not a limit n-partite graph, then K has a single diagonal component.

Proof: Put $K^1 = K_{m_1, m_2, \ldots, m_n}$. Denote by $S_1, S_2, \ldots, S_n$ the maximal independent sets of K^1 with $|S_i| = m_i$ for $1 \leq i \leq n$. For each $i = 1, 2, \ldots, n$, put $V_i = V(K) - S_i$. V_i is the vertex-set of any link $L(x, K)$ such that $x \in S_i$. The cardinality of V_i is equal to $m_1 + m_2 + \cdots + m_n - m_i$, and therefore Proposition (6.1) gives

$$(I_{ij}) \qquad \forall\, i \neq j,\ 1 \leq i,\ j \leq n:\ 2|S_j| \leq |V_i|\ .$$

If $n = 3$, m_1, m_2 and m_3 are necessarily equal to the same integer α (6.2); K^1 is 3-chromatic, therefore S_1, S_2 and S_3 are the three diagonal components of K (3.3). So (i) is proved.

Suppose $n \geq 4$. The chromatic number of K^1 is greater than 3; therefore K has at most two diagonal components. In order to prove (ii) and (iii) we show that K has two diagonal components if and only if K^1 is equal to a limit n-partite graph.

a. Suppose that K has a double diagonal component D_d and a simple diagonal component D_s. Suppose also that a vertex $x_1 \in S_1$ belongs to D_s. If $[y, z]$ is an edge of $L(x_1, K)$, y and z must be in D_d, since the triangle $[x, y, z]$ contains a single vertex of D_s. Therefore all the vertices of $L(x_1, K)$ are in D_d. Since V_1 is the vertex-set of $L(x_1, K)$, we have $V_1 \subset D_d$ and $D_s \subset S_1$.

Consider an index $i \geq 2$ and a vertex $x_i \in S_i$, such that x_i is contained in D_d. If $[y, z]$ is an edge of $L(x_i, K)$, then y and z connot both belong to D_d, for the triangle $[x_i, y, z]$ would have three vertices in D_d. Therefore the vertices of $L(x_i, K)$ belong alternately to D_d and D_s. D_s is entirely contained in V_i for $S_1 \subset V_i$, so we must have

$$|D_s| = |V_i - D_s|.$$

The preceding equality implies $2|D_s| = |V_1|$. The inequality (I_{i1}) gives $|V_i| \geq 2|S_1|$. Since $|D_s| \leq |S_1|$, we obtain $|D_s| = |S_1|$ and $|V_i| = 2|S_1|$ for $i \geq 2$. So there exists α such that $|S_2| = |S_3| = \cdots = |S_n| = \alpha$ and $|S_1| = (n-2)\alpha$, and K^1 is equal to $K\ell(n, \alpha)$. D_d is equal to $S_2 \cup S_3 \cup \cdots \cup S_n$; therefore the cardinality of D_d is equal to $(n-1)\alpha$.

(b) Suppose that $K^1 = K\ell(n, \alpha)$. We prove first that every triangle $t \in T(K)$ contains a vertex of S_s.

Put $t = [x, y, z]$ and suppose that $x \in S_i$. If $i = 1$ the property is proved. If $i \geq 2$ consider the cycle $L(x, K)$. The set of vertices of this cycle is equal to V_i. Two vertices of S_1 cannot be successive on the cycle $L(x, K)$ for S_1 is an independent set. We have the equality $|S_1| = |V_i|/2$ for $K^1 = K\ell(n, \alpha)$; therefore every edge of the cycle $L(x, K)$ must contain a vertex of S_1. In particular $[y, z]$ is an edge of $L(x, K)$; therefore $[x, y, z]$ contains a vertex of S_1.

A triangle of K cannot contain two vertices of S_1 for S_1 is an independent set, so each triangle contains exactly one vertex of S_1. Therefore S_1 is a diagonal component of K. The complement of S_1 is either equal to a double diagonal component or to the union of two simple diagonal components. The second case is excluded, since K has at most two diagonal components.

(6.7) Proof of Theorem 4: Consider a triangulation K such that $K^1 = G$ and an integer $m \geq 2$. There exists a generative m-valuation of K when m is odd (Theorem 2). If K has a double component, then G is equal to a limit n-partite graph $K\ell(n, \alpha)$. The cardinality of that double component is equal to $(n-1)\alpha$ (Proposition (6.6)), so that double component is even, since $K\ell(n, \alpha)$ is eulerian if and only if $(n-1)\alpha$ is even (Remark (6.5.2)). If K has three diagonal components, these diagonal components have the same cardinality (Proposition (6.6)), so they have the same parity. So Theorem 3 implies the existence of a generative m-valuation when m is even.

7. Some cases of triangular imbeddings of $K_{n(m)}$.

Table I (resp. II) has two axes corresponding to the residue classes of the integers n (mod 12) (resp. mod. 6) and m (mod. 6). This table indicates some cases of triangular imbeddings of $K_{n(m)}$ into an orientable surface (resp. non orientable surface). The conventions are the following.

(i) A square is crossed by two diagonal lines when the Euler formula does not allow a triangular imbedding. More precisely, if we have a triangular imbedding of a graph G into a surface S, then the Euler characteristic of S is equal to $|V(G)| - |E(G)|/3$; therefore $|E(G)|$ must be a multiple of 3. If $G = K_{n(m)}$, we have $|E(G)| = m^2n(n-1)/2$, so a square of the Table II is crossed by two diagonal lines if and only if $m^2n(n-1)$ is not a multiple of 6. If S is an orientable surface, the Euler characteristic must be even. We have $|V(G)| - |E(G)|/3 = (6nm - m^2n(n-1))/6$, so a square of the Table I is crossed by two diagonal lines if and only if $6nm - m^2n(n-1)$ is not a multiple of 12.

(ii) A square is crossed by a single diagonal line if the degrees of $K_{n(m)}$ are odd. The triangular imbeddings of such a graph cannot be studied by application of Theorem 4. The degree of a vertex of $K_{n(m)}$ is equal to $m(n-1)$, so a square is crossed by a diagonal line if and only if m is odd and n is even.

(iii) A square with a reference (r) and perhaps an additional note $(n = p?)$ means that a triangular imbedding of $K_{n(m)}$ can be constructed by the application of Theorem 4 when $n \neq p$. The problem is still unsolved when $n = p$. Details are given in the reference (r) (see below.)

(iv) A square with a reference (r?) indicates that the existence of a triangular imbedding of $K_{n(m)}$ is still not solved. The reference (r) gives details.

References to Tables I and II:

(1) We use the triangular imbeddings of K_n which have been constructed for the proof of the map-coloring theorem [10]; for every integer $s \geq 0$, K_{12s+3} and K_{12s+7} are triangularly imbeddable into orientable surfaces; for every integer $s \geq 1$, K_{6s+1} and K_{6s+3} are triangularly imbeddable into non-orientable surfaces. A problem remains in finding a triangular imbedding of $K_{7(m)}$ into a non-orientable surface when $m \geq 2$. $K_{3(2)}$ is a planar graph; it has been verified that $K_{3(3)}$ has no triangular imbedding into a non-orientable surface [1], and a problem remains for $K_{3(m)}$ when $m \geq 4$.

Orientable and triangular imbeddings have already been constructed for $K_{3(m)}$ by G. Ringel and J. W. T. Youngs [11], S. Stahl and A. T. White [13]; for $K_{12s+3(2^\alpha)}$ by B. L. Garman, R. D. Ringeisen and A. T. White [4], and for $K_{12s+7(3^\alpha)}$ and $K_{12s+7(2\cdot 3^\alpha)}$ in [2].

(2) We use the triangular imbeddings of $K_{n(2)}$ into orientable surfaces for $n \equiv 0, 1 \pmod{3}$ which have been announced by G. Ringel and M. Jungerman [12].

(2') We use the triangular imbedding of $K_{n(2)}$ into a non-orientable surface announced by M. Jungerman [8] for $n \equiv 0, 1 \pmod{3}$ and $n \geq 6$. A problem remains for $K_{3(2m)}$ and $K_{4(2m)}$. The first one has already been discussed in the reference (1). The case of $K_{4(m)}$ has been completely solved by M. Jungerman [7]: $K_{4(m)}$ has a triangular imbedding into a non-orientable surface if $m \geq 3$, and $K_{4(2)}$ has no triangular imbedding into an non-orientable surface.

(3) We use the triangular imbeddings of $K_{4s+3(3)}$ into orientable surfaces constructed in [2] and [3].

(4) Orientable and triangular imbeddings of $K_{12s+2(6m)}$ and $K_{12s+5(6m)}$ will exist if they exist for $m = 1$. Non-orientable and triangular imbeddings of $K_{6s+2(6m)}$ and $K_{6s+5(6m)}$ will exist if they exist for $m = 1$, and they will exist for $K_{6s+5(3m)}$ if they exist for $m = 1$ Such imbeddings have been constructed for $K_{n(3)}$ when $n \equiv 0 \pmod{12}$ [9] and when n is a prime-power integer equal to $3 \pmod{4}$ [1].

m \ n	0	1	2	3	4	5	6	7	8	9	10	11
0	(2)	(2)	(4?)	(1) (2) (3)	(2)	(4?)	(2)	(1) (2) (3)	(4?)	(2)	(2)	(3)
1,5				(1)				(1)				
2,4	(2)	(2)		(1) (2)	(2)		(2)	(1) (2)		(2)	(2)	
3				(1) (3)				(1) (3)				(3)

Table I

m \ n	0	1	2	3	4	5
0	(2')	(1) (2') n=7?	(4?)	(1) (2') n=3?	(2')	(4?)
1,5		(1) n=7?		(1) n=3?		
2,4	(2')	(1) (2') n=7?		(1) (2') n=3?	(2')	
3		(1) n=7?		(1) n=3?		(4?)

Table II

References

[1] P. Bayle, Séminaire de théorie des graphes, Centre Universitaire du Mans.

[2] D. Bernard, A. Bouchet, Some cases of triangular imbedding for $K_{n(m)}$, to appear in J. Combinatorial Theory.

[3] A. Bouchet, Cyclabilité des groupes commutatifs d'ordre impair, to appear.

[4] B. L. Garman, R. D. Ringeisen, A. T. White, On the genus of strong tensor products of graphs, Canad. J. Math. 28(1976), 523-532.

[5] J. L. Jolivet, Le joint d'une famille de graphes, Compte-rendu du colloque de l'institut des Hautes Etudes, Universite' Libre de Bruxelles, avril 1973, pp. 319-325.

[6] M. Jungerman, The genus of the symmetric quadripartite graph, J. Combinatorial Theory (B), 19(1975), 181-187.

[7] M. Jungerman, The non-orientable genus of $K_{4(n)}$, abstract 6 in graph Theory Newsletter 5(1975), p. 86.

[8] M. Jungerman, The non-orientable genus of O_n, abstract 7 in Graph Theory Newsletter 5(1975), p. 86.

[9] M. Jungerman, personal communication.

[10] G. Ringel, Map Color Theorem, Grundlehren der mathematischen Wissenchaften, Bd. 209, Springer Verlag, New York, 1974.

[11] G. Ringel, J. W. T. Youngs, Das Geschlecht des vollständige dreifarbaren Graphen, Comment. Math. Helv. 45(1970), 152-158.

[12] G. Ringel, M. Jungerman, The genus of O_n, regular cases, abstract 13 in Graph Theory Newsletter 5(1975), p. 88.

[13] S. Stahl, A. T. White, Genus embeddings for some complete tripartite graphs, Disc. Math. 14(1976), 279-296.

[14] W. T. Tutte, Introduction to the theory of matroids, Elsevier, New York, 1971.

[15] A. T. White, Graphs, Groups and Surfaces, North-Holland, Amsterdam, 1973.

MIXED GRAPHS OF HOMOMORPHISMS

Michael Capobianco and John Molluzzo

St. John's University, Staten Island, N.Y.

Introduction

Certain mappings from one graph into another can themselves be represented by (mixed) graphs. For example, the mapping pictured in figure 1 from the graph $K_{1,3} + x$ to P_3 can be represented by the mixed graph of that figure, or in what we call the compact form shown in figure 2. Here it is to be understood

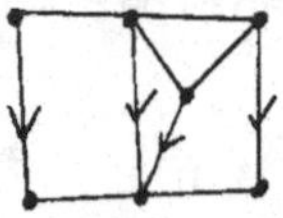

FIG. 1

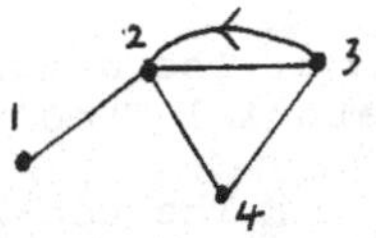

FIG. 2

that points 1, 2, and 4 map into themselves while point 3 maps into point 2. In this paper we will deal exclusively with the compact form, and it should be noted that not all mappings can be pictured in this way. An example is given in figure 3.

FIG. 3

We shall be particularly interested in mappings, , for which the image graph, H, is a homomorphic image of the preimage graph, G i.e., if v_1 and v_2 are adjacent in G then (v_1) and (v_2) are adjacent in H, and furthermore any pair of adjacent points in H has a pair of adjacent preimages in G. It is easy to see that the latter condition always holds true for mappings which can be represented by the compact form. Since it is possible that a single mapping can be represented by more than one mixed graph, we define a <u>normal form</u> for the mixed graph, and present an algorithm for converting any mixed graph to normal form, which simultaneously detects whether the mapping is a homomorphism of the type described above.

2. The Normal Form

Consider any mixed graph, say the one in figure 4.

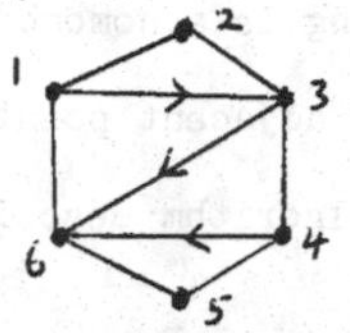

FIG. 4

This represents the mapping pictured in figure 5, which can be represented more

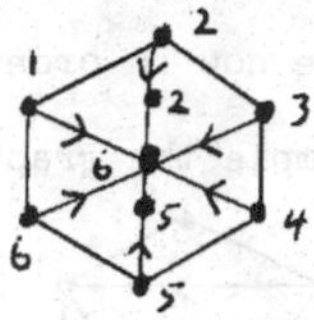

FIG. 5

simply as in figure 6.

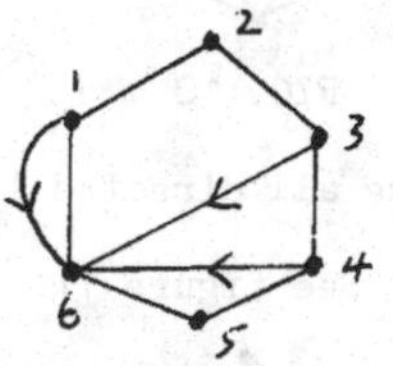

FIG. 6

This is an example of normal form. Note that no dipath of length greater than 1 exists. In other words, each point is adjacent to its image (unless it is its own image). This is the essential feature of the normal form. It should also be made clear that one must be careful about points with outdegree greater than 1. For example, the graph in figure 7 does not even represent a mapping! On the other hand, the one in figure 8 does. In fact, it is the mapping whose norma form is given in figure 9.

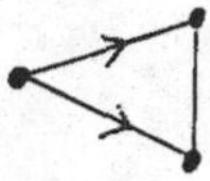

FIG. 7

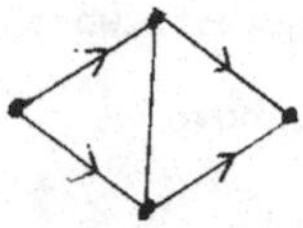

FIG. 8

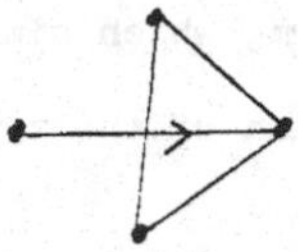

FIG. 9

To check whether the mapping is a homomorphism it is only required that we check to see if each pair of adjacent points in G gets mapped into a pair of adjacent points in H. The algorithm described in the next section does this as the final step.

3. The Algorithm

Some formal definitions are now in order. The original mixed graph we shall denote by M. We use as an example the graph of figure 10.

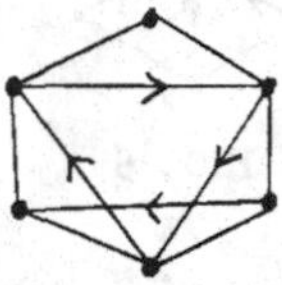

FIG. 10

The graph obtained from M by deleting all directed lines is denoted by G. This is the domain graph of the mapping. See figure 11.

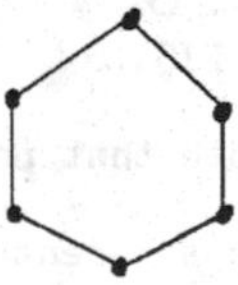

FIG. 11

The digraph obtained from M by deleting all undirected lines we call D; Fig. 12.

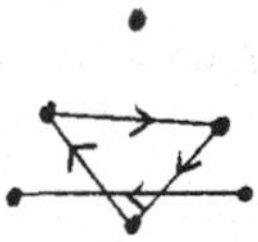

FIG. 12

The image graph we denote by H; figure 13.

FIG. 13

How to obtain H from M is an open question. It is not simply the subgraph induced by the image points. Figure 6 provides a counterexample.

Our algorithm is given by the flow chart in figure 14. $\bar{A}(G)$ is the complement of A(G), and $\bar{m}$ is a matrix defined by the subroutine shown.

4. Concluding Remarks

Homomorphisms help us to study graphs. It is to be hoped that with enough work along the lines of this paper, <u>graphs</u> will help us study homomorphisms. We have obtained some interesting but limited results on connectedness for the "open" form of M. These will hopefully be the subject of a future paper.

<u>References</u>

1. Harary, F., "Graph Theory", Addison-Wesley, Reading 1969

2. Hedetniemi, S., "Homomorphisms of Graphs", University of Michigan Technical Report, Ann Arbor 1965

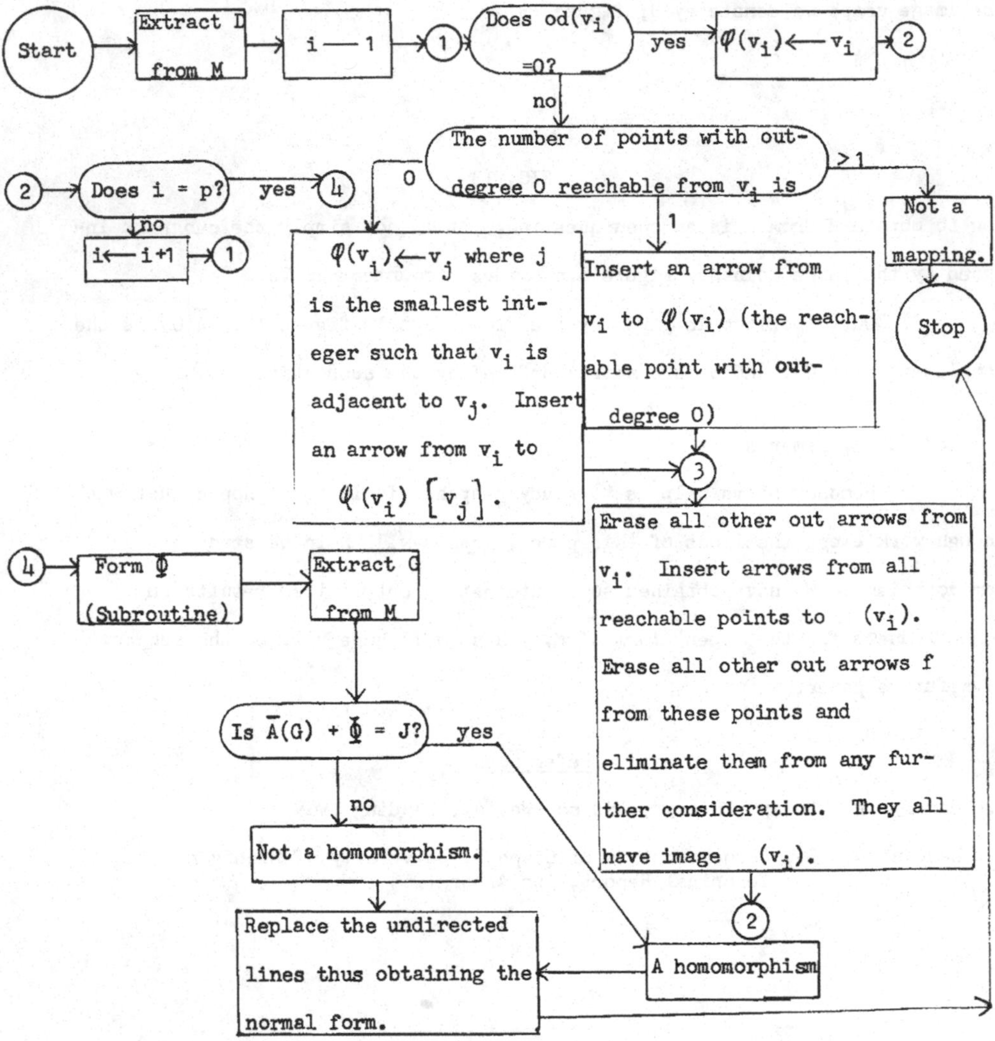

<u>Subroutine</u> <u>for</u> Φ (a pxp matrix $\| \varphi_{ij} \|$)

Let i_1, i_2, ...i_k be the preimages of $\varphi(i)$. For i=1,2,...,p and j=1,2,...p do the following:

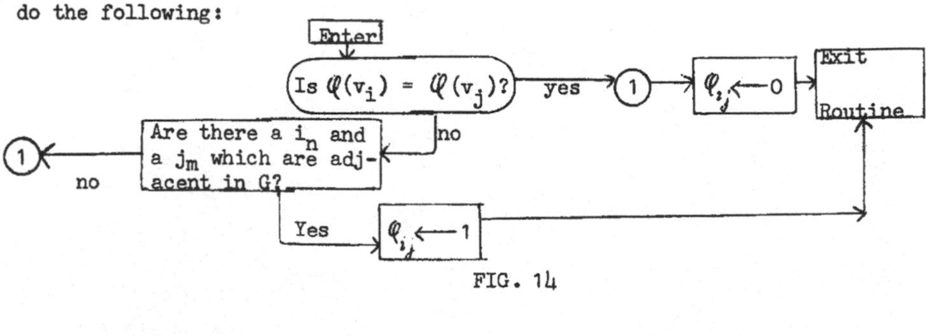

FIG. 14

ON CHROMATIC EQUIVALENCE OF GRAPHS

Chong-Yun Chao and Earl Glen Whitehead, Jr.
Department of Mathematics
University of Pittsburgh
Pittsburgh, Pa. 15260

We show that the cycle on n vertices and each θ-graph are chromatically unique. Also, we prove that a large class of nonisomorphic connected graphs are chromatically equivalent. This class includes the cactus graphs. Combinatorial identities are obtained from the matrices connecting chromatic polynomial bases.

1 INTRODUCTION

The graphs which we consider here are finite, undirected, simple and loopless. Two graphs X and Y are said to be chromatically equivalent if they have the same chromatic polynomial, i.e., $P(X, \lambda) = P(Y, \lambda)$. A graph X is said to be chromatically unique if $P(Y, \lambda) = P(X, \lambda)$ implies that Y is isomorphic to X. There are many graphs which are chromatically equivalent and are nonisomorphic, e.g., two nonisomorphic trees with the same number of vertices. On the other hand, we know that each of the null graphs, N_n, with n vertices is chromatically unique. So is each of the complete graphs, K_n, with n vertices. It seems natural to ask whether we can find other families of chromatically unique graphs. Here, in §2 we show that the cycle, C_n, with n vertices for $n \geq 3$, and the θ-graphs, θ_n, with n vertices for $n \geq 4$ are chromatically unique families of graphs where a θ_n, $n \geq 4$, is a graph consists of two cycles with one edge in common. Each of the wheels, W_n, with n vertices for $n \geq 4$ seems to be chromatically unique. However, we can neither prove nor disprove that.

Our results and computations depend heavily on Whitney's reduction formula in [10] and Whitney's broken cycle theorem in [11]. The formula is

$$P(X, \lambda) = P(X', \lambda) + P(X'', \lambda),$$

or

$$P(X', \lambda) = P(X, \lambda) - P(X'', \lambda)$$

where X' is obtained from X by adding an edge and X" is obtained from X by contracting the endpoints of the added edge to a single vertex. Let X be a graph with n vertices, Y be a subgraph of X such that Y has s edges and p components. We define the rank i and nullity j of Y by $i = n - p$ and $j = s - i = s - n + p$. Then $p = n - i$ and $s = i + j$. Let m_{ij} be the number of subgraphs of X with rank i and nullity j. Then

$$P(X, \lambda) = \sum_{i,j} (-1)^{i+j} m_{ij} \lambda^{n-i} = \sum_i m_i \lambda^{n-i}$$

where $m_i = \sum_j (-1)^{i+j} m_{ij}$. A broken cycle is a cycle without the last edge. Whitney's broken cycle theorem states: The number $(-1)^i m_i$ is equal to the number of subgraphs with i edges in X each of which does not contain all the edges of any broken cycle.

A cactus is a connected graph where any two cycles have no edge in common. By using Whitney's broken cycle theorem, it is easy to see that two cacti with the same number of vertices, the same number of edges and the same number of cycles have the same chromatic polynomial. In §3, we shall point out a class of graphs which includes cacti and which is a family of chromatically equivalent graphs.

Since each of the families of N_n, K_n and C_n are chromatically unique family of graphs, we use each family as a basis for the Z-module $Z[\lambda]$ where Z is the ring of integers. By changing the bases for $Z[\lambda]$, we can obtain many combinatorial identities in §4. Although all of these identities are known (e.g., see [3] and [6]), our method seems to be different.

2. CHROMATICALLY UNIQUE GRAPHS

Let X be a graph with n vertices and $P(X, \lambda)$ be its chromatic polynomial. We state some of the elementary properties of $P(X, \lambda)$ (see [5]): The degree of $P(X, \lambda)$ is n, the coefficient of λ^n in $P(X, \lambda)$ is 1, $P(X, \lambda)$ has no constant term, the terms in $P(X, \lambda)$ alternate in sign, the absolute value of the second coefficient of $P(X, \lambda)$ is the number of edges and the smallest number r such that λ^r has a nonzero coefficient in $P(X, \lambda)$ is the number of components in X. Consequently, X is connected if and only if $r = 1$. Theorem 3 in [5] states that if two graphs X_1 and X_2 "overlap" in a complete graph on k vertices, then the chromatic polynomial of the graph formed by X_1 and X_2 together is

$$\frac{P(X_1, \lambda)P(X_2, \lambda)}{\lambda(\lambda-1)(\lambda-2)\ldots(\lambda-k+1)} . \tag{1}$$

Also, the chromatic polynomial of a n-cycle, C_n, for $n \geq 3$ is

$$P(C_n, \lambda) = (\lambda-1)^n + (-1)^n(\lambda-1). \tag{2}$$

Expand (2), we have

$$P(C_n, \lambda) = \lambda^n - \binom{n}{1}\lambda^{n-1} + \binom{n}{2}\lambda^{n-2} - \ldots \pm \binom{n}{n-2}\lambda^2 \mp \left[\binom{n}{n-1}-1\right]\lambda \tag{3}$$

where $\binom{n}{k}$ are the binomial coefficients.

By using Whitney's reduction formula and by using induction, we also have

$$P(C_n, \lambda) = \begin{cases} \lambda(\lambda-1)(\lambda-2)\left[\sum_{i=0}^{(n-3)/2} (\lambda-1)^{2i}\right], & \text{if } n \text{ is odd,} \\ \lambda(\lambda-1)\left[\sum_{i=0}^{n-2} (-1)^i(\lambda-1)^i\right], & \text{if } n \text{ is even.} \end{cases} \tag{4}$$

(4) also indicates that a cycle of odd (even) length can be

colored by three (two) colors so that adjacent vertices receive distinct colors. In any case, $(\lambda-1)^2$ does not divide $P(C_n, \lambda)$.

Theorem 1. Each cycle C_n, $n \geq 3$, is chromatically unique.

Proof. Let Y be a graph with n vertices such that $P(Y, \lambda) = P(C_n, \lambda)$. Then, by the elementary properties, Y is connected with n edges. If Y had a vertex of degree 1, then, by applying Whitney's reduction formula and by connectness of Y, $P(Y, \lambda) = P(C_n, \lambda)$ would have a factor $(\lambda-1)^2$. That is a contradiction. Hence, every vertex in Y is of degree at least 2. Since Y has only n edges, every vertex is of degree 2 and Y is isomorphic to C_n.

Our theorem 1 can also be proved by using Whitney's broken cycle theorem, i.e., again, let Y be a graph with n vertices such that $P(Y, \lambda) = P(C_n, \lambda)$, then Y is a connected graph with n edges. Since Y has n vertices and n edges, Y is not a tree and Y contains a cycle. Suppose this cycle C is of length k and $k < n$, then, by Whitney's broken cycle theorem, the absolute value of the coefficient of λ in $P(Y, \lambda)$ would be less than $\binom{n}{n-1}-1$. That is a contradiction to (3). Hence, Y is a cycle of length n.

Theorem 2. Each θ-graph θ_n, $n \geq 4$, is chromatically unique.

Proof. Let our θ_n be a graph with n vertices consisting of a p-cycle and a q-cycle with one edge in common where $p + q - 2 = n$, then by (1), we have

$$P(\theta_n, \lambda) = \frac{P(C_p, \lambda)P(C_q, \lambda)}{\lambda(\lambda-1)}$$

By our formula (4), we know that $P(\theta_n, \lambda)$ is divisible by $(\lambda-1)$, but not by $(\lambda-1)^2$. We assume $p \geq q$.

Let Y be a graph with n vertices such that $P(Y, \lambda) = P(\theta_n, \lambda)$. Then Y is a connected graph with $n + 1$ edges. We show that the degree of every vertex in Y is either 2 or 3: (a) We claim that the degree of every vertex in Y $\geq$ 2. Since Y is connected, the degree of every vertex in Y is not zero. Suppose that there existed a vertex with degree 1 in Y, then, by Whitney's reduction formula and by the connectness of Y, $(\lambda-1)^2$ would divide $P(Y, \lambda) = P(\theta_n, \lambda)$. That is a contradiction. (b) We claim that no vertex in Y is of degree $\geq$ 4. Since this connected graph Y has n + 1 edges and n vertices and valency $\geq$ 2 at every vertex, no vertex in Y can have degree > 4, and Y can have at most one vertex of degree 4. Suppose that the degree of a vertex v in Y was 4, then we would have Y consisting of two cycles with v in common, and, by (1),

$$P(Y, \lambda) = \frac{P(C_r, \lambda)P(C_s, \lambda)}{\lambda}$$

where r + s - 1 = n. By our formula (4), $P(Y, \lambda)$ would be divisible by $(\lambda-1)^2$. That is a contradiction. Hence, every vertex in Y is of degree 2 or 3. In fact, since Y is connected with n vertices and n + 1 edges, there are exactly two vertices in Y are of degree 3.

We claim that Y does not contain an isthmus. Suppose the contrary, then we would have

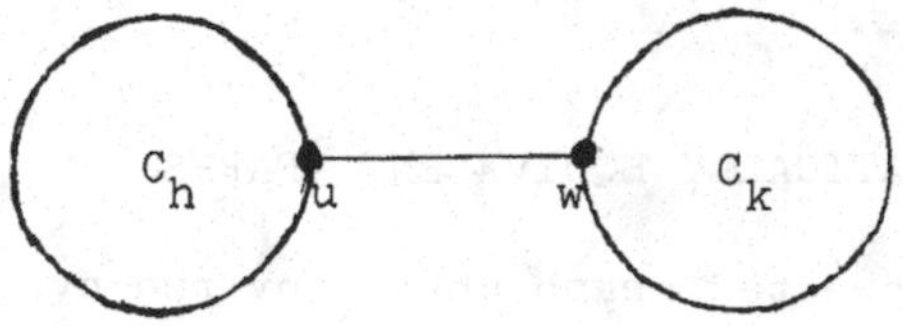

where h + k = n. By Whitney's reduction formula on the edge (u, w), by (1) and by our formula (4), $(\lambda-1)^2$ would divide

$$P(Y, \lambda) = (\lambda-1)\frac{P(C_h, \lambda)P(C_k, \lambda)}{\lambda} .$$

That is a contradiction.

Now, since our graph Y is a connected graph having 2 vertices of degree 3 and n - 2 vertices of degree 2 and having n + 1 edges and containing no isthmus, Y must be:

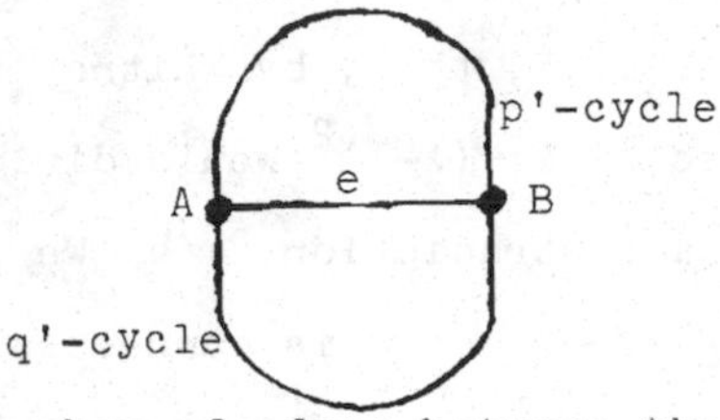

where e is the number of edges between the vertices A and B, i.e., Y consists of two cycles, say, a p'-cycle and a q'-cycle with e edges in common. Say, $p' \geq q'$.

We want to show that e = 1, p' = p and q' = q. There are two cases to be considered: Case 1. e > 1. In turn, there are two subcases: (a) $p' \neq p$, say, $p' < p$. Then, by Whitney's broken cycle theorem, the coefficient of $\lambda^{n-p'}$ in $P(Y, \lambda)$ would be $\binom{n}{p'-1}-1$, and the coefficient of $\lambda^{n-p'}$ in $P(\theta_n, \lambda)$ would be $\binom{n}{p'-1}$. That is a contradiction. (b) p' = p. Then q' > q, i.e., $p' = p \leq q < q'$. By Whitney's broken cycle theorem, the coefficient of λ^{n-q} in $P(Y, \lambda)$ would be greater than the coefficient of λ^{n-q} in $P(\theta_n, \lambda)$ by 1. That is again a contradiction. Case 2. e = 1. If $p' \neq p$, then the case is similar to the Case 1 (a), and we would get a contradiction. Hence, p' = p, then q' = q and Y is isomorphic to θ_n.

3. CHROMATICALLY EQUIVALENT GRAPHS

A cactus [2] is a connected graph where any two cycles have no edge in common. In [1], Geller and Manvel proved that any cactus can be reconstructed. Here, we shall show that all cacti with n vertices, e edges and the same number of cycles of each length are chromatically equivalent. In fact, we shall prove a more general theorem.

In a planar graph, a cycle is said to be a mini-cycle if it is one of the two smaller cycles in every θ-subgraph which contains this cycle, or if it is a cycle which is not contained in any θ-subgraph. A connected planar graph X is said to be forest-like if every pair of cycles have at most one edge in common, and if the dual graph X^d of X is a forest where X^d is obtained from X by replacing each mini-cycle by a vertex, and by joining two vertices in X^d by an edge if and only if the corresponding mini-cycles in X have one edge in common.

Theorem 3. All forest-like connected planar graphs with n vertices, e edges and the same number of mini-cycles of each length are chromatically equivalent.

Proof. Let X and Y be two such graphs. By using Whitney's broken cycle theorem, we show that $P(X, \lambda) = P(Y, \lambda)$.

Consider X, then X^d is a forest. Consider each tree in X^d. Choose any vertex of the first tree, and label it and its corresponding mini-cycle in X by 1. If there are j vertices adjacent to the vertex 1, label these vertices and their corresponding mini-cycles in X by 2, 3,..., j + 1. If there are k unlabelled vertices adjacent to 2, label these k vertices and their corresponding mini-cycles in X by j + 2, j + 3,..., j + k + 1 Since a tree does not contain any cycle, none of 2, 3,..., j + 1 is equal to anyone of j + 2, j + 3,..., j + k + 1. Continue this process until all vertices in the tree and their corresponding mini-cycles are labelled. Repeat the same process to each of the rest of trees. Break the mini-cycle 1 by removing any one of its edges, then break the remaining mini-cycles as follows: Find the mini-cycle with the smallest label which has an edge in common with the cycle to be broken. Break the cycle by removing their common edge. The broken cycles obtained this way have the following property: Any nonmini-cycle will have edges in common with a mini-cycle, and the broken cycle obtained from this mini-cycle will be contained in the nonmini-cycle. Thus, when we apply Whitney's broken cycle theorem to compute the coefficients in $P(X, \lambda)$, we do not need to consider the nonmini-cycles in X.

Apply the same process to Y. Since X and Y have the same number of vertices and edges and the same number of broken mini-cycles of each length, by Whitney's broken cycle theorem, $P(X, \lambda) = P(Y, \lambda)$ and X and Y are chromatically equivalent.

Corollary 3.1. All cacti with the same number of vertices, the same number of edges and the same number of cycles of each length are chromatically equivalent.

Example. Let X and Y be the following graphs:

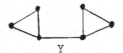

X Y

Then X^d is , and Y^d is . Also, $P(X, \lambda) = P(Y, \lambda) = \lambda(\lambda-1)^3(\lambda-2)^2$.

4. POLYNOMIAL BASES AND IDENTITIES

Let Z be the ring of integers and $Z[\lambda]$ be the Z-module of all polynomials in λ with integral coefficients. Then $P(X, \lambda) \in Z[\lambda]$ for every graph X. Since $1, \lambda, \lambda^2, \ldots, \lambda^n, \ldots$ constitute a basis for $Z[\lambda]$ and since $P(X, \lambda)$ has no constant term and since $\lambda^i = P(N_i, \lambda)$ for $i \geq 1$, $P(X, \lambda)$, for every X, can be uniquely written as a Z-combination of the null-graph-basis, namely, of $\lambda, \lambda^2, \ldots, \lambda^n, \ldots$. In fact, after repeatly applying Whitney's reduction formula by deleting edges in X, we can obtain the coefficients of $P(X, \lambda)$ with respect to the null-graph-basis. In particular, we have

$$(\lambda)_n = \sum_{k=1}^{n} s(n, k)\lambda^k, \quad n = 1, 2, \ldots$$

where $(\lambda)_n = P(K_n, \lambda) = \lambda(\lambda-1)(\lambda-2)\ldots(\lambda-n+1)$, and $s(n, k)$ are the stirling numbers of the first kind. Since 1, $(\lambda)_1, (\lambda)_2, \ldots, (\lambda)_n, \ldots$ also constitute a basis for $Z[\lambda]$ and since the family of complete graphs is a family of chromatically unique graphs, $P(X, \lambda)$, for every X, can also be uniquely written as a Z-combination of the complete-graph-basis, namely, of $(\lambda)_1, (\lambda)_2, \ldots, (\lambda)_n, \ldots$. In particular, we also have

$$\lambda^n = \sum_{k=1}^{n} S(n, k)(\lambda)_k, \quad n = 1, 2, \ldots$$

where $S(n, k)$ are the Stirling numbers of the second kind. $s(n, k)(S(n, k))$ is also called the connection matrix associated with changing from the null-graph-basis (complete-graph-basis) to the complete-graph-basis (null-graph-basis).

Since the $S(n, k)$ is the inverse of the $s(n, k)$ matrix, we have

$$\sum_{j=1}^{i} s(i, j)S(j, k) = \sum_{j=1}^{i} S(i, j)s(j, k) = \delta_{ik}.$$

We shall use the notation $(C)_n$ for $P(C_n, \lambda) = (\lambda-1)^n + (-1)^n(\lambda-1)$, $n \geq 3$. We define $(C)_1 = P(K_1, \lambda) = \lambda$ and $(C)_2 = P(K_2, \lambda) = \lambda(\lambda-1)$. Then one can easily verify that 1, $(C)_1, (C)_2, \ldots, (C)_n, \ldots$ also constitute a basis for $Z[\lambda]$. Again, after repeatedly using Whitney's reduction formula by adding as well as deleting suitable edges in a graph X, we can obtain the coefficients of $P(X, \lambda)$ with respect to the cycle-basis, namely, $(C)_1, (C)_2, \ldots, (C)_n, \ldots$.

We express $(C)_n$ with respect to the null-graph-basis, we have

$$(C)_n = \sum_{k=1}^{n} c(n, k)\lambda^k \tag{5}$$

where, by using $(C)_1 = \lambda$ and $(C)_n = (\lambda-1)^n + (-1)^n(\lambda-1)$, we have

$$c(n, k) = \begin{cases} \delta_{ik}, & \text{for } n = 1, \\ (-1)^{n+k}[\binom{n}{k}-1], & \text{for } k = 1,\ n \geq 2, \\ (-1)^{n+k}\binom{n}{k}, & \text{for } k > 1,\ n \geq 2. \end{cases}$$

We define

$$C(n, k) = \begin{cases} 1, & \text{for } k = 1,\ n \geq 1, \\ \binom{n}{k}, & \text{for } k > 1,\ n \geq 1. \end{cases}$$

Then $\sum_{j=1}^{i} C(i, j)c(j, k) = \sum_{j=1}^{i} c(i, j)C(j, k) = \delta_{ik}$, i.e., $C(n, k)$ is the inverse matrix of $c(n, k)$, and

$$\lambda^n = \sum_{k=1}^{n} C(n, k)(C)_k. \tag{6}$$

By using Stirling numbers, (5) and (6), we also can express $(C)_n$ ($(\lambda)_n$) with respect to the complete-graph-basis (cycle-basis).

Let T_n be a tree with n vertices, then

$P(T_n, \lambda) = \lambda(\lambda-1)^{n-1} = \sum_{k=1}^{n} t(n, k)\lambda^k$, $n \geq 1$, where $t(n, k) = (-1)^{n+k}\binom{n-1}{k-1}$. To express $P(T_n, \lambda)$ with respect to the complete-graph-basis, we have

$$P(T_n, \lambda) = \sum_{k=1}^{n} t(n, k) \sum_{j=1}^{k} S(k, j)(\lambda)_j. \tag{7}$$

The product of matrices in (7) can be written as

$$[t(n, k)][S(n, k)] = [1] \oplus [S(n, k)] \tag{8}$$

where

$$[1] \oplus [S(n, k)] = \begin{bmatrix} 1 & 0\ 0\ 0 \ldots \\ 0 & \\ 0 & \\ 0 & S(n, k) \\ \cdot & \\ \cdot & \\ \cdot & \end{bmatrix}$$

The right side of equation (8) is related to Zykov's product of N_1 and N_{n-1} (see [12]). Multiplying the equation (8) by the matrix $s(n, k)$, we obtain

$$[t(n, k)] = ([1] \oplus [S(n, k)][s(n, k)]. \tag{9}$$

Since $t(i, k) = (-1)^{i+k}\binom{i-1}{k-1}$, (9) can be written as follows: for $1 \leq i, k \leq n$,

$$(-1)^{i+k}\binom{i-1}{k-1} = \sum_{j \geq 1} S(i-1, j-1)s(j, k) \tag{10}$$

which is the identity (6) on p.187 in [3].

The authors wish to thank Professor T. W. Tutte for his helpful comments.

REFERENCES

1. D. Geller and B. Manvel, "Reconstruction of Cacti," Canadian J. Math. 21 (1969), 1354-1360.

2. F. Harary, Graph Theory, Addison-Wesley, Reading, Mass., 1969.

3. C. Jordan, Calculus of Finite Differences, Chelsea, 1947 (Second Ed.).

4. L. A. Lee, On Chromatically Equivalent Graphs, Doctoral Dissertation, George Washington University, 1975.

5. R. C. Read, "An Introduction to Chromatic Polynomials," J. Combinatorial Theory 4, (1968), 52-71.

6. J. Riordan, Combinatorial Identities, Wiley, New York, 1968.

7. G. C. Rota and R. Mullin, "On the Foundations of Combinatorial Theory," B. Harris (ed.) Graph Theory and Its Applications, Academic Press, 1970.

8. N. J. A. Sloane, A Handbook of Integer Sequences, Academic Press, 1973.

9. W. T. Tutte, "On Chromatic Polynomials and the Golden Ratio," J. Combinatorial Theory 9, (1970), 289-296.

10. H. Whitney, "A Logical Expansion in Mathematics," Bull. Amer. Math. Soc. 38 (1932), 572-579.

11. H. Whitney, "The Coloring of Graphs," Ann. of Math. 33 (1932), 688-718.

12. A. A. Zykov, "On Some Properties of Linear Complexes," Amer. Math. Soc. Transl. No. 79 (1952); translated from Mat. Sb. 24, No. 66 (1949), 163-188.

ON GRAPHS HAVING PRESCRIBED CLIQUE NUMBER, CHROMATIC NUMBER, AND MAXIMUM DEGREE

James M. Benedict
Augusta College, Augusta, GA 30904

and

Phyllis Zweig Chinn*
Humboldt State University, Arcata, CA 95521

Denote the clique number of a graph G by $\omega(G)$, the chromatic number by $\chi(G)$ and the maximum degree of any points in G by $\Delta(G)$. Given integers $a, b \geq 1, c \geq 0$, the ordered triple (a, b, c) is called a graphical triple if there is a graph G such that $\omega(G) = a$, $\chi(G) = b$, and $\Delta(G) = c$. We define $p(a, b, c)$ to be the least integer p such that (a, b, c) is graphical via some graph of order p. Clearly $p(a, b, c)$ exists iff (a, b, c) is graphical.

The authors wish to propose the questions: For which triples does $p(a, b, c)$ exist? What is the value of $p(a, b, c)$ when it does exist? This paper answers these questions for several classes of triples and offers two conjectures.

We refer the reader to Behzad and Chartrand [2] for definitions and explanations of symbols.

We first give two results which follow from Brooks' Theorem [4]. If $\Delta(G) < \chi(G)$ then $\Delta(G) + 1 = \chi(G)$ and the only graphical triples are $(a, a, a-1)$, $a \geq 1$, and $(2, 3, 2)$, where $p(a, a, a-1) = a$ and $p(2, 3, 2) = 5$. Thus we can henceforth assume that $b \leq c$.

*Research supported in part by a Towson State College Faculty Research Grant.

Clearly $\omega(G) \le \chi(G)$. Also the only $(1, b, c)$ triple is $(1, 1, 0)$. Thus the discussion is narrowed to examining the existence and value of $p(a, b, c)$ where $2 \le a \le b \le c$.

To help with classification, triples will often be written as $(a, a+j, a+k)$ where $0 \le j \le k$; $a \ge 2$. We put aside for now the general question of existence of $p(a, b, c)$ and investigate some particular cases.

Remark 1. $p(2, 3, c) = c + 3$.

Proof. Obtain the graph G by identifying a vertex of $K_{1,c-2}$ of largest degree with any vertex of C_5 . Then $(2, 3, c)$ is graphical via G so that $p(2, 3, c)$ exists and is no more than $p(G) = 5 + c - 2 = c + 3$.

To show that $p(2, 3, c)$ is at least $c + 3$, let G be any $(2, 3, c)$ graph. Let v be any vertex of G of degree c . Since $\chi(G) = 3$, G must have an odd cycle. If this cycle was C_3 then $\omega(G) \ge 3$. Hence any odd cycle has 5 or more vertices. Let C_{2n+1}, $n \ge 2$, be an odd cycle of G. The vertex v can be adjacent with at most n vertices of the cycle since G has no triangles. This is true whether or not v is a vertex of the cycle. Hence $p(G) \ge c + 1 + n \ge c + 3$. ■

Lemma 2. If $(a,a+j,a+k)$ is graphical then $p(a,a+j,a+k) \ge 2j+a$.

Proof. Say $(a,a+j,a+k)$ is graphical via the graph G, and let any coloring of G be given by $b=a+j$ colors. At most a of the color classes can have a single vertex since these color classes induce a complete subgraph of G. Hence $p(G) \ge 2(b-a)+a = 2j+a$. ■

Lemma 3 . If $(a,a+j),a+k)$ is graphical then $p(a,a+j,a+k) \ge a+k+1$.

Proof. Say $(a,a+j,a+k)$ is graphical via the graph G . Then since $\Delta(G) = a+k$ we have $K_{1,a+k} \subseteq G$ so that $p(G) \geq a+k+1$. ∎

Note that Lemma 3 supercedes Lemma 2 if and only if $k \geq 2j-1$.

Lemma 4. $5j-2 \leq p(2j, 3j, 5j-3) \leq 5j$ for any positive integer j.

Proof. The existence and the upper bound is gained by considering $G = C_5 + C_5 + \cdots + C_5$ where there are j copies of C_5 used in the join operation. We have $\omega(G) = 2j$ and $\chi(G) = 3j$ since both clique number and chromatic number are additive over the join operation. The lower bound is an application of Lemma 3.

Lemma 5. If $p(a, b, c) \leq M$ then $p(a+N, b+N, M+N-1) = M+N$ where N is any positive integer.

Proof. Let (a, b, c) be graphical via G' where $p(G') = M$. Let $G = G' + K_N$ for any positive integer N, and note that $M \geq c+1$. Then $\Delta(G) = \max\{N-1+p(G'), \Delta(G')+N\} = \max\{M+N-1, c+N\} = M+N-1$. It easily follows then that G is an $(a+N, b+N, M+N-1)$ graph of order $M+N$. Then by Lemma 3, $p(a+n, b+N, m+N-1) = M+N$. ∎

Lemma 6. If $p(a, b, c) \leq M$ then $p(a, b, c+N) \leq M+N$ where N is any non-negative integer.

Proof. Let G' be any (a, b, c) graph of order M. Obtain the graph G from G' and $K_{1,N}$, where N is any non-negative integer, by identifying a vertex of G' of largest degree with a vertex of $K_{1,N}$ of largest degree. Then G is an $(a, b, c+N)$ graph of order $M+N$. ∎

Note that if $M = C+1$ then all three of the inequalities used in the statements of Lemmas 5 and 6 are equalities. The next two results essentially show that $p(a, b, c)$ "often" exists and attains $c+1$ in value.

Theorem 7. $p(a, a, a+k) = a+k+1$ for each non-negative integer k.

Proof. Identifying the highest degree points if K_a and $K_{1,k+1}$ gives an $(a, a, a+k)$ graph of order $a+k+1$. The result follows from Lemma 3. ∎

The triples for which $a = 2$ seem special. Moreover the triple $(a, a+1, a+1)$ seems to require separate examination. The omission of these cases allows the next theorem, which is quite general, to follow.

Theorem 8. $p(a, a+j, a+k) = a+k+1$ whenever $a \geq 2j+1$, $j \geq 1$, and $k \geq 3j-1$.

Proof. Let $a-(2j+1) = N \geq 0$ and let $k-(3j-1) = M \geq 0$. By Lemma 4, $p(2j, 3j, 5j-3) \leq 5j$. By Lemma 5, $p(2j+1, 3j+1, 5j)$ $= 5j+1$, so that $p(2j+1+N, 3j+1+N, 5j+N) =$ $p(a, a+j, a+k-M) = a+k-M+1$. The result now follows by Lemma 6 and Lemma 3. ∎

Lemma 9. If $p(a, b, c) \leq M$, then $p(a+N, b, c) \leq M+N$ for all $0 \leq N \leq (b-a)$.

Proof. If (a, b, c) is graphical via G, then $(a+N, b, c)$ is graphical via $G \cup K_N$ plus lines drawn from each point of the K_N to each point of some K_a in G. ∎

The next four remarks give some cases where the value of $p(a, a+1, a+1)$ has been determined or bounded. Note that a bound for one triple gives exact values for an infinite family of other triples using Lemma 5. Figure 1 shows graphs associated with the triples of Remarks 10 and 11. Those associated with Remarks 12 and 13 are essentially generalizations of that for Remark 11. The corresponding construction will not work for the triple (7 , 8 , 8). The equality in Remarks 10 and 11 are seen by considering possible 6 and 8 order graphs respectively.

Remark 10. $p(3, 4, 4) = 7$.

Remark 11. $p(4, 5, 5) = 9$.

Remark 12. $p(5, 6, 6) \leq 11$.

Remark 13. $p(6, 7, 7) \leq 14$.

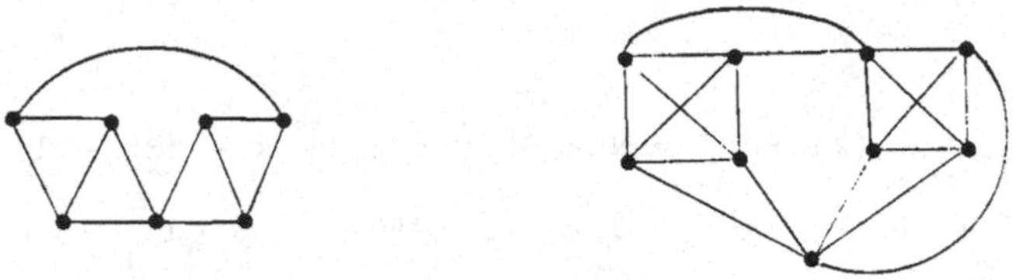

Figure 1 .

Remark 14. If (a, b, c) is graphical via a graph of order M, then $p\left(2a, 2b + \left\lceil \frac{b+1}{2} \right\rceil, 2M + c\right) \leq 5M$.

Remark 14 may be proven using the lexicorgraphic product $C_5[G]$ where G is an (a, b, c) graph of order M.

The situation of triangle-free graphs having arbitrary chromatic number has been discussed by many authors. The following results by Mycielski [11] is particularly helpful in the context of (a, b, c) triples.

Theorem 15. (Mycielski) Given any positive integer n, there is a graph of order $3 \cdot 2^{n-1} - 1$ which has chromatic number $n + 1$ and does not contain any triangles.

The graph G constructed in Mycielski's proof has $\Delta(G) = 3 \cdot 2^{n-2} - 1$. Thus the following corollary is a direct consequence of Theorem 15 and Lemma 6.

Corollary 16. $p(2, b, c)$ exists for all $c \geq 3 \cdot 2^{b-3} - 1$ and furthermore $p(2, b, c) \leq 3 \cdot 2^{b-2} - 1 + M$ where $c - M = 3 \cdot 2^{b-3} - 1$.

Remark 17. $p(a, a+j, c) = c + 1$ for $3 \leq a \leq 2j$ and c sufficiently large.

Note that c is sufficiently large if $c \geq 3 \cdot 2^{j} + a - 4$. The Remark follows from Corollary 16 and Lemma 5.

Knowing particular $(2, b, c)$ graphs having fewer points than those constructed for Theorem 15 would give values for $p(a, a+j, a+k)$ beginning with smaller values of k.

Theorem 18. If (a, b, c) is graphical via a graph of order M, then $p(a, b+1, d) \leq 2M + 1$ where $d = \max\{M, 2c\}$.

The proof of Theorem 18 is a direct extension of Mycielski's proof [11].

To compare the results of Lemma 5 and Theorem 18, consider the information to be gained from knowing that, for example, $p(3,4,4) = 7$. From Lemma 5 comes a family of equalities: $p(4, 5, 7) = 8$, $p(5, 6, 8) = 9$, $p(6, 7, 9) = 10, \ldots$. From Theorem 18, one finds $p(3, 5, 8) \leq 15$ which in turn can be used in Lemma 5 to learn $p(3, 5, 15) = 16$, $p(4, 6, 16) = 17, \ldots$. A second application of Theorem 18 next gives $p(3, 6, 16) \leq 31$ and so on. Notice that beginning with C_5 as a $(2, 3, 2)$ graph and using Theorem 18 shows $p(2, 4, 5) \leq 11$. Now Lemma 5 says; $(3, 5, 11) = 12$ and then Lemma 6 says $p(3, 5, 15) = 16$. Thus two different paths lead one to the same conclusion.

In general when mathematicians have considered the existence of graphs having various chromatic numbers and clique numbers, additional constraints have been imposed and generally the existence of graphs has been of more importance than finding the least possible order of such a graph. The triples (a, b, c) considered in this paper allow one to determine the minimum order graph with prescribed clique and chromatic number, in certain cases. Let $p(a, b)$ denote the least order of any graph G such that $\omega(G) = a$ and $\chi(G) = b$. Then

$$p(a, b) = \min \{p(a, b, c)\} .$$

Conjecture 19 might allow the proof of the Conjecture 20. The second conjecture is probably true even if the first is not.

<u>Conjecture 19</u>. If a, b are positive integers, $a \leq b$, such that there exists c satisfying $p(a, b, c) = c + 1$ and c_0 is the least such positive integer, then $p(a, b, c) = c_0 + 1$ implies that $c = c_0$.

<u>Conjecture 20</u>. If a, b are such that there exists a positive integer c satisfying $p(a, b, c) = c + 1$ and c_0 is the least such positive integer, then $p(a, b) = c_0$.

Varying slightly from the main thesis of this paper, we present the following problem.

<u>Problem 21</u>. Let the triple (a, b, c) be graphical via the graph G of order $p(a, b, c)$. Is it the case that G is necessarily connected? Will there always be a vertex of G of largest degree contained in a subgraph K_a of G?

In the spirit of good mathematics, the participants of the conference at which this paper was presented helped improve this paper. Chvátal has shown that $p(2, 4, 5) = 11$ in [6]. Moreover, he gave us the following $(2, 4, 4)$ graph and stated that

$p(2, 4, 4) = 12$ in [7] .

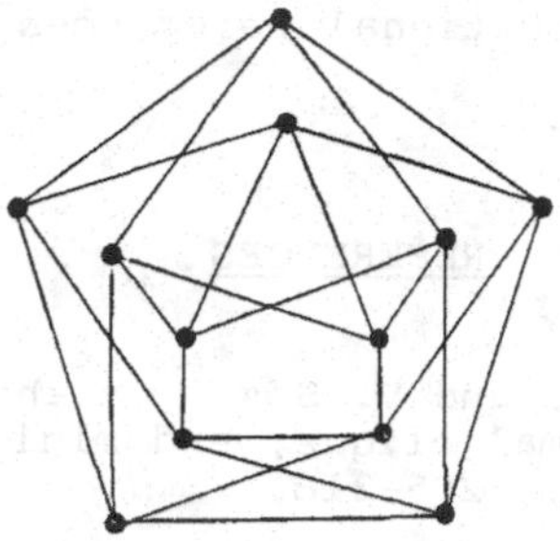

Figure 2.

Bernhart and Stahl were helpful in reducing the proof of Remark 14. Catlin informed us of the following (as yet unpublished) result [5].

Theorem 22. (Catlin) For any graph G the following holds:

$$\chi(G) \le \Delta(G) + 1 - \left\lfloor \frac{(\Delta(G) + 1)}{\max\{4, \omega(G) + 1\}} \right\rfloor .$$

This result yields a criterion for finding non-graphical triples (a, b, c) where $2 \le a \le b \le c$. The "smallest" triple proven to be non-graphical by Catlin's results is $(2, 7, 7)$.

Corollary 23. Let a be an integer exceeding 2 . The triple $(2, 2+j, 2+k)$ is not graphical for all integers $j \ge 5$ and $k \le [(4j-5)/3]$. The triple $(a, a+j, a+k)$ is not graphical for all integers $j \ge a+1$ and $k \le [(j-1)(a+1)/a]$.

Of course if the triple (a, b, c) is not graphical for $2 \le a \le b \le c$, then neither is the triple (a', b, c') for $b \le c' \le c$ and for $2 \le a' \le a$ by Lemmas 6 and 9 . If the triple $(a+N, b+N, M+N-1)$ is not graphical for $N \ge 0$ and $M \ge b$ then Lemma 5 shows $p(a, b, c) \ge M+1$ whenever $b \le c \le M$. This seems to lend validity to Conjecture 20. Similar results follow from Remark 14 and Theorem 18.

The authors request additional references on this topic from the mathematical community.

REFERENCES

1. B. Andrasfai, P. Erdös, and V. Sós, "On the connection between chromatic number, maximal clique, and minimal degree of a graph," Discrete Math. 8(1974), 205-218.

2. M. Behzad and G. Chartrand, Introduction to the Theory of Graphs, Allyn and Bacon, Boston (1972).

3. C. Berge, Graphs and Hypergraphs, North Holland, Amsterdam (1973).

4. R. L. Brooks, "On colouring the nodes of a network," Proc. Cambridge Philos. Soc. 37(1941), 194-197.

5. P. Catlin, "A bound on the chromatic number of graphs." (To appear.)

6. V. Chvátal, "The minimality of the Mycielski graph," Graphs and Combinatorics, Springer-Verlag, Berlin (1974), 243-246.

7. V. Chvátal, "The smallest triangle-free 4-chromatic 4-regular graph," J. Combinatorial Theory 9(1970), 93-94.

8. D. Geller and S. Stahl, "The chromatic number and other functions of the lexicographic product," J. Combinatorial Theory 19B (1975), 87-95.

9. F. Harary, Graph Theory, Addison-Wesley, Reading, Mass., (1969).

10. L. Lovász, "Normal hypergraphs and the perfect graph conjecture," Discrete Math, 2(1974), 253-267.

11. J. Mycielski, "Sur le coloriage des graphs," Colloq. Math. 3(1955), 161-162.

12. D. Seinsche, "On a property of the class of n-colorable graphs," J. Combinatorial Theory 16B (1974), 191-193.

DOMINATION OF UNDIRECTED GRAPHS - A SURVEY

By E. J. Cockayne, University of Victoria,
Victoria, British Columbia, Canada

1. Introduction

Any undefined term or notation in this paper may be found in Harary [11]. A set D of vertices of a graph is a dominating set if each vertex not in D is adjacent to at least one vertex of D. This paper attempts to survey recent results concerning dominating sets in graphs and to suggest ways in which future work might proceed. The theory bears a resemblance to the well known work on independent sets and colourings and several results relate these theories. It should be noted that dominating sets have also been called "externally stable sets" and "absorbants" (see [1]). Miscellaneous applications of the theory are mentioned in [1, 6, 10, 16, 18, 20]. Berge [1, Page 308] has studied domination in directed graphs and applications in game theory.

2. The domination number

In this section we list a number of results on dominating sets, most of them concerning a parameter called the domination number of a graph.

Proposition 1 (Ore [20]) If G has no isolated vertices and D is a minimal dominating set of G then $V(G) - D$ is a dominating set of G.

Proposition 2 (Berge [1, Page 309]) D is an independent dominating set in G if and only if D is a maximal independent set in G.

The domination number $\gamma(G)$ of a graph G is the smallest cardinality of a dominating set of G.

Proposition 3 (Berge [1, Page 304] If G is a (p,q) graph with domination number γ and maximum degree Δ then $p - q \le \gamma \le p - \Delta$.

Theorem 4 (Vizing [22]) If G is a (p,q) graph with domination number γ then $q \le (p - \gamma)(p - \gamma + 2)/2$.

Corollary 4 (Vizing [22]) $\gamma \le p + 1 - \sqrt{1 + 2q}$.

The next results closely resemble those of Nordhaus and Gaddum [19] in the theory of independence.

Theorem 5 (Jaegar and Payan [14]) For any graph G with p vertices

$$\gamma(G)\ \gamma(\overline{G}) \le p$$

and

$$\gamma(G) + \gamma(\overline{G}) \le p + 1 \ .$$

Let $\varepsilon(G)$ equal the maximum number of end edges in a spanning forest of G.

Theorem 6 (J. Nieminen [18]) For any graph G with p vertices

$$\gamma(G) + \varepsilon(G) = p .$$

Corollary 6 (Hedetniemi) For any graph G, $\gamma(G) \le \beta_1(G)$.

Nieminen used the above result to establish upper and lower bounds for $\gamma(G)$ in terms of a simply constructed spanning forest of G.

Meir and Moon [17] have considered a generalisation of the concept of domination in trees. They define a k-packing to be set of vertices P such that for all $x,y \in P$ $d(x,y) > k$ and a k-covering to be a set of vertices C such that for every $x \in V(G)$ there is at least one $y \in C$ such that $d(x,y) \le k$. $P_k(G)$, $C_k(G)$ and the numbers of vertices in a largest k-packing and smallest k-covering respectively. We note that $\gamma(G) = C_1(G)$ and $\beta_0(G) = P_1(G)$. Meir and Moon obtain the following variety of relations involving $C_k(T)$, $P_k(T)$ where T is a tree with p vertices.

Proposition 7 (Meir and Moon [17]) If $p \ge 2$, then $P_1(T) + C_1(T) \le p$.

Corollary 7 (Meir and Moon [17]) If $p \ge 2$, then $1 \le C_1(T) \le \frac{p}{2} \le P_1(T) \le p - 1$

Theorem 8 (Meir and Moon [17]) If $p \ge 1$, then $P_1(T) + 2C_1(T) \ge p + 1$.

Corollary 2 If $p \ge 1$ and $0 \le \lambda \le 2$, then

$$P_1(T) + \lambda C_1(T) \ge \frac{1}{2}(1 + \frac{\lambda}{2})p + \frac{\lambda}{2} .$$

Theorem 9 (Meir and Moon [17]) If α_1, α_2 are positive integers such that $\alpha_1 \ge p/2$, $\alpha_1 + \alpha_2 \le p$ and $\alpha_1 + 2\alpha_2 \ge p + 1$, then there exists a tree T with p vertices such that $P_1(T) = \alpha_1$ and $C_1(T) = \alpha_2$.

Theorem 10 (Meir and Moon [17]) If $p \ge [1/2(k + 3)]$, then

$$P_k(T) \le [2p/(k + 2)] \quad \text{if } k \text{ is even, and}$$

$$P_k(T) \le [(2p - 2)/(k + 1)] \quad \text{if } k \text{ is odd.}$$

Theorem 11 (Meir and Moon [17]) If $p \ge k + 1$, then $C_k(T) \le [p/(k + 1)]$.

Theorem 12 (Meir and Moon [17]) If $p \ge k + 1$, then $P_k(T) + kC_k(T) \le p$.

Theorem 13 (Meir and Moon [17]) If $k \ge 1$, then $P_{2k}(T) = C_k(T)$.

Corollary 13 (Meir and Moon [17]) If $p \ge k + 1$, then $P_k(T) + kP_{2k}(T) \le p$;

If $p > 2k + 1$, then $C_k(T) + 2kC_{2k}(T) \le p$.

3. The domatic number

A much studied parameter in the theory of colouring is the chromatic number $\chi(G)$ of a graph G. Analagously in the theory of domination we define the domatic number $d(G)$ of a graph G to be the largest order of a partition of $V(G)$ into dominating sets of G. Then the following bounds are apparent.

Proposition 14 (Cockayne and Hedetniemi [6]) For any graph G with p vertices

$$d(G) \le p/\gamma(G)$$

$$d(G) \le \delta(G) + 1 .$$

We define a graph to be (domatically) full if $d(G) = \delta(G) + 1$. If G is any of the following graphs, then G is domatically full:

$$K_n,\ C_{3n},\ W_{3n+1},\ \text{a tree},\ \text{maximal outerplanar.}$$

An analogue of Nordhaus and Gaddum's result holds for the parameter $d(G)$.

Theorem 15 (Cockayne and Hedetniemi [6]) For any graph G with p vertices, $d(G) + d(\overline{G}) \le p + 1$ with equality if and only if $G = K_p$ or $G = \overline{K}_p$.

A lower bound for $d(G)$ was established by Jaegar and Payan [14] as a by-product of their elegant proof of theorem 5.

Theorem 16 (Jaegar and Payan [14]) For any graph G, $\gamma(\overline{G}) \le d(G)$.

Combining theorems 15 and 16 we deduce

Corollary 16(a) For any graph G with p vertices, $\gamma(G) + d(G) \le p + 1$ with equality if and only if $G = K_p$ or $G = K_{p+1}$.

Corollary 16(b) (Jaegar and Payan [14]) If G has diameter less than or equal to two and every edge of G is in a triangle then $d(G) \ge 3$.

For any graph G, $\gamma(\overline{G}) \le c(G)$ where $c(G)$ is the cardinality of the smallest clique of G. In view of this inequality and theorem 16 it is natural to ask whether the domatic number is bounded below by the size of the smallest clique. The answer to this question is no, as shown by the example of Fig.1 of [6] in which $c(G) = 4$, $d(G) = 3$. However, it appears that if the clique graph $K(G)$ of G is not too complex, $c(G)$ is in fact less than or equal to $d(G)$.

Theorem 17 (Cockayne and Hedetniemi [6]) If $K(G)$ is a tree or an even length cycle then $c(G) \le d(G)$.

Further examples of graphs satisfying this inequality will be given in the following section.

The final results in this section concern optimal domination.

Theorem 18 (Cockayne and Hedetniemi [4]) The minimum number Q of edges in a graph with p vertices and domatic number d is given by

$$Q = \binom{d}{2}(k + 1) - \binom{d - r}{2}$$

where $\quad p = kd + r \qquad 0 \le r < d .$

A simple class of graphs which achieve this minimum is also constructed.

Theorem 19 The maximum number of edges Q in a graph with p vertices and domatic number at most $k(k \le p)$ is $\binom{p-1}{2} + k - 1$.

Proof Let G consist of a complete graph on $p - 1$ vertices with $k - 1$ of these vertices joined to one additional vertex. $V(G)$ has a partition into k dominating sets and this is optimal by proposition 14, i.e. $d(G) = k$. Hence $Q \ge \binom{p-1}{2} + k - 1$.

If G has $\binom{p-1}{2} + k$ edges, its complement $\overline{G}$ has $p - (k + 1)$ edges. By proposition 3, $\gamma(\overline{G}) \ge k + 1$ hence by theorem 16, $d(G) \ge k + 1$. Thus $Q < \binom{p-1}{2} + k$ and the result follows.

4. Independent Dominating Sets

Proposition 2 infers that sets of vertices of a graph G which are both independent and dominating form cliques in the complementary graph $\overline{G}$. Hence the study of independent dominating sets in graphs is equivalent to studying cliques. Our interest in such sets was motivated by the conjecture of Berge that every incomplete regular graph has at least two disjoint cliques. The next two theorems are concerned with partial results on this conjecture.

Theorem 20 (Cockayne and Hedetniemi [5, 6]) Incomplete regular graphs of degree ≤ 8 have at least two disjoint cliques.

Let L be a subset of a finite group X such that the identity $e \notin L$ and $L = L^{-1}$. The group graph $G(L,X)$ has vertex set X and edge set $\{[x,x\ell] \mid x \in X, \ell \in L\}$. Let $N(L)$ denote the normalizer of L and define

$$\mathcal{G} = \{G(L,X) \mid N(L) \cap (X - L - \{e\} \neq \phi\} .$$

Theorem 21 (Cockayne and Hedetniemi [7]) Any group graph $G(X,L)$ in $\mathcal{G}$ has at least two disjoint cliques.

The line graph $L(G)$ of G has disjoint independent dominating sets if and only if G has disjoint maximal sets of independent edges, i.e. disjoint matchings. Trees and unicyclic graphs having no two disjoint matchings are characterised in [8], [13] respectively. P. J. Slater [21] has characterised trees which have k disjoint maximum matchings.

Graph G is *indominable* if its vertex set may be partitioned into independent dominating sets. For example, complete graphs and their complements, all connected bipartite graphs and complete k-partite graphs and cycles C_{2n}, C_{3n} are all indominable.

Proposition 22 (Cockayne and Hedetniemi [5]) If G is uniquely n-colourable, then G is indominable.

This result implies that maximal outerplanar graphs and n-cubes ([11, Page 23]) are indominable.

Proposition 23 (Cockayne and Hedetniemi [5]) If G is regular and domatically full, then G is indominable.

We are now able to exhibit an additional class of graphs which satisfy the inequality of theorem 17.

Theorem 24 (Cockayne and Hedetniemi [6]) If G or $\overline{G}$ is indominable then $c(G) \leq d(G)$.

5. Computation of Dominating Parameters

The problems of computing $d(G)$ and $\gamma(G)$ for a general graph G are probably in the Cook-Karp class of NP-complete problems (see [15]). Computation of $\gamma(G)$ is a special case of the set-cover problem which is known to be NP-complete. Cockayne, Goodman and Hedetniemi [3] have presented a linear algorithm for computing the dominating number of a tree and hope to generalise this to compute in linear time and storage the smallest weight of a dominating set in a tree with weights assigned to the vertices. Computation of $d(G)$ is a special case of the following problem. Let P,Q be the defining independent sets in a bipartite graph B. A dominating set of B is a subset P' of P such that each vertex of Q is adjacent to at least one vertex of P'. Compute $\tau(B)$, the largest partition of P into dominating sets of B. A branch and bound algorithm for $\tau(B)$ has been obtained by Cockayne and Roberts [10]. Finally Cockayne, Hartnell and Hedetniemi [2] have constructed an algorithm for finding two disjoint matchings in a tree (if they exist). Experimental evidence suggests linearity but no analysis to this effect has been given so far.

6. Topics for further study

In this section we list a variety of areas for further work. Most of the problems mentioned are connected with results given in earlier sections. The following conjecture of Vizing is mentioned in [23]. Let $G \times H$ denote the product of graphs G and H (see [11, Page 22]).

Problem 1 Prove or disprove $\gamma(G \times H) \geq \gamma(G)\,\gamma(H)$.

Problem 2 Is the computation of $\gamma(G)$ for an arbitrary graph G in the Cook-Karp class?

Problem 3 Characterise the class of domatically full graphs.

Problem 4 Study the class of graphs for which $c(G) \leq d(G)$.

Problem 4a Prove or disprove: If $K(G)$ is an odd cycle, then $c(G) \leq d(G)$.

Problem 4b Prove or disprove: If $K(G)$ is bipartite, then $c(G) \leq d(G)$.

Following the development of the theory of colouring we define two properties of graphs analagous to uniquely colourable and critical. (See [11], Chapter 12]) A graph is called uniquely domatic if $d(G) = n$ and G has a unique partition into n dominating sets. A graph is domatically critical if $d(G) = n$ and for each edge x of G, $d(G - x) < n$.

Problem 5 Study the class of uniquely domatic graphs.

Problem 6 Study the class of domatically critical graphs.

Problem 7 Is computation of $d(G)$ for an arbitrary G in the Cook-Karp class?

Problem 8 Relate $d(G)$ to connectivity.

The next topic is concerned with a possible counterpart in the theory of domination to the homomorphism interpolation theorem of colouring theory (see [12]). A dominating set is called indivisable if it is not the union of two dominating sets. Then the largest order of a vertex partition of G into indivisable dominating sets is $d(G)$. We define the adomatic number $ad(G)$ to be the smallest order of a vertex partition into indivisable dominating sets.

Problem 9 Do there exist vertex partitions into indivisable dominating sets of all orders between $ad(G)$ and $d(G)$?

Problem 10 Study the class of indominable graphs.

Problem 11 (Berge's conjecture) Prove or disprove: Any incomplete regular graph has two disjoint cliques.

Problem 11a Prove or disprove: Any incomplete group graph has two disjoint cliques.

Problem 11b Prove or disprove: If G is regular, $G \neq \overline{K}_p$ and G has disjoint cliques then $\overline{G}$ has disjoint cliques.

Problem 12 (Meir and Moon) Study the class of graphs G for which $P_{2k}(G) = C_k(G)$.

Acknowledgement

The author gratefully acknowledges the support of the Canadian National Research Council grant number A7544.

REFERENCES

[1] C. Berge, Graphs and Hypergraphs, North-Holland, Amsterdam, 1973.

[2] E. J. Cockayne, B. L. Hartnell and S. T. Hedetniemi, An Algorithm for two Disjoint Matchings in a Tree (submitted).

[3] E. J. Cockayne, S. Goodman and S. T. Hedetniemi, A Linear Algorithm for the Domination Number of a Tree, Information Processings Letters 4 (1975), 41-44.

[4] E. J. Cockayne and S. T. Hedetniemi, Optimal Domination in Graphs, IEEE Trans. on Circuits and Systems, Vol CAS-22 No. 11 (1975), 855-857.

[5] E. J. Cockayne and S. T. Hedetniemi, Disjoint Independent Dominating Sets in Graphs, Discrete Math (to appear).

[6] E. J. Cockayne and S. T. Hedetniemi, Towards a theory of Domination in Graphs, Networks (to appear).

[7] E. J. Cockayne and S. T. Hedetniemi, On Group Graphs with Disjoint Cliques (submitted).

[8] E. J. Cockayne and S. T. Hedetniemi, Which Trees do not have Disjoint Matchings, Utilitas Math (to appear).

[9] E. J. Cockayne and S. T. Hedetniemi, Disjoint Cliques in Regular Graphs of degree seven or eight, J. Comb. Theory (to appear).

[10] E. J. Cockayne and F. D. K. Roberts, Computation of Dominating Partitions (submitted).

[11] F. Harary, Graph Theory, Addison-Wesley, Reading, Mass., 1969.

[12] F. Harary, S. T. Hedetniemi and G. Prins, An Interpolation Theorem for Graphical Homomorphisms, Portugal Math 26 (1967), 454-462.

[13] B. L. Hartnell, Disjoint Matchings in Unicyclic Graphs, Utilitas Math (to appear).

[14] F. Jaegar and C. Payan, Relations du type Nordhaus-Gaddum pour le Nombre d'absorption d'un graphe simple, C. R. Acad. Sc. Paris, Series A, t. 274, (1972), 728-730.

[15] R. Karp, Reducibility Among combinatorial problems, Complexity of Computer Computations (R. E. Miller and J. W. Thatcher eds.), Plenum Press, New York (1972), 85-104.

[16] C. L. Liu, Introduction to Combinatorial Mathematics, McGraw-Hill, New York, 1968.

[17] J. Moon and A. Meir, Relations between packing and covering numbers of a tree, Pacific. J. Math Vol 61 No. 1 (1975), 225-233.

[18] J. Nieminen, Two bounds for the domination number of a graph, J. Inst. Maths Applics 14 (1974), 183-187.

[19] E. A. Nordhaus and J. W. Gaddum, On Complementary Graphs, American Math. Monthly 63 (1956), 175-177.

[20] O. Ore, Theory of Graphs, American Math. Soc. Colloq. Publ. 38, Providence, R. I., 1962.

[21] P. J. Slater, A constructive characterisation of trees having no k maximum matchings (submitted).

[22] V. G. Vizing, A bound on the External Stability Number of a Graph, Doklady A. N. 164 (1965), 729-731.

[23] V. G. Vizing, The Cartesian product of graphs, Vyč. Sis., 9, 1963, p 30-43.

SOME EXTREMAL PROBLEMS FOR SIMPLE TWO-COMPLEXES

Richard A. Duke
Georgia Institute of Technology
Atlanta, GA 30332

ABSTRACT

A variety of extremal problems are considered for the members of two particular families of simple 2-complexes. We let B_k denote the complex consisting of k 2-cells all sharing one single edge in common and W_k the complex consisting of k 2-cells sharing exactly one vertex in common. The problems treated here include the following questions for the B_k and the corresponding questions for the W_k.

Suppose that for each 2-coloring of the 2-cells of a complex G, either one color contains a monochromatic copy of B_k or the other contains a monochromatic copy of B_ℓ. How many 2-cells must G have? How many vertices? How many vertices and how many 2-cells must G have if one of B_k and B_ℓ must be "faithfully imbedded" in one of the colors? How do these numbers compare with the "generalized Ramsey numbers" for the pairs (B_k, B_ℓ)? What is the smallest integer $f(n,k)$ such that each 2-complex on n vertices having $f(n,k)$ 2-cells contains a copy of B_k?

SOME EXTREMAL PROBLEMS FOR SIMPLE 2-COMPLEXES

1. Introduction. In [3] and [4] the generalized Ramsey numbers were computed for some members of several families of simple 2-complexes or 3-graphs. Here a variety of other extremal problems are considered for the members of two of these families. In certain cases the results are compared with the Ramsey numbers obtained previously. We let B_k denote the 2-complex having exactly k 2-cells all sharing one single edge in common and W_k the 2-complex having k 2-cells all sharing one single vertex in common. In [3] and [4] B_k was called the k-book and W_k the k-wedge.

Following Erdös [6], $K \longrightarrow (P,Q)$ will denote the fact that for every 2-coloring of the 2-cells of the 2-complex K, either the first color contains a monochromatic copy of the 2-complex P or the second color contains a monochromatic copy of Q. Likewise $K \longleftrightarrow (P,Q)$ will denote the fact that in every 2-coloring of the 2-cells of K either P is "faithfully imbedded" in the first color or Q is "faithfully imbedded" in the second color, where a complex A is said to be faithfully imbedded in a single color in the complex B provided that each 2-cell of A is of that color and there are no other 2-cells of either color in the subcomplex of B spanned by the vertices of A. Also following Erdös [5], $f(n,Q)$ will denote the smallest integer m such that every 2-complex on n vertices having at least m 2-cells contains a copy of Q.

2. Problems involving the B_k. It is obvious that, in the above notation, $B_{m+n-1} \longrightarrow (B_m,B_n)$. The first several results show that, in terms of both the number of vertices and the number of 2-cells, B_{m+n-1} is the smallest 2-complex for which this is so.

By the valence of an edge (that is, 1-cell) in a 2-complex Q is meant the number of 2-cells in Q which contain that edge.

The edge of valence k in a book B_k, $k \geq 2$, will be called the <u>spine</u> of B_k. The number of vertices of Q will be denoted by $V(Q)$.

<u>Theorem 1</u>. If Q is a 2-complex for which $Q \longrightarrow (B_m, B_n)$, then Q has at least $m+n-1$ 2-cells, and, when either $m \neq 2$ or $n \neq 2$, the only such 2-complex with exactly $m+n-1$ 2-cells is B_{m+n-1}.

<u>Proof</u>. Suppose that $Q \longrightarrow (B_m, B_n)$, where $m \geq n$. For any coloring of the 2-cells of Q with the colors blue and red, there exists either a blue m-book or a red n-book. Consider the coloring in which each 2-cell is colored blue. This coloring must produce a blue m-book, B_m. Let B_k be the book in Q having the largest number of 2-cells which contains B_m. If $k \geq m+n-1$, then Q has at least $m+n-1$ 2-cells and if, in this case, Q has exactly $m+n-1$ 2-cells, then $Q = B_{m+n-1}$.

If $k < m+n-1$, we may recolor the 2-cells of B_k so that fewer than m are blue and fewer than n are red. This coloring produces another blue m-book. That Q has at least $m+n-1$ 2-cells now follows from the fact that this second m-book can have at most one 2-cell in common with B_k. If this is so and Q has exactly $m+n-1$ 2-cells, then Q must consist entirely of B_k and the second m-book, and these two books must share one 2-cell. Furthermore, $k = m = n \geq 2$. If the common 2-cell is colored red and all of the rest of the 2-cells of Q are colored blue, another blue m-book must be formed. Since no edge other than the spines of B_k and the second m-book can have valence greater than two, it follows that $m = n = 2$. In this case the 2-complex with four vertices and three 2-cells satisfies the conditions of the theorem.

Since it is also clear that $B_{m+n-1} \longleftrightarrow (B_m, B_n)$, the next result is now immediate.

Corollary 2. If $Q \longleftrightarrow (B_m, B_n)$, then Q has at least $m+n-1$ 2-cells, and the only such 2-complex with exactly $m+n-1$ 2-cells is B_{m+n-1}.

If $Q \longrightarrow (B_m, B_n)$, then the smallest number of vertices that Q can have is equal to the generalized Ramsey number $r(B_m, B_n)$ discussed in [4]. It is easy to see that $r(B_m, B_n) \le m + n + 1$. In [4] it was shown that $r(B_m, B_n) \le m + n$ for m and n both even, as well as for $m \equiv 0 \pmod 6$ and $n \equiv 5 \pmod 6$, and for $m \equiv 3 \pmod 6$, $n \equiv 2$ or $5 \pmod 6$. Also if $m \le 4$ or $n \le 4$, it was shown that $r(B_m, B_n) = m + n$ in these cases and $r(B_m, B_n) = m + n + 1$ otherwise.

The next theorem shows that if faithful imbeddings are required, then the minimum number of vertices is $m + n + 1$ in each case.

Theorem 3. If $Q \longleftrightarrow (B_m, B_n)$, then $V(Q) \ge m + n + 1$.

Proof. Suppose $Q \longleftrightarrow (B_m, B_n)$. Since the result is easily obtained when $m \le 2$ or $n \le 2$, we assume that $m \ge n > 2$. As in the proof of Theorem 1, we may also assume that there exists an m-book, B_m^1, and a maximal book, B_k, containing B_m^1, where B_m^1 is now faithfully imbedded and $k < m + n - 1$.

We may color the 2-cells of Q so that there are exactly $n-1$ red 2-cells, all in B_m^1. At most $m-1$ 2-cells of B_k are blue. There must exist another faithfully imbedded blue m-book, B_m^2, contained in a maximal book, B_j. Again if $j \ge m + n - 1$, then $V(Q) \ge m + n + 1$. Assume, therefore, that $j < m + n - 1$. Note that B_m^1 and B_m^2 have at most one 2-cell in common. For if they do share a 2-cell, then the spine of one of them is contained in the other, and they can have at most three vertices in common. It would then follow that $V(Q) \geq 2(m+2) - 3 \geq m + n + 1$. Suppose that B_m^1 and B_m^2 have no 2-cell in common. We may recolor B_m^2 so that exactly $n - 1$ of its 2-cells are red. There must then exist still another faithfully imbedded blue m-book, B_m^3.

In this way we obtain a finite sequence of m-books, B_m^i, no two sharing a common 2-cell. This sequence can be extended until an m-book is found whose spine is on one of the m-books in the sequence or contains the spine of one of them, or until a faithfully imbedded red n-book is formed. In the former case $V(Q) \geq m + n - 1$. If a red n-book, B_n, is obtained, the 2-cells of B_n must be in the union of the m-books forming the sequence. But B_n can share only three vertices with each of the m-books in which it has a 2-cell. It follows that $V(Q) \geq (m+2) + (n+2)-3 = m + n + 1$.

It was shown by Brown, Erdős, and Sós in [1] that $\lim_{n\to\infty} n^{-2} f(n,B_2) = 1/6$. A slight modification of their proof yields the next result.

__Theorem 4__. For each $k \geq 2$, $\lim_{n\to\infty} n^{-2} f(n,B_k) = \frac{k-1}{6}$.

__Proof__. A straightforward counting argument shows that if Q is a 2-complex having n vertices and more than $(k-1)\binom{n}{2}/3$ 2-cells, then some edge of Q has valence at least k.

In [1] the lower bounds for $k = 2$ were obtained when $n \equiv 1$ or $3 \pmod 6$ by considering Steiner triple systems. Such systems yield examples of 2-complexes having n vertices and $n(n-1)/6$ 2-cells in which each edge is on exactly one 2-cell. The lower bounds for the other congruence classes are obtained by deleting up to three vertices from these examples.

For $k > 2$ the lower bounds may be obtained when $n \equiv 1$ or $3 \pmod 6$ by considering 2-complexes whose 2-cells are the triples of a collection of pairwise disjoint Steiner triple systems. Doyen has shown in [2] that for these values of n there exist at least $(n-1)/12$ pairwise disjoint Steiner triple systems. Thus for a given k and for sufficiently large n, $n \equiv 1$ or $3 \pmod 6$, one can construct a 2-complex having n vertices and $(k-1)\binom{n}{2}/3$ 2-cells in which each edge has valence $k - 1$. Again the lower bounds for the other congruence classes are obtained by deleting up to three vertices.

3. Problems involving the W_k. In this section we consider the questions treated in Section 2, but for the W_k rather than the B_k. Here it is clear that $W_{n+m-1} \longrightarrow (W_m, W_n)$, and the first few results parallel those for the B_k almost exactly. The remaining questions, however, seem to be more difficult for the W_k.

Theorem 5. If Q is a 2-complex for which $Q \longrightarrow (W_m, W_n)$, then Q has at least $m + n - 1$ 2-cells, and, when either $m \neq 2$ or $n \neq 2$, the only such 2-complex having exactly $m + n - 1$ 2-cells is W_{m+n-1}.

Proof. The entire argument used for Theorem 1 may be modified in an obvious fashion here. There exists one 2-complex, P, different from W_3, which is such that P has just three 2-cells and $P \longrightarrow (W_2, W_2)$. This is the complex having six vertices, 1, 2, 3, 4, 5, and 6, and the 2-cells (1,2,3), (3,4,5), and (5,6,1).

Corollary 6. If Q is a 2-complex for which $Q \longleftrightarrow (W_m, W_n)$, then Q has at least $m + n - 1$ 2-cells, and, when either $m \neq 2$ or $n \neq 2$, the only such 2-complex having exactly $m + n - 1$ 2-cells is W_{m+n-1}.

The generalized Ramsey number $r(W_m, W_n)$ gives the minimum value of $V(Q)$ among 2-complexes Q for which $Q \longrightarrow (W_m, W_n)$. In [4] it was shown that $r(W_2, W_n) = 2n + 1$ for all $n \geq 2$. The values of $r(W_m, W_n)$ for $m > 2$, $n > 2$, have not been computed. The next result represents a first step in determining the minimum value of $V(Q)$ where $Q \longleftrightarrow (W_2, W_n)$. For $k \geq 2$, the vertex of W_k which is contained in more than one 2-cell of W_k is called its <u>apex</u>.

<u>Theorem 7</u>. Suppose Q is a 2-complex which is such that $Q \longleftrightarrow (W_2, W_n)$ and has as few vertices as possible. Then $n + 2 \leq V(Q) \leq n + 3$, with $V(Q) = 2n + 3$ for $n = 1$ or $n = 3$, and $V(Q) = n + 2$ for $n = 2$.

<u>Proof</u>. Since $W_{n+1} \longleftrightarrow (W_2, W_n)$, it follows that $V(Q) \leq 2n + 3$ for all n. Also $V(Q) \geq V(W_2)$, so $V(Q) = 2n + 3$ when $n = 1$.

Suppose $Q \longleftrightarrow (W_2, W_n)$ for some $n > 1$. For any 2-coloring of Q with the colors red and blue either there is a red 2-wedge faithfully imbedded in Q or a faithfully imbedded blue n-wedge. We may assume that W_n^1 is a faithfully imbedded blue n-wedge in Q. If W_n^1 is contained in a k-wedge of Q for some $k > n$, then $V(Q) \geq 2n + 3$. Otherwise we can recolor exactly one 2-cell of W_n^1 red. Then there exists a red 2-wedge or another n-wedge in Q. In either case there is at least one 2-cell which is not contained in W_n^1. This implies that at least one vertex of Q is not in W_n^1, and hence $V(Q) \geq 2n + 2$. Consideration of the 2-complex on six vertices described in the proof of Theorem 5 shows that for $n = 2$ we have $V(Q) = 2n + 2$.

It remains to be shown that for $n = 3$, $V(Q) > 2n + 2$. Suppose instead that there exists a 2-complex Q with $2n + 2 = 8$ vertices such that each 2-coloring of Q with the colors red and blue produces either a faithfully imbedded red 2-wedge or a faithfully imbedded blue 3-wedge. As above, we may assume that Q contains a faithfully imbedded 3-wedge, W_3^1, which is not contained in any k-wedge of Q for any $k > 3$. Color one 2-cell of W_3^1 red and the rest of the 2-cells of Q blue. There is no red 2-wedge in Q and there must exist a second faithfully imbedded blue 3-wedge, W_3^2, in Q. Since W_3^1 is faithfully imbedded, the apex of W_3^2 is the apex of W_3^1 or is the one vertex of Q not on W_3^1. In the former case, W_3^1 and W_3^2 must share two 2-cells. In the second case, they have no 2-cell in common. If two of the 2-cells of W_3^2 are in W_3^1, then the remaining 2-cell of W_3^2 shares exactly one edge with the third 2-cell of W_3^1, and these two 2-cells can be colored red without forming a red 2-wedge. If W_3^1 and W_3^2 have no 2-cell in common, then either again a 2-cell of W_3^1 shares exactly one edge with some 2-cell of W_3^2 or there are a 2-cell of W_3^1 and a 2-cell of W_3^2 with no vertices in common. In all cases we can recolor the 2-cells of W_3^1 and W_3^2 so that exactly two 2-cells are red and neither a red 2-wedge nor a blue 3-wedge is formed.

Thus there must exist a faithfully imbedded 3-wedge, W_3^3, different from W_3^1 and W_3^2. The apex of W_3^3 is on at least one of W_3^1 and W_3^2, say W_3^1. The apex of W_3^3 must then be the apex of W_3^1, and these two 3-wedges share two 2-cells. If the apex of W_3^2 is not on W_3^1, it must be on W_3^3. But then the vertices of W_3^2 span some 2-cell of W_3^3 which is not in W_3^2. Hence all three must have the same apex. All must share the same two blue 2-cells and the remaining 2-cell of W_3^3 is spanned by the vertices of the red 2-cell of W_3^1 and the red 2-cell of W_3^2. This last 2-cell of W_3^3 may then be recolored red also, without forming a red 2-wedge. Since the 2-cells of these three 3-wedges do not yield a blue 3-

wedge with this coloring, Q contains still another blue 3-wedge. The argument concerning W_3^3 shows that no such blue 3-wedge can exist in the 2-complex spanned by the vertices of the W_3^i, i = 1, 2, 3. It follows that $V(Q) > 2n + 2$.

The next result concerns the minimum number of 2-cells needed to insure the existence of a 2-wedge in a 2-complex having a given number of vertices. Here D_3 denotes the 2-complex having four vertices and three 2-cells, and K_4 will denote the 2-complex having four vertices and four 2-cells.

<u>Theorem 8</u>. If $n \geq 5$, then

$$f(n,W_2) = n + 1 \quad \text{for} \quad n \equiv 0 \pmod 4$$
$$f(n,W_2) = n \quad \text{for} \quad n \equiv 1 \pmod 4$$
$$f(n,W_2) = n - 1 \quad \text{for} \quad n \equiv 2 \text{ or } 3 \pmod 4.$$

<u>Proof</u>. That the given values are lower bounds follows from the existence of 2-complexes consisting of disjoint copies of K_4 or, if $n \equiv 3 \pmod 4$, disjoint copies of K_4 plus a single additional 2-cell disjoint from the copies of K_4.

To show that the given values are upper bounds as well, we proceed by induction on n. To begin, separate consideration is required for each of n = 5, 6, 7, and 8 representing the four congruence classes. We outline the proof for n = 5, the other three being progressively longer, but involving the same ideas.

Suppose Q has five vertices and at least five 2-cells, but no 2-wedge. There is a vertex of Q contained in at least three 2-cells. Since there is no W_2, there must be a 3-book or a copy of D_3. If there were a 3-book, then each additional 2-cell in Q would form a 2-wedge. On the other hand, D_3 plus one more 2-cell yields either a 2-wedge or a copy of K_4. Finally, since there are only five vertices, K_4 plus a fifth 2-cell would produce a 2-book.

Suppose the theorem is true for all n, $5 \leq n < m$, and let Q be a 2-complex having $m \geq 9$ vertices and at least $f(m-4,W_2) + 4$ 2-cells. Suppose Q contains no 2-wedge. Some vertex of Q is contained in at least three 2-cells. Then Q contains either a 3-book or a copy of D_3. If there is a copy of D_3, no 2-cell of Q has vertices in this D_3 as well as vertices in its complement. Thus the $m - 4$ vertices of Q not in this D_3 span a subcomplex having at least $f(m-4,W_2)$ 2-cells, and hence containing a 2-wedge. Suppose Q contains a 3-book, B_3. Any 2-cell containing vertices of B_3 and vertices not in B_3 must also have the spine of B_3 as an edge. Let B_k be the book in Q with the largest number of 2-cells which contains B_3. No 2-cell of Q not in B_k shares a vertex with B_k. Notice that for $m = 9$ this case is impossible and for $m > 9$ we have $k \leq m - 7$, for otherwise there are not enough vertices in the complement of B_k to account for the number of 2-cells not in B_k. The $m - (k+2)$ vertices not in B_k must span at least $f(m-4,W_2) + 4-k$ 2-cells. Since $f(m-4,W_2) + 4 - k \geq m - k - 1 \geq f(m - k - 2,W_2)$, the complement of B_k contains a 2-wedge.

4. Conclusions and questions. The partial results contained in the last several theorems of Section 3 leave some of the general questions which were settled in Section 2 for the B_k, unanswered for the W_k. It is clear that $r(W_m,W_n) \leq 2m + 2n + 1$, and that if $Q \longleftrightarrow (W_m,W_n)$, then $V(Q) \leq 2m + 2n + 1$. It seems likely that for all n, $n \neq 2$, $Q \longleftrightarrow (W_2,W_n)$ implies $V(Q) = 2n + 3$, and it may be that $V(Q) = 2m + 2n + 1$ in general. It seems less probable that $r(W_m,W_n) = 2m + 2n + 1$ for all $m > 2$ and $n > 2$.

Examples such as those used in Theorem 8 show that for all $n \geq 7$, $f(n,W_3) \geq 10n/3$. On the other hand, it can be shown using graph-theoretic extremal results that $f(n,W_3) \leq 5n^2/6$. Professor Paul Erdős conjectures [7] that there is a constant, c_3, such that $f(n,W_3) \leq c_3 n$, and if so, that for each $k \geq 2$, there exists a constant c_k such that $f(n,W_k) \leq c_k n$. For many related results see [1] and [5].

All of the questions considered here for the B_k and W_k could, of course, be asked for the members of the other families discussed in [3] and [4], as well as for other families of 2-complexes. The question of the determination of $\lim_{n\to\infty} n^{-3} f(n,D_3)$ was mentioned by Erdős in [52], and the determination of $f(n,K_4)$ is the first open case of Turan's problem.

REFERENCES

1. W. Brown, P. Erdős, and V. T. Sos, On the existence of triangulated spheres in 3-graphs and related problems. Period, Math. Hungar. 3(1973) 221-228.

2. J. Doyen, Constructions of disjoint Steiner triple systems. Proc. Amer. Math. Soc. 32(1972) 409-416.

3. R. A. Duke and F. Harary, Generalized Ramsey Theory VI: Ramsey numbers for small plexes. J. Austral. Math. Soc., to appear.

4. R. A. Duke, Ramsey numbers of families of 2-complexes. Proc. of the Sixth Southeastern Conference on Combinatories, Graph Theory, and Computing. Congressus Numerantium No. XIV, Utilitas Math., Winnipeg, Man. (1975) 265-277.

5. P. Erdős, Extremal problems on graphs and hypergraphs. Hypergraph Seminar. Springer-Verlag, New York (1974) 75-84.

6. P. Erdős, Problems and results on finite and infinite graphs, to appear.

7. P. Erdős, private communication.

8. P. Erdős, R. J. Faudree, R. H. Schelp and C. C. Rousseau, The size Ramsey number, to appear.

VARIOUS LENGTH PATHS IN GRAPHS

R.J. Faudree and R.H. Schelp
Memphis State University
Memphis, TN 38152

Abstract

A simple graph G on n vertices satisfies property P_i if between every pair of distinct vertices of G, there exists a path with i vertices. In this paper known results and open questions about the relationship between property P_i and property P_j are considered. There are degree conditions, edge conditions and the condition of being the power of a graph which guarantee that a graph is Hamiltonian-connected (satisfies property P_n). Included is a discussion of these conditions and their relationship to property P_i for $i \neq n$.

PRELIMINARIES

All graphs will be finite, undirected and without loops or multiple edges. A graph will be denoted by $G = G(V,E)$, where V represents the set of vertices and E the set of edges. If V has cardinality n, and E has cardinality m, then G is a graph of <u>order</u> n and <u>size</u> m.

If there exists a path containing ℓ vertices connecting u and v in V, then property $P_\ell(u,v)$ will be said to hold. For $2 \leq \ell \leq n$, let S_ℓ be the set of all unordered pairs of distinct u and v in V for which property $P_\ell(u,v)$ holds, and let S_1 be the set of all unordered pairs of vertices which are not connected by any path.

The <u>path length distribution</u> (PLD) of a graph G of order n

and size m is the sequence $(X_1, X_2, \ldots, X_n)$ where $X_i = |S_i|$. Thus $X_1 = 0$ if and only if G is connected, $X_1 = m$, and for all i, $X_i \leq n(n-1)/2$. Usually only connected graphs will be considered. A graph G satisfies <u>property</u> P_i if $X_i = n(n-1)/2$. Thus property P_i holds if and only if there exists a path with i vertices connecting any two distinct vertices of G. The well known property of <u>Hamiltonian-connectedness</u> is just property P_n. A graph is <u>PLD-maximal</u> if property P_i holds for $3 \leq i \leq n$. A complete graph is certainly PLD-maximal but the converse is not true. In fact a graph can be PLD-maximal with only a small number of edges. Any path connecting two vertices must of course be at least as long as the distance between the vertices. A graph G is <u>panconnected</u> if for each pair of vertices u and v, property $P_i(u,v)$ holds for all $i > d(u,v)$, (the distance between u and v).

In this paper the few general results that are known about the path length distribution will be given and some open questions will be posed. Also the path length distribution of special classes of graphs will be discussed. The relationship between and the independence of the various properties P_i will be considered. In particular the following type of question will be considered. Does there exist a graph G for which property P_i does not hold but property P_j holds for all $j \neq i$? An emphasis will be placed on Hamiltonian-connectedness and its relationship to the other properties P_i.

There have been numerous conditions which guarantee that a graph is Hamiltonian-connected. It is now known that some of these conditions also insure that other properties P_i, for $i \neq n$, are satisfied or even in some cases that the graph is panconnected. In most cases it is an open question as to which properties P_i hold. Conditions which imply that a graph is Hamiltonian-connected that fall into the following categories will be discussed: edge conditions, degree conditions, topological conditions, and the condition of being a power of a graph.

PATH LENGTH DISTRIBUTION

The concept of path length distribution was first introduced in a paper by M.F. Capobianco (see [2]). This concept was investigated in a paper by Faudree, Rousseau and Schelp [6], and proofs of several of the results mentioned in this section appear there. A natural question to ask and one that was posed in [2] is the following: Does the PLD of a graph determine the graph up to isomorphism? The answer is no, even for a restricted class of graphs like trees.

Let T be a tree of order n, and thus of size $n-1$. In a tree there exists a unique path between any two vertices. Therefore if $(X_1, X_2, \ldots, X_n)$ is the PLD of T, then $X_1 = 0$, $X_2 = n-1$, and $\sum_{i=2}^{n} X_i = n(n-1)/2$. By direct calculation (see [13], Appendix 3) one can verify for $n \leq 8$ that there do not exist non-isomorphic trees with the same PLD. There is only one tree of diameter 1 and any tree of diameter 2 is a star. Let T be a tree of diameter 3. If u and v are the bicenters of T and $r = d(u) \geq d(v)$, then $X_2 = n-1$, $X_4 = (n-r-1)(r-1)$, $X_i = 0$ for $i > 4$, and $X_3 = n(n-1)/2 - X_2 - X_4$. But if $(n-r-1)(r-1) = (n-s-1)(s-1)$ and both r and s are $\geq n/2$, then $r = s$. Hence any two non-isomorphic trees of diameter 3 have different PLD's. For larger order and diameter the situation is very different as the following result indicates.

THEOREM 1. Non-isomorphic trees of order n and diameter 4 having the same PLD exist for all $n \geq 9$. Moreover, for any N, it is possible to find N trees having the same PLD.

Trees of course have a small number of paths. At the other extreme, PLD-maximal graphs have many paths, but even in this case the PLD does not determine the graph as the following indicates.

THEOREM 2. *If G is a graph of order $n \geq 4$ and if G satisfies P_3 and P_n, then the size m of G is $\geq 2(n-1)$. Moreover, $m = 2(n-1)$ if and only if G is isomorphic to W_n, the wheel on n vertices.*

One can easily verify that any wheel is PLD-maximal. Therefore any graph of order n which has W_n as a subgraph is PLD-maximal. This can be used to prove the following.

THEOREM 3. *For every $n \geq 6$ and for $2n-1 \leq m \leq n(n-1)/2 -2$, there exists non-isomorphic PLD-maximal graphs of order n and size m.*

Therefore just as in the case of trees, there exist non-isomorphic PLD-maximal graphs with the same PLD.

Some graphs of course are determined by their PLD. For example K_n (the complete graph on n vertices), K_n-e (the complete graph with one edge deleted) and W_n are the only PLD-maximal graphs of order n for which the second term in the PLD is equal to $n(n-1)/2$, $n(n-1)/2 -1$ and $2(n-1)$ respectively. Also a path is the only graph on n vertices having a PLD with $X_1 = 0$, $X_2 = n-1$ and $X_n \neq 0$, since it is the only tree of order n and diameter $n-1$. Also one can verify that graphs of order ≤ 5 are determined by their PLD's, and it has already been remarked that this is true for trees of order ≤ 8. The following is a very basic question but probably very difficult.

QUESTION 1. *Find all connected graphs which are determined by their PLD's.*

It would be interesting to just exhibit large classes of graphs which are determined by their PLD.

A more specific question but also a basic one is the following.

QUESTION 2. *Determine all PLD-maximal graphs.*

If a graph G of order n and size m is PLD-maximal then by Theorem 2, $m \geq 2(n-1)$ with equality if and only if $G = W_n$. Therefore W_n is a PLD-maximal graph which loses this property with the deletion of any edge. Such a graph is called a <u>minimal PLD-maximal graph</u>. To determine all PLD-maximal graphs it is sufficient to know the minimal ones.

QUESTION 3. <u>Determine all minimal PLD-maximal graphs or at least give large classes of minimal ones.</u>

The wheel is not the only minimal PLD-maximal graph.

THEOREM 4. <u>If</u> G <u>is a graph of order</u> $n \geq 4$ <u>and size</u> m, <u>then</u> G <u>is PLD-maximal if either</u>

(i) $d(v) \geq (n+2)/2$ <u>for all vertices</u> v <u>in</u> G.

<u>or</u> (ii) $m \geq n(n-1)/2 - (n-4)$.

Therefore by (i) there exists PLD-maximal graphs, and hence minimal ones, which do not have vertices of high degree. By (ii) any minimal PLD-maximal graph on n vertices must have size $m \leq n(n-1)/2 - (n-4)$. A natural question to ask is the following.

QUESTION 4. <u>For which pairs of integers</u> (n,m) <u>do there exist minimal PLD-maximal graphs of order</u> n <u>and size</u> m?

Certainly not all sequences can be the PLD of a graph, so the realizability question arises. It is sufficient to consider only connected graphs.

QUESTION 5. <u>Which sequences</u> $(0, X_2, X_3, \ldots, X_n)$ <u>can be the PLD of a connected graph of order</u> n?

This question appears to be very difficult. One might want to consider a restricted class of graphs. A natural one to consider would be the following.

QUESTION 6. <u>Which sequences</u> $(0,X_1,X_2,\ldots,X_n)$ <u>can be the PLD of a tree of order</u> n?

PROPERTIES P_i

Not too much is known about the relationships that exist between a property P_i and a property P_j. Some observations can be made. Let S be a subset of $\{2,3,\ldots,n\}$. The closure $\bar{S}$ of S is defined to be the set of j in $\{2,3,\ldots,n\}$ such that if G is any graph of order n satisfying P_i for all $i \in S$ then G satisfies P_j. The only graph which satisfies P_2 is a complete graph. Therefore $\overline{\{2\}} = \{2,3,\ldots,n\}$, or more generally if $2 \in S$, then $\bar{S} = \{2,3,\ldots,n\}$. With this notation a question concerning the relationships between various properties P_i can be stated.

QUESTION 7. <u>For each subset</u> S <u>of</u> $\{2,3,\ldots,n\}$ <u>determine the closure</u> $\bar{S}$ <u>of</u> S.

Any PLD-maximal graph which is not a complete graph satisfies property P_i for $i > 2$, but does not satisfy P_2. Define an (r,s,t) graph to be the graph $\bar{K}_s + (K_r \cup K_t)$. The graphical operations <u>join</u> (+) and <u>union</u> ($\cup$) are those defined in [13]. It is easily seen that an $(r,s,1)$ graph does not satisfy P_3 and any $(r,s,2)$ graph does not satisfy P_4. If r is not too small, one can check directly that each of the following graphs satisfy property P_i for $i > 4$ and satisfy either P_3 or P_4.

$$G = (r,r-1,2) \qquad |G| = 2r+1$$
$$G = (r+1,r-1,1) \qquad |G| = 2r+1$$
$$G = (r-1,r-1,2) \qquad |G| = 2r$$
$$G = (r+1,r-2,1) \qquad |G| = 2r.$$

Define $K_r \# K_s$ to be the graph on $r+s-2$ vertices which consists of a complete graph on r vertices and a complete graph on s vertices with precisely two vertices and one edge in common. The two vertices in common to both complete graphs form a cutset. Thus if $r \geq s \geq 3$, then $K_r \# K_s$ satisfies P_i for $3 \leq i \leq r$ but does not satisfy P_i for $i > r$. In particular $K_{n-1} \# K_3$ does not satisfy P_n but does satisfy P_i for $3 \leq i \leq n-1$. One can of course generalize the graph $K_r \# K_s$ and consider the graph determined by any finite number of complete graphs with precisely one edge and two vertices in common to each complete graph. The above results are summarized in the following.

THEOREM 5. _Let_ S _be any subset of_ $\{3,4,\ldots,n\}$. _Then_ $i \in \bar{S}$ _implies_ $i \in S$ _for_ $i = 3, 4$ _and_ n.

This leaves the following question.

QUESTION 8. _Let_ S _be any subset of_ $\{3,4,\ldots,n\}$. _For which_ j _does_ $j \in \bar{S}$ _imply_ $j \in S$? _In particular, does_ $5 \in \bar{S}$ _imply_ $5 \in S$?

Let $t \geq 3$ and $(r_1,r_2,\ldots,r_t)$ be a sequence of positive integers. Define $K_{r_1}*K_{r_2}*\ldots*K_{r_t}$ to be the graph of order $\sum_{i=1}^{t} r_i$ which consists of t disjoint complete graphs of order $r_1,r_2,\ldots,r_t$ respectively. Also each vertex of K_{r_i} is adjacent to each vertex of $K_{r_{i+1}}$ for $1 \leq i < t$, and precisely one vertex of K_{r_1} is adjacent to precisely one vertex of K_{r_t}. For $3 \leq i < t$, there exists no path with i vertices between the vertex in K_{r_1} and the vertex in K_{r_t} which are adjacent. Thus property P_i does not hold for $3 \leq i < t$. Also if $r_i \geq 2$ for $1 < i < t$, then one can easily verify that property P_i holds for $i \geq t$. Note that if $r_1 = r_t = 1$ and $r_i = 2$ for $1 < i < t$, then the order of $K_{r_1}*K_{r_2}*\ldots*K_{r_t}$ is $2t-2$.

The above example along with the generalization of the graph

$K_r \# K_s$ give the following result.

THEOREM 6. Let $n \geq 6$ and $2 \leq t \leq n$.

(i) For any $t \leq n/2$, there exists a graph $G = G(t)$ of order n which satisfies property P_i if and only if $i > t$.

(ii) For any $t \geq 3$, there exists a graph $G = G(t)$ of order n, which satisfies property P_i if and only if $3 \leq i \leq t$.

All of the examples considered so far indicate a positive answer to the following question.

QUESTION 9. If S is a subset of $\{4,5,\ldots,n\}$, then is $\bar{S}$ an interval (i.e. $\bar{S} = \{i: r \leq i \leq s\}$ for some r and s)?

A very basic question is the following.

QUESTION 10. Give an example of a subset S of $\{3,4,\ldots,n\}$ such that $S \neq \bar{S}$.

If no such example exists, then all the previous questions about the closure of a set are answered.

The number of edges needed to insure that a property P_i holds was first investigated by Ore. He proved the following (see [20]).

THEOREM 7. If G is a graph of order $n \geq 4$ and size $m \geq (n-1)(n-2)/2 +3$, then G is Hamiltonian-connected. If $m < (n-1)(n-2)/2 +3$, then G need not be Hamiltonian-connected.

In fact a stronger result holds.

THEOREM 8. If G is a graph of order $n \geq 4$ and size $m \geq (n-1)(n-2)/2 +3$, then G is PLD-maximal.

This theorem can be used to prove a result concerning the minimum number of edges needed in a graph to give property P_i. Let $f_{nk}(i)$ denote the minimum size of a graph G or order n with every vertex of degree at least k which will insure that G satisfies property P_i. For notational convenience let $n_i = (n-i)(n-i-1)/2$ throughout the remainder of this section.

THEOREM 9. <u>For</u> $n \geq 4$, $f_{n1}(i) = n_1+2$ <u>if</u> $3 \leq i \leq n-1$. <u>Also</u> $f_{n1}(2) = n_0$ <u>and</u> $f_{n1}(n) = n_1+3$.

<u>Proof</u>. The result is obvious for $i = 2$, and the result for $i = n$ is just Theorem 7. Since any graph with a vertex of degree 1 cannot satisfy property P_i for $i \geq 3$, $f_{n1}(i) > n_1+1$ for $i \geq 3$. In a graph of order n and size n_1+2, any two vertices are commonly adjacent to at least one vertex, thus $f_{n1}(3) = n_1+2$.

For small values of n one can check the result directly, so assume $n \geq 8$ and that the theorem is true for all graphs of order less than n. Let G be a graph of order n and size $m \geq n_1+2$. If v is a vertex of G of degree $\leq n-3$, then $G - v$ has order $n - 1$ and size at least n_2+3. Therefore $G - v$ is PLD-maximal by Theorem 8. The vertex v has degree at least 2. This implies that G satisfies property P_i for $3 \leq i \leq n$. If $d(v) \geq n-2$ for all vertices v of G, then $m \geq n(n-2)/2 = n_1+n/2 -1$. Since $n \geq 8$, G is PLD-maximal. This completes the proof.

If one does not allow any vertices of degree 1, then the number of edges need to insure property P_i can be reduced, as the following theorem of Lewin (see [16]) indicates.

THEOREM 10. <u>For</u> $n \geq 9$, $f_{n2}(i) = n_2+4$ <u>if</u> $5 \leq i \leq n-2$. <u>Also</u> $f_{n2}(2) = n_0$, $f_{n2}(3) = n_1+2$, $f_{n2}(4) = n_1+5$, $f_{n2}(n-1) = n_2+6$ <u>and</u> $f_{n2}(n) = n_1+3$.

Also included in [16] are the values of $f_{n2}(i)$ for small values of n. Hence the values $f_{n1}(i)$ and $f_{n2}(i)$ are known.

QUESTION 11. <u>Determine</u> $f_{nk}(i)$ <u>for</u> $k > 2$.

HAMILTONIAN-CONNECTEDNESS

In this section evidence will be given which indicates that the following statement holds.

QUESTION 12. <u>Property</u> P_n <u>implies property</u> P_i <u>for each</u> i, $n/2+1 \leq i \leq n$, <u>in a graph of order</u> n.

Direct calculations in graphs of small order, randomly generated graphs that are not to large, and graphs that are commonly used as counterexamples have failed to produce any example which contradicts the above statement. Another source of evidence is to consider conditions which imply property P_n and investigate what other properties P_i are also implied by these conditions. Hamiltonian and Hamiltonian-connected graphs have been studied extensively, and there are many conditions that are known to imply Hamiltonian-connectedness. Several of these conditions will be discussed below along with the additional properties P_i they are known to imply.

A condition on the number of edges in a graph G which insures that the graph is Hamiltonian-connected was given in Theorem 7. By Theorem 8 this same condition implies that the graph is in fact PLD-maximal. In [19], Ore gave a condition on the sum of degrees of non-adjacent vertices which implies that a graph is Hamiltonian. In [20], a strengthened version of this condition was shown to imply Hamiltonian-connectedness. In fact the last condition also implies much more as the following result indicates.

THEOREM 11 ([7]). <u>If</u> G <u>is a graph of order</u> $n \geq 5$ <u>with</u> $d(x)+d(y) > n$ <u>for each pair of distinct non-adjacent vertices</u> x <u>and</u> y, <u>then</u> G <u>satisfies property</u> P_i <u>for</u> $5 \leq i \leq n$.

Degree conditions which imply that a graph is Hamiltonian were given in [21] by Pósa, in [1] by Bondy, and in [4] by Chvátal. Just as in the case of the Ore condition, each of these conditions were strengthened to give a condition which implies Hamiltonian-connectedness. These results can be found in [17], [18] and [23] respectively. A different type of condition was considered by Chvátal and Erdös in [5], where they proved the following result. If G is a s-connected graph containing no independent set of s vertices, then G is Hamiltonian-connected. It would be interesting to know if the conditions mentioned above also imply other properties P_i as in the case of the Ore condition.

Powers of connected graphs have many paths of many lengths. For example, Karaganis in [15] and Sekanina in [22] proved that the cube of a connected graph is Hamiltonian-connected. By an induction proof one can prove the following stronger result. If G is a connected graph of order n, then G^3 is panconnected, and thus G^3 satisfies property P_i for $n/3 < i \leq n$. Recently there have been several interesting results proved on the square of a connected graph.

The well-known Nash-Williams and Plummer conjecture that the square of a block is Hamiltonian was proved by Fleischner in [11]. In a companion paper the following was proved.

THEOREM 12 ([10]). *The square of a connected bridgeless DT-graph (every edge is incident to a vertex of degree 2) is Hamiltonian-connected.*

Shortly thereafter a short but ingenious proof of the following result was published.

THEOREM 13 ([13]). *The square of a block is Hamiltonian-connected.*

Just as in several other cases, the conditions in the previous two theorems which imply Hamiltonian-connectedness also imply much more.

THEOREM 14 ([8]). *Let* G *be a block or a bridgeless DT-graph. Then*

G^2 is panconnected.

More than any other theorem, the following one proved by Fleishner leads one to believe that the statement that property P_n implies property P_i for $n/2 + 1 \leq i \leq n$ is true.

THEOREM 15 ([12]). The square of a graph is Hamiltonian-connected if and only if it is panconnected.

Let G be a plane graph and let $V(G)$, $E(G)$, and $F(G)$ denote the set of vertices, edges and faces, respectively. Two distinct vertices (edges, faces) of G are adjacent if they share a common edge (vertex, edge). A vertex and an edge, a vertex and a face, or an edge and a face, are adjacent if they are incident. The entire graph of G, denoted $e(G)$, is the graph with vertex set $V(G) \cup E(G) \cup F(G)$ with two vertices of $e(G)$ adjacent if and only if they are adjacent in G. Hamiltonian properties of entire graphs were first discussed by Mitchem, and then the following was proved by Hobbs and Mitchem.

THEOREM 16 ([14]). The entire graph of a bridgeless connected plane graph is Hamiltonian.

Fitting the pattern of the previous results, a stronger result was proved.

THEOREM 17 ([9]). The entire graph of a bridgeless connected plane graph is panconnected.

It would be interesting if an analogue of the theorem of Fleishner on squares of graphs could be proved and the following question answered.

QUESTION 13. Is the entire graph of a plane graph Hamiltonian-connected if and only if it is panconnected?

One would really like to prove that property P_n implies property P_i for $n/2 + 1 \leq i \leq n$. In this section some evidence has been given to substantiate that this fact is true. But property P_n seems to be a difficult property to get hold of, so one might aim for a less ambitious goal. In particular one could consider the following weaker version of Question 12.

QUESTION 14. *Does any reasonable condition on a graph G of order n that implies G is Hamiltonian-connected also imply that G satisfies property P_i for $n/2 + 1 \leq i \leq n$?*

REFERENCES

[1] J. Bondy, Properties of graphs with constrains on degrees. *Studia. Sci. Math. Hungar.* 4 (1969) 473-475.

[2] M.F. Capobianco, Statistical inference in finite populations having structure. *Trans. New York Acad. Sci. Ser. II* 32 (1970) 401-413.

[3] G. Chartrand, A.M. Hobbs, H.A. Jung, S.F. Kapoor, and C. St. J. A. Nash-Williams, The square of a block is Hamiltonian connected. *J. Comb. Theory Ser. B* 16 (1974) 290-292.

[4] V. Chvátal, On Hamilton's Ideals. *J. Comb. Theory Ser. B* 12 (1972) 163-168.

[5] V. Chvátal and P. Erdös, A note on Hamiltonian circuits. *Discrete Math.* 2 (1972) 111-113.

[6] R.J. Faudree, C.C. Rousseau, and R.H. Schelp, Theory of path length distributions I. *Discrete Math.* 6 (1973) 35-53.

[7] R.J. Faudree and R.H. Schelp, Path connected graphs. *Acta Math. Acad. Scient. Hung.* 25 (1974) 313-319.

[8] _____________________, The square of a block is strongly path connected. *J. of Comb. Theory Ser. B* 20 (1976) 47-61.

[9] _____________________, The entire graph of a bridgeless connected plane graph is panconnected. *J. Lond. Math. Soc.* (2) 12 (1975) 59-66.

[10] H. Fleishner, On spanning subgraphs of a connected bridgeless graph and their application to DT-graphs. *J. Comb. Theory Ser. B* 16 (1974) 17-28.

[11] __________, The square of every two-connected graph is Hamiltonian. *J. Comb. Theory Ser. B* 16 (1974) 29-34.

[12] ___________, In the square of graphs, Hamiltonianity and pancyclicity, Hamiltonian connected and panconnectedness are equivalent concepts. (To appear).

[13] F. Harary, Graph Theory. Addison-Wesley, Reading (1969).

[14] A.M. Hobbs and J. Mitchem, The entire graph of a bridgeless connected plane graph is Hamiltonian. (To appear).

[15] J.J. Karaganis, On the cube of a graph. Canad. Math. Bull. 11 (1969) 295-296.

[16] M. Lewin, On path-connected graphs. (To appear).

[17] D.R. Lick, n-Hamiltonian connected graphs. Duke Math. J. 37 (1970) 387-392.

[18] ________, A sufficient condition for Hamiltonian connectedness. J. Comb. Theory 8 (1970) 444-445.

[19] O. Ore, Note on Hamiltonian circuits. Amer. Math. Monthly 67 (1960) 55.

[20] ______, Hamiltonian connected graphs. J. Math. Pures. Appl. 42 (1963) 21-27.

[21] L. Pósa, A theorem on Hamiltonian lines. Maggar Tud. Akad. Mat. Kutato Int. Kozl. 8 (1963) 355-361.

[22] M. Sekanina, On an ordering of the set of vertices of a connected graph. Publ. Fac. Sci. Univ. Brno. 412 (1960) 137-142.

[23] J.E. Williamson, A note on Hamiltonian-connected graphs. (To appear).

ORTHOGONAL GROUPS OVER GF(2) AND RELATED GRAPHS

J. Sutherland Frame
Michigan State University
East Lansing, MI 48824

Abstract

Regular graphs are considered, whose automorphism groups are permutation representations P of the orthogonal groups in various dimensions over GF(2). Vertices and adjacencies are defined by quadratic forms, and after graphical displays of the trivial isomorphisms between the symmetric groups S_2, S_3, S_5, S_6 and corresponding orthogonal groups, a 28-vertex graph is constructed that displays the isomorphism between S_8 and $O_6^+(2)$. Explored next are the eigenvalues and constituent idempotent matrices of the (-1,1)-adjacency matrix A of each of the orthogonal graphs, and the commuting ring R of the rank three permutation representation P of its automorphism group. Formulas are obtained for splitting into its irreducible characters $\chi^{(i)}$ the permutation character χ of P, by expressing the class sums P_λ of P in terms of the identity matrix and the (0,1)-matrices H and K obtained from the adjacency matrix $A = H - K$.

ORTHOGONAL GROUPS OVER GF(2) AND RELATED GRAPHS

1. INTRODUCTION

Interesting relationships exist between graphs, their adjacency matrices, and their automorphism groups. The orthogonal groups $O_n(2)$ over the two element field GF(2) are the automorphism groups of certain "orthogonal graphs" described by J.J. Seidel in his 1973 paper "On two-graphs and Shult's characterization of symplectic and orthogonal geometries over GF(2)." [3]. The orthogonal groups are defined by the invariance of standard quadratic and bilinear forms. The known isomorphisms between the symmetric groups of degrees 2,3,5,6 and 8 with corresponding orthogonal groups is displayed graphically in Section 2. Relationships are explored in Section 3 between the constituent idempotent matrices of the graph adjacency matrix and the commuting ring of the rank three permutation matrix representation P of the orthogonal automorphism group G, noting that the class sums of P are in this ring. These relationships are used in Section 4 to compute some characters of the two non-trivial irreducible constituents of P in terms of the adjacencies in the graph, involving the double coset decompositions of class sums.

For $s = +1$ or -1, let $Q_m^s(x)$ denote the two non-equivalent standard quadratic forms on the vector space $V = V(2m,2)$:

$$Q_m^+(x) = \sum_{i=1}^{2m} x_{2i-1}x_{2i} \, ; \qquad Q_m^-(x) = x_1^2 + x_2^2 + Q_m^+(x) \tag{1.1}$$

Both quadratic forms determine the same non-degenerate bilinear form over GF(2)

$$B_m(x,y) = Q_m^s(x) + Q_m^s(y) + Q_m^s(x+y) \tag{1.2}$$

Triples of distinct vectors x,y,z of V such that

$$B_m(x,y) + B_m(y,z) + B_m(z,x) = 0 \qquad (1.3)$$

form the triple set Δ^s_{2m} of the orthogonal two-graph, studied by E. Shult, J.J. Seidel, and others, whose vertex set Ω^s_{2m} consists of vectors x of V such that $Q^s_m(x) = 0$. As required in any two-graph, each 4-subset of the vertex set contains an even number of triples from the triple set. The derived graph of Ω^s_{2m} with respect to $z = 0$ is the regular orthogonal graph O^s_{2m} with vertex set $\Omega^s_{2m}\backslash\{0\}$, in which two vertices x,y are adjacent if they form a triple with 0 in Ω^s_{2m}. Since $Q^s_m(x) = Q^s_m(y) = B_m(x,0) = B_m(y,0) = 0$, the conditions for adjacency in O^s_{2m} can be written:

$$Q^s_m\ (x+y) = 0, \text{ whenever } x \text{ and } y \text{ are adjacent in } O^s_{2m} \qquad (1.4)$$

The two-graphs Ω^+_{2m} and Ω^-_{2m} both have as their automorphism group the symplectic group $Sp(2m,2)$ known to be isomorphic with the orthogonal group $O(2m+1,2)$. Each class of this group is represented in one or both of its maximal orthogonal subgroups $O^s(2m,2)$ which are automorphism groups of the orthogonal graphs O^s_{2m}.

2. ISOMORPHISMS BETWEEN SYMMETRIC AND ORTHOGONAL GROUPS

Both the two-graph Ω^+_2, with three vertices $(0,0),(1,0),(0,1)$ from $V(2,2)$ satisfying $Q^+_1(v) = v_1v_2 = 0$, and the two-graph Ω^-_4 with the six vertices

$$(0000),(1011),(0111),(0010),(0001),(1111) \qquad (2.1)$$

from $V(4,2)$ satisfying $Q^-_2(v) = v_1^2 + v_2^2 + v_1v_2 + v_3v_4 = 0$ have vacuous triple sets satisfying (1.3). Hence their automorphism groups $Sp(2,2)$ and $Sp(4,2)$, known to be isomorphic with $O(3,2)$ and $O(5.2)$ respectively, are also isomorphic with S_3 and S_6 and their subgroup

$O^+(2,2)$ and $O^-(4,2)$ which fix the vertex $z = 0$ are isomorphic with S_2 and S_5.

The less obvious isomorphism between $O^+(6,2)$ and S_8 can be exhibited graphically as follows. The 28 vectors v_{ij} of $V(6,2)$ such that $Q_3^+(v_{ij}) = 1$ are precisely the sums of distinct pairs chosen from the eight vectors

(2.2)

(000000), (101100), (011100), (001011), (000111), (110010), (110001), (111111)

denoted x_i, $i = 0,1,\cdots, 7$. Consider the 28 vertex graph Γ with vertices v_{ij} labeled by vector sums $v_{ij} = x_i + x_j \pmod 2$, where u and v are adjacent if $Q_3^+(u+v) = 1$. Then u and v are adjacent if and only if they share a vector x_i. The 28 vertices of Γ correspond to unordered digit pairs $ij, (i,j = 0,1,\cdots,7)$ with vertex ij incident to the 14 vertices ik and kj. Clearly Γ has S_8 for its automorphism group, so S_8 is isomorphic with $O^+(6,2)$.

If J denotes the matrix with all entries 1, the matrix $J-I$ interchanges the sums $x_0 + x_i$ and $x_7 + x_i$, $i = 1,\cdots 6$, and corresponds to the permutation (07) of S_8. We can embed S_8 in $O^+(6,2)$ by constructing the matrix Z of order eight that maps the row vectors $x_i + x_j$ into $x_{i+1} + x_{j+1}$, $i,j = 0,1,\cdots 7$, where $x_8 = x_0$. Then $J-I$ and Z generate S_8.

$$J-I = \begin{bmatrix} 0&1&1&1&1&1\\ 1&0&1&1&1&1\\ 1&1&0&1&1&1\\ 1&1&1&0&1&1\\ 1&1&1&1&0&1\\ 1&1&1&1&1&0 \end{bmatrix} \qquad Z = \begin{bmatrix} 0&0&0&1&0&1\\ 0&1&0&0&1&0\\ 1&0&0&1&0&1\\ 0&1&0&0&0&0\\ 0&0&1&0&1&0\\ 0&0&0&1&0&0 \end{bmatrix}$$

(07) (01234567)

3. EIGENVALUES AND IDEMPOTENTS OF THE ADJACENCY MATRIX

Any n-vertex graph can be described either (1) by an $n \times n$ symmetric incidence matrix such that $K_{ij} = 1$ if vertices are adjacent and $K_{ij} = 0$ otherwise, or (2) by an $n \times n$ non-incidence matrix H such that $H_{ij} = 1$ if v_i and v_j are not adjacent and $H_{ij} = 0$ otherwise, or (3) by the (-1,1)-adjacency matrix $A = H - K$ used advantageously by Seidel and others [3]. It is assumed that K, H and A have diagonal entries 0, and we note that $J = I + H + K$ is the $n \times n$ matrix with all entries 1.

For a regular graph each vertex is adjacent to k other vertices and non-adjacent to $h = n - 1 - k$ others. Hence

$$HJ = hJ, KJ = kJ, AJ = a_0 J \quad \text{with} \quad a_0 = h - k \tag{3.1}$$

Thus the matrix $E_0 = J/n$ is a constituent idempotent matrix of rank 1 for the eigenvalues h of H, k of K and a_0 of A.

Theorem 3.1. The matrices H and K for the orthogonal graphs O^s_{2m} satisfy the product relation

$$HK = (k/2)H + (h/2)K = KH \tag{3.2}$$

Proof: Let v_1 and v_2 be two vertices of O^s_{2m} which are both adjacent to some vertex v_0 but non-adjacent to each other. Without loss of generality we can choose

$$v_0 = (X,0,0),\ v_1 = (X,1,0),\ (v_2 = X,0,1) \tag{3.3}$$

where X denotes a vertex of O^s_{2m-2}. Just half the k vertices $v_j = (Y_j, a_j, b_j)$ adjacent to v_1 are non-adjacent to v_2, and half the h vertices v_j non-adjacent to v_0 are adjacent to v_1 so $\sum K_{1j}H_{j2} = k/2$ and $\sum H_{0j}K_{j1} = h/2$, and the coefficients of H and K in $KH = HK$ are $k/2$ and $h/2$ respectively. Since

$HK - (k/2)H - (h/2)K$ has non-negative entries, and its product with J vanishes by (3.1), it is O, and (3.2) is proved.

<u>Theorem 3.2</u>. The adjacency matrix A satisfies the minimal equation

$$(A - a_0 I)(A^2 - a_0 A - nI) = O, \qquad a_0 = h - k \tag{3.4}$$

<u>Proof</u>: The relations $H + K = J - I$ and $J^2 = nJ$ imply

$$A^2 = (H - K)^2 = (J - I)^2 - 4HK \tag{3.5}$$

$$= (n - 2)J + I - 2kH - 2hk$$

$$= a_0 A + (n - 2)J + I - (h + k)(H + K)$$

$$A^2 = a_0 A + nI - J \tag{3.6}$$

Since $AJ = a_0 J$, by (3.1), equation (3.4) follows from (3.6). Q.E.D.

Hence A has three eigenvalues a_0, a_1, a_2, such that

$$a_1 + a_2 = a_0 = h - k, \quad -a_1 a_2 = n = h + k + 1 \tag{3.7}$$

Let $E_0 = J/n$, E_1, and E_2 denote the constituent idempotent matrices of A. The rank and trace of E_i is the multiplicity of the corresponding eigenvalue a_i of A, h_i of H, or k_i of K, and it is the dimension d_i of one of the three irreducible constituents D^i of the rank 3 permutation representation P of the automorphism group $G = O^s(2m, 2)$ of the orthogonal graph O^s_{2m}.

Since $J = nE_0 + O \cdot E_1 + O \cdot E_2$, the eigenvalues h_i of $H = (J - I + A)/2$ and k_i of $K = (J - I - A)/2$ for E_1 and E_2 are

$$h_i = (-1 + a_i)/2, \quad k_i = (-1 - a_i)/2, \quad k = 1,2 \tag{3.8}$$

The values of $h = h_0$ and $k = k_0$ are found from (3.7):

$$h = -(1-a_1)(1-a_2)/2 = -2h_1 h_2 \tag{3.9}$$

$$k = -(1+a_1)(1+a_2)/2 = -2k_1 k_2$$

Since the traces of A^2, A, I are $n(n-1), 0, n$ respectively, the idempotent matrix E_2 and its trace d_2 are

$$E_2 = \frac{(A-a_0I)(A-a_1I)}{(a_2-a_0)(a_2-a_1)}, \quad d_2 = \frac{n(n-1+a_0a_1)}{(a_2-a_0)(a_2-a_1)} \tag{3.10}$$

Setting $a_0 = a_1 + a_2$ and $n = -a_1a_2$ from (3.7) yields

$$d_2 = a_2(-1+a_1^2)/(a_2-a_1) = (na_1+a_2)/(a_1-a_2) \tag{3.11}$$

$$d_1 = (na_2 + a_1)/(a_2 - a_1) \tag{3.12}$$

We denote by d the differences

$$d = (a_2 - a_1)/2 = h_2 - h_1 = k_1 - k_2 = h_1k_2 - h_2k_1 \tag{3.13}$$

Values of the parameters a_i, h_i, k_i, d for the orthogonal groups are found as follows. We set $M = 2^{m-1}s$, $s = +1$ or -1. Then among the $2^{2m} = 4M^2$ vectors x in the vector space $V = V(2m,2)$ there are $n+1 = 2M^2+M$ vectors including 0 for which $Q_m^s(x) = 0$. Hence the graph O_{2m}^s has $n = (2M-1)(M+1)$ vertices $\neq 0$. Each is non-adjacent to $h = M^2$ vertices and adjacent to $k = (M+2)(M-1)$. The two factors a_i of $-n$ with sum $a_0 = h - k = 2 - M$ are

$$a_1 = 1+M, \quad a_2 = 1 - 2M, \quad M = 2^{m-1}s, \quad s = \pm 1 \tag{3.14}$$

From (3.8) we obtain h_i and k_i:

$$h_1 = M/2, \quad h_2 = -M, \quad k_1 = -M/2 - 1, \quad k_2 = M - 1 \tag{3.15}$$

The change of basis from I, H, K to E_0, E_1, E_2 is expressed by a 3×3 matrix

$$S = \begin{bmatrix} 1 & h & k \\ 1 & h_1 & k_1 \\ 1 & h_2 & k_2 \end{bmatrix} = \begin{bmatrix} 1 & M^2 & (M+2)(M-1) \\ 1 & M/2 & -(M+2)/2 \\ 1 & -M & M-1 \end{bmatrix} \tag{3.16}$$

such that

$$[I,H,K] = [E_0,E_1,E_2][S \times I] \tag{3.17}$$

Adding the first two columns of S to the third yields the column vector $[n,0,0]^T$, so

$$\det S = n(h_2-h_1) = nd \tag{3.18}$$

The eigenvalue vector $[a_0,a_1,a_2]^T$ for A is the difference between columns 2 and 3 of S. Clearly (3.17) implies

$$[E_0,E_1,E_2] = [I,H,K][S^{-1} \times I] \tag{3.19}$$

but we shall compute S^{-1} in another way in terms of S^T.

<u>Theorem 3.3</u>. The transpose and inverse of the matrix S in (3.16) are related by the equations

$$\text{diag}\{1,h,k\}\; nS^{-1} = S^T\; \text{diag}\{1,d_1,d_2\} \tag{3.20}$$

<u>Proof</u>: We have $n\,E_0 = J = I + H + K$, and by (3.10) and (3.6),

$$\begin{aligned} n\,E_2/d_2 &= (A-a_0I)(A-a_1I)/(n-1+a_0a_1) \qquad (3.21)\\ &= (n\,I-J-a_1A+a_0a_1I)/(-1+a_1^2)\\ &= I + (H+K+a_1H-a_1K)/(1-a_1^2)\\ &= I + H/(1-a_1) + K/(1+a_1)\end{aligned}$$

$$\begin{aligned} n\,E_2/d_2 &= I + h_2H/h + k_2K/k \qquad (3.22)\\ n\,E_1/d_1 &= I + h_1H/h + k_1K/k\end{aligned}$$

$$n[E_0,E_1,E_2] = [I,H,K]\begin{bmatrix} I & d_1I & d_2I \\ I & d_1h_1I/h & d_2h_2I/h \\ I & d_1k_1I/k & d_2k_2I/h \end{bmatrix} \tag{3.23}$$

Equations (3.16), (3.19) and (3.23) imply (3.20). Q.E.D.

We repeat a theorem proved in 1937 [1].

Theorem 3.4. If $\det S = nd$, $d = h_2 - h_1$, then

$$n\,h\,k/d_1 d_2 = d^2 \tag{3.24}$$

Proof: Multiply (3.20) by S, take determinants and divide by $n^2 d_1 d_2$.

For the orthogonal groups we have $d = h_2 - h_1 = -3M/2$.

$$n\,S^{-1} = \begin{bmatrix} 1 & d_1 & d_2 \\ 1 & d_1 h_1/h & d_2 h_2/k \\ 1 & d_1 k_1/k & d_2 k_2/k \end{bmatrix} = \begin{bmatrix} 1 & 4(M^2-1)/3 & (2M-1)(M+2)/3 \\ 1 & 2(M^2-1)/3M & -(2M-1)(M+2)/3M \\ 1 & -2(M+1)/3 & (2M-1)/3 \end{bmatrix} \tag{3.25}$$

We verify that $d_1 d_2 d^2 = n\,h\,k = (M^2-1)(2M-1)(M+2)M^2$.

4. CLASS SUMS AND IRREDUCIBLE CHARACTERS

The commuting ring R of the rank 3 permutation representation P of the automorphism group G of the orthogonal graph O^s_{2m} is determined by the adjacency matrix A. It is spanned by I, A, A^2 or by I, H, K or by E_0, E_1, E_2. The sum of the g_λ elements in class C_λ of G commutes with each element of G, so its matrix $P_\lambda = P(C_\lambda)$ belongs to R. If g is an element of C_λ the trace χ_λ of $P(g)$ is the number of vertices fixed by g, and P_λ has trace $g_\lambda \chi_\lambda$. Let $g \in C_\lambda$ map h^*_λ vertices each into a non-adjacent vertex and k^*_λ vertices into an adjacent vertex. Then

$$n = \chi_\lambda + h^*_\lambda + k^*_\lambda = 1 + h + k \tag{4.1}$$

$$P_\lambda = g_\lambda[\chi_\lambda I/n + h^*_\lambda H/nh + k^*_\lambda K/nk] \tag{4.2}$$

since $I, H,$ and K have respectively n, $n\,h$, and $n\,k$ entries equal to 1.

We now seek expressions for the values χ_λ^i for class C_λ of the characters χ^i of the irreducible constituents D^i of the permutation representation P of G_1, in terms of the graph adjacency relations determined by the double coset coefficients h_λ^* and k_λ^*, [2].

<u>Theorem 4.1</u>. The irreducible characters χ_λ^i are given by

$$[1,\chi_\lambda^{(1)},\chi_\lambda^{(2)}] = [\chi_\lambda,h_\lambda^*,k_\lambda^*]S^{-1} \tag{4.3}$$

where the matrix nS^{-1} is defined in (3.25).

<u>Proof</u>: The coefficients of E_i in the expansion of P_λ are the class multipliers $g_\lambda\chi_\lambda^i/d_i$. Hence by (4.2)

$$\begin{aligned} P_\lambda/g_\lambda &= [I,I\chi_\lambda^{(1)}/d_1,I\chi_\lambda^{(2)}/d_2][E_0,E_1,E_2]^T \\ &= n^{-1}[I\chi_\lambda, Ih_\lambda^*/h,I\,k_\lambda^*/k][I,H,K]^T \end{aligned} \tag{4.4}$$

Expressing $[I,H,K]$ in terms of the E_i by (3.17) yields

$$[1,\chi_\lambda^{(1)}/d_1,\chi_\lambda^{(2)}/d_2] = n^{-1}[\chi_\lambda,h_\lambda^*/h,k_\lambda^*/k]S^T \tag{4.5}$$

When we multiply both sides by $\mathrm{diag}[1,d_1,d_2]$ and apply formula (3.20) of Theorem 3.3, equation (4.3) results.

<u>Theorem 4.2</u>. The irreducible characters χ_λ^i can be expressed in terms of either of the two adjacency numbers h_λ^* or k_λ^* by the formulas

$$-\chi_\lambda^{(1)}d = h_2(a_1-\chi_\lambda) + h_\lambda^* = k_2(a_1+\chi_\lambda) - k_\lambda^* \tag{4.6a}$$

$$\chi_\lambda^{(2)}d = h_1(a_2-\chi_\lambda) + h_\lambda^* = k_1(a_2+\chi_\lambda) - k_\lambda^* \tag{4.6b}$$

<u>Proof</u>: We first eliminate $\chi_\lambda^{(2)}$ and k_λ^* from (4.3) by multiplying on the right by $[-d_1k_1/k,1,0]^T$. Applying (3.25) we then obtain

$$-d_1k_1/k + \chi_\lambda^{(1)} = [\chi_\lambda,h_\lambda^*,k_\lambda^*][1-k_1/k,h_1/h-k_1/k_1,0]^T d_1/n \tag{4.7}$$

From (3.8), (3,9) and (3.11) we obtain the relations

$$k_1/k = 1/(1+a_2),\ h_1/h = 1/(1-a_2),\ dd_1 = a_1(1-a_2^2)/2 \qquad (4.8)$$

where $d = (a_2-a_1)/2$. We multiply (4.7) by d and obtain

$$a_1h_2 + \chi_\lambda^{(1)} d = [\chi_\lambda, h_\lambda^*][-a_1a_2h_2, a_1a_2]^T/n \qquad (4.9)$$

Since $n = -a_1a_2$ and $d = a_2 - a_1$, equation (4.9) reduces to the first equality in (4.6a). To interchange the roles of h_λ^* and k_λ^* we note that $A = H - K$, so we change signs of a_i and interchange h's and k's. To obtain (4.6b) from (4.6a) we interchange subscripts 1 and 2.

Example. We illustrate Theorem 4.2 by computing the irreducible characters χ_3^i for the class C_3 of the group $G = O^-(2m,2)$ containing the matrix

$$P(g) = \begin{bmatrix} 0 & 1 \\ 1 & 1 \end{bmatrix} \oplus I_{2m-2} \qquad (4.10)$$

which fixes the $\chi_3 = (M+1)(M-2)/2$ row vectors $(0,0,X)$ of O_{2m}^-, and permutes cyclically the $(M+1)M/2$ triples of mutually non-adjacent row vectors $(1,0,Y),(0,1,Y),(1,1,Y)$. Here $h_\lambda^* = 3(M+1)M/2$ and $k_\lambda^* = 0$, so we use the second parts of equations (4.6) with $d = -3M/2$, $a_1 = 1 + M$, $a_2 = 1 - 2M$, $k_1 = -(M+2)/2$, $k_2 = M - 1$, and obtain

$$\chi_3^{(1)} = -k_2(a_1+\chi_3)/d = -(M-1)(M+1)M/2d = (M^2-1)/3 \qquad (4.11a)$$

$$\chi_3^{(2)} = k_1(a_2+\chi_3)/d = -(M+2)(M-5)M/4d = (M+2)(M-5)/6 \qquad (4.11b)$$

We check that $\chi_3 = 1 + \chi_3^{(1)} + \chi_3^{(2)}$. For $G = O^-(8,2)$ we have $M = -8$, $n = 119$, $\chi_3 = 35$, $h_3^* = 84$, $k_3^* = 0$, $d_1 = 84$, $\chi_3^{(1)} = 21$, $d_2 = 34$, $\chi_3^{(2)} = 13$.

REFERENCES

1. Frame, J. S., "The Degrees of the Irreducible Representations of Simply Transitive Permutation Groups," Duke Math. Journal 3, (1937), 8-17.

2. Frame, J. S., "Group Decomposition by Double Coset Matrices," Bull. Amer. Math. Soc. 54, (1948) 740-755.

3. Seidel, J. S., "On Two-Graphs and Shult's Characterization of Symplectic and orthogonal Geometries Over GF(2)," T.H. - Report 73-WSK-02, Technological University, Eindhoven, The Netherlands.

DISTANCE MATRIX POLYNOMIALS OF TREES

R. L. Graham
Bell Laboratories
Murray Hill, New Jersey

and

L. Lovász
Eötvös Lorand University
Budapest, HUNGARY

Abstract

For a finite undirected tree T with n vertices, the distance matrix $D(T)$ of T is defined to be the n by n symmetric matrix whose (i, j) entry is d_{ij}, the number of edges in the unique path from i to j. Denote the characteristic polynomial of $D(T)$ by

$$\Delta_T(x) = \det (D(T)-xI) = \sum_{i=0}^{n} \delta_k(T)x^k .$$

In this talk we describe exactly how the coefficients $\delta_k(T)$ depend on the structure of T. In contrast to the corresponding problem for the adjacency matrix of T, the results here are surprisingly difficult, requiring the use of a number of interesting auxiliary results.

1. For a finite undirected tree T with n vertices, the distance matrix $D(T)$ of T is defined to be the n by n symmetric matrix whose (i, j) entry is d_{ij}, the number of edges in the unique path from i to j. Denote the characteristic polynomial of $D(T)$ by

(1) $$\Delta_T(x) = \det(D(T)-xI) = \sum_{k=0}^{n} \delta_k(T)x^k .$$

In [1], the problem of relating the values of the coefficients $\delta_k(T)$ to the structural properties of T was initiated. In this note, we announce a complete solution to this problem. Proof of the assertions are rather complicated and will be given elsewhere.

For an acyclic graph (i.e., forest) F, let $N_F(T)$ denote the number of subgraphs of T which are isomorphic to F.

Theorem 1. There exist unique coefficients $A_F^{(k)}$ depending only on F and k so that for all trees T with n vertices,

(2) $$\delta_k(T) = (-2)^n \sum_F A_F^{(k)} N_F(T)$$

where F ranges over all forests having no isolated points and at most $k+1$ edges.

The uniqueness of the $A_F^{(k)}$ follows from a recent result in [4]. This may be compared with the corresponding theorem for the adjacency matrix characteristic polynomial

$$\det(A(T)-xI) = \sum_{k=0}^{n} \alpha_k(T)x^k$$

where $A(T) = (a_{ij})$ with

$$a_{ij} = \begin{cases} 1 & \text{if } d_{ij} = 1, \\ 0 & \text{otherwise.} \end{cases}$$

In this case, it is well known [6] that

$$\alpha_k(T) = \begin{cases} (-1)^t N_{tP_1}(T) & \text{if } k = n-2t, \\ 0 & \text{otherwise.} \end{cases}$$

where tP_1 denotes the forest consisting of t disjoint edges.

The exact values of the $A_F^{(k)}$ are given as follows. For a forest F which is the disjoint union of trees having $n_1, n_2, \ldots,$ and n_r vertices, define $\pi(F)$ to be the integer $n_1 n_2 \ldots n_r$. Let $\mathfrak{F}_n$ denote the set of forests having no isolated vertices and exactly k edges. By convention, $\mathfrak{F}_0$ will consist of only the empty forest F^* having no vertices and $\pi(F^*) = N_{F^*}(T) = 1$ for all T.

<u>Theorem 2</u>. For all trees T with n vertices

$$\delta_k(T) = (-1)^{n-1} 2^{n-k-2} \left\{ \sum_{F \in \mathfrak{F}_{k+1}} a_F \pi(F) N_F(T) + \sum_{F \in \mathfrak{F}_k} b_F \pi(F) N_F(T) + \sum_{F \in \mathfrak{F}_{k-1}} c_F \pi(F) N_F(T) \right\}$$

where a_F, b_F, c_F are recursively determined as follows:

(i) $a_{F*} = b_{F*} = 0$, $c_{F*} = -4$

(ii) If F is a <u>tree</u> T' with $n' \geq 1$ vertices and distance matrix (d'_{ij}) then

$$a_{T'} = \frac{1}{n'} \sum_{i<j} d'_{ij}(2-d'_{ij})$$

$$b_{T'} = \frac{4}{n'} \sum_{i<j} (2-d'_{ij})$$

$$c_{T'} = \frac{-4}{n'}$$

(iii) If F is the disjoint union of forests F_1 and F_2 then

$$a_F = a_{F_1} + a_{F_2}$$

$$b_F = b_{F_1} + b_{F_2}$$

$$c_F = c_{F_1} + c_{F_2} + 4$$

As an immediate corollary, taking $k = 0$ we have

$\mathfrak{F}_1 = \{P_1\}$, $a_{P_1} = \frac{1}{2}$, $b_{P_1} = 2$, $c_{P_1} = -2$, $\pi(P_1) = 2$ and

$$(3) \quad \delta_0(T) = (-1)^{n-1}2^{n-2}(\tfrac{1}{2}\cdot 2\cdot N_{P_1}(T) + 0) = (-1)^{n-1}(n-1)2^{n-2}$$

independent of the structure of T (see [2], [3]).

An interesting result used in the proof of Theorem 1 is the following.

Lemma. The inverse $D(T)^{-1} = (d^*_{ij})$ of $D(T)$ is given by

$$d^*_{ij} = \frac{(2-d_i)(2-d_j)}{2(n-1)} + \begin{cases} +\frac{1}{2}a_{ij} & \text{if } i \neq j \\ -\frac{1}{2}d_i & \text{if } i = j \end{cases}$$

where d_i denotes the degree of ith vertex of T.

It is natural to expect that similar results hold for general connected graphs G. It is known [5], for example, that if G has blocks G_j, $j \in J$, then letting cof X denote the sum of the cofactors of a matrix X we have

$$(a) \quad \text{cof } D(G) = \prod_{j \in J} \text{cof } D(G_j)$$

$$\text{(b)}\quad \det\ D(G) = \sum_{j\in J} \det D(G_j) \prod_{i\neq j} \operatorname{cof}\ D(G_i)$$

Of course, (3) follows at once from (a) and (b) since all blocks of a tree are isomorphic to P_1 .

It is still not known whether $\Delta_T(x)$ determines T . This is not the case for general graphs G (see [1]) and almost never the case for $\det(A(T)-xI)$ (see [7]) .

REFERENCES

1. M. Edelberg, M. R. Garey and R. L. Graham. On the distance matrix of a tree, Discrete Math, 14 (1976), 23-29.

2. R. L. Graham and H. O. Pollak, On the addressing problem for loop switching, Bell Sys. Tech. Jour. 50 (1971), 2495-2519.

3. R. L. Graham and H. O. Pollak, Embedding graphs in squashed cubes, Springer Lecture Notes in Math., Vol. 303 (1973), 99-110.

4. R. L. Graham and E. Szemerédi, On subgraph number independence in trees, Jour. Comb. Th. (to appear).

5. R. L. Graham, A. J. Hoffman and H. Hosoya, On the distance matrix of a directed graph (to appear).

6. A. Moshowitz, The characteristic polynomial of a graph, Jour. Comb. Th. (B), 12 (1972), 177-193.

7. Allen. J. Schwenk, Almost all trees are cospectral, New Directions in the Theory of Graphs, Acad. Press, New York (1973), 275-307.

ODD CYCLES AND PERFECT GRAPHS

Don Greenwell
Marshall Space Flight Center
Huntsville, ALA. 35812

Abstract

A simple proof (using theorems of Hajnal and Tutte) of a characterization of odd cycles due to Melnikov and Vising is given.

1. Introduction. The purpose of this paper is to present a simple proof of the following characterization of odd cycles, first conjectured by Toft [5] and later proved by Melnikov and Vising [4]:

Theorem 1: A graph G is an odd cycle if and only if $|V(G)| = 2n+1$, $\alpha(G) = n$ and $\alpha(G - v - w) = n$ for all $v, w \in V(G)$.

Theorem 1 also gives a characterization of complements of odd cycles by replacing the independence number with its complementary notion, the clique number. It seems therefore that this characterization might be a useful tool in attacking the strong perfect graph conjecture.

A graph is perfect if the clique number and chromatic number, of each induced subgraph, are equal. The Berge conjecture (strong perfect graph conjecture) is that odd cycles and complements of odd cycles are the only irreducible forbidden induced subgraphs of perfect graphs. Lovász [3] proved the following characterization of perfect graphs:

Theorem 2: A graph G is perfect if and only if $\alpha(G')\cdot\omega(G') \geq |V(G')|$ for every induced subgraph G' of G.

It is easy to see from Theorem 2 that an irreducible non-perfect graph G has $\alpha(G)\cdot\omega(G) + 1$ vertices and that removing $\omega(G)$ vertices from G will not lower the independence number. Thus if $\omega(G) = 2$, Theorem 1 will give that G is an odd cycle, and if $\alpha(G) = 2$ we see that G is the complement of an odd cycle.

2. Proof of Theorem 1. A graph is α-critical if $\alpha(G - e) > \alpha(G)$ for every edge e. Hajnal [2] (see also [1, p. 290]) proved:

Theorem 3: In any α-critical graph G, without isolated vertices, every independent set S satisfies $|\Gamma(S)| \geq |S|$, where $\Gamma(S)$ is the set of vertices of G adjacent to vertices of S.

Tutte [6] has shown that the above condition is equivalent to the graph having a certain spanning subgraph.

Theorem 4: Every independent set S of G satisfies $|\Gamma(S)| \geq |S|$ if and only if G contains a spanning subgraph whose connected components are odd cycles or complete graphs on two vertices.

The proof now follows: Let G be a graph with $2n+1$ vertices and with $\alpha(G) = \alpha(G - v - w) = n$ for any $v, w \in V(G)$. Let G' be an α-critical subgraph of G. If v is an isolated vertex of G' we would have $n - 1 = \alpha(G' - v) \geq \alpha(G - v) = n$ which is impossible. Therefore Theorems 3 and 4 can be used to obtain a spanning subgraph G'' of G whose components are odd cycles or complete graphs on two vertices. Since G has an odd number of vertices there must be at least one odd cycle in G''. If G'' contained more than one odd cycle it would have to contain at least three of them. But if G'' contained three odd cycles it could not contain n independent vertices which is impossible since $\alpha(G'') \geq \alpha(G) = n$. Thus G'' contains exactly one odd cycle. If G'' contains a complete graph on two vertices v and w as a component then $\alpha(G) = \alpha(G - v - w) \leq \alpha(G'' - v - w) = n - 1$ which is impossible. Thus G'' is just an odd cycle. If there is an edge e of G not in G'' then e will split G'' into an even cycle and an odd cycle. Let v and w be the two vertices on the even cycle which are adjacent to the end vertices of e. We would have $\alpha(G - v - w) < n$ which is impossible. Thus G is an odd cycle.

The author would like to thank L. Lovász for many useful discussions concerning this problem.

REFERENCE

1. Berge, C., Graphs and Hypergraphs, North Holland, (1973).

2. Hajnal, A., A theorem on k-saturated graphs, Canad. Math. J., 17(1965), 720-724.

3. Lovász, L., A characterization of perfect graphs, J. Combinatorial Theory, 13(1972), 95-98.

4. Melnikov, L.S., V.G. Vising, Solution of Toft's problem, Diskret. Analiz., 19(1971), 11-14.

5. Toft, A., Combinatorial Theory and its Applications III, (P. Erdos, A. Renyi and Vera T. Sos, eds.) North Holland Publishing Company, 1970, 1193.

6. Tutte, W.T., The 1-factors of oriented graphs, Proc. Amer. Math. Soc., 4(1953), 922-931.

IMBEDDINGS OF METACYCLIC CAYLEY GRAPHS

Jonathan L. Gross *
Columbia University
New York, N.Y. 10027

Abstract

Any group whose commutator subgroup and commutator quotient group are both cyclic is called metacyclic. The Cayley graph corresponding to any generating set for a metacyclic group is called a metacyclic Cayley graph. Whereas the earliest work on Cayley graph imbeddings concentrates mainly on planarity, recent work of A.T. White and others shows that higher genus imbeddings are accessible, especially for abelian groups. Since metacyclic groups are an especially tractable kind of nonabelian groups, this paper begins a development of an imbedding theory for non-abelian Cayley graphs by considering the metacyclic Cayley graphs. The author has previously used voltage graphs (the duals of the current graphs of Gustin, Ringel, and Youngs) to show that a special subclass of metacyclic groups has toroidal Cayley graphs. In this paper, toroidal Cayley graphs are constructed for two additional subclasses of metacyclic groups.

*The author is an Alfred P. Sloan Fellow. His research was partially supported by NSF Contract MPS74-05481-A01 at Columbia University.

1. Introduction

Let G be a group and X a set of generators for G. The (right) Cayley color graph for G and X has the elements of the group G as its set of vertices. For each generator x and each group element g there is an edge (x,g) from the vertex g to the vertex gx. Thus, the outdegree of each vertex (also the indegree) is the cardinality of the generating set X. The underlying graph is simply called a Cayley graph.

Let H be a subgroup of G. The (right) Schreier coset graph for G, H, and X has as its set of vertices the right cosets of H in G. An edge (H_i, x) runs from the coset H_i to the coset $H_i x$. Thus, a Schreier coset graph is a generalization of a Cayley graph. The underlying graph for a Schreier coset graph is called a Schreier graph.

From their algebraic description one may correctly anticipate that Cayley graphs and Schreier graphs admit automorphism groups that may be useful in determining their graph-theoretic properties. Gross and Tucker [10] give necessary and sufficient topological conditions for a graph to be a Cayley graph or a Schreier graph that may also be useful in the determination of graph-theoretic properties.

Nearly all known graph imbeddings either have been or can be obtained using Cayley graphs. Sometimes, adjacency modifications are needed, as in Ringel [15] or in Gross [6]. Many interesting graphs, such as the Petersen graph, are Schreier graphs but not Cayley graphs, so attention toward Schreier graphs would also be appropriate. The following result of Gross [7] indicates the generality of Schreier graphs.

Theorem. Every connected regular graph of every degree is a Schreier graph.

An odd degree regular graph with a 1-factor, e.g. the Petersen graph, is also a Schreier graph, if the doubling convention is permitted.

The graphs in this paper often have multiple edges and self-adjacencies. The above theorem holds for such graphs.

The author wishes to thank S.J. Lomonaco for several helpful conversations regarding metacyclic groups.

2. Metacyclic groups

A group is called metacyclic if both its commutator subgroup and its commutator quotient group are cyclic. Zassenhaus [20,p.174] proves that every metacyclic group has a presentation of the form

$$S, T : S^m = T^n = E, T^{-1}ST = S^r$$

where $r^n \equiv 1$ modulo m and $\gcd(r-1,m) = 1$. The group so presented is denoted ((m,n,r)).

Using the relation $T^{-1}ST = S^r$ repeatedly, it is possible to rewrite every commutator as a power of the generator S in the group ((m,n,r)), from which it follows that the commutator subgroup is cyclic. The commutator quotient group is cyclic of order n because $\gcd(r-1,m) = 1$ and $S^{-1}T^{-1}ST = S^{r-1}$. For additional discussion, the reader is referred to Coxeter and Moser [4].

Any Cayley graph for a metacyclic group is called a metacyclic Cayley graph. The present reason for interest in metacyclic Cayley graphs is that metacyclic groups are among the best understood of the nonabelian groups.

Since $ST = TS^r$ in the metacyclic group ((m,n,r)), it is possible to rewrite any word in the generators S and T as an equivalent word of the form T^jS^i, where $0 \leq j < n$ and $0 \leq i < m$.

3. Voltage graphs

Recognizing that the current graph constructions of Ringel and Youngs were combinatorial equivalents of the topological constructions called branched covering spaces, Gross and Alpert [8,2] unified and generalized the theory. Later, Gross [5] introduced the terminology voltage graph for the dual of a current graph. For the same reasons that topologists study both homology and cohomology, it is useful to have both current graph theory and voltage graph theory available for combinatorialists.

Imbeddings of metacyclic Cayley graphs are constructed here using voltage graphs. Gross and Tucker [10,11] explain this construction in somewhat greater detail than [5], whose understanding depends on prior acquaintance with current graphs.

A branched covering space is simply a topological abstraction of a Riemann surface. Alexander [1] proves that every orientable closed surface is a branched covering space of the sphere. Figure 1 shows an imbedding of a Cayley graph for the metacyclic group ((7,3,2)) in the torus is obtained from a voltage graph in the sphere.

The generators used for the Cayley graph of Figure 1 are T and TS, both of which have order three in the group ((7,3,2)). It follows that the monogons of the voltage graph imbedding with net voltages T and TS on their respective boundaries are covered by triangles in the derived imbedding. Since the exterior digon of

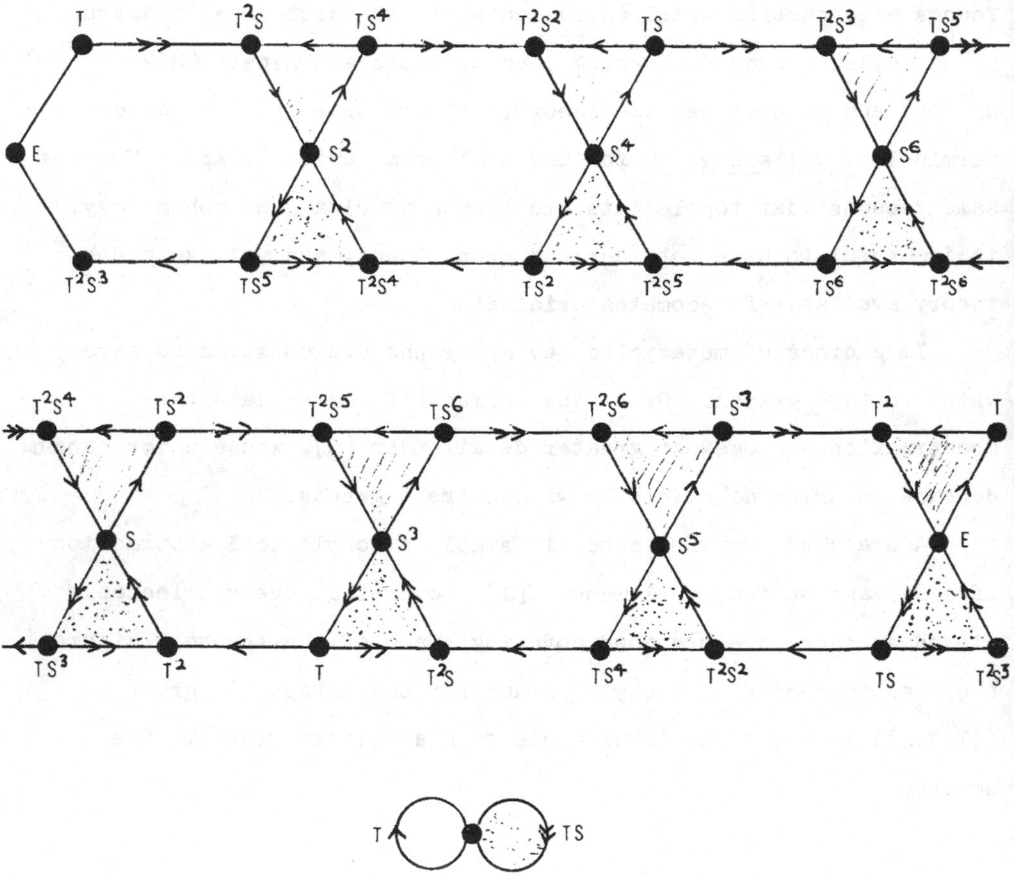

Figure 1. A toroidal imbedding of a Cayley graph for the metacyclic ((7.3.2)) obtained from a voltage graph in the sphere.

the voltage graph imbedding has net voltage T^2S on its boundary, and the group element T^2S also has order 3, that digon is covered by hexagons. These are special instances of the following rule.

General Principle. An s-sided polygon of the imbedded voltage graph with net voltage of order w in the group G lifts to $|G|/w$ ws-sided polygons in the derived imbedding.

4. On the genus of a group

The genus of a group is the smallest genus of any Cayley graph for that group. Recent interest in this subject is largely due to White [18, 19], who obtained a general upper bound for the genus of a group, the genus of cartesian products of even order cyclic groups, the genus of dicyclic groups, and more.

Some other highlights of this subject are the classification by Maschke [14] of all planar finite groups, the proof by Levinson [13] that the genus of an infinite group is either zero or infinite, the proof by Babai [3] that the genus of a group cannot be exceeded by the genus of any of its subgroups, and the computation by Jungerman and White [12] of the genus of a substantial majority of the finite abelian groups. A related result of Stahl [16] is that every finite abelian group of order four or more has a self-dual orientable imbedding. Such self-dual imbeddings do not ordinarily realize the genus of the group.

Special attention has been directed toward nonabelian groups by Terry, Welch, and Youngs [17] who used them on Case 0 of the Heawood problem and by Himelwright [11a], who calculated the genus of certain Hamiltonian groups.

5. Two new classes of toroidal metacyclic groups

It is proved by Gross [5] that every metacyclic group $((m,n,-1))$ has genus one. Again we use Maschke's classification of finite planar groups for assurance of nonplanarity. The voltage graphs in Figure 2 prove that every metacyclic group $((m,3,r)$ and every metacyclic group $((m,6,r))$ is toroidal.

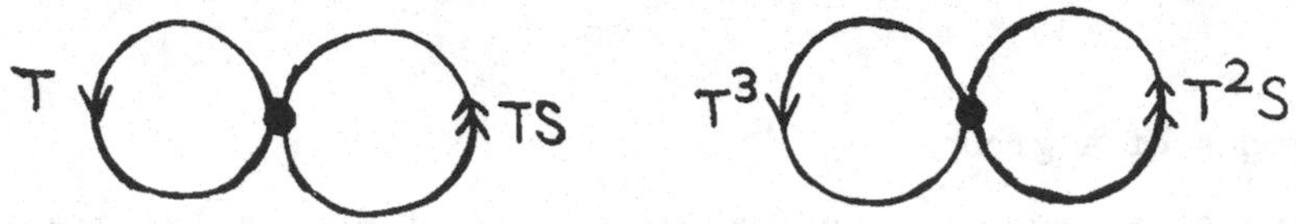

Figure 2. The voltage graph on the left yields a toroidal imbedding for $((m,3,r))$, the voltage graph on the right for $((m,6,r))$.

The covering graphs resulting from the lefthand and righthand voltage graphs in Figure 2 both have mn vertices and $2mn$ edges. Since the net boundary voltages T, TS, and T^2S all have order three in the metacyclic group $((m,3,r))$, the lefthand derived imbedding has

$$mn/3 + mn/3 + mn/3 = mn$$

faces, yielding an orientable surface of euler characteristic

$$mn - 2mn + mn = 0$$

that is, the torus. Since the net boundary voltages T^3, T^2S, and T^5S have orders 2, 3, and 6, respectively, in the group $((m,6,r))$, the righthand derived imbedding has

$$mn/2 - mn/3 + mn/6 = mn$$

faces, yielding another surface of characteristic zero. The number of faces in both instances is computed using the General Principle of Section 3.

Many metacyclic groups have genus greater than one. The basic trick in proving such a statement is obtaining a lower bound on the girth or at least an upper bound on the number of small circuits.

References

1. J.W. Alexander, Note on Riemann spaces, Bull. Amer. Math. Soc. 26 (1920), 370-372.

2. S.R. Alpert and J.L. Gross, Components of branched coverings of a current graph, J. Combinatorial Theory, to appear in 1976.

3. L. Babai, Chromatic number and subgraphs of Cayley graphs, These Conference Proceedings.

4. H.S.M. Coxeter and W.O.J. Moser, Generators and Relations for Discrete Groups, 3rd ed., Ergebnisse der Mathematik und ihrer Grenzgebiete, Vol. 14 (Springer, Berlin, 1972).

5. J.L. Gross, Voltage graphs, Discrete Math. 9 (1974), 239-246.

6. J.L. Gross, The genus of nearly complete graphs - Case 6, Aequationes Math. 13 (1975), 243-249.

7. J.L. Gross, Every connected regular graph of even degree is a Schreier coset graph, J. Combinatorial Theory, to appear.

8. J.L. Gross and S.R. Alpert, The topological theory of current graphs, J. Combinatorial Theory 17 (1974), 218-233.

9. J.L. Gross and T.W. Tucker, Quotients of complete graphs: revisiting the Heawood map-coloring problem, Pacific J. Math. 55 (1974), 391-402.

10. J.L. Gross and T.W. Tucker, Generating all graph coverings by permutation voltage assignments, to appear.

11. W. Gustin, Orientable embeddings of Cayley graphs, Bull. Amer. Math. Soc. 69 (1963), 272-275.

11a. P. Himelwright, On the genus of Hamiltonian groups, Specialist Thesis, Western Michigan Univ. 1972.

12. M. Jungerman and A.T. White, On the genus of finite abelian groups, Abstract No. 9, Graph Theory Newsletter May, 1976.

13. H. Levinson, On the genera of graphs of group presentations, Ann. New York Acad. Sci. 175 (1970), 277-284.

14. W. Maschke, The representation of finite groups, Amer. J. Math. 18 (1896), 156-194.

15. G. Ringel, Map Color Theorem, Springer, New York, 1974.

16. S. Stahl, Self-dual embeddings of graphs, Ph.D. Dissertation, Western Michigan Univ. 1975.

17. C.M. Terry, L.R. Welch, and J.W.T. Youngs, The genus of K_{12s}, J. Combinatorial Theory 2 (1967), 43-60.

18. A.T. White, On the genus of a group, _Trans. Amer. Math. Soc._ 173 (1972), 203-214.

19. A.T. White, _Graphs, Groups and Surfaces_, North-Holland, Amsterdam, 1973.

20. H. Zassenhaus, _The Theory of Groups_ (2nd ed.), Chelsea, New York, 1958.

ON THE CHROMATIC INDEX AND THE COVER INDEX OF A MULTIGRAPH

Ram P. Gupta
Ohio State University
Columbus, OH 43210

Abstract

The problem of determining bounds for the chromatics index and the cover index of a multigraph is considered. Results stated include (I) a bound for the cover index which is analogous to the well-known bound for the chromatic index due to Shannon, (II) a bound for the chromatic index (and analogous bound for the cover index) which establishes conjectures by Berge and Bosak.

1. Let G be a <u>multigraph</u> (undirected, finite, without loops) with set of vertices $V(G)$ and set of edges $E(G)$. Let $F \subseteq E(G)$. The set of edges F is called a <u>matching</u> [<u>cover</u>, resp.] if for any $x \in V(G)$, F contains at most [at least, resp.] one edge incident with x. The <u>chromatic index</u> of G, denoted by $\mu(G)$, is the minimum number k such that $E(G)$ can be expressed as a (disjoint) union of k matchings. The <u>cover index</u> of G, denoted by $\kappa(G)$, is the maximum number k such that $E(G)$ can be expressed as a disjoint union of k covers.

In this article, we consider the problem of determining bounds for the chromatic index $\mu(G)$ and the cover index $\kappa(G)$ in terms of other simpler parameters of G.

2. The <u>degree</u> of a vertex x, denoted by $d_G(x)$, is the number of edges incident with x. Let $\Delta(G) = \Delta = \max_x d_G(x)$ and $\delta(G) = \delta = \min_x d_G(x)$ be the <u>maximum degree</u> and the <u>minimum degree</u> in G respectively. Then, evidently, we must have

(1) $$\mu(G) \geq \Delta(G) \text{ and } \kappa(G) \leq \delta(G) .$$

3. Let $S \subseteq V(G)$. Let $E(S)$ denote the set of edges with both ends in S and $E'(S)$ denote the set of edges with at least one endpoint in S. If $|S| = 2r + 1$, where r is a positive integer, then it is obvious that any matching of G can contain at most r edges from $E(S)$ and any cover of G must contain at least $r + 1$ edges from $E'(S)$. We introduce the following parameters for G:

<u>Definition</u>: $$t_r(G) = \max_{\substack{S \subseteq V(G) \\ |S| = 2r+1}} \left[\frac{|E(S)|}{r}\right]^* , \text{ and}$$

$$t_r^*(G) \quad \min_{\substack{S \subseteq V(G) \\ |S| = 2r+1}} \left[\frac{|E'(S)|}{r+1}\right] , \quad r = 1, 2, 3, \ldots$$

[Here, $|S|$ denotes the cardinality of the set S. For any number x, $[x]$ is the greatest integer p with $p \leq x$ and $[x]^*$ is the smallest integer q with $q \geq x$.]

From the remark preceding the definition, the following lower bounds for the chromatic index and upper bounds for the cover index are obvious:

"For any multigraph G,

(2) $\mu(G) \geq \max \{\Delta(G), t_1(G), t_2(G), \ldots, t_r(G)\}$, $r = 1, 2, \ldots$;

(3) $\kappa(G) \leq \min \{\delta(G), t_1^*(G), t_2^*(G), \ldots, t_r^*(G)\}$, $r = 1, 2, \ldots$.

4. In order to establish upper bounds for $\mu(G)$ and lower bounds for $\kappa(G)$, we first reformulate the problem in a more general way.

Let G be any multigraph. Let k be any nonnegative integer and let J_k denote a set of k distinct elements called 'colors'. Any mapping

$$\sigma : E(G) \to J_k$$

is called a k-coloration of G. If $\lambda \in E(G)$ and $\sigma(\lambda) = \alpha$ then λ is called an α-edge. For any $x \in V(G)$, let $\nu(x,\sigma)$ be the number of distinct colors α such that there exists at least one α-edge incident with x. Then, for any k-coloration σ of G, we have

(4) $\nu(x,\sigma) \leq \min \{k, d_G(x)\}$ for all $x \in V(G)$.

It may be observed that the chromatic index $\mu(G)$ can be defined as the minimum number k such that there exists a k-coloration σ of G satisfying $\nu(x,\sigma) = d_G(x)$ for all $x \in V(G)$, and the cover index $\kappa(G)$ can be defined as the maximum number k such that there exists a k-coloration σ of G satisfying $\nu(x,\sigma) = k$ for all $x \in V(G)$.

We formulate the following general problem: "Given a multigraph G and any nonnegative integer k, find a k-coloration σ of G for which $\sum_{x \in V(G)} \nu(x,\sigma)$ is maximum possible."

A more specific and apparently more useful problem is the following: "Determine conditions on G and/or k under which there exists a k-coloration σ of G satisfying

$$\nu(x,\sigma) = \min \{k, d_G(x)\} \text{ for all } x \in V(G)."$$

A few such conditions have already been obtained which yield important bounds for $\mu(G)$ and $\kappa(G)$. The following two propositions are given in [7].

Proposition 1: If G is bipartite, then for any nonnegative integer k, there exists a k-coloration σ of G satisfying

$$\nu(x,\sigma) = \min\{k, d_G(x)\} \quad \text{for all} \quad x \in V(G).$$

From the above proposition, we easily obtain (by taking $k = \Delta(G)$ and $k = \delta(G)$, respectively) the following two theorems.

Theorem 1A: If G is bipartite, then $\mu(G) = \Delta(G)$.

Theorem 1B: If G is bipartite, then $\kappa(G) = \delta(G)$.

Theorem 1A is a well-known result due to König [9] and Theorem 1B is due to the author [6].

For $x \in V(G)$, let $m_G(x)$ be the minimum number m such that x is joined to any other vertex in G by at most m edges. Let $m(G) = \max_x m_G(x)$. $m(G)$ is called the multiplicity of G. We have

Proposition 2: For any multigraph G and any nonnegative integer k, there exists a k-coloration σ of G satisfying:

$$\nu(x,\sigma) \geq \min\{k - m_G(x), d_G(x)\} \quad \text{if} \quad d_G(x) \leq k$$

$$\geq \min\{k, d_G(x) - m_G(x)\} \quad \text{if} \quad d_G(x) \geq k$$

for all $x \in V(G)$.

For even a stronger statement than Proposition 2, see [7, Theorem 2.2].

From Proposition 2, we easily deduce the following theorems:

Theorem 2A: For any multigraph G, $\mu(G) \leq \Delta(G) + m(G)$.

Theorem 2B: For any multigraph G, $\kappa(G) \geq \delta(G) - m(G)$.

Theorem 2A is well-known in literature as Vizing's Theorem [12] and was also proved independently by the author [3, Chapter 2], [4]. Theorem 2B is due to the author [3, Chapter 1], [5].

5. We state below some additional propositions and derive their consequences. For the sake of brevity, the proofs of these propositions, which are lengthy, are omitted here and will be published separately. First we introduce some further notations and terminology.

As above, G is any multigraph, k any nonnegative integer, J_k a set of k 'colors' and $\sigma : E(G) \to J_k$ any k-coloration of G. For any vertex $x \in V(G)$, let $C_r(x,\sigma)$ denote the set of colors α such that there are at least r α-edges incident with x, $r = 1, 2, 3$. Let $\overline{C}(x,\sigma) = J_k - C_1(x,\sigma)$. Finally, let $D(x,\sigma) = \overline{C}(x,\sigma) \cup C_2(x,\sigma)$. We may note that $\nu(x,\sigma) = |C_1(x,\sigma)|$.

The following lemma, required later, is quite trivial.

<u>Lemma</u> 1: <u>For any vertex</u> $x \in V(G)$,

(a) $|D(x,\sigma)| \geq |k - d_G(x)|$ <u>if</u> $C_3(x,\sigma) = \emptyset$; <u>and</u>

(b) $|D(x,\sigma)| \geq |k - d_G(x)| + 2$ <u>if</u> $C_3(x,\sigma) = \emptyset$ <u>and</u>

$\nu(x,\sigma) < \min\{k, d_G(x)\}$.

In the following, x_0 is any vertex such that $\nu(x_0,\sigma) < \min\{k, d_G(x_0)\}$. We say that σ is <u>improvable at</u> x_0 if there exists another k-coloration ρ of G such that $\nu(x_0,\rho) \geq \nu(x_0,\sigma) + 1$ and $\nu(x,\rho) \geq \nu(x,\sigma)$ for all $x \in V(G)$. Let α be any color such that $\alpha \in C_2(x_0,\sigma)$. [Note that our assumption on x_0 is equivalent to the assumption that $C_2(x_0,\sigma) \neq \emptyset$ and $\overline{C}(x_0,\sigma) \neq \emptyset$.] Let $\lambda_i = [x_0,x_i]$, $i = 1, 2$, be any two α-edges incident with x_0. The vertices x_1 and x_2 defined by λ_1 and λ_2 need not be distinct. Under these notations, we have

<u>Proposition</u> 3: <u>Either</u> (a) σ <u>is improvable at</u> x_0 ;

<u>or</u> (b) <u>the vertices</u> x_1 <u>and</u> x_2 <u>are distinct and</u>

(i) $D(x_0,\sigma)$, $D(x_1,\sigma)$, $D(x_2,\sigma)$ <u>are mutually disjoint, and</u>

(ii) $C_3(x_0,\sigma)$, $C_3(x_1,\sigma)$, $C_3(x_2,\sigma)$ <u>are empty.</u>

Since $D(x_i,\sigma) \subseteq J_k$, $i = 0, 1, 2,$ if condition (b) in the above proposition holds, then we get $\sum_{i=0}^{2} |D(x_i,\sigma)| \leq |J_k|$. Hence, from Proposition 3, using Lemma 1, we obtain:

Corollary 3: Either (a) σ is improvable at x_o ;

or (b) the vertices x_1 and x_2 are distinct and $\sum_{i=0}^{2} |k - d_G(x_i)| + 2 \leq k$.

We now derive some bounds for $\mu(G)$ and $\kappa(G)$ from Corollary 3. Let $S \subseteq V(G)$. S is called coherent if the subgraph of G induced by S is connected. We have:

Theorem 3A (O. Ore [10]): For any multigraph G ,

$$(5) \qquad \mu(G) \leq \max \{\Delta, \max [\frac{d_G(x) + d_G(y) + d_G(z)}{2}]\}$$

where the inner maximum is taken over all coherent, 3-subsets $\{x,y,z\} \subseteq V(G)$.

Proof: Let k be the right hand side of (5). Let σ be a k-coloration of G such that $\sum_x \nu(x,\sigma)$ is maximum possible. Let, if possible, $x_o \in V(G)$ be such that $\nu(x_o,\sigma) < \min \{k, d_G(x_o)\} = d_G(x_o)$. Let x_1 and x_2 be defined as above. Then, by corollary 3, since σ cannot be improvable at x_o , we have

$$\sum_{i=0}^{2} |k - d_G(x_o)| + 2 \leq k ,$$

or

$$\sum_{i=0}^{2} (k - d_G(x_o)) + 2 \leq k , \text{ since } k \geq \Delta ,$$

which implies

$$k < [\frac{d_G(x_o) + d_G(x_1) + d_G(x_2)}{2}] .$$

But this contradicts our choice of k . Hence, we must have $\nu(x,\sigma) = d_G(x)$ for all $x \in V(G)$ so that by definition, $\mu(G) \leq k$. This proves the theorem.

Theorem 3B (C. E. Shannon [11]): For any multigraph G ,

$$(6) \qquad \mu(G) \leq [\frac{3\Delta}{2}] .$$

The proof follows immediately from Theorem 3A.

Just as Theorems 3A and 3B, from Corollary 3, we obtain:

Theorem 3C: For any multigraph G,

$$\kappa(G) \geq \min \{\delta, \min [\frac{d_G(x) + d_G(y) + d_G(z) + 1}{4}]\} \tag{7}$$

where the inner minimum is taken over all coherent, 3-subsets $\{x,y,z\} \subseteq V(G)$.

Theorem 3D: For any multigraph G,

$$\kappa(G) \geq [\frac{3\delta + 1}{4}]. \tag{8}$$

The bounds (7) and (8) for the cover index $\kappa(G)$ are analogous to the Ore's bound (5) and Shannon's bound (6) for the chromatic index $\mu(G)$, respectively.

It is easy to show that the bounds (6) and (8) are both attainable for each value of Δ and δ respectively. In fact, for any $\Delta \geq 0$, let G_Δ denote the multigraph with three vertices in which two pairs of vertices are joined by $[\frac{\Delta}{2}]$ edges each and the third pair of vertices is joined by $[\frac{\Delta}{2}]^*$ edges. Then G_Δ is a multigraph with maximum degree Δ and chromatic index $[\frac{3\Delta}{2}]$. Also, for any $\delta \geq 0$, let G'_δ denote the multigraph with three vertices in which two pairs of vertices are joined by $[\frac{\delta}{2}]^*$ edges each and the third pair of vertices is joined by $[\frac{\delta}{2}]$ edges. Then, G'_δ is a multigraph with minimum degree δ and cover index $[\frac{3\delta + 1}{4}]$.

Remark 1: C. Berge [1, Chapter 12] has shown that the Shannon's Theorem 3B can be deduced from the Vizing - Gupta Theorem 2A. I do not know if Theorem 3D can similarly be deduced Theorem 2B.

6. V. G Vizing (1965) has shown that the multigraph G_Δ are essentially the only multigraphs for which the equality in (6) can hold. More precisely:

"If G is a multigraph with maximum degree $\Delta \geq 4$ and if G does not contain G_Δ as a subgraph then $\mu(G) \leq [\frac{3\Delta}{2}] - 1$."

C. Berge [1] made the following conjecture.

<u>Conjecture</u>: "<u>If</u> $G \not\supseteq G_s$, <u>where</u> $4 \leq s \leq \Delta$, <u>then</u>

(9) $$\mu(G) \leq [\frac{3\Delta}{2}] - [\frac{\Delta}{s}] \text{ ."}$$

The above conjecture which generalizes Vizing's result was established by Berge [1] when both Δ and s are even, and by Bosak [2] when Δ is even and s is odd

The following proposition provides bounds for the chromatic index that are stronger than (9) and also analogous bounds for the cover index.

Let G be any multigraph, k any nonnegative integer and σ any k-coloration of G . Let $x_o \in V(G)$ be a vertex such that $\nu(x_o,\sigma) < \min\{k, d_G(x_o)\}$. Let $\lambda_i = [x_o,x_i]$, $i = 1, 2$ be any two α-edges incident with x_o for some $\alpha \in C_2(x_o,\sigma)$.

Under these hypothesis, we have

<u>Proposition</u> 4: <u>If</u> $k \neq d_G(x_o)$, <u>then,</u>

<u>either</u> (a) σ <u>is improvable</u> at x_o ;

<u>or</u> (b) <u>the vertices</u> x_1 <u>and</u> x_2 <u>are distinct and</u>

(i) $D(x_o,\sigma)$, $D(x_1,\sigma)$, $D(x_2\sigma)$ <u>are pairwise disjoint</u>,

(ii) $C_3(x_o,\sigma)$, $C_3(x_1,\sigma)$, $C_3(x_2,\sigma)$ <u>are empty</u>

(iii) <u>for any</u> $\beta \in \overline{C}(x_i,\sigma)$, <u>there exists a</u> β-<u>edge</u> $[x_{i-1}, x_{i+1}]$, $i = 0, 1, 2$

(iv) <u>for any</u> $\gamma \in C_2(x_i,\sigma)$, <u>there exist</u> γ-<u>edges</u> $[x_i,x_{i-1}]$, $[x_i,x_{i+1}]$, $i = 0, 1, 2$

(v) for any $\delta \in J_k - \bigcup_{i=0}^{2} D(x_i,\sigma)$, $i = 0, 1, 2$, <u>there exists a</u> δ-<u>edge</u> $[x_j,x_k]$ <u>for some</u> $0 \leq j < k \leq 2$.

(<u>the subscripts in</u> (iii) <u>and</u> (iv) <u>are all modulo</u> 3);

<u>or</u> (c) <u>there exist vertices</u> x_3 <u>and</u> x_4 <u>such that</u> $\{x_o, x_1, x_2, x_3, x_4\}$ <u>is a coherent</u> 5-<u>subset of</u> $V(G)$ <u>and</u>

(i) $D(x_o,\sigma)$, $D(x_1,\sigma)$, ... , $D(x_4,\sigma)$ <u>are mutually disjoint</u>,

(ii) $C_3(x_o,\sigma)$, $C_3(x_1,\sigma)$, ... , $C_3(x_4,\sigma)$ <u>are empty</u>.

From the above proposition, we obtain

<u>Theorem</u> 4A: <u>For any multigraph</u> G ,

$$(10) \qquad \kappa(G) \geq \min \{\delta - 1,\ t_1^*(G),\ \min_T [\frac{\sum_{x \in T} d_G(x) + 1}{6}]\}$$

<u>where the inner minimum is taken over all coherent</u>, 5-<u>subsets</u> $T \subseteq V(G)$.

<u>Proof</u>: Let k be the right hand side of (10). Let σ be an k-coloration of G such that $\sum_x \nu(x,\sigma)$ is maximum possible. Let, if possible, $x_o \in V(G)$ be such that $\nu(x_o,\sigma) < \min\{k, d_G(x_o)\} = k$. Let x_1 and x_2 be defined as above. Since $k \leq \delta - 1$, $k < d_G(x_o)$. Hence, at least one of the conditions (a), (b) or (c) in Proposition 4 must hold. By our own choice of σ , (a) cannot hold. Assume (b) holds. Let $S = \{x_o, x_1, x_2\}$. Then, counting the number of edges with at least one endpoint in S , we find $|E'(S)| \leq 2k - 1$. This implies $t_1^*(G) \leq [\frac{|E'(S)|}{2}] \leq k - 1$ which contradicts our choice of k . Assume (c) holds. Then, by Lemma 1,

$$\sum_{i=0}^{4} |k - d_G(x_i)| + 2 \leq k$$

or

$$\sum_{i=0}^{4} (d_G(x_i) - k) + 2 \leq k \text{ , since } k < d_G(x_i) ,$$

which implies

$$k > \frac{\sum_{i=0}^{4} d_G(x_i) + 1}{6} \geq [\frac{\sum_{i=0}^{4} d_G(x_i) + 1}{6}]$$

again contradicting our choice of k . Hence, we must have $\nu(x,\sigma) = k$ for all $x \in V(G)$ so that by definition, $\kappa(G) \geq k$. This proves the theorem.

<u>Theorem</u> 4B: <u>For any multigraph</u> G ,

$$(11) \qquad \kappa(G) \geq \min \{[\frac{5\delta + 1}{6}],\ t_1^*(G)\} .$$

<u>Proof</u>: If $\delta = 0$ or 1 , then $\kappa(G) = \delta$. Otherwise, the proof follows immediately from Theorem 4A.

From Proposition 4, we also obtain the following:

<u>Theorem</u> 4C: <u>For any multigraph</u> G ,

$$(12) \qquad \mu(G) \leq \max \{\Delta + 1,\ t_1(G),\ \max_T [\frac{\sum_{x \in T} d_G(x) + 2}{4}]\}$$

where the inner maximum is taken over all coherent 5-subsets $T \subseteq V(G)$.

Theorem 4D: For any multigraph G ,

(13) $$\mu(G) \leq \max \{[\frac{5\Delta + 2}{4}], t_1(G)\} .$$

The details of the proofs of Theorems 4C and 4D are similar to that of Theorems 4A and 4B and are left to the reader.

Remark 2: It can easily be seen that the bound on the chromatic index given by Theorem 4D is much stronger than that conjectured by Berge [(9) above]. Hence, Berge's conjecture is established. We also note that (13) had also been conjectured by Bosak [2].

Remark 3: Since $t_1^*(G) \leq [\frac{3\Delta}{2}]$ always, bound (13) on the chromatic index is an extension of Shannon's bound (6). Similarly, bound (11) on the cover index is an extension of bound (8).

Remark 4: It may be observed that Proposition 4 is an extension of Proposition 3. Proposition 4, in turn, has been extended 'one step further' from which we obtain:

Theorem 5A: For any multigraph G ,

(14) $$\mu(G) \leq \max \{[\frac{7\Delta + 4}{6}], t_1(G), t_2(G)\} .$$

Theorem 5B: For any multigraph G ,

$$\kappa(G) \geq \min \{[\frac{7\delta + 1}{8}], t_1^*(G), t_2^*(G)\} .$$

The details of the proofs are too lengthy to be included here.

7. We make the following conjectures:

Conjecture I: For any multigraph G ,

$$\mu(G) \leq \max \{[\frac{(2r + 3)\Delta + 2r}{2r + 2}], t_1(G), \ldots , t_r(G)\}, r = 1, 2, \ldots .$$

<u>Conjecture</u> II: <u>For any multigraph</u> G ,

$$\kappa(G) \geq \min \{[\frac{(2r+3)\delta+1}{2r+4}], t_1^*(G), \ldots, t_r^*(G)\}, \ r = 1, 2, \ldots .$$

Let $t(G) = \max_{r \geq 1} t_r(G)$ and $t^*(G) = \min_{r \geq 1} t_r^*(G)$. Then,

<u>Conjecture</u> III: <u>For any multigraph</u> G ,

$$\mu(G) = t(G) \quad \underline{\text{if}} \quad t(G) \geq \Delta + 1 ,$$

$$\Delta \leq \mu(G) \leq \Delta + 1 \quad \underline{\text{otherwise}}.$$

<u>Conjecture</u> IV: <u>For any multigraph</u> G ,

$$\kappa(G) = t^*(G) \quad \underline{\text{if}} \quad t^*(G) \leq \delta - 1$$

$$\delta - 1 \leq \kappa(G) \leq \delta \qquad \underline{\text{otherwise}}.$$

<u>Remark</u> 5: It is easily seen that Conjectures III and IV follow from Conjectures I and II respectively.

<u>Remark</u> 6: The conjectures I and II have been proved for $r = 1$ and $r = 2$. Also, from Theorems 2A and 2B, these conjectures are trivially true for graphs (without multiple edges).

The conjectures seem extremely plausible and we hope that these will be proved soon.

<u>Remark</u> 7: The problem of characterization of multigraphs G for which the equality $\mu(G) = \Delta(G)$ and/or $\kappa(G) = \delta(G)$ hold seems extremely difficult as, for instance, it is well-known that the Four Color Conjecture will be proved if it can be shown that for any nonseparable, trivalent planar graph G , $\mu(G) = \Delta(G) = 3$ and $\kappa(G) = \delta(G) = 3$.

REFERENCES

1. C. Berge, Graphes et Hpergraphes. Dunod, Paris (1970). (eng. Tr. Graphs and Hypergraphs, North-Holland 1973.)

2. J. Bosák, Chromatic Index of finite and infinite graphs. Czechoslavak Math. J. 22(1972), 272-290.

3. R. P. Gupta, Studies in the Theory of Graphs. Thesis, Tata Inst. Fund. Res., Bombay (1967).

4. R. P. Gupta, The chromatic index and the degree of a graph. Notices Amer. Math. Soc. 13, No. 6 (1966) Abstract GGT-429.

5. R. P. Gupta, A theorem on the cover index of an s-graph. Notices Amer. Math. Soc. 13, No. 6 (1966) Abstract 637-14.

6. R. P. Gupta, A decomposition theorem for bipartite graphs. Theorie des Graphes, Rome I.C.C. (P. Rosenstiehl, Ed.)

7. R. P. Gupta, On decompositions of a multigraph into spanning subgraphs. Bull Am. Math. Soc. 80 (1974).

8. J. Fiamcik and E. Jcovic, Coloring the edges of a multigraph. Arch. Math. 21(1970).

9. D. König, Theorie der endlichen und unendlichen graphen. Leipzig (1936); reprinted by Chelsea, New York (1950).

10. O. Ore, The Four Color Problem, Academic Press, New York (1967).

11. C. E. Shannon, A theorem on coloring the lines of a network, J. Math. and Phys. 28 (1949).

12. V. G. Vizing, On an estimate of the chromatic class of a p-graph (Russian) Diskret Analiz, 3 (1964).

13. V. G. Vizing, Critical graphs with a given chromatic class (Russian) Diskret, Analiz, 5 (1965).

14. V. G. Vizing, On the chromatic class of a multigraph, Cybernetics, 1, No. 3 (1965).

PRUNING AND DEPTH FIRST SEARCH

Gary Haggard
University of Maine at Orono
Orono, Maine 04473

Abstract

This paper contrasts two algorithms for finding a fundamental set of cycles for a connected graph. The first algorithm based on a PRUNING technique is known to have complexity $O(\beta(G)|V|^2)$. The second algorithm based on a DEPTH FIRST SEARCH is shown to have complexity $O(\beta(G)|V|)$.

I. INTRODUCTION

From its beginnings graph theory has been concerned with finding effective algorithms for determining whether or not a graph has a certain property. Some of the most important concerns in graph theory deal with properties that have never been adequately characterized. Among the more notable cases are the 4-Color Conjecture and the Hamiltonian Cycle Problem. For particular graphs, questions of this sort can always be answered eventually because one can do an exhaustive search of the possibilities. This is only of theoretical interest in most cases since graphs with as few as 20 vertices admit too many possibilities for completing - in a reasonable amount of time - a search through all possible candidates to determine the presence or absence of a particular property.

For the properties of graphs which have reasonable algorithmic determination, one is interested in measuring, in some way, the relative effectiveness of the particular procedure being used. Since the input of an algorithm for graphs consists of the vertex set and the edge set of the graph, we would like to describe graph theoretic algorithms by functions which indicate the necessary amount of processing of the inputs. We can then measure a graph theoretic algorithm by means of the function which describes it. Obviously an algorithm which is described by a function which is linear in its inputs will have a much wider applicability than an algorithm which is described by a function for which the size of the input sets occurs as exponents, i.e. $2^{**}(\ell|V|+k|E|)$. For example, if we assume that one unit of time equals one millisecond, then an input of size 1000 can be processed in one second if the algorithm is described by the function $f(n) = n$. If, however, the function which describes the algorithm is $g(n) = 2^{**}n$, then only a maximum input of size 9 can be processed in a second. Extrapolating these functions to a time of one hour, the algorithm described by function f can process an input of size 3.6 × 10**6 while the algorithm described by function g can only process an input of size 21.

We say a function f is O(g(n)) if there exists a constant c such that $f(n) \le cg(n)$ for all but some finite (possibly empty) set of nonnegative values for n. If f is O(n**2) we say f is of order n**2. Additional information about the complexity of algorithms can be found in [1, p. 12].

Unless explicitly defined, graph theoretic terms will conform to the definitions in [2].

An important reason for studying the complexity of graph theoretic algorithms is that these algorithms have applications in other disciplines. Increasingly efficient algorithms allow analysis of cases once viewed as intractable. For instance, electrical engineering is concerned with finding special bases for the circuit space of graphs which represent electrical networks [9, p. 132]. Specifically, it is necessary to find a basis for the circuit space for a graph G in the following way:

(1) Find a spanning tree T for G.

(2) For each edge $e \in E(G) - E(T)$ identify the unique cycle in $T \cup \{e\}$ and include each such cycle in the basis.

A basis for the circuit space of G which is found in this way is called a <u>fundamental set of cycles for G</u>. For 3-connected planar graphs a set of fundamental cycles can be found by using the boundaries of the faces of its planar embedding as elements of the basis. For a connected graph G, the number of cycles in a fundamental set of cycles is

$$\beta(G) = |E(G)| - |V(G)| + 1.$$

In this paper we will show how a method for finding a fundamental set of cycles for a graph G called PRUNING [6] as modified in [3] has complexity $O(|V(G)|^2\beta(G))$. This analysis was essentially done in [3]. We next show that another method based on a Depth First Search has complexity $O(|V(G)|\beta(G))$.

For the remainder of this paper we will assume that all graphs are connected.

II. PRUNING

To find a fundamental set of cycles for a graph G, the following algorithm could be used.

ALGORITHM 1.

STEP 1. Find a spanning tree T for G.

STEP 2. For each edge $e \in E(G) - E(T)$ remove the edges of $T \cup \{e\}$ which are not in the cycle of $T \cup \{e\}$. Identify the cycle which is now distinguished.

For each $e \in E(G) - E(T)$ we can implement STEP 2 by iteratively finding vertices of degree 1 in $T \cup \{e\}$ and deleting the corresponding end edge. The process of removing the edges of $T \cup \{e\}$ which are not in the cycle is called PRUNING. In FIGURE 1 we indicate the result of each required pass through the vertex set needed to identify the cycle in T_0. The graph T_i for i = 0, 1, 2, 3 is the result of pass i through the vertex set.

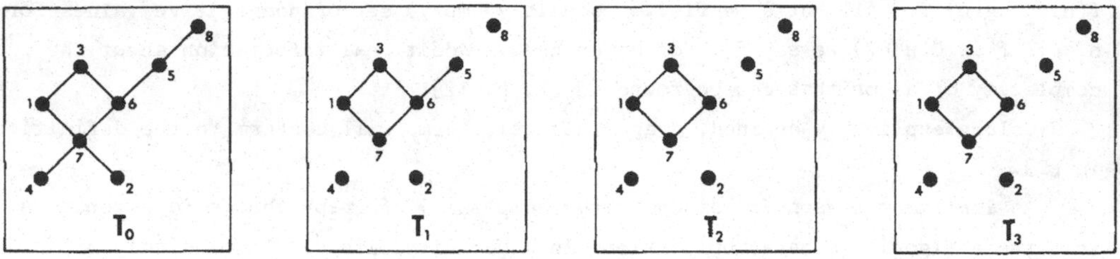

FIGURE 1

The third pass through the vertex set is required to verify that the deletion of the edge (5, 6) did not create a new end vertex. One can see the limitations of this iterative procedure by considering the number of passes required to isolate the cycle in the graph pictured in FIGURE 2.

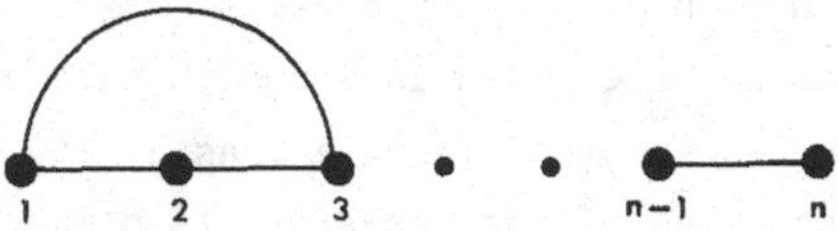

FIGURE 2

We can modify this procedure so that only a single pass through the vertex set is required. In FIGURE 3 we label by S_i where $i \in \{1,2,\cdots,8\}$ the result of the modifications triggered by vertex i.

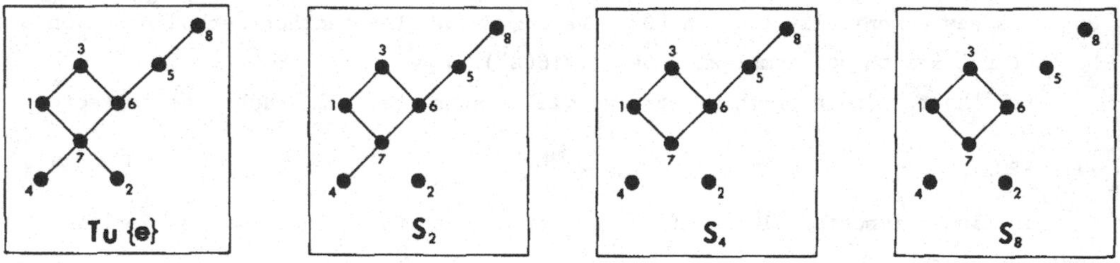

FIGURE 3

When vertex 8 is encountered both the edge (8, 5) and the edge (5, 6) are deleted. Each time an end vertex is encountered edges are deleted until there is one fewer end vertex in the graph. This is effected by continually asking if the other end of the end edge being deleted becomes an end vertex by this process. If a new end vertex is created, the deletion process is repeated for this new end edge. If no end vertex is created, the scan of the vertex set of the graph is continued.

We assume $T \cup \{e\}$ is represented by an incidence matrix where T is a spanning tree for G and $e \in E(G) - E(T)$. We further assume that there is an array called DEGREE which contains the degree of the vertex i in DEGREE (i) for $i = 1, 2, \cdots, |V(G)|$.

PRUNING

STEP 1. Set $I \leftarrow 1$.

STEP 2. Set $J \leftarrow I$.

STEP 3. If DEGREE (J) = 1 GO TO STEP 4 ELSE GO TO STEP 7.

STEP 4. FIND THE OTHER END OF THE EDGE WITH ONE END J AND CALL THIS OTHER END K.

STEP 5. DELETE THE EDGE (J, K) FROM THE GRAPH $T \cup \{e\}$ AND DECREMENT BY 1 BOTH DEGREE (J) AND DEGREE (K).

STEP 6. SET $J \leftarrow K$ AND GO TO STEP 3.

STEP 7. SET $I \leftarrow I + 1$. IF $I \leq |V(G)|$ GO TO STEP 2 ELSE STOP.

It is important to prove that PRUNING deletes all the non-cycle edges of $T \cup \{e\}$. The result is obvious if $T \cup \{e\}$ is itself the cycle. It is equally obvious if $(v, w) \in E(G) - E(T)$ and either v or w has degree 1. Suppose therefore that $(v, w) \in E(G) - E(T)$ and that both DEGREE (v) and DEGREE (w) are greater than 1. Also assume that v is farther from the cycle than w. Let $u_1, u_2, \cdots, u_k$ be all the vertices of $T \cup \{e\}$ of degree 1 which are joined to v by paths which do not contain vertices of the cycle. Further suppose $u_1 \leq u_2 \leq \cdots \leq u_k$. Let P be the unique path from v to u_k. When I = k in PRUNING, no vertex of P can have degree greater than 2 for if $s \in V(P)$ and DEGREE (s) > 2 there must be a path from s to an end vertex $u \neq u_k$ since s is not on the cycle. The path v - s - u must be a path from v to an end vertex having no vertex of the path contained in the cycle. Therefore, either $u \in \{u_1, u_2, \cdots, u_{k-1}\}$ or u is created by the deletion process initiated at some of the end vertices $u_1, u_2, \cdots, u_{k-1}$. The first is impossible because PRUNING eliminates the end edges incident with $u_1, u_2, \cdots, u_{k-1}$ before I = k. The second is also impossible since each time STEP 4 is entered, edges are deleted until there is one fewer end vertex in the graph. Therefore, no end vertex is created in the deletion process begun at any of $u_1, u_2, \cdots, u_{k-1}$. Hence, no vertex of P has degree greater than 2. It now follows that P and then at least (v, w) are deleted before the testing condition used in STEP 3 is false.

For $e \in E(G) - E(T)$ let C_e be the number of edges in the cycle in $T \cup \{e\}$. To determine the complexity of PRUNING one observes that STEP 4 is the only step which depends on the whole vertex set for its execution. None of the other steps requires more than a constant number of operations. None of the other steps is executed more than $|V|$ times. Thus, the other steps can be executed $O(|V|)$ times. STEP 4 requires $O(|V|)$ operations each time it is executed and it will be executed $|V| - C_e$ times when e is added to T. Therefore an estimate for the complexity of PRUNING for each $e \in E(G) - E(T)$ is

$$k|V| + \ell(|V| - C_e)|V|$$

where k, ℓ are constants. This expression is dominated by $\ell(|V| - C_e)|V|$. Since there are $\beta(G)$ edges in E(G) - E(T), the total complexity of PRUNING is dominated by

$$O((\sum_{e \in E(G)-E(T)}(|V| - C_e))|V|) \approx O(\beta(G) \cdot |V| \cdot \text{average cycle size})$$

The remainder of STEP 2 of ALGORITHM 1 has complexity at most $O(\beta(G)|V|^2)$ where $|V|^2$ represents the number of operations needed to search the incidence matrix for the cycle isolated and $\beta(G)$ represents the number of times this will be done. It is known [5, p. 197] that STEP 1 has complexity bounded by

$$O(\max\{|V| \log_2|V|, |E|\}).$$

If we estimate the average cycle size as $|V|/2$, we get the following expression as a bound on the complexity of this algorithm for finding a fundamental set of cycles for G

$$O(\max\{|V| \log_2|V|, |E|\}) + O(\beta(G) \frac{|V|}{2} |V|) + O(\beta(G)|V|^2).$$

This expression is dominated by the terms involving $\beta(G)|V|^2$ so this algorithm has complexity $O(\beta(G)|V|^2)$. This analysis is essentially the analysis presented in [3].

III. DEPTH FIRST SEARCH

One feature of the PRUNING method that should disturb the reader is that the incidence matrix representation of a graph forces many operations to have an $O(|V|^2)$ complexity. Obviously the representation of a graph by an incidence matrix has nothing essential to do with the structure of the graph and may be making the algorithm more complex than it need be. This, in fact, is what has happened. Using an alternate data structure comprised of a set of adjacency lists - one for each vertex - and a new search method called a depth first search [7] reduces the complexity of finding a fundamental set of cycles for a graph by a power of $|V|$.

The purpose of presenting this different data structure and an introduction to an application of a Depth First Search (DFS) is not only to offer an improved algorithm for finding a fundamental set of cycles but also to introduce to a wider audience a method of proof that has had an important role in recent developments in finding improved graph theory algorithms [4].

For a graph G a set of adjacency lists is a set of linked lists, one for each vertex of G. An adjacency list for a vertex v of G is a linked list which contains a listing of all the other vertices of G adjacent to v. We see an example for the random graph - FIGURE 4.

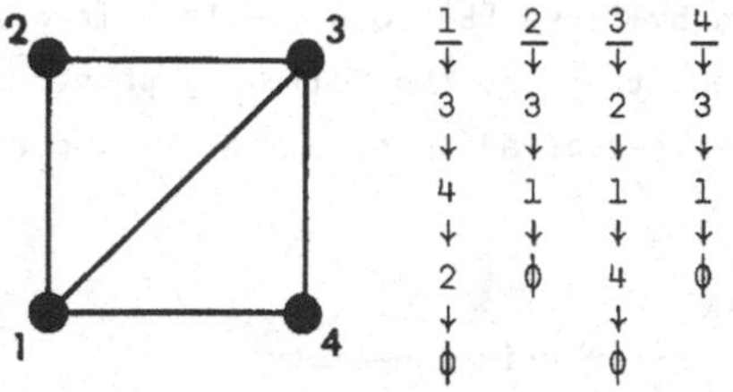

FIGURE 4

Since each edge (v, w) generates two entries, one in the list for v and one in the list for w, we process each edge twice while processing the adjacency lists. To avoid this repetition we include an additional linkage between the two occurrences of information about the same edge. After processing an edge we can use this additional linkage to update the lists by deleting both references to that edge. For example, suppose the edge (1, 3) has been processed and that we want to update the adjacency lists. We will adjust the adjacency lists for 1 and 3 as indicated in FIGURE 5.

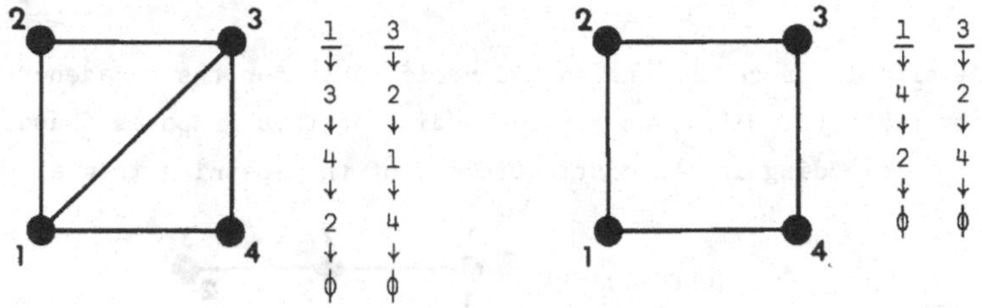

FIGURE 5

With any search procedure which systematically examines all possibilities and backtracks when dead ends are found, it is important to devise a way of avoiding the reexamination of previously considered cases. For example, if we wish to number the vertices of a graph in some way in order to make subsequent processing more reasonable, then the following recursive strategy can be used. Pick a vertex v. Distinguish v in some way. Pick a vertex adjacent to v that has not been examined, say w. If w is not a distinguished vertex, proceed from v to w along this edge. If w is already distinguished, pick another unexamined vertex adjacent to v and proceed as before. If we have proceeded along the edge (v, w), we distinguish w and begin the search anew from w. When all the vertices adjacent to w have been examined, return to v and proceed to examine the other unexamined vertices adjacent to v. This search procedure explores each edge of the graph exactly once. Further, if the vertex t is distinguished when the search is emanating from w, and t is encountered later when the search is emanating from the vertex x, we avoid the reexamination of cases emanating from t since t has already been distinguished and all possibilities emanating from t were examined when t was first encountered. This strategy is called a Depth First Search (DFS) since we try to examine as many different vertices as possible before the backtracking process is initiated. The DFS is an important variant of the search devised by Tarry [8] to solve labyrinthes.

We can more formally present DFS by the following procedure. Assume G is connected and is represented by a set of adjacency lists. Also assume vertex v is coded as 1.

```
DFS:  SEARCH (v)
      DO   while adjacency list for v is non-empty.
           Let w be the next adjacency on the list for v.
           Delete both references to (v, w) on the adjacency lists for G.
           If w has not been encountered previously, then
                BEGIN
                     Code w with a value 1 more than the value given to the
                          last vertex coded.
                     Start SEARCH from w.
                END
      END
END SEARCH
```

As an example of this procedure one should verify that for the adjacency lists given for the random graph (FIGURE 4) we get the coding of this graph as found in FIGURE 6. Implicit with this coding is the distinguishing of the spanning tree also pictured in FIGURE 6.

CODE (1) = 1
CODE (2) = 3
CODE (3) = 2
CODE (4) = 4

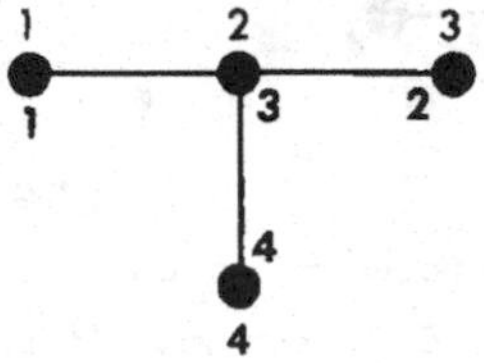

FIGURE 6

In addition to the spanning tree edges, the random graph contains the edges (1, 4) and (1, 2). From the coding given the vertices by the DFS both of these edges have induced on them a direction from a higher coded vertex to a lower coded vertex. In the case of (1, 4) we get [4, 1] and in the case of (1, 2) we get [2, 1]. Since DFS produced a spanning tree, each of these additional edges will generate a fundamental cycle when joined to this spanning tree. We need to expand the coding procedure for new vertices slightly. Suppose w is adjacent to v and that the DFS proceeding from v encounters w as an uncoded vertex. In addition to coding w as indicated in DFS we need also to attach to w the name v in an array as BACKCODE (w) = v. Now if an edge (u, v) is encountered and CODE (u) < CODE (v) we can use this additional field to proceed back down the tree along the unique path from v to u, and thus find the fundamental cycle associated with (u, v) in a number of operations that is proportional to the number of edges in this cycle. For example, with [4, 1] the first edge will be (4, BACKCODE (4)). Since BACKCODE (4) ≠ 1 we then add the edge

(BACKCODE (4), BACKCODE(BACKCODE (4))).

Since $BACKCODE^2(4) = 1$ we add (1, 4) to the edges found to determine the fundamental cycle (4 - 3 - 1). This procedure is formalized as ALGORITHM 2. Suppose G is connected and represented by adjacency lists. Further assume CODE (v) = 1 and BACKCODE (v) = 0.

```
ALGORITHM 2
        DFS - FC (v)
          DO   while the adjacency list for v is non-empty.
               Let w be the next adjacency on the list for v.
               Delete both references to (v, w) on the adjacency lists for G.
               If w has not been encountered
                  THEN
                     BEGIN
                        Code w with a value 1 more than that given to the
                            last vertex coded.  Set  BACKCODE(w) = v .
                        Start DFS - FC from w.
                     END
                  ELSE
                     BEGIN
                        Use BACKCODE to find the fundamental cycle of G
                            associated with the edge [v, w].
                     END
               END
        END DFS - FC
```

The correctness of this algorithm follows from the argument given in [7]. The search procedure has complexity $O(\max\{|V|, |E|\})$ as determined in [7]. The only part of DFS - FC left to analyze is the block with uses BACKCODE to find a fundamental cycle. Clearly, this block requires $O(C_e)$ operations for $e \in \underline{S} = E(G) - E(T)$ - remember C_e is the number of edges in the cycle in $T \cup \{e\}$. Thus, the total complexity of the algorithm is

$$O(\max\{|V|, |E|\}) + O(\sum_{e \in S} C_e)$$

Using the average estimate for the size of a typical cycle, we get

$$\sum_{e \in S} C_e \approx \beta(G) \text{ (average cycle size).}$$

Now taking $|V|/2$ as the average cycle size, we get

$$O(\max\{|V|, |E|\}) + O(\beta(G)(\text{average cycle size}))$$
$$\approx O(\max\{|V|, |E|\}) + O(\beta(G)|V|/2)$$

as an estimate for the complexity of DFS - FC. Since the $\beta(G)|V|/2$ term dominates this expression, we conclude ALGORITHM 2 has complexity $O(\beta(G)|V|/2)$ which is an obvious improvement over the estimate for the complexity of ALGORITHM 1 which was $O(\beta(G)|V|^2)$.

1. A. Aho, J. Hopcroft and J. Ullman. The Design and Analysis of Computer Algorithms. Addison Wesley, Reading, Mass. (1974).

2. M. Behzad and G. Chartrand. Introduction to the Theory of Graphs. Allyn and Bacon, Boston (1962).

3. C. Gotlieb and D. Cornell. Algorithms for finding a fundamental set of cycles for an undirected linear graph. Comm. ACM 10 (1967), 780-783.

4. J. Hopcroft and R. Tarjan. Efficient algorithms for graph manipulation (Algorithm 447). Comm. ACM 16 (1973), 372-378.

5. A. Nijenhuis and H. Wilf. Combinatorial Algorithms. Academic Press, New York (1975).

6. R. Read. Teaching Graph Theory to a Computer, Recent Progress in Combinatorics. Academic Press, New York (1969).

7. R. Tarjan. Depth first search and linear graph algorithms. Siam J. Comput. 1 (1972), 146-160.

8. G. Tarry. Le probleme des labyrinthes. Nouvelles Ann. de Math. 14 (1895), 187.

9. L. Weinberg. Network Analysis and Synthesis. McGraw-Hill, New York (1962).

GRAPHS AND THEIR DEGREE SEQUENCES: A SURVEY[1]

S. Louis Hakimi
Northwestern University
Evanston, IL 60201
and
Edward F. Schmeichel
University of Southern California
Los Angeles, CA 90007

ABSTRACT

Some recent work on the relationship between an integer sequence and various graph theoretic properties which a realization of the sequence may or must possess is surveyed.

I. INTRODUCTION

Our terminology and notation will be standard (see Harary [26]) except as indicated. We will consider primarily finite, undirected graphs without loops or multiple edges having p vertices and q edges. In certain places we will also discuss multigraphs in which multiple edges are allowed, but not loops.

A sequence of positive integers $\pi = (d_1, d_2, \ldots, d_p)$ is called <u>graphical</u> if there exists a graph G having π as the degrees of its vertices, and such a graph G is called a <u>realization</u> of π. Graphical sequences were characterized by Havel [27], Erdös and Gallai [18], and Hakimi [23]. The condition given by Erdös and Gallai is the following: <u>A sequence</u> $\pi = (d_1, d_2, \ldots, d_p)$ <u>with</u> $d_i \geq d_{i+1}$ <u>is graphical if and only if</u> $\sum_{i=1}^{p} d_i$ <u>is even and</u>

$$\sum_{i=1}^{k} d_i \leq k(k-1) + \sum_{i=k+1}^{p} \min(k, d_i), \quad \underline{\text{for}}\ 1 \leq k \leq p-1.$$

The corresponding condition for multigraphs was given by Senior [48] and Hakimi [23]. Our purpose in this paper is to survey some recent results dealing with the relationship between a graphical integer sequence $\pi = (d_1, d_2, \ldots, d_p)$ and various graph theoretical properties (e.g., connectivity, transversability, planarity, etc.) which a realization of π may or must possess. In particular, let π be a graphical sequence and let P be a graph theoretic property. Let $\mathscr{R}(\pi)$ denote the collection of all non-isomorphic realizations of π. Our survey will divide naturally into two sections: In the next section, we will deal with the question of when <u>at least one</u> member of $\mathscr{R}(\pi)$ possesses property P (i.e., when is π <u>potentially</u> P-graphical). In Section III, we deal with the question of when <u>every</u> member of $\mathscr{R}(\pi)$ possesses property P (i.e., when is π <u>forcibly</u> P-graphical).

---

[1]This work is supported by the Air Force Office of Scientific Research, Air Force Systems Command, under Grant AFOSR 71-2103 (D).

All integer sequences π discussed in the sequel are assumed to be graphical unless explicitly stated otherwise.

II. ON THE EXISTENCE OF A GRAPH IN $\mathcal{R}(\pi)$ WITH A SPECIFIED PROPERTY

Our first results deal with sequences graphical with prescribed connectivity. A complete characterization of sequences graphical with prescribed vertex connectivity, originally conjectured by Hakimi [25], was first proved by Wang and Kleitman [50]. Their proof was constructive, containing an explicit algorithm for obtaining the realization.

Theorem 1. The sequence $\pi = (d_1 \geq d_2 \geq \ldots \geq d_p)$ is graphical with vertex connectivity $d_p = k$ if and only if

(1)
$$\sum_{i=1}^{k-1} d_i \leq \tfrac{1}{2} \sum_{i=1}^{p} d_i + \binom{k-1}{2} - (p-k).$$

Moreover, π is graphical with vertex connectivity $d_p - 1$ if (1) is not satisfied.

The corresponding problem for vertex connectivity in multigraphs was studied by Hakimi [25], and will not be given here.

A graph G is called maximally edge-connected if for all $v, w \in V(G)$, there exist min(deg v, deg w) edge disjoint paths from v to w. Wang [52] obtained the following result:

Theorem 2. A sequence $\pi = d_1 \geq d_2 \geq \ldots \geq d_p \geq 2$ has a realization which is maximally edge-connected.

Since a maximally edge-connected realization of π above is a priori d_p edge-connected, the last two theorems completely characterize sequences graphical with prescribed edge-connectivity. Such a characterization was obtained earlier by Edmonds [17] and Wang and Kleitman [51]. Also Chou and Frank [11] have studied sequences realizable as maximally edge-connected multigraphs.

A characterization of sequences realizable as "maximally vertex-connected" graphs appears very difficult.

We next consider results dealing with sequences graphical with prescribed factors. The following theorem (essentially conjectured by Grünbaum [22] and Rao and Rao [41]) was first proved independently by Kundu [30] and Kleitman and Wang [29]. The latter gave a constructive proof. Lovász [32] had previously given an elegant proof of a slightly less general theorem.

Theorem 3. Let $\pi = (d_1, d_2, \ldots, d_p)$ and $(k_1, k_2, \ldots, k_p)$ be graphical sequences, with $k_i \leq d_i$ and $|k_i - k_j| \leq 1$, for all i and j. Then there exists a realization G of π containing a factor F with $d_G(v_i) = d_i$ and $d_F(v_i) = k_i$, $1 \leq i \leq p$ if and only if the sequence $(d_1 - k_1, \ldots, d_p - k_p)$ is graphical.

The result may fail to hold if the condition $|k_i - k_j| \leq 1$ is omitted. For example, let $\pi = (7,7,7,4,4,3,3,3)$ and let the k_i-sequence be $(4,4,4,3,3,3,3,3,2)$. It is easy to check that $\{d_i - k_i\}$ is a graphical sequence. But π has exactly one realization, and it is easily verified that the k_i-sequence cannot be the vertex degree sequence of any edge-deleted subgraph of this realization. As another example, let $\pi = (13,13,9,9,7,7,7,7,7,6,3,3,2,2,1,1)$ and let the k_i-sequence be $(7,7,7,7,7,7,7,7,0,0,0,0,0,0,0,0)$. It is easily checked that both these sequences as well as $\{d_i - k_i\}$ are graphical. However, no realization of π contains K_8 as an induced subgraph.

Rao and Rao [41] explored the important question of when a sequence π is graphical with a connected k-factor. Their result is the following:

Theorem 4. Let $\pi = (d_1 \geq d_2 \geq \ldots \geq d_p)$ be graphical with a k-factor (i.e., the conditions in Theorem 3 are satisfied with $k_i = k$, for all i). Then π is graphical with a connected k-factor if and only if

$$\sum_{i=1}^{r} d_i < r(p-r-1) + \sum_{i=p-r+1}^{p} d_i, \qquad \text{for all } r < p/2.$$

The next group of results deal with sequences having realizations with specified cycle or path structure. A characterization of sequences having a Hamiltonian realization is already contained in Theorems 3 and 4, because a Hamiltonian cycle is a connected 2-factor.

A modification of the technique used to prove Theorem 4 will yield the following result:

Theorem 5. The sequence $\pi = (d_1 \geq d_2 \geq \ldots \geq d_p)$ has a realization with a Hamiltonian path joining a vertex of degree d_i to one of degree d_j (with $i < j$ both chosen to be the largest indices corresponding to the prescribed degrees d_i and d_j) if and only if the sequence

(i) $$(d_1-2,\ldots d_{i-1}-2, d_i-1, d_{i+1}-2,\ldots d_{j-1}-2, d_j-1, d_{j+1}-2,\ldots d_p-2)$$

is graphical, and

(ii) $$\sum_{i=1}^{k} d_i < \begin{cases} k(p-k-2) + \sum_{i=p-k}^{p} d_i, & \text{if } k < \frac{p-1}{2} \text{ and } i > k,\ j > p-k \\ k(p-k-1) + \sum_{i=p-k+1}^{p} d_i, & \text{if } k < \frac{p}{2} \text{ and either } i > k, j \leq p-k \text{ or } i \leq k, j > k \\ k(p-k) + \sum_{i=p-k+2}^{p} d_i, & \text{if } k < \frac{p+1}{2} \text{ and } i,j \leq k. \end{cases}$$

Details will appear in Schmeichel and Hakimi [47]. The following corollary follows easily from this theorem.

Corollory. The sequence $\pi = (d_1 \geq d_2 \geq \ldots \geq d_p)$ has a realization with a Hamiltonian path if and only if

(i) $$(d_1-2,\ldots,d_{p-2}-2, d_{p-1}-1, d_p-1)$$ is graphical, and

(ii) $$\sum_{i=1}^{r} d_i < r(p-r-2) + \sum_{i=p-r}^{p} d_i, \quad \text{for all } r < \frac{p-1}{2}.$$

(Rao and Rao [41] had already noted that if π has a realization with a 2-factor, then π has a realization with a Hamiltonian path.)

A graph G is called Hamiltonian connected if for each pair of vertices $v,w \in V(G)$, there exists a Hamiltonian path in G joining v to w. We conjecture the following characterization of sequences having a Hamiltonian connected realization:

Conjecture. A sequence $\pi = (d_1 \geq d_2 \geq \ldots \geq d_p)$ is realizable as a Hamiltonian connected graph if and only if

(i) π is realizable as a Hamiltonian graph, and

(ii) $$\sum_{i=1}^{r} d_i < r(p-r) + \sum_{i=p-r+2}^{p} d_i, \quad \text{for all } r < \frac{p+1}{2}, \text{ and}$$

(iii) $d_p \geq 3$.

A graph with p vertices is called pancyclic if it contains cycles of all lengths ℓ, $3 \leq \ell \leq p$. The problem of characterizing sequences having pancyclic realizations seems extremely difficult. In fact, the minimum number of edges required in a pancyclic graph on p vertices is not even known (although Bondy [7] has shown that this number is asymptotic to $p + \log_2 p$).

The problem of which sequences are graphical with specified girth (i.e., the length of the shortest cycle) also seems exceedingly difficult. One need only consider the effort that has been expended on finding minimal cubic graphs of given girth to appreciate this [49]. One might expect that the general problem for girth four might be approachable, but even with this restriction the problem remains difficult. In fact, even the related problem of when a sequence is graphical as a bipartite(bicolorable)graph has never been satisfactorily answered.

A connected graph is called k-cyclic if it contains exactly k distinct cycles. For example, a tree is 0-cyclic. Boesch and Harary [5] gave a characterization of sequences realizable as 1-cyclic graphs. Beineke and Schmeichel [3] extended the above result by characterizing sequences realizable as a k-cyclic graph for k = 2,3,4. The general characterization for arbitrary k seems out of reach.

We next consider the problem of when a sequence π has a planar realization. Although a full characterization seems hopelessly difficult, some interesting partial results are available.From Euler's classical formula, a necessary condition for a sequence $\pi = (d_1, d_2, \ldots, d_p)$ to be planar graphical is that

(2) $$\sum_{i=1}^{p} d_i \leq 6(p-2)$$

We call π an Euler sequence if (2) holds. We call an Euler sequence maximal (resp., nonmaximal) if equality (resp., strict inequality) holds in (2).

The authors have attempted to determine which "nearly regular" Euler sequences have planar realizations. In particular we have studied sequences $d_1 \geq d_2 \geq \ldots \geq d_p$ in which $d_1 - d_p \leq 2$. For brevity of notation, let us denote a monotone sequence $\underbrace{a_1 \ldots a_1}_{e_1} \underbrace{a_2 \ldots a_2}_{e_2} \ldots \underbrace{a_r \ldots a_r}_{e_r}$ by $a_1^{e_1} a_2^{e_2} \ldots a_r^{e_r}$. Hawkins, Hill, Reeve and Tyrell [28] have determined completely which regular, Euler sequences have planar realizations.

Theorem 6. Every regular, Euler sequence π has a planar realization except for 4^7 and 5^{14}.

A complete proof appears in Owens [37].

The following three theorems all appear in Schmeichel and Hakimi [46]. The proofs make use on a profound theorem of Grünbaum [21, p. 272].

Theorem 7. Every Euler sequence $\pi = (d_1 \geq \ldots \geq d_p)$, with $d_1 - d_p = 1$, has a planar realization except for $5^{10}4^1$, $5^{12}4^1$, $6^1 5^{12}$, and $6^1 5^{14}$.

Theorem 7 contains the answer to an interesting question, which Grünbaum and Motzkin [20] attribute to Coxeter: For which $p \geq 12$ does $6^{p-12}5^{12}$ have a planar realization? It was conjectured by R. H. Fox that the answer is all even $p \geq 12$ (see Ringel [43] and Alpert and Gross [1]). As the above theorem indicates, however, the correct answer is every $p \geq 12$ except $p=13$. This was apparently first proved (in dual form) by Grünbaum and Motzkin [20], and later explicitly by E. Etourneau [19] and others.

The problem of determining which Euler sequences π with $d_1 - d_p = 2$ have planar realizations is considerably more complex. The following two theorems, however, leave only a small number of undecided sequences. The authors conjecture that none of these undecided sequences have a planar realization.

Theorem 8. Every maximal Euler sequence $\pi = (d_1 \geq \ldots \geq d_p)$ with $d_1 - d_p = 2$ has a planar realization except for $5^1 4^4 3^1$, $5^3 3^3$, $5^4 4^1 3^2$, $6^1 4^6$, $5^5 4^2 3^1$, $6^1 5^2 4^5$, $5^7 4^1 3^1$, $5^9 3^1$, $6^1 5^{10} 4^1$, $7^1 5^{13}$, and $7^1 6^1 5^{13}$, and possibly the following undecided cases:

$$7^1 6^2 5^{13}$$
$$7^k 6^1 5^{k+12}, \quad \text{for } k = 2,3,5,7$$
$$7^k 5^{k+12}, \quad \text{for } k = 3,5,7,9.$$

Theorem 9. Every nonmaximal Euler sequence $\pi = (d_1 \geq d_2 \geq \ldots \geq d_p)$ with $d_1 - d_p = 2$ is planar graphical except for $4^5 2^1$, $5^5 3^3$, $5^{11} 3^1$, $7^1 5^{15}$, and $6^{p-7} 4^7$ for $p > 7$, and possibly the following undecided cases: $5^{13} 3^1$, $7^1 5^{17}$, $7^3 5^{17}$.

Various attempts have been made to find stronger necessary conditions for a sequence to have a planar realization than Euler's condition (2). The usual form of these conditions is an upperbound on the partial sums $\sum_{i=1}^{k} d_i$ of the sequence (see Bowen [9] and Chvátal [12], for example). We now give a slightly weakened version of a theorem due to the authors which generalizes and improves results of this sort. We should point out, however, that none of the available results-including ours-substantially strengthens Euler's condition (see Schmeichel and Hakimi [46]).

Theorem 10. If $d_1 \geq d_2 \geq \ldots \geq d_p$ is planar graphical, then

$$\sum_{i=1}^{k} d_i \leq \begin{cases} 6(k-2) + \sum_{i=k+1}^{3k-4} \min(3,d_i) + \sum_{i=3k-3}^{p} \min(2,d_i), & \text{for } 3 \leq k \leq \frac{p+4}{3} \\ 6(k-2) - 2\,\omega(2) - 4\,\omega(1), & \text{for } k=p \end{cases}$$

where $\omega(i)$ is the number of times i occurs in the sequence $(d_1,d_2,\ldots,d_p)$

A graph is called self-complimentary if it is isomorphic to its compliment. Sequences which have self-complimentary realizations have only recently been fully characterized. The necessity portion of the following theroem is due to Sachs [44] and Ringel [42], while the more difficult sufficiency part is due to Clapham and Kleitman [15], who gave a constructive proof.

Theorem 11. A sequence $\pi = (d_1 \geq d_2 \geq \ldots \geq d_p)$ is realizable as a self-complimentary graph if and only if

(i) $p \equiv 0$ or 1 (modulo 4), and

(ii) $d_i + d_{p-i+1} = p-1$, for $1 \leq i \leq \left\lceil \frac{p+1}{2} \right\rceil$, and

(iii) $d_{2i} = d_{2i-1}$, for $1 \leq i \leq \left\lceil \frac{p}{4} \right\rceil$.

The question of when π has a unique realization up to isomorphism (i.e. when $\mathcal{G}(\pi)$ contains exactly one member) remains unsolved in general. The corresponding problem for multigraphs is of substantial interest in organic chemistry, where it is related to the problem of the generation of isomers. The multigraph problem was solved by Hakimi [24], while Senior [48, 24] answered the corresponding (and more difficult) problem for connected multigraphs.

III. CONDITIONS IMPLYING EVERY GRAPH IN $\mathcal{G}(\pi)$ HAS A SPECIFIED PROPERTY

The opening results in this section give conditions on π which guarantee that every graph in $\mathcal{G}(\pi)$ has a specified cycle or path structure.

An early result of this type was the following theorem of Dirac [16]: If $\pi = (d_1,d_2,\ldots,d_p)$, and if $d_i \geq p/2$ for $1 \leq i \leq p$, then every graph in $\mathcal{G}(\pi)$ has a Hamiltonian cycle. Dirac's theorem was successively generalized by Ore [36], Pósa [39], Bondy [8], and Chvátal [13]. Chvátal's result is the following:

Theorem 13. Let $\pi = (d_1 \le d_2 \le \ldots \le d_p)$, with $p \ge 3$. If

(3) $$d_k \le k < p/2 \ \text{implies} \ d_{p-k} \ge p-k,$$

then every graph in $\mathcal{R}(\pi)$ is Hamiltonian.

Although it is possible to sharpen Chvátal's theorem (e.g., see Berge [2, p.204]), Chvátal indicated a sense in which Theorem 13 is a best possible result. A sequence $\pi' = (d_i' \le d_2' \le \ldots \le d_p')$ is said to majorize the sequence $\pi = (d_1 \le d_2 \le \ldots \le d_p)$ if $d_i' \ge d_i$, for $1 \le i \le p$. Chvátal observed that if a sequence π fails to satisfy (3), then π is majorized by a sequence π' which has only a nonhamiltonian realization. Explicitly, if π fails to satisfy (3) for some $k < p/2$, then π is majorized by the sequence $\pi(p,k) = k^k\ (p-k-1)^{p-2k}\ (p-1)^k$ which has only the nonhamiltonian realization $G(p,k) = K_k + (K_{p-2k} \cup \bar{K}_k)$.

Theorem 13 does not fully characterize forcibly Hamiltonian sequences. For example, the sequence $2^1 3^2 4^5$, and the infinite class of sequences $(2t)^{2t}\ (p-2t-1)^{p-2t}$ (due to Nash-Williams [35]) are each forcibly Hamiltonian, and yet none satisfies condition (3) in Theorem 13. We conjecture, however, that every sequence π is either forcibly Hamiltonian or has a nonhamiltonian realization in which the removal of k vertices of largest degree will leave $>k$ components, for some $k < p/2$. On the other hand, there are sequences π which are not forcibly Hamiltonian (and which therefore fail to satisfy (3) for at least one $k < p/2$) but which cannot be realized as edge deleted subgraphs of $G(p,k)$ (defined in the last paragraph) for any k. An example of such a sequence is $2^1 3^6 4^1$.

We now give a more general, "local" version of Theorem 13, which also happens to be an immediate consequence of it.

Theorem 14. Let G be a graph with p vertices, and let $\{v_1, v_2, \ldots, v_r\} \subseteq V(G)$, with say $d_i = \deg_G(v_i)$. (We assume the v_i are labeled so that $d_1 \le d_2 \le \ldots \le d_r$). If

$$d_k \le k + (p-r) \ \text{implies} \ d_{r-k} \ge p-k, \ \text{for} \ k < r/2,$$

the subgraph of G induced by $\{v_1, v_2, \ldots, v_r\}$ is Hamiltonian.

Theorem 13 is just the special case r=p. Theorem 14 is also a best possible result in the same sense as above.

The following three theorems each come from a straight-forward application of Theorem 13. Each can be generalized in the same way that Theorem 14 generalizes Theorem 13. Moreover, each is a best possible result in the same way that Theorem 13 was. (After the statement of each theorem, the corresponding "majorizing graph" is indicated.)

Theorem 15 (Chvátal [13]). Let $\pi = (d_1 \le d_2 \le \ldots \le d_p)$. If

$$d_k \le k-1 < \frac{p-1}{2} \ \text{implies} \ d_{p-k+1} \ge p-k$$

then every graph in $\mathscr{G}(\pi)$ has a Hamiltonian path.

(Majorizing graph: $K_{k-1} + (K_{p-2k+1} \cup \bar{K}_k)$).

Theorem 16 (Lick [31]). Let $\pi = (d_1 \le d_2 \le \ldots \le d_p)$. If

$$d_k \le k+1 < \frac{p+1}{2} \text{ implies } d_{p-k-1} \ge p-k$$

then every graph in $\mathscr{G}(\pi)$ is Hamiltonian-connected.

(Majorizing graph: $K_{k+1} + (K_{p-2k-1} \cup \bar{K}_k)$).

We call a graph n-Hamiltonian if the removal of n or fewer vertices leaves a Hamiltonian graph.

Theorem 17 (Chvátal [13]). Let $\pi = (d_1 \le d_2 \le \ldots \le d_p)$ and $0 \le n \le p-3$. If

$$d_k \le k+m < \frac{p+m}{2} \text{ implies } d_{p-k-m} \ge p-k$$

for $0 \le m \le n$, then every graph in $\mathscr{G}(\pi)$ is n-Hamiltonian.

(Majorizing graph: $K_{k+n} + (K_{p-2k-n} \cup \bar{K}_k)$).

The following theorem - possibly the most difficult result of this type - was conjectured by Bondy [7] and Chvátal [14], and proved by Schmeichel and Hakimi [45]. It is once again a best possible result in the above sense.

Theorem 18. Let $\pi = (d_1 \le d_2 \le \ldots \le d_p)$, with $p \ge 3$. If π satisfies condition (3) in Theorem 13, then every graph in $\mathscr{G}(\pi)$ is either pancyclic or bipartite. If a graph in $\mathscr{G}(\pi)$ is bipartite, it contains cycles of every even length.

In [6], Bondy gave an essentially best possible condition for a sequence π to be forcibly k-vertex-connected graphical (although a slightly sharper version appears in Berge [2, p. 170]). Bondy's theorem, which generalizes an earlier result of Chartrand, Kapoor, and Kronk [10], is the following:

Theorem 19. Let $\pi = (d_1 \le d_2 \le \ldots \le d_p)$. If

(4) $$d_i \le i+k-2 \text{ implies } d_{p-k+1} \ge p-i,$$

for $1 \le i \le \left[\frac{p-k+1}{2}\right]$, then every graph in $\mathscr{G}(\pi)$ is k-vertex-connected.

Boesch [4] proved that Theorem 19 is a best possible result in the usual sense. The "majorizing graph" is $K_{k-1} + (K_i \cup K_{p-i-k+1})$, which has connectivity k-1.

None of the results in Theorems 15 through 19 fully characterize sequences which forcibly possess the particular property. For example, $7^5 6^6 4^2 3^1 2^1$ is a forcibly 1-connected sequence which fails to satisfy (4) for k=1. Still the problem of characterizing forcibly 1-connected sequences, unlike the other characterization problems suggested by the last few theorems, seems quite tractable.

In [8], Bondy and Chvátal present a unified approach to a wide variety of problems concerning sequences which are forcibly graphical with a certain property. Their method is inspired by a proof of Ore [36], which they observed actually yields the following result: If v and w are nonadjacent vertices in a p-vertex graph with deg v + deg w ≥ p, and if G + (v,w) is Hamiltonian, then G itself is Hamiltonian. The implications of this simple observation are really quite surprising in their scope, including rather simple proofs of Theorems 13 through 17 and Theorem 19. The interested reader is referred to [8] for details of their approach.

IV. ADDITIONAL REMARKS

There has been a considerable amount of research on the vertex degree sequences of directed graphs. An excellent reference for the available results is Berge [2]. We want to mention one recent key result which does not appear in [2].

Theorem 20 (Meyniel [33]): *Let G be a strongly connected digraph on p vertices. If for any nonadjacent vertices v,w we have deg v + deg w ≥ 2p-1, then G contains a directed Hamiltonian cycle.*

This generalizes a result of Ghouila-Houri [2, p. 196].

Nash-Williams [34] suggests that a great deal more may be true. In essence, he asked the following question: Let G be a directed graph on p vertices. If the outdegree sequence and indegree sequence (arranged in a nondecreasing way) both satisfy Chvátal's condition (3), must G contain a directed Hamiltonian circuit?

One other direction for future work seems worthy of mentioning. It seems reasonable that one may associate a "degree" to each edge of a graph and develop a new theory of degree sequences. Such a theory was initiated by Patrinos and Hakimi [38], who associated an unordered pair of integers with each edge representing the degrees of its end vertices.

REFERENCES

[1] S. R. Alpert and J. L. Gross, Graph embedding problems, *Amer Math. Monthly* 82 (1975), 835-837.

[2] C. Berge, *Graphs and Hypergraphs*, North-Holland, Amsterdam, 1973.

[3] L. W. Beineke and E. F. Schmeichel, On degrees and cycles in graphs, *J. Networks* (to appear).

[4] F. T. Boesch, The strongest monotone degree condition for n-connectedness of a graph, *J. Combinatorial Theory Ser. B.* 16 (1974), 162-165.

[5] F. T. Boesch and F. Harary, Unicyclic realizability of a degree list, *J. Networks* (to appear).

[6] J. A. Bondy, Properties of graphs with constraints on degrees, *Studia Sci. Math. Hungar.* 4 (1966), 473-475.

[7] J. A. Bondy, Pancyclic graphs I, *J. Combinatorial Theory* 11 (1971), 80-84.

[8] J. A. Bondy and V. Chvátal, A method in graph theory (to appear).

[9] R. Bowen, On the sums of valences in planar graphs, *Can. Math. Bull.* 9 (1966), 111-114.

[10] G. Chartrand, S. F. Kapoor, and H. V. Kronk, A sufficient condition for n-connectedness of a graph, *Mathematika* 15 (1968), 51-52.

[11] W. Chou and H. Frank, Survivable communication networks and terminal capacity matrix, IEEE Trans. Circuit Theory CT-17 (1970), 192-197.

[12] V. Chvátal, Planarity of graphs with given degrees of vertices, Nieuw. Arch. Wisk. 17 (1969), 47-60.

[13] V. Chvátal, On Hamilton's ideals, J. Combinatorial Theory Ser. B. 12 (1972), 163-168.

[14] V. Chvátal, New directions in Hamiltonian graph theory, in "New Directions in the Theory of Graphs" (Proceedings of the 1971 Ann Arbor Graph Theory Conference, F. Harary, ed.) Academic Press, New York, 1973.

[15] C. R. J. Clapham and D. J. Kleitman, The degree sequences of self-complimentary graphs, J. Combinatorial Theory Ser. B. 20 (1976), 67-74.

[16] G. A. Dirac, Some theorems on abstract graphs, Proc. London Math. Soc. (3) 2 (1952), 69-81.

[17] J. Edmonds, Existence of k-edge connected ordinary graphs with prescribed degrees, J. of Research of the National Bureau of Standards, 68B (1964), 73-74.

[18] P. Erdős and T. Gallai, Graphs with prescribed degrees of vertices (Hungarian) Mat. Lapok 11 (1960), 264-274.

[19] E. Etourneau, Existence and connectivity of planar graphs having 12 vertices of degree 5 and n-12 vertices of degree 6, Infinite and Finite Sets, Colloq. Math. Soc. J. Bolyai, (A. Hajnal et al, ed.) North-Holland, 1975.

[20] B. Grünbaum and T. Motzkin, The number of hexagons and the simplicity of geodesics on certain polyhedra, Can. J. Math. 15 (1963), 744-751.

[21] B. Grünbaum, Convex Polytopes, Wiley-Interscience, New York, 1967.

[22] B. Grünbaum, Problem No. 2, Combinatorial Structures and Their Applications (Proceedings of the Calgony Conference, June, 1969, Richard Guy, ed.),Gordon and Breach, New York, 1971.

[23] S. L. Hakimi, On realizability of a set of integers as degrees of the vertices of a linear graph, I, J. SIAM 10 (1962), 496-502.

[24] S. L. Hakimi, On the realizability of a set of integers as degrees of the vertices of a linear graph, II, J. SIAM 11 (1963), 135-147.

[25] S. L. Hakimi, On the existence of graphs with prescribed degrees and connectivity, J. SIAM Appl. Math. 26 (1974), 154-164.

[26] F. Harary, Graph Theory, Addison-Wesley, Reading, Mass., 1969.

[27] V. Havel, A remark on the existence of finite graphs (Hungarian), Casopis Pest. Mat. 80 (1955), 477-480.

[28] A. Hawkins, A. Hill, J. Reeve, and J. Tyrell, On certain polyhedra, Math. Gazette 50 (1966), 140-144.

[29] D. J. Kleitman and D. L. Wang, Algorithms for constructing graphs and digraphs with given valences and factors, Discrete Math. 6 (1973), 79-88.

[30] S. Kundu, The k-factor conjecture is true, Discrete Math. 6 (1973), 367-376.

[31] D. R. Lick, A sufficient condition for Hamiltonian connectedness, J. Combinatorial Theory 8 (1970), 444-445.

[32] L. Lovász, Valencies of graphs with 1-factors, Periodica Math. Hungar. 5(2) (1974), 149-151.

[33] M. Meyniel, Une condition suffisante d'existence d'un circuit Hamiltonian dans un graphe oriente, J. Combinatorial Theory Ser. B 14 (1973), 137-147.

[34] C. St. J. A. Nash-Williams, Unexplored and semi-explored territories in graph theory, in "New Directions in the Theory of Graphs" (Proceedings of the 1971 Ann Arbor Graph Theory Conference, F. Harary, ed.) Academic Press, New York, 1973.

[35] C. St. J. A. Nash-Williams, On Hamiltonian circuits in finite graphs, Proc. Amer. Math. Soc. 17 (1966), 466-467.

[36] O. Ore, Note on Hamilton circuits, Amer. Math. Monthly 67 (1960), 55.

[37] A. B. Owens, On the planarity of regular incidence sequences, J. Combinatorial Theory 11 (1971), 201-212.

[38] A. Patrinos and S. L. Hakimi, Relations between graphs and integer-pair sequences, to appear in Discrete Mathematics.

[39] L. Posa, A theorem concerning Hamilton lines, Magyar Tud. Akas. Mat. Kutáto Int. Közl 7 (1962), 225-226.

[40] S. B. Rao and A. R. Rao, Existence of triconnected graphs with prescribed degrees, Pacific J. Math. 33 (1970), 203-207.

[41] A. R. Rao and S. B. Rao, On factorable degree sequences, J. Combinatorial Theory Ser. B. 13 (1972), 185-191.

[42] G. Ringel, Selbst Komplementäre Graphen, Arch. Math. 14 (1963), 354-358.

[43] G. Ringel, Map Color Theorem, Spring-Verlag, New York, 1974.

[44] H. Sachs, Über Selbst Komplementäre Graphen, Publ. Math. Debrecen 9 (1962), 270-288.

[45] E. F. Schmeichel and S. L. Hakimi, Pancyclic graphs and a conjecture of Bondy and Chvátal, J. Combinatorial Theory Ser. B. 17 (1974), 22-34.

[46] E. F. Schmeichel and S. L. Hakimi, On planar graphical degree sequences, SIAM J. Applied Math. (to appear).

[47] E. F. Schmeichel and S. L. Hakimi, On the existence of a traceable graph with prescribed vertex degrees (in preparation).

[48] J. K. Senior, Partitions and their representative graphs, Amer. J. Math. 73 (1951), 663-689.

[49] W. T. Tutte, A family of cubical graphs, Proc. of Comb. Phil. Soc. 43 (1947), 459-474.

[50] D. L. Wang and D. J. Kleitman, On the existence of n-connected graphs with prescribed degrees ($n \geq 2$), Networks 3 (1973), 225-239.

[51] D. L. Wang and D. J. Kleitman, A note on n-edge-connectivity, unpublished note.

[52] D. L. Wang, Construction of a maximally edge-connected graph with prescribed degrees, Stud. Appl. Math. 55(1) (1976), 87-92.

THE BICHROMATICITY OF A TREE

Frank Harary*, Derbiau Hsu, and Zevi Miller
The University of Michigan
Ann Arbor, Michigan

Abstract

The bichromaticity $\beta(B)$ of a bipartite graph B is defined as the maximum order of a complete bipartite graph onto which B is homomorphic. It is specified that no two points of different colors can be sent to the same point under a homomorphism. Exact formulas are obtained for the bichromaticity of a tree and for that of even cycles.

1. Bichromaticity

Our object is to study for bipartite graphs B the invariant corresponding to the achromatic number $\psi(G)$ of an arbitrary graph G which was introduced in [2] . This was defined as the maximum order p of a complete graph K_p onto which G is homomorphic. Analogously, the bichromaticity $\beta(B)$ is the maximum order $p = r+s$ of a complete bigraph $K_{r,s}$ onto which B is homomorphic, no two points in B of different colors being sent to the same point.

An elementary homomorphism will mean, as usual, the identification of two points in the same color class. A homomorphism of a bigraph will then be a sequence of elementary homomorphisms. A

---

*Research supported in part by grant AF 73-2502 from the Air Force Office of Scientific Research.

homomorphism $h: B \to K_{r,s}$ will be called <u>bicomplete</u>. We always take $r \le s$.

Let B denote a connected bigraph with color classes C and D so that its point set is $V = C \cup D$. Without loss of generality, we stipulate that $|C| \ge |D|$. We then call C the <u>majority</u> of B, and D the <u>minority</u>, and we write $\mu = \mu(B) = |C|$. Furthermore, whenever we say that a bipartite graph is <u>complete</u>, we will mean that it is a complete bigraph. For other notation and terminology, we follow the book [1].

2. Bichromaticity of a Tree

We now derive two simple inequalities involving these numbers which will be useful in establishing the bichromaticity of a tree.

<u>Lemma 1a</u>. If $h: B \to K_{r,s}$ is a bicomplete homomorphism of a bigraph B having q lines onto $K_{r,s}$, then

(1) $$rs \le q, \quad \text{and}$$

(2) $$r + s \ge \mu + 1 .$$

<u>Proof</u>: The first inequality (1) is immediate since the number of edges in the image of a homomorphism is at most the number of edges in its domain. For (2), we define the homomorphism h' by identifying all points in the minority of the two color classes C and D. Then $h'(B) = K_{1,\mu}$ so that $\beta(B) \ge \mu + 1$.

The second Lemma deals with the case that B is a tree T, as a step in determining a formula for $\beta(T)$.

<u>Lemma 1b</u>. If $h: T \to K_{r,s}$ is a bicomplete homomorphism of a tree T such that $\beta(T) = r + s$, then $r \le 2$ or $r = s = 3$.

Proof: By (1) and the fact that T is a tree,

$$rs \le q = p - 1,$$

so $s \leq (p-1)/r$. But by (2) ,

$$r + s \geq \mu + 1 ,$$

so that $(p-1)/r + r \geq \mu + 1$. As we obviously have $\mu \geq p/2$, it follows that $p(2-r)/2r \geq 1 - r + 1/r$. We now show that $r \geq 4$ leads to a contradiction as follows. If $r \geq 3$, the above inequality becomes, after routine manipulation,

$$\frac{p}{r} \leq 2 - \frac{2}{2-r} + \frac{2}{r(2-r)} .$$

But we also have

$$(3) \qquad s \leq \frac{p-1}{r} \leq 2 - \frac{2}{2-r} + \frac{2}{r(2-r)} - \frac{1}{r} .$$

For $r \geq 4$, this gives $s \leq 2 + 2/(r-2) \leq 3$, contradicting the convention $r \leq s$.

We now examine the case $r = 3$. Substituting into (3), we obtain $s \leq 3$. By our convention, we then have $r = s = 3$. This leaves $r \leq 2$ as required.

Remark: It is possible to characterize those trees for which $r = s = 3$ in the preceeding lemma as follows.

For such a tree T , there is a homomorphism $h: T \to K_{3,3}$ and $\beta(T) = 6$. The tree T must have at least 10 points since otherwise $q(T) \leq 8$. But Lemma 1a gives $q(T) \geq r \cdot s = 9$, a contradiction. Now since $p(T) \geq 10$, we have $\mu(T) \geq 5$. Therefore, $\mu(T) \geq 6$ or $\mu(T) = 5$. If $\mu(T) \geq 6$, we derive a contradiction by noticing that Lemma 1a and $\beta(T) = 6$ imply the absurdity:

$$6 = \beta(T) \geq \mu + 1 \geq 7 .$$

Thus $\mu(T) = 5$. By definition of μ , this forces $p \leq 10$; hence $p = 10$.

We now need only consider the trees T having exactly 10 points. We may immediately eliminate those for which $\mu(T) \geq 6$ by the preceding paragraph. Then we note that for such trees T, the maximum degree satisfies the inequality $\Delta(T) \leq 3$. For a point v of degree greater than 3 will have image $h(v)$ with degree at least one less than v. This can only happen if $q(h(T)) < q(T) = 9$. But $h(T) = K_{3,3}$, contradiction. We now show that T must have at least two points of degree 3. Since $p(T) = 10$ and $h(T) = 6$, we may view the homomorphism h as a sequence of four elementary homomorphisms. The maximum number of points not fixed by a sequence of four elementary homomorphisms is 8, two new points for each elementary homomorphism. But this leaves at least two points fixed by h. These fixed points will now be shown to have degree 3. A fixed point v of T under h has degree in T at least as large as its degree in $h(T)$ the homomorphic image of T. In this image, $\deg_{h(T)}(V) = 3$ since $h(T) = K_{3,3}$, hence $\deg_T(v) \geq 3$. But the previous argument showing $\Delta(T) \leq 3$ forces $\deg_T(v) = 3$. Since these considerations hold for all fixed points, we have at least two points of degree 3, as asserted.

Summarizing, we have shown that the trees T for which $r = s = 3$ in the lemma must satisfy the following properties:

$$p = 10,\ \mu = 5,\ \Delta \leq 3, \text{ and}$$

T has at least two points of degree 3.

A check of Appendix 3 in [1] listing all the trees with ten points shows that only sixteen satisfy these properties. Of those sixteen, we have verified that only the seven shown in Figure 1 have the property $r = s = 3$.

We are now ready to determine th bichromaticity of an arbitrary tree T.

<u>Theorem 1</u>. For any tree T, $\beta(T) = \mu + 1$.

<u>Proof</u>: By Lemma 1b, if $r + s = \beta(T)$ is the maximum order of a complete bigraph $K_{r,s}$ onto which T is homomorphic, then $r = 3$ and $s = 3$, or $r \leq 2$. We begin by considering the first possibility.

As shown in the remark following Lemma 1b, $r = 3$ and $s = 3$ implies $\mu(T) = 5$. But then $\mu(T) + 1 = 5 + 1 = 6 = r + s = \beta(T)$, so the theorem is proved in this case. We now pass to a consideration of the case $r \leq 2$.

We may decompose the point set $V(T) = C \cup D$ uniquely into two color classes C and D. As above, we again take $|C| \geq |D|$.

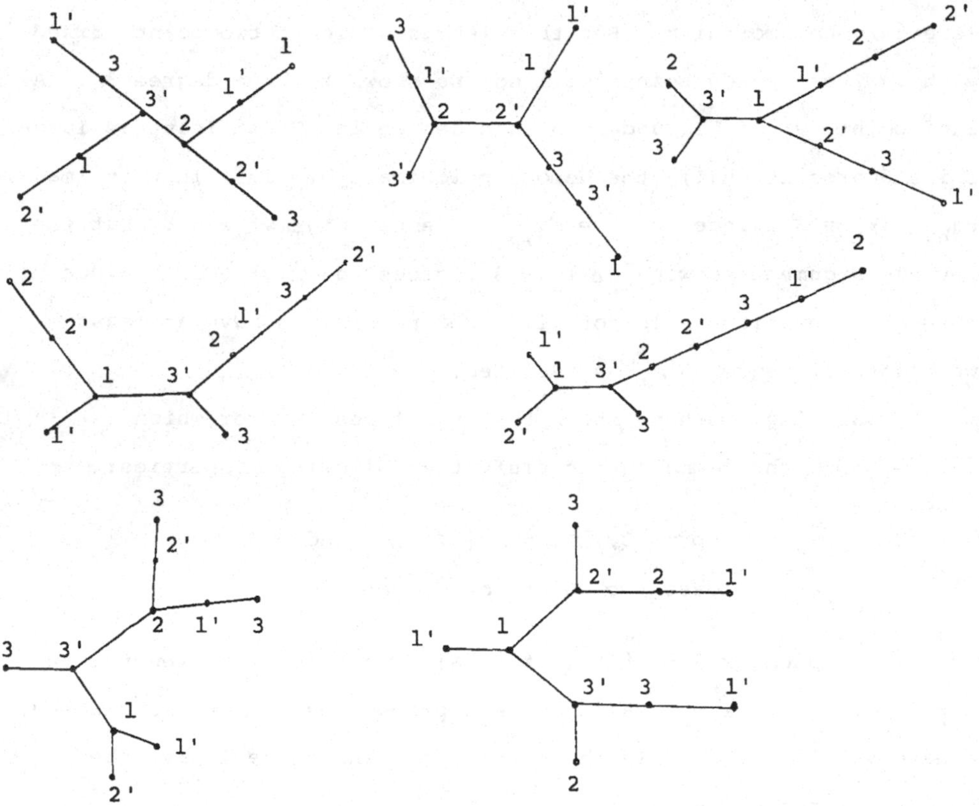

FIGURE 1. The seven trees with $r = 3$. Each has $\beta = 6$ and $K_{3,3}$ as a homomorphic image.

We note first that tight bounds can be given for $\beta(T)$ as follows. Let $h: T \to K_{r,s}$ be a bicomplete homomorphism for which $r + s = \beta(T)$. By Lemmas 1a and 1b and the fact that $s \le \mu$, it follows that $\mu + 1 \le \beta(T) \le \mu + 2$. It only remains to show that $\beta(T) < \mu + 2$.

Suppose to the contrary that $\beta(T) = \mu + 2$. Then since $s \le \mu$ and $r \le 2$, we must have $s = \mu$ and $r = 2$. Thus the homomorphism h sends the points of D onto two points v and w, but h leaves the points of C fixed.

Our plan is to show that in order to derive a contradiction to $\beta(T) = \mu + 2$, it will suffice to prove that C must contain at least one endpoint e of T. For if this holds, let x_0 be the point adjacent to e and partition set D so that $X_1 = \{x \in D/h(x) = v\}$ and $X_2 = \{x \in D/h(x) = w\}$. Then since $D = X_1 \cup X_2$ and $x_0 \in D$, either $x_0 \in X_1$ or $x_0 \in X_2$. Without loss of generality we assume $x_0 \in X_1$. Then the endpoint e is not adjacent to any point of X_2, so e is not adjacent to w in $h(T)$, contradicting the completeness of $h(T)$.

To prove that C does indeed contain an endpoint of T, suppose to the contrary that all the endpoints of T are in D. As usual, define $T^{(k)}$ inductively as the subtree obtained from $T^{(k-1)}$ by deleting all its endpoints, with $T^{(0)} = T$. We find it convenient to take $T^{(k)}$ as empty when $T^{(k-1)}$ is either K_2 or K_1. Also write $U^{(k)}$ for the set of endpoints of $T^{(k)}$ and let N be the smallest integer such that $T^{(N)} = \emptyset$.

Let $k_0 \le N - 1$ be the smallest index such that $|U^{(k_0)}| > |U^{(k_0+1)}|$. Since by assumption $U^{(0)} \subset D$, it follows that

$$U^{(j)} \subset D \quad \text{for } j \text{ even}, \; j \le N-1,$$

$$U^{(j)} \subset C \quad \text{for } j \text{ odd}, \; j \le N-1.$$

Furthermore,

$$D = \bigcup_{\substack{j \text{ even} \\ j \le N-1}} U^{(j)} \quad \text{and} \quad C = \bigcup_{\substack{j \text{ odd} \\ j \le N-1}} U^{(j)} \quad ,$$

so that

$$|D| = \sum_{\substack{j \text{ even} \\ j \le N-1}} |U^{(j)}| \quad \text{and} \quad |C| = \sum_{\substack{j \text{ odd} \\ j \le N-1}} |U^{(j)}| \, .$$

Thus as for all k, $|U^{(k)}| \ge |U^{(k+1)}|$, and since $|U^{(k_0)}| > |U^{(k_0+1)}|$, we obtain

$$|D| = \sum_{\substack{j \text{ even} \\ j \le N-1}} |U^{(j)}| > \underset{\substack{j \text{ odd} \\ j \le N-1}}{\quad} |U^{(j)}| = |C| \, ,$$

a contradiction of our convention.

The theorem is easily specialized to handle paths.

<u>Corollary 1a</u>. The bichromaticity of the path P_n of order n is given by

$$(4) \qquad \beta(P_n) = \left[\frac{n+3}{2}\right] \, .$$

3. <u>Bichromaticity of Even Cycles</u>.

In this section, we determine the bichromaticity of the even cycles.

<u>Lemma 2a</u>. If $h: C_{2n} \to K_{r,s}$ is a bicomplete homomorphism of an even cycles such that $\beta(C\ \ ) = r + s$, then $r \le 2$.

We only sketch the proof, since it is very similar to that of Lemma 1b. We have $p(C_{2n}) = q(C_{2n}) = 2n$. By the inequalities of Lemma 1a, it can be verified that there are just three possibilities: $r \le 2$, $r = s = 3$, or $r = 3$ and $s = 4$,

The latter two possibilities may be eliminated as follows. We will show that the only cycle C_{2n} for which we can have $r = s = 3$ in Lemma 2a is C_{10}. By formula (1), $q(C_{2n}) \geq q(K_{3,3}) = 9$. Thus $2n \geq 10$. On the other hand, if $2n \geq 12$, then formula (2) gives $\beta(C_{2n}) \geq 7$. But this contradicts $\beta(C_{2n}) = r+s = 3+3 = 6$, so that $2n \leq 10$. But now an exhaustive listing of all possible bipartite labelings of C_{10} shows that C_{10} is not homomorphic to $K_{3,3}$.

Similarly, if $r = e$ and $s = 4$ we conclude $C_{2n} = C_{12}$ and again an exhaustive verification shows that C_{12} is not homomorphic to $K_{3,4}$. Thus the two possibilities involving $r = 3$ are eliminated and the Lemma is proved.

Theorem 2. The bichromaticity of the even cycle C_{2n} is given by

$$(5) \qquad \beta(C_{2n}) = \begin{cases} 2 + n & \text{if } n \text{ is even} \\ 1 + n & \text{if } n \text{ is odd} \end{cases} .$$

Proof: As in the case of trees, we first give tight bounds for $\beta(C_{2n})$. Since $\mu(C_{2n}) = n$, Lemma 1a gives $\beta(C_{2n}) \geq n+1$. On the other hand, Lemma 2a and $s \leq \mu$ give $\beta(C_{2n}) \leq n+2$. Thus

$$(6) \qquad n + 1 \leq \beta(C_{2n}) \leq n + 2 .$$

Consider first the case that n is even. Proceed along the cycle in either of the two possible directions and number the points traversed $v_1, v_2, \ldots, v_{2n}$ successively. We partition $V(C_{2n}) = C \cup D$ so that $C = \{v_k | k \text{ odd}\}$ and $D = \{v_k | k \text{ even}\}$. We now describe a particular bicomplete homomorphism $h': C_{2n} \to K_{2,n}$. This will prove (5) for n even by definition of $\beta(C_{2n})$. For the action of h' on D, identify all points v_j such that $j \equiv 2 \pmod 4$ onto one point v, and identify all points v_j such that $j \equiv 0 \pmod 4$ onto another point w. For the action of h' on C, leave all points fixed. Clearly $h'(C_{2n}) = K_{2,n}$ as required.

Now we have the case that n is odd. By equation (6), it is only necessary to show that $\beta(C_{2n}) \neq n + 2$. Suppose to the contrary that $\beta(C_{2n}) = n + 2$. By Lemma 2a, for any bicomplete homomorphism $h: C_{2n} \to K_{r,s}$ we must have $r \leq 2$. This together with $s \leq n$ and the condition $\beta(C_{2n}) = n + 2$ implies the existence of another homomorphism $h'': C_{2n} \to K_{2,n}$. We may assume without loss of generality that C and D are again defined as alternating points of the cycle. Since $r = 2$ and $s = n$ by the existence of h'', one of the sets C or D is left pointwise fixed by h''. We may assume without loss of generality that this set is C. Then D must be sent by h'' onto two points v and w. We may partition D as $x_1 \cup x_2$, where

$$X_1 = \{x \in D / h''(x) = v\} \text{ and}$$

$$X_2 = \{x \in D / h''(x) = w\} .$$

As above, number the points of C_{2n} successively $v_1, v_2, \ldots, v_{2n}$. Since n is odd, there exists an index j, $1 \leq j \leq 2n$, such that v_j and v_{j+2} are both in X_1 or both in X_2. Without loss of generality, assume both are members of X_1. By convention, the index $j + 2$ will be read modulo $2n$. Thus in $h''(C_{2n})$, the point v_{j+1} is not adjacent to w, contradicting the completeness of $h''(C_{2n})$ and proving (5) for n odd.

4. Unsolved Problems

We conclude by stating some open questions concerning the bichromaticity of other bigraphs.

1. Although we have an explicit formula and algorithm for calculating $\beta(T)$ for a tree, the general problem for a connected bigraph B remains open. Several examples have led us to believe

that

$$\beta(B) = \mu + \delta(B) - x$$

where x is a non-negative integer, "small" compared with the minimum degree $\delta(B)$.

2. Given a bigraph B, it would be interesting to find $\beta(B \times K_2)$ in terms of $\beta(B)$ since then the bichromaticity of a cube $\beta(Q_n)$ follows immediately. More generally, the behavior of β under other binary operations on two connected bigraphs B_1 and B_2 is also open.

3. Uniquely colorable graphs were defined in [3] as graphs G such that there is a unique partition of $V(G)$ into $n = \chi(G)$ subsets determined by a homomorphism of G onto K_n. The corresponding concept is also meaningful for the achromatic number $\psi(G)$ defined in [2].

However, for the bichromaticity of a bipartite graph, the situation is a bit different. Clearly connected bigraphs are uniquely 2-colorable in the sense of [3]. We now ask whether the parameters r and s in a bicomplete homomorphism $h: B \to K_{r,s}$ for which $\beta(B) = r + s$ are uniquely determined. For instance $\beta(C_6) = 4$ according to Theorem 2. But this number can arise via homomorphisms

$$h_1: C_6 \to K_{1,3} \quad \text{or} \quad h_2: C_6 \to K_{2,2} .$$

4. Lemma 1a states that $\beta \geq i + \mu$ for any bigraph B. Then Theorems 1 and 2 assure that equality holds for trees and for even cycles C_{2n} with n odd. It is natural to ask for a characterization of bigraphs for which $\beta = 1 + \mu$.

REFERENCES

[1] F. Harary, Graph Theory. Addison-Wesley, Reading, Mass. (1969).

[2] F. Harary and S. T. Hedetniemi, The achromatic number of a graph. J. Combinatorial Theory 8(1970) 154-161.

[3] F. Harary, S. T. Hedetniemi and R. W. Robinson, Uniquely colorable graphs. J. Combinatorial Theory 6(1969) 264-270; 9(1970) 221.

ON DIAMETER STABILITY OF GRAPHS

Jehuda Hartman and Izhak Rubin
University of California
Los Angeles, California 90024

Abstract

A graph G is called an (ℓ,d)-graph if the removal of at least ℓ edges from G is required in order that the resulting graph would have a diameter larger than d. G is called ℓ-distance stable if the removal of at least ℓ edges is required to increase the distance between any two nonadjacent vertices of G. Some properties of (ℓ,d)-graphs, and ℓ-distance stable graphs are derived. In particular, a Mengerian type characterization of ℓ-distance stable graphs is obtained.

1. Introduction. A communication network is modelled here by a graph whose vertices correspond to centers dispatching and receiving information, and edges correspond to communication links. The connectivity of the graph is associated with the reliability of the network, whereas the diameter of the graph serves as a measure of the maximal message delay over the network. In this paper we assume that it is required to construct the network in such a way, that the maximal message delay will not exceed a given upper bound, even if certain communication links fail. We consider properties of graphs satisfying this condition.

All graphs considered in this paper are undirected without loops and multiple edges, and the terminology follows [3]. By V(G) and E(G) we denote the set of vertices and the set of edges of G, respectively. The degree, deg(v), of $v \in V(G)$ is the number of vertices adjacent to v.

The distance $d_G(x,y)$ between two vertices $x,y \in V(G)$ is the length of the shortest path in G joining x and y, and the diameter of G, d(G), is defined as

$$d(G) \triangleq \max_{x,y \in V(G)} d_G(x,y) .$$

A pair of vertices $x,y \in V(G)$ such that $d_G(x,y) = d(G)$, is called a diametrical pair. A pair of nonadjacent vertices $x,y \in V(G)$ will be called an ℓ-distance pair if

$$d_G(x,y) = d_{G-E}(x,y) ,$$

$\forall\ E\subset E(G)$ such that $|E| \le \ell-1$.

An edge $e\in E(G)$, will be called <u>cyclic</u> if there exists a cycle in G containing e . To each cyclic edge we assign a natural number, g(e) , which is the length of the shortest cycle containing e . If e is a bridge then $g(e) \overset{\Delta}{=} \infty$. The <u>girth</u> of G , girth(G) , is defined as

$$\text{girth}(G) \overset{\Delta}{=} \min_{e\in E(G)} g(e) .$$

A graph G with $|V(G)| \ge k+1$ is called <u>k-connected</u> (<u>k-edge-connected</u>) if between any pair x,y of distinct vertices of G , there are at least k vertex (edge) disjoint x,y-paths in G . It is obvious that a k-connected (k-edge-connected) graph cannot be disconnected by removing less than k vertices (edges) from the graph. The converse is also true, hence, a graph is k-connected (k-edge-connected) if and only if $|V(G)| \ge k+1$ and it is impossible to disconnect G by removing k-1 or fewer vertices (edges) from G , (Menger-Whitney theorem, See [3], Chapter 5). A graph G is defined to be an <u>(ℓ,d)-stable graph</u> (with respect to edges) if

$$d(G-E) \le d ,$$
$$\forall\ E\subset E(G) \text{ such that } |E| \le \ell-1 .$$

(ℓ,d)-stable graphs with respect to vertices are defined similarly. In this paper we deal only with (ℓ,d)-stable graphs with respect to edges, and call them simply <u>(ℓ,d)-graphs</u>. The <u>ℓ-diameter-stable</u> graphs defined in [1] are (ℓ,d)-graphs with diameter d . Note that an (ℓ,d)-graph must be in particular ℓ-edge-connected, but the converse is not true. A graph is called <u>ℓ-distance-stable</u> if at least ℓ edges are to be removed from G in order to increase the distance between any pair of non-adjacent vertices of G . It is clear that ℓ-distance-stable graphs are ℓ-diameter stable; in the next section we show that the converse is false. Some properties of ℓ-distance-stable graphs were also obtained in [4], where they are called ℓ-geodetically-line-connected graphs.

The above definitions together with some results, first appeared in [1]. The present paper extends the results of [1]. Some examples of (ℓ,d)-graphs may be found in [5],[6],[7] and [8], where the main problem is to find the minimal number of edges that an (ℓ,d)-graph with diameter λ and n vertices must have, and to construct classes of extremal graphs in this sense. We are concerned here with the properties of (ℓ,d)-graphs in general and some special (ℓ,d)-graphs, in particular ℓ-diameter stable and ℓ-distance stable graphs.

2. <u>Some Results for (ℓ,d)-graphs</u>. The following two lemmas are readily observed

<u>Lemma 1</u>: If between any pair of vertices of G there are at least ℓ edge disjoint paths not longer than d , then G is an (ℓ,d)-graph.

Lemma 2: If for any pair of nonadjacent vertices $x,y \in V(G)$, there are at least ℓ edge disjoint x,y-paths of length $d_G(x,y)$, then G is ℓ-distance stable.

As will be proved later, Lemma 2 is also a necessary condition for ℓ-distance-stability.

We now construct for any arbitrary integer $\ell,d \geq 2$ an ℓ-diameter-stable graph, with diameter d.

Let $H_1,H_2,\ldots,H_{d-1}$ be d-1 disjoint copies of the complete ℓ-vertex graph K_ℓ and denote the vertices of H_j by $v_{1_j},v_{2_j},\ldots,v_{\ell_j}$ $(j = 1,2,\ldots,d-1)$. Join v_{i_j} to $v_{i_{j+1}}$ by an edge, for all $i = 1,2,\ldots,\ell$ and $j = 1,2,\ldots,d-1$. Then join a new vertex u_1 adjacent to all vertices of H_1, and a vertex u_2 adjacent to all vertices of H_{d-1}. We denote the resulting graph by H.

Any pair of vertices in H is joined by at least ℓ disjoint paths, each not longer than d. By Lemma 1 H is an (ℓ,d)-graph. Since $d(H) = d$, H is an ℓ-diameter stable graph. It is easy to see that H is not an ℓ-distance-stable graph. Note that the previous construction can be used to obtain ℓ-diameter-stable graphs with diameter d, having exactly n vertices $(n \geq (d-1)\ell+2)$. Simply join to H $n-(d-1)\ell-2$ vertices each of them adjacent to all vertices of H_1.

ℓ-distance-stable graph with diameter d are obtained as follows. Take d-1 disjoint copies of the complement of K_ℓ $H_1,H_2,\ldots,H_{d-1}$, where the vertices of H_j are denoted by $v_{1_j},v_{2_j},\ldots,v_{\ell_j}$ $(j = 1,2,\ldots,d-1)$. Join v_{m_j} to $v_{n_{j+1}}$ for all $1 \leq m, n \leq \ell$ and $j = 1,2,\ldots,d-1$. Then join new vertices u_1 and u_2 adjacent to all vertices of H_1 and H_{d-1} respectively. By Lemma 2, the resulting graph H is an ℓ-distance stable graph with diameter d.

Another class of (ℓ,d)-graphs having n vertices may be obtained by a simple application of a result obtained in [2]. There the function $g(n,d)$ $(n-1 > d > 2)$ is defined as the least integer r such that if the degree of every vertex of G $(|V(G)| = n)$ is greater or equal to r, then $d(G) \leq d$. Using the results of [2] if all vertices of G satisfy

$$\deg(v) \geq \begin{cases} [\frac{n}{t}] + \ell-1 & \text{if } d = 3t-4 \\ [\frac{n-1}{t}]+\ell-1 & \text{if } d = 3t-3 \\ [\frac{n-2}{t}]+\ell-1 & \text{if } d = 3t-2 \end{cases}$$

then G is an (ℓ,d)-graph.

The following property of ℓ-diameter stable graphs is quite obvious.

Theorem 1: Let G be an ℓ-diameter stable graph, $(\ell \geq 2)$, then

$$g(e) \leq d(G)+1 \;,\; \forall\, e \in E(G) \;, \tag{1}$$

and this result is best possible.

Proof: If G is an ℓ-diameter stable graph, $(\ell \geq 2)$, then G is in particular

2-edge-connected and hence all its edges are cyclic. Let $e = ab$ be an edge of G and assume $g(e) > d(G)+1$. Then $d(G-e) \geq d_{G-e}(a,b) > d(G)$ contradicting the assumption that G is an ℓ-diameter stable graph.

To show that (1) cannot in general be improved we construct for any arbitrary integer $d \geq 2$ an ℓ-diameter stable graph with at least one edge e such that $g(e) = d(G)+1$. Let G_1 and G_2 be distinct ℓ-diameter-stable graphs with diameters d_1 and d_2 respectively, and let x_i, y_i be a diametrical pair of G_i $(i = 1,2)$. The graph G' generated from G_1 and G_2 by identifying the vertex $x_1 \in V(G_1)$ with the vertex $x_2 \in V(G_2)$ is an ℓ-diameter-stable graph with diameter $d = d_1 + d_2$. Let us assume in addition that G_i contains a triple x_i, y_i, z_i such $d_i = d_{G_i}(x_i,y_i) = d_{G_i}(x_i,z_i) = d_{G_i}(y_i,z_i)$, $i = 1,2$. Finally define the graph $G = G' + y_1y_2$, obtained from G' by joining the vertices y_1 and y_2 by an edge y_1y_2. G is clearly an ℓ-diameter stable graph which contains an edge y_1y_2 such that $g(y_1y_2) = d(G)+1$. ∎

In particular we conclude from Theorem 1 that the girth of an ℓ-diameter stable graph is upper bounded by $d(G)+1$.

One easily can prove a similar theorem for (ℓ,d)-stable graphs, $(g(e) \leq d+1)$

Equality (1) of Theorem 1 does not yield a sufficient condition on a graph G to be an ℓ-diameter stable graph. Take for instance the graph H, composed of 3 5-cycles C_1, C_2, C_3 with vertices v_i^k $1 \leq i \leq 5$, $1 \leq k \leq 3$, such that $C_1 \cap C_2 = v_2' v_3'$, $C_1 \cap C_3 = v_4' v_5'$ and $C_2 \cap C_3 = \phi$. Although every edge of H is contained in a cycle of length $\leq d(H)+1$, $d(H-v_3'v_4') = 6 > d(H) = 5$.

Next we prove a theorem that will be used later.

<u>Theorem 2</u>: A pair of nonadjacent vertices $x,y \in V(G)$ is an ℓ-distance pair if and only if there are at least ℓ edge disjoint x,y-paths of length $d_G(x,y)$ in G.

<u>Proof:</u> Clearly, if there are at least ℓ edge disjoint x,y-paths of length $d_G(x,y)$ in G, then x,y is an ℓ-distance pair. The converse may be proved by applying the digraph version of Menger's Theorem for edge connectivity [4]. We present here a constructive proof for the case $\ell = 2$. The same type of proof works for $\ell > 2$, but will be too lengthy to present here.

First define the following sets

$$A_i(x) \stackrel{\Delta}{=} \{y \in V(G) : d_G(x,y) = i\} \quad 0 \leq i \leq d(G),$$

and

$$E_i(x) \stackrel{\Delta}{=} \{e \in E(G) : ab = e, \ a \in A_i(x), \ b \ A_{i+1}(x)\} \ 0 \leq i \leq d(G)-1.$$

Since x,y is a 2-distance pair there are obviously at least two x,y-paths of length $d_G(x,y)$ in G. We shall prove that one can always find at least two edge disjoint such paths. If $d_G(x,y) = 2, 3$ the proof follows immediately. Assume therefore $d_G(x,y) > 3$, and let $d_G(x,y) = d_0$. If there are no two edge disjoint x,y-paths of length d_0 in G, there must be a maximal integer $i (1 < i < d_0)$ so that there are two x,y-paths P_1 and P_2 of length d_0 with edge disjoint subpaths from x to a vertex x_i, in $A_i(x)$. (note that P_1 and P_2 may have common edges in $E_m(x)$ only if $m \geq i$). If $i = 0$ x,y is not a 2-distance pair, and $i = 1$ is impossible in a graph without multiple edges. Let $e \in E_i(x)$ be an edge contained both in P_1 and P_2, with end vertices x_i and x_{i+1}, $(x_i \in A_i(x)$, $x_{i+1} \in A_{i+1}(x))$. Since x,y is a 2-distance pair, there must be a vertex $y_i \in A_i(x)$ and a y_i,y-path of length d_0-i not containing e. Otherwise, $d_G(x,y) < d_{G-e}(x,y)$. Denote this path by P_3'. By definition there exists an x,y_i-path P_4' of length i. If $P_1(P_2)$ and P_4' are edge disjoint, define $P = P_3' \cup P_4'$. $P_1(P_2)$ and P are two x,y-paths of length d_0, such that if there exists $e_1 \in E_m(x)$ and $e_1 \in P_1 \cap P (e_1 \in P_2 \cap P)$ then $m > i$. This contradicts the maximality of i.

If P_4', P_1 and P_4', P_2 have common edges let $k < i$ be the greatest integer such that there is an edge $e_2 \in E_k(x)$ and $e_2 \in P_1 \cap P_4'$ or $e_2 \in P_2 \cap P_4'$, $e_2 = x_k x_{k+1}$, $x_k \in A_k(x)$ and $x_{k+1} \in A_{k+1}(x)$. Define a new path P' composed of P_1 resp. P_2 from x to x_k and of P_4' from x_k to y_i. As before the paths P_2 resp. P_1 and P' contradict the maximality of i. This completes the proof. ■

Since any diametrical pair of vertices in an ℓ-diameter stable graph must be an ℓ-distance pair, the following necessary condition for ℓ-diameter stable graphs is obtained by applying Theorem 2.

<u>Theorem 3</u>: If G is an ℓ-diameter stable graph and x,y is a diametrical pair of vertices in G then there are at least ℓ edge disjoint x,y-paths of length $d(G)$ in G.

Note that the statement in Theorem 3 cannot serve as a sufficient condition for ℓ-diameter stability, as can be seen for instance from a graph which is a cycle of even length. Such a cycle is not 2-diameter stable, but satisfies the condition of Theorem 3.

The following is an equivalent definition of an (ℓ,d)-graph.

A graph G is an (ℓ,d)-graph if and only if for any two distinct vertices x,y and any $\ell-1$ disjoint edges $e_1,e_2,\ldots,e_{\ell-1}$, there exists an x,y-path of length not exceeding d, which avoids $e_1,e_2,\ldots,e_{\ell-1}$.

To show the equivalence of the definitions let G be an (ℓ,d)-graph, then by removing $e_1,e_2,\ldots,e_{\ell-1}$ from G the diameter of the resulting graph $G-\{e_1,\ldots,e_{\ell-1}\}$ does not exceed d, consequently for any pair of distinct vertices x,y there is an x,y-path of length $\leq d$ avoiding $e_1,e_2,\ldots,e_{\ell-1}$.

To show the sufficiency of the condition, let $a_1, a_2, \ldots, a_k$ $(k < \ell)$ be a minimal set of edges of G whose elimination increases the diameter of the resulting graph. Therefore, $\exists$ $x, y \in V(G)$ such that $d_{G-A}(x,y) > d$, where $A = \{a_1, a_2, \ldots a_k\}$. Choose $\ell-k-1$ other distinct edges $a_{k+1}, \ldots, a_{\ell-1}$. According to the condition in the theorem there exists an x,y-path of length $\le d$, avoiding $a_1, a_2, \ldots, a_{\ell-1}$ contrary to our assumption. Hence G is an (ℓ,d)-graph.

We conclude this section by stating a necessary and sufficient condition for a graph to be ℓ-distance stable.

Theorem 4: A graph G is ℓ-distance stable if and only if between any pair of non-adjacent vertices $x, y \in V(G)$ there are at least ℓ edge disjoint paths of length $d_G(x,y)$.

Proof: Sufficiency was stated in Lemma 2. Necessity is derived from Theorem 2, since a graph is ℓ-distance stable if and only if any pair of vertices in G is an ℓ-distance pair. ∎

There is an analogy between Theorem 4 characterizing ℓ-distance stable graphs and the "edge" version of the Menger-Whitney Theorem ([3], Chapter 5) derived by Ford and Fulkerson, Elias, Feinstein, Shanon and A. Kotzig. A similar result for ℓ-distance-stable graphs with respect to vertices appears in [4].

The rest of this paper is devoted to $(2,d)$-graphs.

3. Some Classes of $(2,d)$-graphs. The following theorem gives a sufficient condition for a graph to be a $(2,d)$-graph.

Theorem 5: If for some integer $m \ge 0$, we have for a graph G,

(1) $g(e) \le 3 + m$, $\forall$ $e \in E(G)$.

(2) any pair of vertices $x, y \in V(G)$, such that $d_G(x,y) \ge d-m$ is joined by at least two edge disjoint paths of length not exceeding d;

Then G is a $(2,d)$-graph.

Proof: Let $x, y \in V(G-e)$. If $d_G(x,y) \ge d-m$, then by (2) $d_{G-e}(x,y) \le d$. Assume therefore, $d_G(x,y) \le d-m-1$; by (1), $d_{G-e}(x,y) \le 3 + m-1 + d-m-2 = d$. ∎

Theorem 6 is a consequence of Theorem 5 $(m=0)$ and Theorem 3.

Theorem 6: Let G be a graph such that $g(e) = 3$ $\forall$ $e \in E(G)$. Then G is 2-diameter stable graph if and only if every pair of diametrical vertices in G is joined by at least two edge disjoint paths of length $d(G)$.

By Theorem 6, the following class of graphs with arbitrary diameter, is a class of 2-diameter stable graphs. Take a cycle with even number of vertices labeled $v_1, v_2, \ldots, v_{2n}$. To each pair of adjacent vertices $v_i, v_{i+1 \pmod{2n}}$ $(i = 1,2,\ldots,2n)$ join a vertex w_i, such that w_i is adjacent to v_i and to $v_{i+1 \pmod{2n}}$ $(i = 1,2,\ldots,2n)$. The resulting graph H_{2n} is of diameter $n+1$, and is

2-diameter-stable. The graph H_{2n} shows that Lemma 1 is not a necessary condition for stability if $\ell = 2$, unless the diameter is less than 4.

Theorem 7: A graph G with diameter 2 or 3 is 2-diameter-stable if and only if there are at least two edge disjoint paths of length not exceeding $d(G)$, between any pair of vertices of G.

Proof: Sufficiency follows from Lemma 1. If $d(G) = 2$, necessity follows Theorems 1 and 3. If $d(G) = 3$ then for any pair of vertices $x,y \in V(G)$ such that $d_G(x,y) = 1,3$, the necessity again is derived from Theorems 1 and 3. Therefore, let $x,y \in V(G)$ be two vertices such that $d_G(x,y) = 2$, and denote the shortest x,y-path by $P_1 (P_1 = xzy)$. Since G is 2-diameter stable there exists, an x,y-path $P_2 \neq P_1$ and $2 \leq |P_2| \leq 3$. If $|P_2| = 2$ then P_1, P_2 are vertex disjoint. If all x,y-paths ($\neq P_1$) are of length 3, and none of them is disjoint from P_1, take such a path $P_3 = xaby$, and assume without loss of generality that $xz = xa$. There must be an x,y-path P_4 of length 3 ($P_4 = xcdy$) such that $xc \neq xz$, otherwise $d_{G-xc}(x,y) > 3$, contradicting the 2-diameter stability of G. In this case P_3 and P_4 are edge disjoint x,y-paths of length 3. ∎

Conclusions: This note presents a preliminary study of diameter stability. It suggests a variety of open problems. We mention only a few of them.

(1) Find a characterization of (ℓ,d)-graphs and of ℓ-diameter stable graphs.

(2) Given integers $\ell, d \geq 3$, find a family of (ℓ,d)-graphs (ℓ-diameter stable graphs) for which the condition of Lemma 1 does not hold.
(In other words, prove that Lemma 1 is not a necessary condition for a graph to be an (ℓ,d)-graph.)

(3) Obtain the minimal number of edges that an ℓ-diameter stable graph on n vertices must have.

(4) Theorem 5 is apparently not a necessary condition, find a counter example.

References:

1. J. Hartman and I. Rubin, "On Diameter Stability of Communication Networks", Proc. 1976 Conf. on Information Sciences and Systems, The Johns Hopkins Univ.

2. J. W. Moon, "On the Diameter of a Graph", Michigan Math. J., 12 (1965), pp. pp. 349-351.

3. F. Harary, Graph Theory, Addison-Wesley, Reading, Mass., 1969.

4. R. C. Entringer, D. E. Jackson and P. J. Slater, "Geodetic Connectivity of Graphs", to appear.

5. U. S. R. Murty, "On Critical Graphs of Diameter 2", Math. Mag., 41 (1968), pp. 138-140.

6. U. S. R. Murty, "On Some Extremal Graphs", Acta. Math. Acad. Sci. Hungar., 19 (1968), pp. 69-74.

7. B. Bollobás, "Graphs of Given Diameter", Proc. Coll. at Tihany (Hungar.), 1968, pp. 29-36.

8. B. Bollobás and P. Erdös, "An Extremal Problem of Graphs with Diameter 2", to appear.

Acknowledgement: This work was supported by the Office of Naval Research under Contract N00014-75-C-0609 and by the National Science Foundation under Grant ENG75-03224.

GENERALIZED PSEUDOSURFACE EMBEDDINGS OF GRAPHS AND ASSOCIATED BLOCK DESIGNS

Clare E. Heidema
Syracuse University
Syracuse, New York 13210

Abstract

In this paper we discuss recent results concerning the embedding of graphs into topological spaces which differ from surfaces at only finitely many points, called generalized pseudosurfaces. Such graph embeddings can be associated with certain classes of block designs. We will review the correspondences between graph embeddings and block designs, and then we will discuss some implications of these relationships to the constructing of new block designs as well as to the constructing and analyzing of proper graph embeddings.

1. INTRODUCTION

The study of block designs has flourished primarily because of their applications to scheduling problems, to the theory of design of experiments as applied in statistics and to coding theory, but also because block designs arise in many mathematical contexts; e.g., in connection with Latin squares, finite projective geometries, finite Euclidean geometries, difference sets, graphs and hypergraphs. There has been a recent resurgence of interest in the study of graph embeddings due primarily to the Ringel-Youngs solution [24] to the Heawood map-coloring theorem [14].

A relationship between certain block designs, two-fold triple systems, and triangular complete graph embeddings was first noted by Heffter [15], [16] who in 1891 made the first significant contribution to the solution of the Heawood map-coloring theorem. In 1897 he explored using graph embeddings in the construction of triple systems. Emch [6] employed triangular graph embeddings associated with designs as an aid in calculating the automorphism groups of the designs. The work of Heffter and Emch seems to have received little attention, possibly because very few explicit graph embeddings were known at that time. Now, however, recent work on graph embedding problems has provided some techniques for generating specific types of embeddings and, consequently, an abundance of graph embeddings. The relationship between block designs and graph embeddings is receiving a deserved re-examination.

We will survey recent results in the study of the relationships between graph embeddings and block designs and will note some implications of these relationships.

2. GENERALIZED PSEUDOSURFACES

A <u>surface</u> is a compact, connected Hausdorff space in which every point has an open neighborhood homeomorphic to an open disk; that is, a compact, connected 2-manifold. If $\mathcal{S}$ is a surface, then the classification theorem for surfaces (see Massey [22, p.29]) says that $\mathcal{S}$ is homeomorphic to a sphere or to a connected sum of tori or to a connected sum of projective planes. A surface is <u>orientable</u> if it is homeomorphic to a sphere or to a connected sum of tori. Since a connected sum of k $(k \geq 1)$ tori can be realized by attaching k handles to a sphere, we denote by S_k the orientable surface obtained by attaching k $(k \geq 0)$ handles to a sphere. The <u>genus</u> of the orientable surface S_k is k. A non-orientable surface is one which is homeomorphic to a connected sum of projective planes and as such can be realized by attaching cross-caps to a sphere. We denote by N_k the non-orientable surface which results from attaching k $(k \geq 1)$ cross-caps to a sphere. The genus of the non-orientable surface N_k is k.

Another topological invariant for a surface $\mathcal{S}$, the <u>Euler characteristic</u> $\epsilon(\mathcal{S})$, can be defined by

$$\epsilon(S_k) = 2 - 2k, \quad k \geq 0$$

and

$$\epsilon(N_k) = 2 - k, \quad k \geq 1 .$$

We will be interested in embeddings of graphs into surfaces and into topological spaces which fail to be surfaces at only finitely many points. Such spaces can be realized as certain identification spaces, and have been called pinched manifolds or pseudomanifolds or pseudosurfaces. We adopt the terminology of White [28].

Let $\mathcal{S}_1, \ldots, \mathcal{S}_k$ be pairwise disjoint surfaces and let $\{X_1, \ldots, X_t\}$ be a collection of pairwise disjoint subsets of $\bigcup_{i=1}^{k} \mathcal{S}_i$ with $1 < |X_j| < \infty$ for all $j = 1, \ldots, t$. We define an equivalence relation $\sim$ by

$$x \sim y \text{ if and only if } x = y \text{ or } x, y \in X_j \text{ for some } j = 1, \ldots, t .$$

If under $\sim$ the identification space $\bigcup_{i=1}^{k} \mathcal{S}_i/\sim$ is a connected topological space, then it is called a generalized pseudosurface or more precisely a k-component pseudosurface. Clearly, a generalized pseudosurface depends only on the underlying surfaces $\mathcal{S}_1,\ldots,\mathcal{S}_k$ and the set $\{X_1,\ldots,X_t\}$; therefore, we denote it by $\mathcal{P}\left(\bigcup_{i=1}^{k} \mathcal{S}_i ; \{X_1,\ldots,X_t\}\right)$. A 1-component pseudosurface is called simply a pseudosurface.

The points at which a generalized pseudosurface fails to be a surface are called singular points, and they are precisely the t points which result by identifying the points of X_j for each $j = 1,\ldots,t$. The singular point associated with X_j is said to have multiplicity $|X_j|$. Surfaces are, of course, pseudosurfaces with no singular points.

The components of a k-component pseudosurface are the spaces $\mathcal{S}_i/\sim$ for $i = 1,\ldots,k$. Clearly, each component is a pseudosurface with singular points resulting from identifying the points of $X_j \cap \mathcal{S}_i$ for all $j = 1,\ldots,t$ such that $|X_j \cap \mathcal{S}_i| > 1$. A pseudosurface is orientable if its underlying surface is orientable, and a generalized pseudosurface is orientable if all of its components are orientable.

Let $\mathcal{P}$ be a generalized psuedosurface $\mathcal{P}\left(\bigcup_{i=1}^{k} \mathcal{S}_i ; \{X_1,\ldots,X_t\}\right)$; we define the Euler characteristic of $\mathcal{P}$ by

$$\epsilon(\mathcal{P}) = \sum_{i=1}^{k} \epsilon(\mathcal{S}_i) - \sum_{j=1}^{t} (|X_j|-1) .$$

Generalized pseudosurfaces, unlike surfaces, cannot be characterized by orientability and Euler characteristic. It is possible to have two non-homeomorphic generalized pseudosurfaces with the same underlying surfaces and the same Euler characteristic. However, the Euler characteristic formula (see Theorem 1 below) holds and, in fact, motivated the above definition of Euler characteristic for generalized pseudosurfaces.

3. GRAPH EMBEDDINGS

Unless otherwise specified, $\Gamma = (V, \mathcal{E})$ denotes a connected graph (no loops and no multiple edges) with vertex set V and edge set $\mathcal{E}$. An <u>embedding</u> of a graph Γ into a generalized pseudosurface $\mathcal{P}$ is a homeomorphism of a topological realization of Γ into $\mathcal{P}$ such that each singular point of $\mathcal{P}$ is the image of some vertex of Γ. Therefore, the image of an embedding homeomorphism, which we denote $\Gamma(\mathcal{P})$, is a subspace of $\mathcal{P}$ containing all the singular points of $\mathcal{P}$ and homeomorphic to Γ. We usually refer to this subspace as the embedding of Γ into $\mathcal{P}$. The components of $\mathcal{P} - \Gamma(\mathcal{P})$ are the <u>regions</u> of the embedding. If every region of an embedding is a 2-cell (that is, homeomorphic to an open disk), then the embedding is said to be a <u>proper</u> or <u>2-cell embedding</u>.

The number of regions of a proper embedding of a connected graph in a generalized pseudosurface is finite and is completely determined by the graph and the generalized pseudosurface. This fact is the essence of the famous Euler characteristic formula, which is well-known in the case that the embedding is into a surface (see Massey [22, p.29]) and was established by Petroelje [23] in the case that the embedding is into a pseudosurface. To show that the formula also holds in the case that the embedding is into a generalized pseudosurface is an easy consequence of these earlier results and the definition of Euler characteristic.

<u>Theorem 1</u>. (Euler Characteristic Formula) Suppose Γ has ν_0 vertices, ν_1 edges and has a proper embedding into a generalized pseudosurface $\mathcal{P}$ with ν_2 regions. Then

$$\nu_0 - \nu_1 + \nu_2 = \varepsilon(\mathcal{P}) .$$

Because surfaces become more complicated (that is, acquire more handles or crosscaps) as the Euler characteristic decreases, it is natural to define a surface embedding of a graph to be <u>simplest</u> if it is into a surface of maximum Euler characteristic among all surfaces into which the graph will embed. However, for generalized pseudosurface embeddings, this definition is not useful because it is not possible to choose a space of maximum Euler characteristic. For example, if a given graph embeds into a generalized pseudosurface of Euler characteristic ϵ, then by attaching a sphere to this space at a point associated with a vertex of the graph, one finds that the graph also embeds into a generalized pseudosurface of Euler characteristic $\epsilon + 1$. A solution to this difficulty is to define a generalized pseudosurface embedding of a graph to be <u>simplest</u> if it is into a k-component pseudosurface $\mathcal{P}$ for which the number $\epsilon(\mathcal{P}) - k$ is a maximum.

If $\Gamma(\mathcal{P})$ is a simplest embedding, then $\epsilon''(\Gamma)$ denotes the <u>general pseudocharacteristic</u> of Γ defined as $\epsilon(\mathcal{P})$. We say simply pseudocharacteristic (characteristic) when we restrict consideration to pseudosurface (surface) embeddings and denote this by $\epsilon'(\Gamma)(\epsilon(\Gamma))$.

A well-known characterization theorem of Youngs [29] states that a surface embedding of a connected graph is a simplest embedding if and only if it is a maximal (in terms of the number of regions) proper embedding. Petroelje [23] has shown that an analogous result holds for orientable pseudosurface embeddings. The corresponding result for generalized pseudosurface embeddings holds when we ignore superfluous embeddings; that is, embeddings in which some region is bounded by 0 or 2 edges.

The quest for a combinatorial and ultimately algebraic approach to graph embeddings began with MacLane's [19] characterization of planar graphs. Graver [8] extended MacLane's result to non-planar graphs by introducing a combinatorial substitute for the genus of a graph, which he calls the <u>embedding index</u>. The main result of [8] is that for any graph Γ, $\iota(\Gamma) = 2 - \epsilon'(\Gamma)$ where $\iota(\Gamma)$ is the embedding index of Γ.

Edmonds [5] announced a combinatorial definition for embedding graphs into orientable surfaces suggesting a useful method of constructing such embeddings. Since it appears that Heffter was aware of this method and used it in his attack on Case 7 of the Heawood conjecture, it is known as the Heffter-Edmonds embedding technique. An extension of Edmonds' theorem can be established for proper embeddings of graphs into generalized pseudosurfaces as follows. A pseudorotation on a connected graph Γ is a map that assigns to each vertex $v \in V$ a permutation σ_v of $V(v) = \{u \in V : \{u,v\} \in \mathcal{E}\}$. Let $D = \{(u,v) : \{u,v\} \in \mathcal{E}\}$ and define $\Sigma : D \to D$ by $\Sigma((u,v)) = (v, \sigma_v(u))$. Then Σ is a permutation of D and the orbits of Σ describe oriented regions of a proper embedding of Γ.

Theorem 2. For a connected graph Γ, every pseudorotation σ on Γ determines a proper embedding of Γ into an orientable generalized pseudosurface. Conversely, given a proper embedding of Γ into an oriented generalized pseudosurface and with no region bounded by 0 edges, there is a pseudorotation on Γ which determines this embedding.

Another combinatorial approach to graph embeddings is the association of a block design to each graph embedding. In this case, however, the impetus for studying the relationships between block designs and graph embeddings is as much to learn about block designs from the corresponding graph embeddings as to learn about graph embeddings from the block designs.

4. BLOCK DESIGNS

A balanced incomplete block design (BIBD) with positive integral parameters b, v, r, k and λ, called a (b,v,r,k,λ)-design, consists of a finite set V of v objects, called vertices, and a set $\mathcal{B}$ of b not necessarily distinct k-subsets of V, called blocks, satisfying (i) each vertex is in exactly r blocks, and (ii) each pair of distinct vertices occurs in exactly λ blocks. Simple counting arguments can be used to establish the following well-known relations on the five parameters for a BIBD (see Hall [13, p.101]):

$$vr = bk \quad \text{and} \quad \lambda(v-1) = r(k-1)$$

or equivalently

$$r = \frac{\lambda(v-1)}{k-1} \quad \text{and} \quad b = \frac{\lambda v(v-1)}{k(k-1)} .$$

(Note that the definition of a BIBD forces $v,k \geq 2$) . Thus, it is clear that if a BIBD exists, then its parameters are determined by v, k and λ ; therefore, a (b,v,r,k,λ)-design may more simply be called a (v,k,λ)-design.

In an attempt to construct experimental designs related to BIBD's and with possible values of v, b, r and k which are not related as above, Bose and Nair [3] introduced the notion of a <u>partially balanced incomplete block design</u> (PBIBD). Let $\lambda_1 > \lambda_2 > \ldots > \lambda_m \geq 0$ be integers. Then an m-class PBIBD with parameters $b,v,r,k,\lambda_1,\ldots,\lambda_m$, called a $(b,v,r,k;\ \lambda_1,\ldots,\lambda_m)$-design, consists of a finite set V of v objects (vertices) and a set $\mathcal{B}$ of b k-subsets of V (blocks) with $k < v$ satisfying (i) each vertex is in exactly r blocks where $r < b$, (ii) each pair of vertices occurs in exactly λ_1 or $\lambda_2 \cdots$ or λ_m blocks, (iii) for each $x \in V$ the number of vertices $y \neq x$ which are <u>i^{th}-associates</u> of x is a constant n_i , $i = 1,\ldots,m$, (by i^{th}-associate we mean that the pair $\{x,y\}$ occurs in exactly λ_i blocks) and (iv) for any $x,y \in V$ if x,y are i^{th}-associates, then the number of vertices which are both j^{th}-associates of x and k^{th}-associates of y is a constant p^i_{jk} . This last condition is an attempt to restore some of the balance to the experiment which is lost when one λ cannot be found for all pairs $\{x,y\}$.

5. CORRESPONDENCES

A triple system is a $(v,3,\lambda)$-design. Of special interest here are Steiner triple systems $(\lambda = 1)$ and twofold triple systems $(\lambda = 2)$. One correspondence between graph embeddings and block designs, announced by Alpert [1], associates certain triangular graph embeddings and twofold triple systems. To each block of a twofold triple system, we associate a triangle whose vertices are labeled by the objects in that block. Since $\lambda = 2$, each pair of vertices appears exactly twice so that the standard identification procedure of combinatorial topology yields a 2-manifold whose connected components are triangulated surfaces. Recall that in this identification procedure, edges which are labeled by the same pair of objects will be identified in the appropriate way but no other identifications are made. Now, if we identify identically labeled vertices, the result is a triangular embedding of the complete graph K_v into a generalized pseudosurface.

Theorem 3. (Alpert) The above correspondence is a bijection between twofold triple systems and triangular embeddings of complete graphs into generalized pseudosurfaces.

Special cases of this theorem are determined when we classify some twofold triple systems. Two blocks B_1 and B_2 of a twofold triple system are adjacent if $|B_1 \cap B_2| = 2$. Using this notion of adjacency to construct in the usual way the adjacency graph of a system, we say that a twofold triple system is connected if its adjacency graph is connected. A vertex x is simple if the subgraph of the adjacency graph constructed by considering only blocks containing x is connected, and a twofold triple system is simple if all of its vertices are simple. A twofold triple system is orientable if there exists an assignment $\{x,y,z\} \mapsto (x,y,z)$ or (x,z,y) on $\mathcal{B}$ such that if $\{x,y,w\}, \{x,y,z\} \in \mathcal{B}$, then $\{x,y,w\} \mapsto (x,y,w)$ and $\{x,y,z\} \mapsto (x,z,y)$ or $\{x,y,w\} \mapsto (x,w,y)$ and $\{x,y,z\} \mapsto (x,y,z)$.

Theorem 4. There is a bijective correspondence between

a) connected twofold triple systems and triangular embeddings of complete graphs into pseudosurfaces;

b) simple twofold triple systems and triangular embeddings of complete graphs into surfaces;

c) orientable twofold triple systems and triangular embeddings of complete graphs into orientable generalized pseudosurfaces.

The adjacency graph of a twofold triple system is the dual graph of its corresponding graph embedding. If the adjacency graph is bipartite, then the twofold triple system decomposes into two Steiner triple systems. (Such a twofold triple system is said to be decomposable.) Also, any two Steiner triple systems with the same vertex set V can be composed to form a decomposable twofold triple system with vertex set V. Thus

Theorem 5. There is a bijective correspondence between pairs of Steiner triple systems on v objects and triangular embeddings with bipartite dual of K_v into generalized pseudosurfaces.

Let Γ be a regular, non-complete, non-empty graph and let $x,y \in V$ be adjacent (non-adjacent) vertices. Define $p^1(x,y)$ ($p^2(x,y)$) to be the number of vertices adjacent to both x and y. If both p^1 and p^2 are constant functions, then Γ is said to be strongly regular. Associating triangles to blocks of a $(b,v,r,3;2,0)$-design and making identifications as above (since a pair of vertices occurs in 2 or 0 blocks, every edge will again have exactly one mate), the result will be a triangular embedding of a strongly regular graph into a space whose connected components are generalized pseudosurfaces. The following results of White [28] are analogous to Theorems 3 and 5.

Theorem 6. The above correspondence is a bijection between (b,v,r,3;2,0)-designs and triangular embeddings of strongly regular graphs of order v into spaces with generalized pseudosurface components.

Theorem 7. There is a bijective correspondence between pairs of (b,v,r,3;1,0)-designs on the same vertex set and triangular embeddings with bipartite dual of strongly regular graphs of order v into spaces with generalized pseudosurface components.

6. CONSTRUCTION OF DESIGNS AND ANALYZING OF GRAPH EMBEDDINGS

An abundance of literature exists on the construction of BIBD's and in particular on constructions of triple systems. See for example [2], [4], [11], [12], [17], [18] and [26]. A close examination of these constructions in the case of twofold triple systems reveals that, with the exception of the Bhattacharya constructions, the resulting designs are disconnected. Bhattacharya's constructions of twofold triple systems on $6s+6$ and $6s+4$ objects use Bose's method of symmetrically repeated differences and give connected, non-orientable designs which, in general, are not simple. Therefore, as noted by Alpert [1], the triangular embeddings of complete graphs K_n, $n \equiv 0,3,4,7 \pmod{12}$, into orientable surfaces (see for example [10], [25] and [27]) all correspond to new twofold triple systems. Also, the triangular embeddings of complete graphs K_n, $n \not\equiv 2 \pmod 3$ and $n \neq 7$, into non-orientable surfaces (see [10], [21], [25] and [27]) in general correspond to new twofold triple systems. Note that these embeddings together with Theorem 3 give another proof that the necessary conditions for the existence of a (v,3,2)-design are in fact sufficient. Bhattacharya's [2] constructions gave the first proof of this fact; and, using different methods, Hanani [11] showed that the necessary conditions are also sufficient for the existence of a $(v,3,\lambda)$-design with any choice of λ.

Close scrutiny of schemes used in the construction of triangular embeddings of complete graphs may reveal interesting properties of the corresponding block designs. For example, consider the following current graph with currents from $\mathbb{Z}_{19}$ used to construct an orientable embedding of K_{19} .

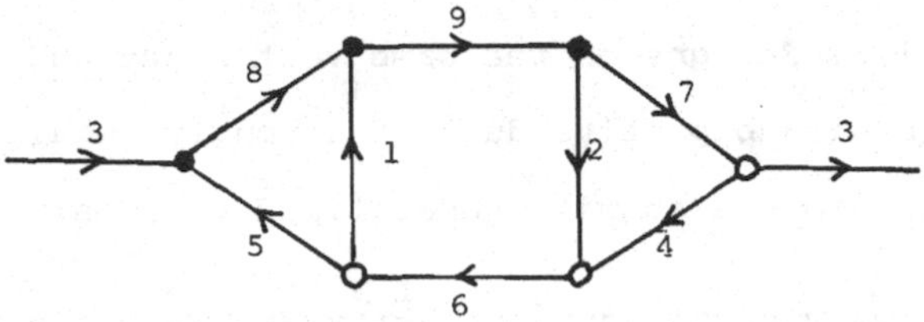

For easier presentation, the graph is drawn in the plane (instead of the quotient manifold S_2) with the usual convention that a solid vertex has edges ordered clockwise and a hollow vertex has edges ordered counterclockwise. The dual of this current graph embedded in S_2 is a voltage graph (see Gross [9]):

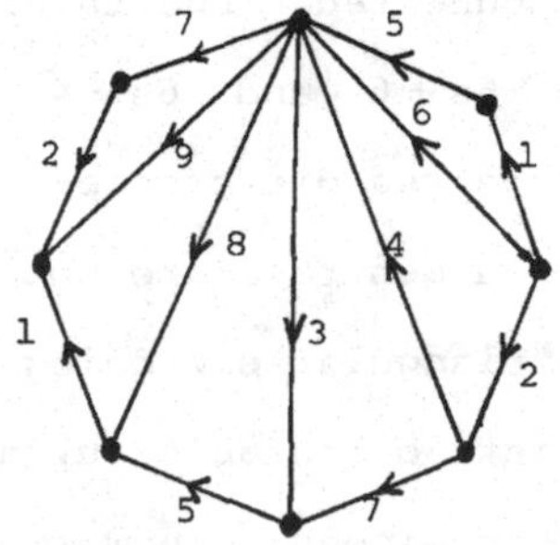

and the derived graph is a 19-fold covering of this (pseudo)graph. The derived embedding is a triangular embedding of K_{19} into S_{20} . Since the base space, in this case a bouquet of 9 circles triangularly embedded in S_2 , does not have bipartite dual, the covering space K_{19} embedded into S_{20} does not. Therefore, the corresponding twofold triple system is indecomposable and can be found immediately from the voltage graph. Blocks are cyclically generated (addition modulo 19) by those corresponding to the 6 triangles in the voltage graph; namely, $\{0,1,6\}$, $\{0,2,6\}$, $\{0,4,7\}$, $\{0,3,8\}$, $\{0,8,9\}$ and $\{0,7,9\}$. Similarly, one shows that for $n \equiv 7 \pmod{12}$ the index 1 published solutions for orientable triangular embeddings of K_n correspond to indecomposable twofold triple systems.

Consideration of the twofold triple systems corresponding to triangular surface embeddings of a complete graph may answer negatively the question of whether such embeddings are unique; that is, are the duals of any two triangular surface embeddings of a given complete graph isomorphic? For example, it is possible to find both a decomposable and an indecomposable simple twofold triple system on 9 vertices; therefore, K_9 has triangular embeddings into N_5 both with and without bipartite duals. The same is true for non-orientable triangular embeddings of K_{15} and orientable triangular embeddings of K_{19}.

White [28] uses known triangular embeddings of strongly regular graphs and voltage graph theory to construct many new PBIBD's. In particular using the orientable triangular embeddings for $K_{3(m)}$ from Garman, Ringeisen and White [7], all having bipartite dual, he finds $(2m^2,3m,2m,3;2,0)$-designs each splitting into two $(m^2,3m,m,3;1,0)$-designs. Also, using results in [7] which indicate a doubling of strongly regular graphs and a construction of triangular embeddings with bipartite duals from given such embeddings, White finds infinite collections of new PBIBD's occurring in triples, one $(b,v,r,3;2,0)$-design and two $(b/2,v,r/2,3;1,0)$-designs.

Observing that the group divisible $(b,v,r,3;2,0)$-designs correspond to triangular embeddings of strongly regular graphs $K_{n(m)}$ with $v = n \cdot m$ and employing a result of Hanani [12] which states that a group divisible $(b,v,r,3;2,0)$-design with vertex set partitioned into n groups of m vertices exists if and only if $3 \mid mn(n-1)$, White proves the following graph embedding theorem.

Theorem 8. For $n > 2$, $\varepsilon''(K_{n(m)}) = \frac{mn(6+m-mn)}{6}$ if and only if $3 \mid mn(n-1)$.

An easy consequence of results from design theory is

Theorem 9. Triangular embeddings of K_n into orientable generalized pseudosurfaces exist if and only if $v \equiv 0,1 \pmod 3$ and $v \neq 1,6$.

This theorem follows immediately from Theorem 4(c) and the corresponding result of Mendelsohn [20] for orientable twofold triple systems.

We have seen a rich interplay between the theory of designs and the theory of graph embeddings which contributes to both. Recent advances in the theory of graph embeddings have made possible alternate methods of solving problems in the theory of designs. The topological viewpoint of graph embedding theory may shed light on the isomorphism questions of design theory. Results from the theory of designs have been useful both in analyzing existing graph embeddings and in constructing new embeddings.

REFERENCES

1. S. R. Alpert, Twofold Triple Systems and Graph Embeddings. J. Combinatorial Theory (A) 18 (1975), 101-107.
2. K. N. Bhattacharya, A Note on Two-fold Triple Systems. Sankhyā 6 (1943), 313-314.
3. R. C. Bose and K. R. Nair, Partially Balanced Incomplete Block Designs. Sankhyā 4 (1939), 337-372.
4. J. Doyen, Construction of Disjoint Steiner Triple Systems. Proc. Amer. Math. Soc. 32 (1972), 409-416.
5. J. R. Edmonds, A Combinatorial Representation for Polyhedral Surfaces. Notices Amer. Math. Soc. 7 (1960), 646.
6. A. Emch, Triple and Multiple Systems; Their Geometric Configurations and Groups. Trans. Amer. Math. Soc. 31 (1929), 25-42.
7. B. L. Garman, R. D. Ringeisen and A. T. White, On the Genus of Strong Tensor Products of Graphs. Submitted for publication.
8. J. E. Graver, The Imbedding Index of a Graph. Submitted for publication.

9. J. L. Gross, Voltage Graphs. Discrete Math. 9 (1974), 239-246.

10. R. K. Guy and J. W. T. Youngs, A Smooth and Unified Proof of Cases 6, 5 and 3 of the Ringel-Youngs Theorem. J. Combinatorial Theory (B) 15 (1973), 1-11.

11. H. Hanani, The Existence and Construction of Balanced Incomplete Block Designs. Ann. Math. Stat. 32 (1961), 361-386.

12. H. Hanani, Balanced Incomplete Block Designs and Related Designs. Discrete Math. 11 (1975), 255-369.

13. M. Hall, Combinatorial Theory, Ginn-Blaisdell, Waltham, Massachusetts (1967).

14. P. J. Heawood, Map Colour Theorem. Quart. J. Math. 24 (1890), 332-338.

15. L. Heffter, Über das Problem der Nachbargebiete. Math. Ann. 38 (1891), 477-508.

16. L. Heffter, Über Triplesysteme. Math. Ann. 49 (1897), 101-112.

17. F. K. Hwang and S. Lin, A Direct Method to Construct Triple Systems. J. Combinatorial Theory (A) 17 (1974), 84-94.

18. E. S. Kramer, Indecomposable Triple Systems. Discrete Math. 8 (1974), 173-180.

19. S. MacLane, A Combinatorial Condition for Planar Graphs. Fund. Math. 28 (1937), 22-32.

20. N. S. Mendelsohn, A Natural Generalization of Steiner Triple Systems. Proc. Sci. Res. Council Atlas Sympos. No. 2 Oxford (1969), 323-338.

21. E. M. Landesman and J. W. T. Youngs, Smooth Solutions in Case 1 of the Heawood Conjecture for Non-orientable Surfaces. J. Combinatorial Theory (B) 13 (1972), 26-39.

22. W. S. Massey, Algebraic Topology: An Introduction. Harcourt, Brace and World, Inc., New York (1967).

23. W. Petroelje, Imbedding Graphs on Pseudosurfaces. Specialist Thesis, Western Michigan University, Kalamazoo, Michigan (1971).

24. G. Ringel and J. W. T. Youngs, Solution of the Heawood Map-Coloring Problem. Proc. Nat. Acad. Sci. U.S.A. 60 (1969), 343-353.

25. G. Ringel, Map Color Theorem. Springer-Verlag, Berlin (1974).

26. J. VanBuggenhaut, On the Existence of 2-designs $S_2(2,3,v)$ Without Repeated Blocks, Discrete Math. 8 (1974), 105-109.

27. A. T. White, Graphs, Groups and Surfaces. North-Holland, Amsterdam (1973).

28. A. T. White, Block Designs and Graph Imbeddings. Western Michigan University Mathematics Report #41, Kalamazoo, Michigan (1975).

29. J. W. T. Youngs, Minimal Imbeddings and the Genus of a Graph J. Math. Mech. 12 (1963), 303-315.

POWERS OF GRAPHS, LINE GRAPHS, AND TOTAL GRAPHS

Arthur M. Hobbs
Department of Mathematics
Texas A & M University
College Station, Texas 77843

This paper contains a brief review of some published results on powers of graphs, line graphs, and total graphs. This review, together with a new characterization of total graphs, indicates that many of the theorems on Hamiltonian cycles and other properties of these graphs almost certainly can be proved in a more general setting. The presentation concludes by indicating a possible direction of research on this problem.

The first section of this paper consists of a partial survey and history of the subject of Hamiltonian cycles in the squares, cubes, line graphs, and total graphs of graphs, together with a unification of the formulations of these theorems The second section includes a similar survey of the characterizations of such graphs. The space allotted for this paper is not sufficient to do justice even to these limited areas of graph theory, and besides I am aiming at a specific conclusion in each section. Hence very many interesting and worthwhile results have been omitted (in particular, I have skipped all results relating to pancyclic and panconnected graphs, although these are certainly very important). I have even had to omit some results which are very closely related to the work I present here. I want to apologize now to all of the fine mathematicians I am thereby slighting.

I use the terminology of Behzad and Chartrand [1], but the class of graphs is here assumed to include those with parallel edges. I denote a trail or walk by its sequence of vertices. Given a graph G, the <u>subdivision</u> $S(G)$ of G is a graph obtained from G by replacing each edge of G with a path of length two in such a way that no vertex of $S(G) - V(G)$ has degree different from two in $S(G)$ The nth power G^n of a graph G is a graph with no parallel edges and with $V(G^n) = V(G)$ such that two vertices of G^n are adjacent if and only if their distance in G is in $\{1, 2, \ldots, n\}$. We call G^2 the square of G and G^3 the cube of G. The line graph $L(G)$ of G is the graph whose vertices are the edges of G such that two vertices of $L(G)$ are adjacent if and only if they are edges incident with the same vertex of G. The total graph $T(G)$ is the graph whose vertices are the edges and vertices of G in which two vertices of $T(G)$ are adjacent if and only if they are adjacent vertices in G, edges of G incident with the same vertex of G, or an edge and a vertex of G such that the edge is incident to the vertex in G. It is not difficult to show that $T(G)$ is isomorphic to $(S(G))^2$. Given a graph G, $V_i(G)$ is the set of vertices of G of degree i. Given a subgraph S of G, $G - S = (G - E(S)) - V_0(G - E(S))$, where $E(S)$ is the set of edges of S. Given a vertex v of G, the v-fragments of G are the subgraphs of G of the form $\langle V(H) \cup \{v\} \rangle$, where H is a component of $G - v$. $|A|$ denotes the number of elements in set A.

I. HAMILTONIAN CYCLES

One of the major attacks on the Hamiltonian cycle problem was begun in about 1960 with work on line graphs and powers of graphs. In this work, a graph-valued function f acting on graphs is specified, and conditions are placed on the graphs in the domain of f to guarantee that their images are Hamiltonian. Most of the work I have done in the past eight years has been in this area, and I wish now to report a few of my observations and one direction in which I will be working in the next few months.

The subject of powers of graphs was introduced independently by I. C. Ross and F. Harary [24], who named it, and by M. Sekanina [25], when he was solving a problem in topology. Translating Sekanina's results, he proved (and independently later Karaganis proved):

THEOREM 1 (Sekanina [25], Karaganis [18]): *If G is a connected graph, then G^3 is Hamiltonian connected.*

I will discuss this theorem further later on. Sekanina asked [26], in effect, which squares of graphs are Hamiltonian. His student, F. Neuman [22], determined for any tree and any two distinct vertices u and v of the tree, whether or not the square of the tree contains a Hamiltonian u,v-path. A consequence of Neuman's work, which was later proved more directly and simply by Harary and Schwenk [14], is the following theorem:

THEOREM 2: The square T^2 of a tree T with at least three vertices is Hamiltonian if and only if $T - V_1(T)$ is a path.

Let e be an Eulerian circuit in a spanning Eulerian subgraph of a graph G. A Hamiltonian cycle of $f(G)$ follows e if the sequence of vertices of the Hamiltonian cycle can be formed from e by omitting terms from e. In an algorithm constructing the Hamiltonian cycle h from e so that h follows e, we say the algorithm visits each term of e which is not omitted in the formation of h. An edge b of a graph G is doubled if b is replaced by two edges in parallel which join the ends of b. Given sequences a and b, $(a), (b)$ is the sequence formed by listing all of a followed by all of b. A sequence c is a section of the sequence a if there are sequences s and t such that $a = (s), (c), (t)$. A trail is a u,v-trail if its first vertex is the vertex u and its last vertex is the vertex v.

The following algorithm can be used to find a Hamiltonian cycle in the square of a tree T such that $T - V_1(T)$ is a path. Give the vertices along the path $T - V_1(T)$ the labels x_1 through x_n. Double every edge of T to form T^*. Let e be an Eulerian circuit in T^* formed by starting on x_1, passing through all vertices of degree one in T which are adjacent to vertices x_i with i even as e progresses through T^* from x_1 to x_n, and then passing through all vertices of degree one in T which are adjacent to vertices x_i with i odd as e progresses through T^* from x_n back to x_1. Then a Hamiltonian cycle in T^2 is found by the following algorithm:

ALGORITHM I

1) Visit the first vertex of e.

2) After visiting a vertex z, the next vertex z' to be visited should be

as far along e as possible subject to the conditions:

a) the section of e from z to z' shall not have length more than two; and

b) if the next vertex z" after z in e is appearing in e for the last time, and if z" has not been visited, z' shall be z".

3) The last vertex of the Eulerian circuit shall be visited.

It is trivial to check that Algorithm I actually does produce a Hamiltonian cycle in T^2.

An <u>EFS-subgraph</u> of a graph G is a connected spanning subgraph of G which is the edge-disjoint union of an Euler graph E', not necessarily connected, and a forest F' each of whose trees has the property that removing its vertices of degree one results in the empty graph or a path. In an EFS-subgraph S, we call the Euler subgraph E' of S the <u>Euler part</u> of S and always denote it by E', and we call the forest F' of S the <u>forest part</u> of S and always denote it by F'. An EFS-subgraph S is <u>doubled</u> if every edge of F' is doubled.

In 1971, Herbert Fleischner kindly consented to do some joint work with me; we investigated Hamiltonian cycles in the squares of graphs and found:

THEOREM 3 ([9]): <u>If</u> G^2 <u>is Hamiltonian, then there exists an EFS-subgraph</u> S <u>of</u> G <u>such that, if</u> S* <u>is formed from</u> S <u>by doubling the edges of</u> F', <u>then there is an Eulerian circuit</u> e <u>in</u> S* <u>and a Hamiltonian cycle</u> h <u>in</u> G^2 <u>such that</u> h <u>follows</u> e. <u>Further, no two successive vertices of</u> e <u>are omitted in forming</u> h.

This theorem is only necessary; many EFS-subgraphs exist whose squares are not Hamiltonian. Nevertheless, the most productive efforts of mathematicians in finding Hamiltonian cycles in the squares of graphs have been those aimed at finding a suitable EFS-subgraph and the Eulerian circuit described in Theorem 3, and then using some variant of Algorithm I to produce the required Hamiltonian cycle in G^2.

In 1966, Plummer [23] and Nash-Williams [21] independently conjectured that the square of any block with at least three vertices is Hamiltonian. Walther [27] in 1969 and Fleischner [6] in 1970 independently investigated the Plummer -

Nash-Williams conjecture in the special case of cubic graphs and found:

THEOREM 4: <u>If G is a cubic connected graph with a two-factor, then G^2 is Hamiltonian</u>.

COROLLARY: <u>Every two-connected cubic graph has a Hamiltonian square</u>.

The proof of Theorem 4 in both of these papers is by induction. However, each contains the following idea: Let the two-factor of the cubic graph G have k cycles and consider it to be the Euler part of an EFS-subgraph S. Choose $k - 1$ additional edges from G so that the two-factor together with these edges is connected, and consider those additional edges to be the edges of the forest part of S. Choose an arbitrary Eulerian circuit e in the doubling of S. Then use Algorithm I to follow e. It is easy to check that the resulting sequence is a Hamiltonian cycle in G^2.

In early 1970, I proved [15] that if G is an Eulerian graph with the property that $G - V_2(G)$ is a forest, then any Eulerian circuit in G can be followed by a simple modification of Algorithm I to produce a Hamiltonian cycle in G^2. At first I thought that if G is an Eulerian graph whose square is Hamiltonian, then G must contain an Eulerian circuit which can be followed by an algorithm similar to Algorithm I to yield a Hamiltonian cycle in G^2. However, in 1974 I found several counter-examples to this notion. The graph G shown in the Figure is the simplest such counter-example; its square is Hamiltonian, but every Hamiltonian cycle h in G^2 requires at least one edge to be doubled before an Eulerian circuit can be found which is followed by h.

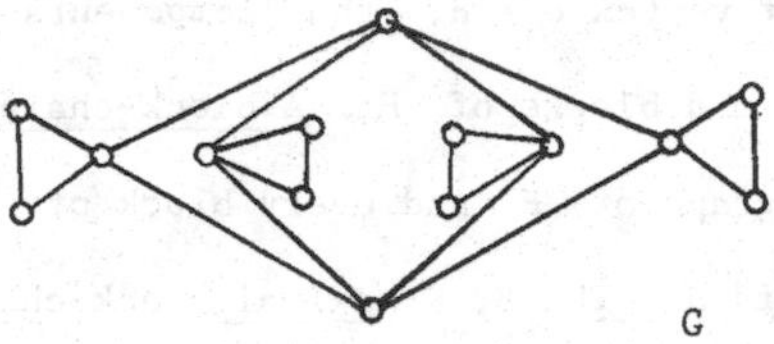

G

While working on my PhD thesis in 1971, I found [16] that the EFS-subgraph required in a cactus by Theorem 3 is precisely the cactus itself if the square of the cactus is Hamiltonian. In the EFS-subgraph, the cycles of the cactus constitute E' and the bridges of the cactus are the edges of F'. I was able to develop an algorithm [17] which follows an arbitrary Eulerian circuit in the doubling of the cactus if the square of the cactus is Hamiltonian, and which concludes that the square is not Hamiltonian if that is the case.

The most difficult and penetrating result obtained so far in the study of squares of graphs is Herbert Fleischner's proof [7, 8] that the square of a block is Hamiltonian. Fleischner made the following initial observation (though he saw it in slightly different terms): If it is true for any two successive vertices of an Eulerian circuit in a graph G that one of them has degree two in G, then Algorithm I can be used to follow the Eulerian circuit to produce a Hamiltonian cycle in G^2. Hence he first studied the total graph of G, which is $(S(G))^2$. I think his first major achievement in this work lay in his <u>believing</u> that any bridgeless connected graph contains an EFS-subgraph, and (even more remarkable) one in which every tree of the forest part is a path. Fleischner called such an EFS-subgraph an <u>EPS-graph</u>. His second major achievement is that he <u>proved</u> the EPS-graph can be found in any bridgeless connected graph. The essential new idea in his proof requires some definitions. Let K be a cycle of a block G, and let $H = G - K$. An <u>end block</u> of H is a block B of H which contains at most one cut vertex of H; thus components of H without cut vertices are also regarded as end blocks of H. A <u>block-chain of</u> H is a block-chain J such that J is a subgraph of H and every block of J is a block of H. Let B be an end block of H. The B, K-<u>maximal block-chain</u> J of H is the block-chain of H with the following properties:

1) B is a subgraph of J;
2) $G - J$ is a block; and
3) no smaller block-chain J' of H has the properties that B is a subgraph of J' and $G - J'$ is a block.

The existence of the B, K-maximal block-chain is easily shown. It is also clear that proofs by induction in two-connected graphs can be made by using such a block-chain. It was necessary to show that a small block has an EPS-graph as a spanning subgraph, and that EPS-graphs can be fitted together at cut vertices of the B, K-maximal block-chain to form a larger EPS-graph. The fitting together caused some difficulties, but Fleischner overcame them by adding the condition that given any two vertices u and v of a block G, then G contains an EPS-graph in which u meets no edge of F' and v meets at most one such edge. With this special EPS-graph, called a [u, v]-EPS-graph, the fitting together at the cut vertices of the block-chain became relatively easy. Thus Fleischner finally proved [7] that every bridgeless connected graph contains an EPS-graph as a spanning subgraph. Doubling this EPS-graph and using a variant of Algorithm I, he proved that the total graph of any bridgeless connected graph is Hamiltonian.

In a second paper [8] , Fleischner used the existence of the [u, v]-EPS-graph in any block, together with a lovely new structural theorem about minimally two-connected graphs, to prove by induction that the square of a block is Hamiltonian.

Fleischner's proof that the square of a block is Hamiltonian does not use for blocks an EFS-subgraph and an Eulerian circuit in its doubling which is followed by a Hamiltonian cycle in the square of the EFS-subgraph. In our first joint paper [9], in addition to showing Theorem 3, he and I showed that the required EFS-subgraph for a block is in fact an EPS-graph.

Theorem 3 also covers total graphs. However, Fleischner and I [10] recently improved Theorem 3 in this case to the following:

THEOREM 5: The total graph of a graph G is Hamiltonian if and only if G contains an EPS-graph.

We went on in [10] to investigate which graphs contain EPS-graphs and thus have Hamiltonian total graphs. Additional work in powers of graphs, while interesting and extensive, is omitted for lack of space.

Harary and Nash-Williams [13] developed a characterization of graphs whose line graphs are Hamiltonian. This work is included here because of its similarity to the other theorems in this section.

THEOREM 6 (Harary and Nash-Williams [13]): L(G) is Hamiltonian if and only if there is a circuit in G which includes at least one end vertex of each edge of G.

We can introduce an algorithm here also in which an Eulerian circuit is followed by the Hamiltonian cycle in L(G). First, subdivide every edge of G. Let the subdivided subgraph formed from the circuit named in the theorem be the Euler part E' of an EFS-subgraph S, and form the forest part by including in

F' for each vertex in S(G) - V(G) not in E' an edge from that vertex to a vertex of E'. Form S* from S by doubling each edge in F', and find an arbitrary Eulerian circuit e in S*. Eliminate every vertex of G from e; the resulting sequence is a Hamiltonian cycle in L(G) which follows e.

We have seen that if f(G) is the square of G, the total graph of G, or the line graph of G, then f(G) has a Hamiltonian cycle only if there is a subgraph in G which can be modified to allow an Eulerian circuit which is followed by a Hamiltonian cycle in f(G). However, we have not shown the equivalent result for cubes of graphs, since the published proofs [25] and [18] of Theorem 1 do not have the algorithmic character that we have found for Theorems 2 through 6. I have therefore developed an algorithm for finding a Hamiltonian path joining a specified two vertices in the cube of a tree. This algorithm uses the method of finding a Hamiltonian path which follows an Eulerian trail in the graph produced by doubling some of the edges of the tree. The proof of the algorithm strongly resembles the published proofs [25] and [18] of Theorem 1.

ALGORITHM II

Let T be a tree and let u and v be distinct vertices of T such that there is a u-fragment of T not containing v if T has more than two vertices and u and v are adjacent. Form T* from T by doubling every edge of T except those on the path joining u and v. Let e be an Eulerian trail in T* from u to v. Visit the vertices of T as follows:

1) Visit u.
2) After visiting vertex z, the next vertex z' to be visited must be as far as possible along e, subject to:
   a) z' must be within distance three of z in e;
   b) z' cannot have been previously visited; and
   c) if an unvisited vertex is being passed by e for the last time, it must be visited.
3) Stop when no unvisited vertices are within distance three of the last visited vertex. The sequence of vertices in the order visited is a Hamiltonian u,v-path in T^3.

PROOF: It is trivial that the claimed result is correct if T has only two vertices. Suppose u and v are adjacent, and suppose T has at least three vertices. Let T' be the union of the u-fragments of T^* which do not include v. Because uv is a bridge of T^*, e has only one section in T'. Let e' be the section of e in T', and let e'' be formed from e' by removing its last term. Let x be the last term of e''; then e'' is an Eulerian u, x-trail in T'. By induction, the algorithm using e'' finds a Hamiltonian u,x-path h' in $(T')^3$. Similarly, let T'' be the union of the v-fragments of T^* which do not include u and let f' be the section of e in T''. Form f'' by removing the first term from f'. If T'' is non-empty, let y be the first vertex in f''. By induction, the algorithm using f'' finds a Hamiltonian y,v-path h'' in $(T'')^3$. If T'' is empty, let $h'' = v$. Since x and y are at distance three in T^* and since the vertices between them in e are u and v, the algorithm visits y immediately after visiting x, if y exists. If y does not exist, then v is visited immediately after x is visited because v is the last term of e. Thus, the algorithm finds $(h'), (h'')$, which is a Hamiltonian u,v-path in T^3.

Next suppose u and v are not adjacent in T. Let the path joining u and v in T be $u = x_0, x_1, \ldots, x_n = v$. For each i, let H_i be the union of the x_i-fragments of T^* which do not contain u or v. Let A_i be the x_i-fragment of T^* which contains v, but let A_n be the empty graph. Since the section of e from the first appearance of u to the last appearance of x_1 in e is an Eulerian trail from u to x_1 in $T^* - A_1$, by the first part of this proof the algorithm finds a Hamiltonian u,x_1-path in $(T^* - A_1)^3$. Suppose the algorithm finds a Hamiltonian u,x_{i-1}-path in $(T^* - A_{i-1})^3$. If H_i is empty, the next vertex visited by the algorithm is x_i. Otherwise, the next vertex y' visited by the algorithm is at distance one or two from x_i and is in H_i. If the distance of y' from x_i is one, the algorithm finds a Hamiltonian path in the cube of H_i from y' to x_i. If the distance of y' from x_i is two, let y'' be the vertex immediately preceding y' on e. Then in the union B' of y''-fragments which do not include x_i, the section of e contained in B' includes an Eulerian trail from y' to y'', and so the algorithm finds a Hamiltonian path in the cube of B' from y' to y''. Further, $H_i - B'$ includes an Eulerian trail which is a section of e from y'' to x_i and which includes all of the vertices of $H_i - B'$, and y'' is adjacent to x_i. Hence the algorithm finds a Hamiltonian path in $(H_i - B')^3$ from y'' to x_i. Thus the algorithm finds a Hamiltonian u,x_i-path in $(T^* - A_i)^3$ for every $i \in \{1, 2, \ldots, n\}$. Since $v = x_n$, it follows that the algorithm finds a Hamiltonian u,v-path in T^3.

Clearly if u and v are chosen to be adjacent and a last step is added to the algorithm to place u after v in the sequence formed by the algorithm, then the modified algorithm finds a Hamiltonian cycle in T^3 which follows an Eulerian circuit in the doubling of T. Thus the above theorems and algorithms prove:

THEOREM 7: Given a graph G, and given that $f(G)$ is G^2, G^3, $L(G)$, or $T(G)$, then if $f(G)$ is Hamiltonian, there is a spanning subgraph of G or $S(G)$ which, by doubling certain edges, admits of an Eulerian circuit e, and there is an algorithm which produces a Hamiltonian cycle in $f(G)$ which follows e.

II. CHARACTERIZATIONS

There is a second group of theorems which again reveal a common form in squares, cubes, line graphs, and total graphs of graphs. It is easy enough to find G^2, G^3, $L(G)$, and $T(G)$ if G is known, but given a graph H, how can we find out if there is a graph G for which H is one of these images of G? The answer to this question appears in the form of four theorems:

THEOREM 8 (Mukhopadhyay [20] and Geller [11]): H is the square of a graph G if and only if there are n complete subgraphs $C_1, \ldots, C_n$ of H and an ordering $v_1, \ldots, v_n$ of the vertices of H such that $H = \bigcup_{i=1}^{n} C_i$ and, for all $\{i, j\} \subseteq \{1, 2, \ldots, n\}$,

1) $v_i \in V(C_i)$, and

2) $v_j \in V(C_i)$ if and only if $v_i \in V(C_j)$.

THEOREM 9 (Escalante, Montejano, and Rojano [5], modified for cubes of graphs only): Let the vertices of H be ordered $v_1, v_2, \ldots, v_n$. Then H is the cube of a graph if and only if there are complete subgraphs $C_1, \ldots, C_n$ of H

such that:

1) $v_i \in V(C_i)$ for all i;

2) $v_i \in V(C_j)$ if and only if $v_j \in V(C_i)$; and

3) $v_i v_j$ is an edge of H if and only if one of the following is true:

   a) $v_i \in V(C_j)$, or

   b) there is a v_k such that $v_i \in V(C_k)$ and $v_k \in V(C_j)$, or

   c) there are vertices v_k and v_m such that $v_i \in V(C_k)$, $v_k \in V(C_m)$, $v_m \in V(C_j)$, and the subgraph of H generated by $V(C_k) \cup V(C_m)$ is complete.

THEOREM 10 (Krausz [19] and Beineke [3]): H is the line graph of a graph G if and only if there are edge-disjoint complete subgraphs $C_1, \ldots, C_n$ of H such that $H = \bigcup_{i=1}^{n} C_i$ and, for every $v \in V(H)$, $|\{C_i : v \in V(C_i)\}| \leq 2$.

The fourth theorem is a characterization of total graphs. Unfortunately, the characterization of total graphs by Behzad [1], although very interesting, is not of the form of the preceding three theorems. Because Behzad's theorem requires several definitions to make it intelligible, it is omitted for lack of space. In order to preserve the form of Theorems 8 through 10 in the total graph characterization, I have used the fact that the total graph of G is $(S(G))^2$ and the characterization of squares of graphs given in Theorem 8.

A node is a vertex of degree different from two. A path is suspended if every internal vertex of the path has degree two in G but the end vertices of the path are nodes. A cycle is suspended if exactly one of its vertices is a node. To prove Theorem 11 we need the following easily proved lemma.

LEMMA: If a connected graph F is not a cycle, then there is a graph G such that $F = S(G)$ if and only if every suspended path and cycle in F is of even length.

THEOREM 11: Let H be a connected graph which is neither K_3 nor the total graph of a cycle. Then there is a graph G such that $H = T(G)$ if and only if there is an ordering $v_1, \ldots, v_n$ of the vertices of H and a sequence $C_1, \ldots, C_n$ of complete subgraphs of H such that $H = \bigcup_{i=1}^{n} C_i$ and

1) $v_i \in V(C_i)$ <u>for all</u> i;

2) $v_j \in V(C_i)$ <u>if and only if</u> $v_i \in V(C_j)$; <u>and</u>

3) <u>let</u> $C_{i_1}, \ldots, C_{i_r}$ <u>be a sequence of members of</u> $\{C_1, \ldots, C_n\}$ <u>with</u> $r \geqq 2$ <u>such that</u>

   a) <u>no two of</u> $C_{i_1}, \ldots, C_{i_r}$ <u>have the same subscript except that</u> $i_1 = i_r$ <u>is possible if</u> $r \geqq 3$;

   b) $\{v_{i_k}, v_{i_{k+1}}\} \subseteq V(C_{i_k}) \cap V(C_{i_{k+1}})$ <u>for all</u> $k \in \{1, \ldots, r-1\}$;

   c) $|V(C_{i_1})| \neq 3$ <u>and</u> $|V(C_{i_r})| \neq 3$; <u>and</u>

   d) $|V(C_{i_k})| = 3$ <u>for all</u> $k \in \{2, \ldots, r-1\}$.

<u>Then</u> r <u>is odd</u>.

PROOF: If $|V(H)| = 1$, the theorem is trivial. Suppose $|V(H)| \geqq 2$. Since a graph is connected if and only if its total graph is connected, we may suppose that H is a connected graph which is neither K_3 nor the total graph of a cycle.

Suppose there is a graph G such that $H = T(G)$. It is known that $T(G)$ is isomorphic to $(S(G))^2$. Hence the existence of the ordering $v_1, \ldots, v_n$ of $V(H)$ and the sequence $C_1, \ldots, C_n$, as well as properties (1) and (2) follow from Theorem 8. Let $C_{i_1}, \ldots, C_{i_r}$ be a sequence with the properties (a) through (d) of part (3). Then $v_{i_1}, \ldots, v_{i_r}$ is a trail in $S(G)$ by (3b). By (3d), the vertices of the trail other than v_{i_1} and v_{i_r} have degree two in $S(G)$. By (3a), the trail is either a path or a cycle. If it is a path, the ends of the path are distinct nodes by (3c), so the path is a suspended path in $S(G)$ joining two nodes. Since the path is in $S(G)$, it has even length, and so r is odd. If the trail is a cycle, the subgraph of $S(G)$ induced by the cycle is a block of $S(G)$ which is a cycle. Hence it contains an even number of edges and so r is odd.

Suppose H satisfies conditions (1) through (3). By (1) and (2), there is a graph F such that $H = F^2$. Let $v_{i_1}, \ldots, v_{i_r}$ be a suspended path or cycle in F. Then C_{i_1} and C_{i_r} do not have three vertices and C_{i_k} does have three vertices if $k \in \{2, \ldots, r-1\}$. By the construction which proves Theorem 8,

$\{v_{i_k}, v_{i_{k+1}}\} \subseteq V(C_{i_k}) \cap V(C_{i_{k+1}})$ for every $k \in \{1, \ldots, r-1\}$. Finally, since $v_{i_1}, \ldots, v_{i_r}$ is a path or a cycle in F, no two of $C_{i_1}, \ldots, C_{i_r}$ have the same subscript except that $i_1 = i_r$ is possible if $r \geq 3$. Thus conditions (a) through (d) of (3) are satisfied by $v_{i_1}, \ldots, v_{i_r}$ and $C_{i_1}, \ldots, C_{i_r}$, so that r is odd. We conclude that the suspended path or cycle is of even length. By the Lemma, there is a graph G such that $F = S(G)$. Therefore, $H = (S(G))^2 = T(G)$.

We see in Theorems 8 through 11 the following common elements: In each case there is a set of complete subgraphs of H whose union is H, and certain intersection properties are specified for these complete subgraphs which guarantee that there is a graph G for which H is G^2, G^3, $L(G)$, or $T(G)$.

III. CONCLUSIONS

The theorems and algorithms described in Sections I and II suggest that G^2, G^3, $L(G)$, and $T(G)$ are actually all special cases of a more general class of graphs about which we can hope to prove similar theorems. We would assume graph H has a collection of complete subgraphs $C_1, \ldots, C_n$ whose union includes all of the vertices of H. Then we want to describe conditions on the intersections of the subgraphs C_i of H which, when satisfied, allow us to find a spanning subgraph S of H, a doubling S^* of S, an Eulerian circuit e in S^*, and an algorithm which produces a Hamiltonian cycle h in H such that h follows e. We have already seen such conditions in the cases of squares, cubes, and total graphs of graphs; we can hope other similar conditions can be found which lead to Hamiltonian cycles in a more general class of graphs than those discussed in the preceding sections. Another fact which bolsters our hope is the following theorem found by Chartrand, Polimeni, and Stewart [4], and independently by Sumner [28], which shows the common structures of squares, cubes, line graphs, and total graphs of graphs:

THEOREM 12: <u>Let</u> $f(G)$ <u>be</u> G^2, G^3, $L(G)$, <u>or</u> $T(G)$. <u>Then</u> $f(G)$ <u>includes a one-factor if and only if</u> $f(G)$ <u>has an even number of vertices</u>.

It is possible that I have been misled by the similarity of Theorem 6 to Theorems 1 through 5 and the two algorithms, and of Theorem 10 to Theorems 8, 9, and 11. It may be useless to direct the efforts discussed in the previous paragraph to include line graphs. At any rate, I think it would be most profitable to try modifications of the cube and square of graph characterizations first, and to use the line graph characterization as a guide later in the study. Hopefully, the effort suggested here would yield general results which have the reported theorems for squares, cubes and total graphs as special cases, and which cover many other kinds of graphs.

REFERENCES

1. M. Behzad, A characterization of total graphs, Proceedings, Amer. Math. Soc. 26 (1970) 383-389.
2. M. Behzad and G. Chartrand, Introduction to the Theory of Graphs, Allyn and Bacon, Inc., Boston (1971).
3. L. W. Beineke, Characterizations of derived graphs, J. Combinatorial Theory 9 (1970) 129-135.
4. G. Chartrand, A. D. Polimeni, and M. J. Stewart, The existence of 1-factors in line graphs, squares, and total graphs, Indagationes Mathematicae (A) 35 (1973) 228-232.
5. F. Escalante, L. Montejano, and T. Rojano, Characterization of n-path graphs and of graphs having nth root, J. Combinatorial Theory (B) 16 (1974) 282-289.
6. H. Fleischner, Über Hamiltonsche Linien im Quadrat kubischer und pseudokubischer Graphen, Mathematische Nachrichten 49 (1971) 163-171.
7. H. Fleischner, On spanning subgraphs of a connected bridgeless graph and their application to DT-graphs, J. Combinatorial Theory (B) 16 (1974) 17-28.
8. H. Fleischner, The square of every two-connected graph is Hamiltonian, J. Combinatorial Theory (B) 16 (1974) 29-34.
9. H. Fleischner and A. M. Hobbs, A necessary condition for the square of a graph to be Hamiltonian, J. Combinatorial Theory (B) 19 (1975) 97-118.
10. H. Fleischner and A. M. Hobbs, Hamiltonian total graphs, Mathematische Nachrichten 68 (1975) 59-82.
11. D. P. Geller, The square root of a digraph, J. Combinatorial Theory 5 (1968) 320-321.
12. S. Goodman and S. Hedetniemi, Sufficient conditions for a graph to be Hamiltonian, J. Combinatorial Theory (B) 16 (1974) 175-180.
13. F. Harary and C. St. J. A. Nash-Williams, On Eulerian and Hamiltonian graphs and line graphs, Canad. Math. Bull. 8 (1965) 701-709.
14. F. Harary and A. Schwenk, Trees with Hamiltonian squares, Mathematika, London 18 (1971) 138-140.
15. A. M. Hobbs, Some Hamiltonian results in powers of graphs, J. Res. Nat. Bur. Standards Sect. B 77B (1973) 1-10.
16. A. M. Hobbs, Maximal Hamiltonian cycles in squares of graphs, J. Combinatorial Theory (B), to appear.
17. A. M. Hobbs, Hamiltonian squares of cacti, J. Combinatorial Theory (B), to appear.
18. J. J. Karaganis, On the cube of a graph, Canad. Math. Bull. 11 (1968), 295-296
19. J. Krausz, Demonstration nouvelle d'une Theoreme de Whitney sur les Reseaux, Mat. Fiz. Lapok 50 (1943) 75-85.
20. A. Mukhopadhyay, The square root of a graph, J. Combinatorial Theory 2 (1967) 290-295.
21. C. St. J. A. Nash-Williams, Problem No. 48, Theory of Graphs (P. Erdös and G. Katona, ed.), Academic Press, New York (1969).
22. F. Neuman, On a certain ordering of the vertices of a tree, Casopis Pest. Mat. 89 (1964) 323-339.
23. M. D. Plummer, Private communication, 26 April 1971.
24. I. C. Ross and F. Harary, The square of a tree, Bell System Tech. J. 39 (1960) 641-647.

25. M. Sekanina, On an ordering of the vertices of a graph, Casopis Pest. Mat. 88 (1963) 265-282.
26. M. Sekanina, Problem No. 28, Theory of Graphs and its Applications (M. Fiedler, ed.), Academic Press, New York (1964), 164.
27. H. Walther, Über eine Anordnung der Knotenpunkte kubischer Graphen, Matematicky Casopis 19 (1969) 330-333.
28. D. P. Sumner, Graphs with 1-factors, Proc. Amer. Math. Soc. 42 (1974) 8-12.

THE CARTESIAN PRODUCT OF TWO GRAPHS IS STABLE

D A. Holton
University of Melbourne
Parkville, Victoria 3052, Australia

and

J. Sims
Linacre College
Oxford, England OX1 1SY

Abstract

If G and H are two connected non-trivial graphs, not necessarily of finite order, then we show that the cartesian product $G \times H$, is stable in the sense of Sheehan [4]. Moreover except when $G = P_2$ and H is a certain restricted class of prime graphs, any edge of $G \times H$ may be removed to give the stability, i.e. $G \times H$ is completely stable.

Consideration is given to the case where $G \times H$ is not connected. In the case of finite graphs $G \times H$ is stable unless one of G or H is totally disconnected and the other is not stable. For non-finite graphs the situation is not as clear. We give examples of cartesian products of non-finite graphs which are not stable.

1. Introduction

Throughout all graphs will be without loops, multiple edges or directed edges, but, unless specifically stated otherwise, they may be on a non-finite vertex set. All graph theoretical terms can be found in Behzad and Chartrand [1] although we use "cartesian product" where they use simply "product", we use G_e for $G - e$, and $u \sim v$ for u adjacent to v.

Sheehan introduced the concept of a stable graph in [4]. If there exists $e \in E(G)$ such that $\Gamma(G_e) \leq \Gamma(G)$, where $\Gamma(G)$ is the automorphism group of the graph G, then G is said to be stable. If $\Gamma(G_e) \leq \Gamma(G)$ for all $e \in E(G)$, then G is completely stable. In [4] Sheehan discussed, among other things, the stability of trees and unicyclic (monocyclic) graphs, both of which he was able to characterise. Here we determine those cartesian products of graphs which are stable.

If $e = \{u, v\}$ is an edge of G, then we use end $[e]$ to denote the set of endvertices u, v of e. This and the definition of stability give us the following characterisation of stability.

<u>Lemma 1.1</u>. <u>Let</u> G <u>be a graph and let</u> $e \in E(G)$. <u>Then</u> G <u>is stable at</u> e <u>if and only if</u> end $[e]$ <u>is fixed by</u> $\Gamma(G_e)$.

Of fundamental importance in the work of this paper is the concept of section sets. These were introduced in Sims and Holton [5] where they were used to prove that all cartesian products are semi-stable. Suppose $V(G) = \{v_i : i \in I\}$ and $V(H) = \{w_j : j \in \bar{I}\}$ where I and $\bar{I}$ are some index sets, then we let $G^{(v_k, w_j)} = \langle (v_i, w_j) : i \in I \rangle$ for some $k \in I$ and some $j \in \bar{I}$ be a <u>section</u> of G in the product $G \times H$. Similarly we define a section of H as $H^{(v_i, w_k)} = \langle (v_i, w_j) : j \in \bar{I} \rangle$ for some $i \in I$ and some $k \in \bar{I}$. Then the <u>section sets</u> of G and H in the product $G \times H$ are defined by

$$E_G = \bigcup_{j \in I} E(G^{(v_i, w_j)}) \quad \text{for some } i \in I$$

and
$$E_H = \bigcup_{k \in I} E(G^{(v_k, w_\ell)}) \quad \text{for some } \ell \in \bar{I}.$$

If G and H are prime graphs then E_G and E_H are the equivalence classes of Sabidussi [3]. It is clear that E_G and E_H are independent of the choice of i and ℓ, that $E_G \cap E_H = \emptyset$, $E_G \cup E_H = E(G \times H)$ and for G, H non-trivial and connected, $E_G \neq \emptyset$ and $E_H \neq \emptyset$.

We now have the following seven important properties of section sets of a connected composite graph $G \times H$.

P1: An edge of each section set is incident with every vertex.

P2: If e_1 and e_2 are opposite sides of a 4-cycle, then e_1 and e_2 belong to the same section set.

P3: If e_1 and e_2 are adjacent edges such that $e_1 \in E_G$ and $e_2 \in E_H$, then there is a 4-cycle without diagonals containing e_1 and e_2 in $G \times H$.

P4: Any maximal connected subgraph T with $E(T) \subseteq E_G$ (or E_H) is isomorphic to G (or H).

P5: If the graph K of Figure 1.1 is a subgraph of $G \times H$, then all edges of K are in the same section set of $G \times H$.

P6: If C is a cycle in $G \times H$ we cannot have $E(C) = E(P_G) \cup E(P_H)$ where P_G and P_H are paths whose edges are contained in E_G and E_H respectively.

P7: If T is an induced subgraph of $G \times H$ such that the set of edges of $G \times H$ incident with exactly one vertex of T is contained in E_H, then $T \cong G \times D$ where D is an induced subgraph of H. Also, $E(T) \cap E_G$ is a section set of T and if $v \notin V(T)$, $v \in V(G \times H)$, $N_{G \times H}[v] \cap V(T)$ is contained in one section of D in T.

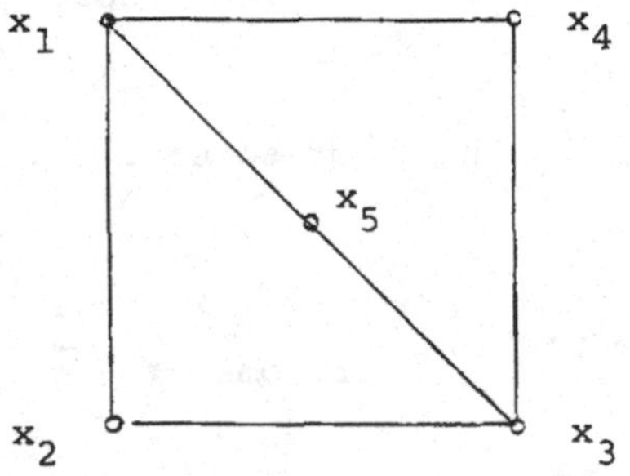

Figure 1.1

Properties P1 through P6 are proved in [5]. The proof of P7 is straightforward and so we omit it.

These properties are basically those used by Dörfler [2] in establishing the reconstructability of the cartesian product, but the recognition of these properties reduces the work of that paper.

We also need to define the projections p_G and p_H on the vertices and edges of $G \times H$ by

$$p_G(v_i\ , w_j) = v_i$$

$$p_G\{(v_i\ , w_j)\ , (v_k\ , w_m)\} = \begin{cases} \{v_i\ , v_k\} & \text{if } i \neq k \\ v_i & \text{if } i \neq k \end{cases}$$

and
$$p_H(v_i\ , w_j) = w_j$$

$$p_H\{(v_i\ , w_j)\ , (v_k\ , w_m)\} = \begin{cases} \{w_j\ , w_m\} & \text{if } j \neq m \\ w_j & \text{if } j = m\ . \end{cases}$$

2. THE STABILITY OF CONNECTED CARTESIAN PRODUCTS

In this section we prove that the cartesian product of two connected non-trivial graphs is stable. But we first need to describe the perturbation edge set of G .

Let G be a non-trivial connected graph and let $e \in E(G)$. If there exists a component M of G_e which is isomorphic to $P_2 \times L$ for some graph L and end $[e] \cap M$ is contained in one of the sections of L in M, then we call e a perturbation edge of G. The set of all perturbation edges of G is called the *perturbation edge set* of G and we denote this set by $\overline{s}(G)$.

We now prove a basic lemma.

Lemma 2.1. *Let* G *and* H *be non-trivial, connected graphs with* $e \in E(G \times H)$, $f \in E(\overline{G \times H})$. *Then* $(G \times H)_e + f$ *is composite if and only if there is an isomorphism* g *mapping* $(G \times H)_e + f$ *onto* $G \times H$ with $(\text{end}[f])^g = \text{end}[e]$ and $(\text{end}[e])^g = \text{end}[f]$, and either

(i) $G = P_2$ *and* $p_H e \in \overline{s}(H)$,

or (ii) $H = P_2$ *and* $p_G e \in \overline{s}(G)$,

or (iii) $G \times H = P_2 \times P_3$ *and* e *is incident with two vertices of degree* 3.

Proof:

($\Leftarrow$) If G and H are non-trivial, connected graphs and $(G \times H)_e + f \cong G \times H$, then $(G \times H)_e + f$ is composite.

($\Rightarrow$) With G, H, e, f as specified above, assume that $(G \times H)_e + f \cong K \times L$ for some non-trivial graphs K and L. $(G \times H)_e$ is connected, as a non-trivial connected cartesian product has edge connectivity at least 2. Thus $K \times L$ is connected, as f is defined to be incident with two vertices of $V(G \times H)$ and so both K and L are connected.

Let $J = (G \times H)_e + f$, so $J \cong K \times L$. As K and L are connected and non-trivial, section sets E_K and E_L exist for the graph J with properties P1 to P7. Let

$$V(G) = \{v_i , i \in I\} \quad \text{and} \quad V(H) = \{w_j , j \in \overline{I}\}$$

where I and $\overline{I}$ are index sets and, exchanging G and H if necessary, we may assume that

$$e = \{(v_1, w_1), (v_1, w_2)\} \quad \text{where} \quad w_1 \sim w_2 \quad \text{in } H .$$

By assumption, $f \neq \{(v_1, w_1), (v_1, w_2)\}$. Now

$$V(J) = V(G \times H)$$

$$E(J) = \Big\{\{(v_i, w_j), (v_k, w_\ell)\}: v_i = v_k \text{ and } w_j \sim w_\ell$$

$$\text{or } v_i \sim v_k \text{ and } w_j = w_\ell, i, k \in I, j, \ell \in \overline{I}\Big\} \cup \{f\} \setminus \Big\{\{(v_1, w_1), (v_1, w_2)\}\Big\}.$$

Note that this labelling of J is not the natural labelling of $J \cong K \times L$, and so is not related to the definition of E_K and E_1 for J.

We proceed to find a contradiction to properties P1 to P7 of the section sets, unless the conditions of the lemma hold. We will assume vertices and edges exist where they are implied by previous assumptions; otherwise the required contradiction is obtained.

As G is non-trivial, there exists $v_2 \in V(G): v_2 \sim v_1$ in G. f cannot be incident simultaneously with one of (v_1, w_1) and (v_2, w_2), and one of (v_1, w_2) and (v_2, w_1), as then f is an edge of $G \times H$. So without loss of generality, we may assume that f is not incident with either of (v_1, w_1) and (v_2, w_2). Hence there is no 4-cycle in J containing $\{(v_1, w_1), (v_2, w_1)\}$ and

$\{(v_2, w_1), (v_2, w_2)\}$, and so by P3, these two edges belong to the same section set, E, say, where E is either E_K or E_L. So

$$\{(v_1, w_1), (v_2, w_1)\} \in E \; . \; . \; . \; . \; . \; . \; . \quad (1)$$

and

$$\{(v_2, w_1), (v_2, w_2)\} \in E \; . \; . \; . \; . \; . \; . \; . \quad (2) \; .$$

By P1, there is an edge in the other section set, $\bar{E}$, say, incident with (v_1, w_1). Either

1: there exists $w_3 \neq w_2 : w_3 \sim w_1$ and $\{(v_1, w_1), (v_1, w_3)\} \in \bar{E}$,

or 2: there exists $v_3 \neq v_2 : v_3 \sim v_1$ and $\{(v_1, w_1), (v_3, w_1)\} \in \bar{E}$,

as, by assumption, f is not incident with (v_1, w_1).

Case 1 (see Figure 2.1).

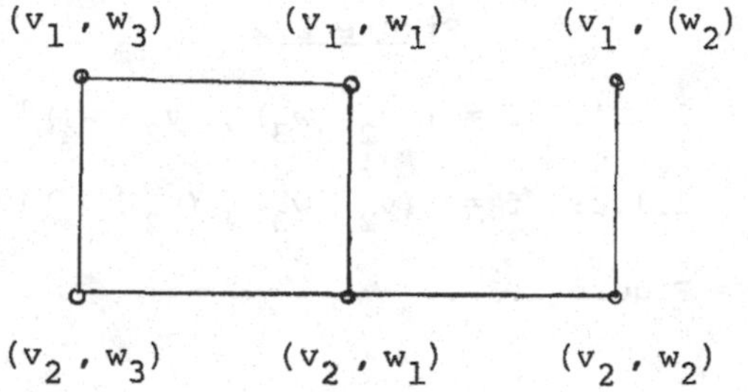

Figure 2.1

There exists $w_3 \sim w_1 : \{(v_1, w_1), (v_1, w_3)\} \in \bar{E} \; . \; . \; . \; . \quad (3)$.

By P_2 and (3), $\{(v_2, w_1), (v_2, w_3)\} \in \bar{E} \; . \; . \; . \; . \quad (4)$.

Then by P3 and (2), there is a 4-cycle in J containing $\{(v_2, w_1), (v_2, w_2)\}$ and $\{(v_2, w_1), (v_2, w_3)\}$. In view of the structure of $G \times H$, the presence of the extra edge f and the assumption that f is not incident with (v_2, w_2), the fourth vertex of this 4-cycle may be

1.1: (v_2, w_4) for $w_4 \in V(H)$; $w_4 \neq w_1, w_3$ and $w_4 \sim w_2$,

or 1.2: (v_1, w_2) ,

or 1.3: (v_3, w_2) for $v_3 \in V(G)$: $v_3 \neq v_1$ and $v_3 \sim v_2$.

<u>Case 1.1</u> (see Figure 2.2).

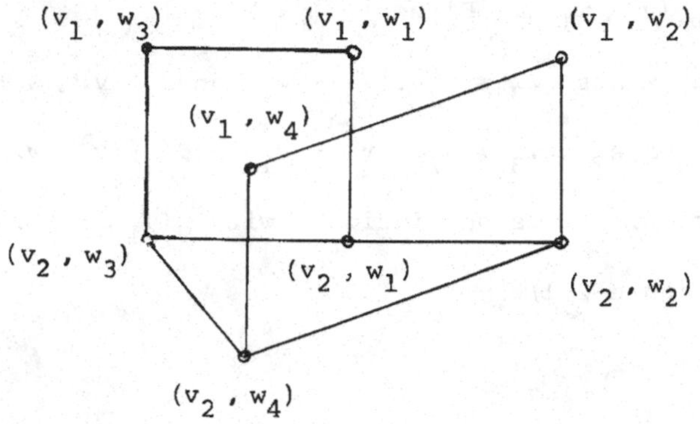

<u>Figure 2.2</u>

Either 1.1.1: $f \neq \{(v_2, w_3), (v_2, w_4)\}$,

or 1.1.2: $f = \{(v_2, w_3), (v_2, w_4)\}$.

<u>Case 1.1.1</u> (see Figure 2.3).

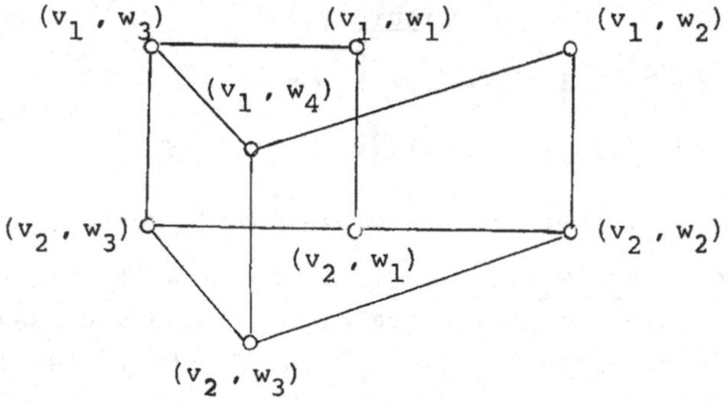

<u>Figure 2.3</u>

In this Case, $w_3 \sim w_4$

By P2 and (2), $\{(v_2, w_3), (v_2, w_4)\} \in E$, so by P2,

$$\{(v_1, w_3), (v_1, w_4)\} \in E \; . \; . \; . \; . \; . \quad (5).$$

By P2 and (4), $\{(v_2, w_2), (v_2, w_4)\} \in \overline{E}$ so by P2,

$$\{(v_1, w_2), (v_1, w_4)\} \in \overline{E} \; . \; . \; . \; . \; . \quad (6).$$

By P3, (3) and (5), there is a 4-cycle in J containing $\{(v_1, w_1), (v_1, w_3)\}$ and $\{(v_1, w_3), (v_1, w_4)\}$, and by P3, (5) and (6), there is a 4-cycle in J containing $\{(v_1, w_2), (v_1, w_4)\}$ and $\{(v_1, w_3), (v_1, w_4)\}$. As $f \notin E(G \times H)$, f cannot be incident with both one of (v_1, w_1) and (v_1, w_4), and one of (v_1, w_2) and (v_1, w_3). So one of the 4-cycles above cannot contain f. Thus either there exists $w_5 \neq w_2, w_3 : w_5 \sim w_1$ and $w_5 \sim w_4$, or there exists $w_6 \neq w_1, w_4 : w_6 \sim w_2$ and $w_6 \sim w_3$. Then one of $\langle (v_2, w_i), \; i = 1, \ldots, 5 \rangle$ and $\langle (v_2, w_i), \; i = 1, \ldots, 4, 6 \rangle$ exists in J and is forbidden by P5, using (2) and (4). So this Case leads to a contradiction.

<u>Case 1.1.2</u> $f = \{(v_2, w_3), (v_2, w_4)\}$, so $w_3 \not\sim w_4$ in H and $\{(v_1, w_3), (v_1, w_4)\} \notin E(J)$ (see Figure 2.2).

Either 1.1.2.1: there exists $v_3 \neq v_1 : v_3 \sim v_2$ in G,

or 1.1.2.2: there exists $v_4 \neq v_2 : v_4 \sim v_1$ in G,

or 1.1.2.3: $G = P_2$.

<u>Case 1.1.2.1</u> (see Figure 2.4).

By P2 and (4), $\{(v_3, w_1), (v_3, w_3)\} \in \overline{E}$, and by P2 and (2), $\{(v_3, w_1), (v_3, w_2)\} \in E$. So by P3, there is a 4-cycle in J containing the latter two edges. As $f = \{(v_2, w_3), (v_2, w_4)\}$, the only possibility is that there exists $w_5 \neq w_1, w_4 : w_5 \sim w_3$ and $w_5 \sim w_2$. Then $\langle (v_2, w_i), i = 1, \ldots, 5 \rangle$ is forbidden by P5, (2) and (4). So this Case gives a contradiction.

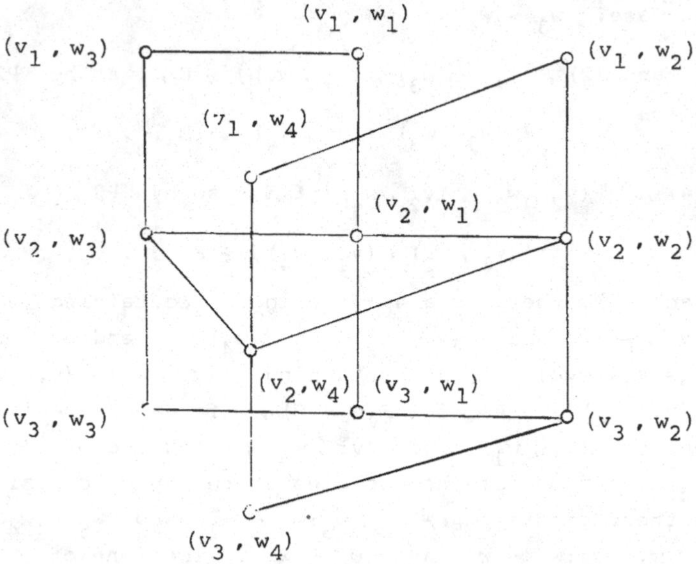

Figure 2.4

Case 1.1.2.2 (see Figure 2.5).

By P2 and (3), $\{(v_4 , w_1) , (v_4 , w_3)\} \in \bar{E}$. Suppose that $\{(v_4 , w_1) , (v_4 , w_2)\} \in E$. Then by P3, there exists a 4-cycle in J containing the latter two edges. So there exists $w_5 \neq w_1 , w_4 : w_5 \sim w_3$ and $w_5 \sim w_2$. But then $\langle (v_2 , w_i) , \ i = 1 , \ldots , 5 \rangle$ is forbidden by P5, using (2) and (4). So we must have

$$\{(v_4 , w_1) , (v_4 , w_2)\} \in \bar{E} \quad \ldots\ldots \quad (7) .$$

Then, using (1), (2) and (7), by P6 applied to $\langle (v_i , w_j) , i = 1 , 2 , 4 ; j = 1 , 2 \rangle$, $\{(v_1 , w_2) , (v_4 , w_2)\} \in E$. But then, by P3 and (7), there is a 4-cycle in J containing $\{(v_4 , w_1) , (v_4 , w_2)\}$ and $\{(v_1 , w_2) , (v_4 , w_2)\}$. This is a contradiction, as $\{(v_1 , w_1) , (v_1 , w_2)\} \notin E(J)$, and no vertex but (v_1 , w_1) can complete the 4-cycle.

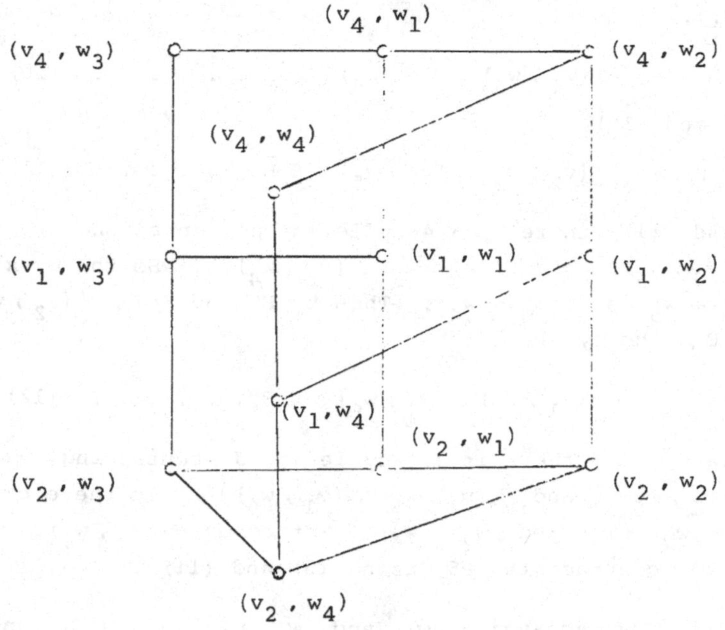

Figure 2.5

Case 1.1.2.3 (see Figure 2.2).

$G = P_2$. By P2 and (2) ,

$$\{(v_2, w_3), (v_2, w_4)\} \in E \text{ . . . . (8)}$$

and by P2 and (1), $\{(v_1, w_3), (v_2, w_3)\} \in E$ . . . . (9) .

Also, $w_2 \not\sim w_3$, as $w_2 \sim w_3$ implies $(v_2, w_2) \sim (v_2, w_3)$ and so by P3, $\{(v_2, w_1), (v_2, w_3)\}$ and $\{(v_2, w_1), (v_2, w_2)\}$ would belong to the same section set, contradicting (2) and (4) .

Consider the edge $\{(v_1, w_1), (v_1, w_3)\}$: either there is an edge of $\overline{E}$ incident with (v_1, w_1) or (v_1, w_3) or there is not. Then using (3) and P4 , either

1.1.2.3.1: there exists $w_5 \neq w_1, w_2, w_4 : \{(v_1, w_3), (v_1, w_5)\} \in \overline{E}$,

or 1.1.2.3.2: there exists $w_5 \neq w_2, w_3, w_4 : \{(v_1, w_1), (v_1, w_5)\} \in \overline{E}$,

or 1.1.2.3.3: $\overline{E} = E_{P_2}$.

Case 1.1.2.3.1.

$$\{(v_1, w_3), (v_1, w_5)\} \in \overline{E} \ . \ . \ . \ . \ . \ . \ . \quad (10) \ .$$

By P2 and (10),

$$\{(v_2, w_3), (v_2, w_5)\} \in \overline{E} \ . \ . \ . \ . \ . \ . \ . \quad (11) .$$

By P3, (8) and (11), there is a 4-cycle in J containing $\{(v_2, w_3), (v_2, w_5)\}$ and $\{(v_2, w_3), (v_2, w_4)\}$. So there exists $w_6 \neq w_3 : w_6 \sim w_5$ and $w_6 \sim w_4$. Then by P2 and (8), $\{(v_2, w_5), (v_2, w_6)\} \in E$, so by P_2,

$$\{(v_1, w_5), (v_1, w_6)\} \in E \ . \ . \ . \ . \ . \ . \ . \quad (12) \ .$$

By P3, (10) and (12), there is a 4-cycle in J containing $\{(v_1, w_3), (v_1, w_5)\}$ and $\{(v_1, w_5), (v_1, w_6)\}$, so there exists $w_7 \neq w_4, w_5 : w_7 \sim w_3$ and $w_7 \sim w_6$. But then $\langle(v_2, w_i), i = 3, \ldots, 7\rangle$ contradicts P5 using (8) and (11).

Case 1.1.2.3.2. By exchanging w_1 and w_3, w_2 and w_4, an argument similar to the previous Case gives a contradiction.

Case 1.1.2.3.3. $\overline{E} = E_{P_2}$.

We will show that $p_H e \in \overline{s}(H)$, and exhibit an isomorphism between J and $G \times H$ which maps end[f] onto end[e] and vice versa, giving the required result.

By P2 and (4),

$$\{(v_2, w_2), (v_2, w_4)\} \in \overline{E} \ . \ . \ . \ . \ . \ . \ . \quad (13) \ .$$

As $\overline{E} = E_{P_2}$, P4 and (13) imply that

$$\{(v_1, w_2), (v_2, w_2)\} \in E \ . \ . \ . \ . \ . \ . \ . \quad (14) \ .$$

H is connected, so if $w_j \in V(H)$, there is either a $w_j - w_1$ path or a $w_j - w_2$ path in $H_{\{w_1, w_2\}}$. Let this path be $w_j, w_j^1, \ldots, w_j^m$ where w_j^m is either w_1 or w_2. By (1) or (14),

$$\{(v_1, w_j^m), (v_2, w_j^m)\} \in E,$$

so applying P2 repeatedly,

$\{(v_1,w_j^{m-1}),(v_2,w_m^{m-1})\},\ldots,\{(v_1,w_j^1),(v_2,w_j^1)\},\{(v_1,w_j),(v_2,w_j)\} \in E$

i.e., $\{(v_1,w_j),(v_2,w_j)\} \in E$ for all $w_j \in V(H)$.

So all edges of J incident with exactly one vertex of $M = \langle (v_1, w_j), w_j \in V(H)\rangle$ are in E . Thus, by P7, $M \cong P_2 \times N$ for some graph N. By (1), (2) and (14), (v_1, w_1) and (v_1, w_2) lie in the same section of N in M . So, as $M \cong H_{\{w_1,w_2\}}$, $\{w_1,w_2\} \in \bar{s}(H)$.

By P2 and (13), $\{(v_1, w_2), (v_1, w_4)\} \in \bar{E}$, so, using (3), (v_1, w_1) corresponds to (v_1, w_3), and (v_1, w_2) to (v_1, w_4) in the two sections of N in M. Let

$$\bar{M} = \langle (v_2, w_j), w_j \in V(H)\rangle ,$$

so $\bar{M}_{\{(v_2,w_1),(v_2,w_2)\},\{(v_2,w_3),(v_2,w_4)\}} \cong M \cong P_2 \times N$

where the first isomorphism is via $(v_2, w_j) \to (v_1, w_j), w_j \in V(H)$.

Define $g : V(J) \to V(G \times H)$ as follows: g fixes $N^{(v_1,w_3)}$ and $N^{(v_2,w_1)}$, g maps $N^{(v_1,w_1)}$ onto $N^{(v_2,w_3)}$ and $N^{(v_2,w_3)}$ + $\{(v_2, w_3), (v_2, w_4)\}$ onto $N^{(v_1,w_1)}$ + $\{(v_1, w_1), (v_1, w_2)\}n$ in the natural manner. Thus, $(v_1, w_1)^g = (v_2, w_3)$, $(v_2, w_3)^g =$ (v_1, w_1), $(v_1, w_2)^g = (v_2, w_4)$ and $(v_2, w_4)^g = (v_1, w_2)$. Then g is an automorphism of

$$(G \times H)_{\{(v_1,w_1),(v_1,w_2)\},\{(v_2,w_1),(v_2,w_2)\}} \cong P_2 \times P_2 \times N ,$$

as g maps one factor of P_2 onto the other; g fixes $\{(v_2, w_1), (v_2, w_2)\} \in E(N^{(v_2,w_1)})$; and $\{(v_2, w_3), (v_2, w_4)\}^g =$ $\{(v_1, w_1), (v_1, w_2)\}$. So g is an isomorphism from J to $G \times H$ with the required property that $(\text{end}[f])^g = \text{end}[e]$ and vice versa.

This completes Case 1.1.2.3, and so Case 1.1.

The remaining Cases follow by similar arguments to those of Case 1.1. We omit the lengthy details here. A complete proof is given in Sims [6]. ∎

As a result of the previous lemma we can produce the following theorem which tells us which edges e of a cartesian product may be removed to give $\Gamma(G_e) \le \Gamma(G)$.

Theorem 2.2. Let G and H be non-trivial, connected graphs and let $e \in E(G \times H)$. Then $G \times H$ is not stable at e if and only if either

(i) $G = P_2$ and $p_H e \in \bar{s}(H)$,

or (ii) $H = P_2$ and $p_G e \in \bar{s}(G)$,

or (iii) $G \times H = P_2 \times P_3$ and e is incident with two vertices of degree 3.

Proof:

(⇒) Suppose $G \times H$ is not stable at e. Then, by Lemma 1.1, end[e] is not fixed by $g \in \Gamma = \Gamma((G \times H)_e)$. Define f by $\text{end}[f] = (\text{end}[e])^g$, so $f \in E(\overline{G \times H})$. Then g is an isomorphism from $G \times H$ to $(G \times H)_e + f$. By Lemma 2.1 the required result holds.

(⇐) It follows by insepction of $P_2 \times P_3$ that if $e \in E(P_2 \times P_3)$ and e is incident with two vertices of degree 3, then $P_2 \times P_3$ is not stable at e. As (ii) is obtained from (i) by interchanging G and H, it suffices to prove that if $G = P_2$ and $\{w_1, w_2\} \in \bar{s}(H)$, $G \times H$ is not stable at $e = \{(v_1, w_1), (v_1, w_2)\}$ for $v_1 \in V(G)$. We do this by exhibiting an element of Γ which does not fix end[e].

Let $G = P_2$ and $\{w_1, w_2\} \in \bar{s}(H)$. Then there exists a component M of $H_{(w_1,w_2)}$ which is isomorphic to $P_2 \times L$ for some graph L, and the vertex set $\{w_1, w_2\} \cap V(M)$ is contained in one of the sections of L in M.

Let $V(L) = \{y_i, i \in I_L\}$ so we may label M

$$V(M) = \{(x_1, y_i), (x_2, y_i), i \in I_L\}$$

where $w_1 = (x_2, y_1)$, say, and

$$E(M) = \{\{(x_1, y_i), (x_2, y_i)\}, i \in I_L, \{(x_k, y_i), (x_k, y_j)\} : k - 1, 2, y_i \sim y_j \text{ in } L\}.$$

Define $g: V(P_2 \times H) \to V(P_2 \times H)$ as follows, where $V(P_2) = \{v_1, v_2\}$:

$$(v_1, (x_2, y_i))^g = (v_2, (x_1, y_i)),$$

$$(v_2, (x_1, y_i))^g = (v_1, (x_2, y_i)), \quad i \in I_L,$$

and $\qquad v^g = v$ otherwise, for $v \in V(P_2 \times H)$.

Either $H_{\{w_1, w_2\}}$ is disconnected, so $w_2 \notin V(M)$, or $H_{\{w_1, w_2\}}$ is connected, so $w_2 = (X_2, y_n)$ for some $n \in I_L$, by the definition of $\overline{s}(H)$. In the former case, no vertex except (v_2, w_2) of $\langle V((P_2 \times H)_e) \setminus V(P_2 \times M)\rangle$ is adjacent to a vertex of $P_2 \times M$, $N_{(P_2 \times H)_e}[(v_2, w_2)]$ is fixed by g; and adjacency is preserved in $P_2 \times M \cong P_2 \times P_2 \times L$ as g excanges the two factors of P_2. So g preserves adjacency in $(P_2 \times H)_e$, i.e., $g \in \Gamma$. However, $(v_1, w_1) \sim (v_1, w_2)$ in $P_2 \times H$, and

$$(v_1, w_1)^g = (v_2, (x_1, y_1)) \not\sim (v_1, w_2) = (v_1, w_2)^g,$$

so $g \notin \Gamma(P_2 \times H)$.

In the latter case, i.e. $H_{\{w_1, w_2\}}$ is connected,

$$(P_2 \times H)_e \cong P_2 \times P_2 \times L + \{(v_2, (x_2, y_1)), (v_2, (x_2, y_n))\};$$

$\{(v_2, (x_2, y_1)), (v_2, (x_2, y_n))\}$ is fixed by g; and g exchanges the two factors of P_2. So g preserves adjacency in $(P_2 \times H)_e$, i.e. $g \in \Gamma$. However, $(v_1, w_1) = (v_1, (x_2, y_1)) \sim (v_1, (x_2, y_n)) = (v_1, w_2)$ in $P_2 \times H$ and

$$(v_1, (x_2, y_1))^g = (v_2, (x_1, y_1)) \not\sim (v_2, (x_1, y_n)) = (v_1, (x_2, y_n))^g$$

in $P_2 \times H$, as $w_1 \not\sim w_2$ in M. So $g \notin \Gamma(P_2 \times H)$.

Thus, in either case, $\Gamma = \Gamma((P_2 \times H)_e) \not\leq \Gamma(P_2 \times H)$, as required. ∎

The result that all cartesian products of connected, non-trivial graphs are stable now follows.

Theorem 2.3. All connected composite graphs are stable.

Proof: If neither G nor H is P_2, then $G \times H$ is stable by theorem 2.2. If $G \times H \simeq P_2 \times P_3$, then $G \times H$ is stable by inspection. If one of G, H is P_2 and the other is not P_3, then $G \times H$ is stable since $\bar{s}(P_2) = \emptyset$. ∎

3. COMPLETELY STABLE CONNECTED CARTESIAN PRODUCTS

In this section we show that if G and H are connected, non trivial graphs then $G \times H$ is completely stable unless $G(H)$ is P_2 and $H(G)$ is prime with $\bar{s}(H)$ $(\bar{s}(G)) \neq \emptyset$.

We need some preliminary results.

Theorem 3.1. Let G, H be non-trivial, connected graphs with $e \in E(G \times H)$. Then $(G \times H)_e$ is prime.

Proof: Interchanging G and H if necessary we may assume that $e = \{(v_1, w_1), (v_1, w_2)\}$ for some $v_1 \in V(G)$, $w_1, w_2 \in V(H)$. We now assume that $(G \times H)_e$ is not prime and produce a contradiction. Let $J = (G \times H)_e$ and assume that $J \simeq K \times L$ for some non-trivial graphs K and L.

As the edge-connectivity of $G \times H$ is at least two for connected graphs G and H, J is connected, so both K and L are connected.

Thus section sets E_K and E_L exist for the graph J, with $E(J) = E_K \cup E_L$ and $E_K \cap E_L = \emptyset$, together with properties P1 to P7.

As G is non-trivial, there exists v_2: $v_2 \sim v_1$ in G. As $(v_1, w_1 \not\sim (v_1, w_2)$ in J, there is no 4-cycle in J containing $\{(v_1, w_1), (v_2, w_1)\}$ and $\{(v_2, w_1), (v_2, w_2)\}$, or $\{(v_2, w_1), (v_2, w_2)\}$ and $\{(v_1, w_2), (v_2, w_2)\}$, so by P3, all three edges are in the same section set, E, say. So

$$\{(v_1, w_1), (v_2, w_1)\} \in E \quad \dots\dots\dots \quad (1),$$

$$\{(v_2, w_1), (v_2, w_2)\} \in E \quad \dots\dots\dots \quad (2),$$

and

$$\{(v_1, w_2), (v_2, w_2)\} \in E \quad \dots\dots\dots \quad (3).$$

By P1, there is an edge in the other section set, $\bar{E}$, say, incident with (v_1, w_1). Either

1: there exists $v_3 \neq v_2 : v_3 \sim v_1$ and $\{(v_1, w_1), (v_3, w_1)\} \in \bar{E}$,

or 2: there exists $w_3 \neq w_2 : w_3 \sim w_1$ and $\{(v_1, w_1), (v_1, w_3)\} \in \bar{E}$.

Case 1.

Assume $\{(v_1, w_1), (v_3, w_1)\} \in \bar{E}$ . . . . (4) .

There is no 4-cycle in J containing $\{(v_1, w_1), (v_3, w_1)\}$ and $\{(v_3, w_1), (v_3, w_2)\}$ or $\{(v_3, w_1), (v_3, w_2)\}$ and $\{(v_1, w_2), (v_3, w_2)\}$ so by P3 and (4), all three edges belong to $\bar{E}$. Then $\langle (v_i, w_j),\ i = 1, 2, 3;\ j = 1, 2\rangle$ is forbidden by P6, using (1), (2) and (3). So Case 1 leads to a contradiction.

Case 2.

Assume $\{(v_1, w_1), (v_1, w_3)\} \in \bar{E}$ . . . . . . (5) .

By P2 and (5), $\{(v_2, w_1), (v_2, w_3)\} \in \bar{E}$ . . . . . . (6),

so by P3 and (2), there is a 4-cycle in J containing $\{(v_2, w_1), (v_2, w_2)\}$ and $\{(v_2, w_1), (v_2, w_3)\}$. So there exists $w_4 \neq w_1 : w_4 \sim w_3$ and $w_4 \sim w_2$.

By P2 and (2), $\{(v_2, w_3), (v_2, w_4)\} \in E$, so by P2, $\{(v_1, w_3), (v_1, w_4)\} \in E$. Then by P3 and (5), there is a 4-cycle in J containing $\{(v_1, w_1), (v_1, w_3)\}$ and $\{(v_1, w_3), (v_1, w_4)\}$. So there exists $w_5 \neq w_2, w_3 : w_5 \sim w_1$ and $w_5 \sim w_4$. Then $\langle (v_2, w_i),\ i = 1, \ldots, 5\rangle$ is forbidden by P5, using (2) and (6). So Case 2 lead to a contradiction. This completes the proof. ∎

Adding an edge to a connected composite graph also gives a prime graph:

<u>Corollary 3.2</u>. *Let G and H be non-trivial, connected graphs with $e \in E(\overline{G \times H})$. Then $G \times H + e$ is prime.*

We also need the following result concerning perturbation edge sets.

Lemma 3.3. If G is a non-trivial, connected graph with $\bar{s}(G) \neq \emptyset$, then G is prime.

Proof: Let $\bar{s}(G) \neq \emptyset$. So there exists $e \in E(G)$ such that there is a component M of G_e which is isomorphic to $P_2 \times L$ for some graph L. If $M \neq G_e$, then e is a bridge of G. Hence, since every connected composite graph has edge connectivity at least 2, G is prime.

On the other hand, if $M = G_e$, $G = M + e$, where M is composite and connected. Hence, by Corollary 3.2, G is again prime. ■

Finally then we are able to determine all completely stable composite graphs. The proof of the theorem follows directly from Theorem 2.2 and Lemma 3.3.

Theorem 3.4. Let G be a connected composite graph. Then, unless $G \cong P_2 \times Q$ where Q is a prime graph with $\bar{s}(Q) \neq \emptyset$, G is completely stable.

It is now not too difficult to prove the following theorem. The proof may be found in [6].

Theorem 3.5. Let G and H be finite graphs. $G \times H$ is not stable if and only if one of G and H is totally disconnected and the other is not stable.

But in the general case, where both G and H may have trivial and non-trivial components, we have no characterisation for the stability of $G \times H$. However, it is possible to construct G and H in this case, so that $G \times H$ is not stable. One such construction is given below.

Let H_0 be any non-trivial graph and let G be any graph with K_1 as an component. Collect all the components of all the graphs $(H_0)_e$ with $e \in E(H_0)$ and repeat this process with H_0 replaced by $(H_0)_e$, for all $e \in E(H_0)$. Continue inductively. Then form $G \times C_0$ where C_0 is any component obtained from H_0 in the inductive process above. Repeat the same process on $G \times C_0$. Then form $G \times C_1$ where C_1 is any one of the components formed from $G \times C_0$. Repeat this process inductively also. Then if H is the union of all the components produced above, from $G \times H$. The product is not stable.

REFERENCES

[1] M. Behzad and G. Chartrand, Introduction To The Theory of Graphs, Allyn and Bacon, Boston, (1971).

[2] W. Dörfler, Some results on the reconstruction of graphs, Coll. Math. Soc. János Bolyai, 10, North Holland, Amsterdam, (1975), 361-383.

[3] G. Sabidussi, Graph Multiplication, Math. Zeit., 72, (1960), 446-457.

[4] J. Sheehan, Fixing subgraphs, J. Comb. Th. 12(B), (1972), 226-244.

[5] J. Sims and D. A. Holton, Stability of cartesian products, to appear.

[6] J. Sims. Stability of the cartesian product of graphs, M.Sc. Thesis, University of Melbourne, 1976.

A NOTE ON A GENERALIZED REGULARITY CONDITION

Joan P. Hutchinson* and Sue H. Whitesides
Tufts University
Medford, MA 02155

Abstract

For $k \geq 1$ let G be a graph with the property that for every set of k vertices of G, the number of vertices adjacent to these k vertices is a constant $d \geq 0$. For $k \geq 3$ we prove that these graphs are either complete on k+d vertices or possibly strongly regular for certain values of k and d.

In this paper we study finite, simple, connected graphs which satisfy a regularity condition on their vertex subsets of size k. We show that in many cases graphs satisfying our conditions are strongly regular.

By a k-set of a graph G, we mean a set of k vertices of G. We define the concepts of k-set degree and k-set regularity as follows. For $k \geq 1$, the degree of a k-set of a graph G is the number of vertices of G adjacent to each of the k given vertices. We say that G is an R(k,d) graph if the degree of every k-set is a constant, $d \geq 0$. (To avoid trivial cases, we always assume that an R(k,d) graph has at least k+d vertices.) R(k,0) graphs are simply graphs with maximum degree less than k, but for $d \geq 1$ we may think of R(k,d) graphs as satisfying a generalized regularity condition, as R(1,d) graphs are regular of degree d.

R(2,1) graphs have been characterized by the Friendship theorem [7,10,13], which states that in a graph with the property that every pair of vertices has a unique common neighbor, there is a vertex adjacent to all other vertices. Thus a graph satisfying the "friendship" condition is a collection of K_3's joined at a common vertex; such graphs are the R(2,1) graphs.

R(2,d) graphs with d>1 have been studied extensively [6] and relate to (v,k,λ) designs [1] and to Ramsey theory [11]; they are also

AMS(MOS) Subject Classification (1970) Primary 05C99.

*Address after June 1, 1976, Smith College, Northampton, MA 01060.

known as (v,k,λ) graphs where $\lambda=d$.

Clearly for each k and d, the complete graph on k+d vertices is an R(k,d) graph. A natural problem is to determine the structure of R(k,1) graphs for $k\geq 3$; in the first theorem we show that these are always complete.

In general we consider R(k,d) graphs for which $k\geq 3$. If $v_1,\ldots,v_k$ are distinct vertices of a graph G, we denote their common neighbors by $f(v_1,\ldots,v_k)$. Thus for G an R(k,d) graph, $f(v_1,\ldots,v_k)$ is a set of d additional and distinct vertices, each adjacent to each of $v_1,\ldots,v_k$. The following two properties of R(k,d) graphs will be used in the subsequent theorems.

(i) Every vertex of an R(k,d) graph has degree at least k+d-1.
Proof: Certainly no set of k-1 vertices $v_1,\ldots,v_{k-1}$ includes all the neighbors of a vertex x since $f(x,v_1,\ldots,v_{k-1})$ is a set of d additional neighbors. By choosing $v_1,\ldots,v_{k-1}$ all to be neighbors of x, we see that the degree of x is at least k+d-1.

(ii) An R(k,d) graph is complete if and only if it has k+d vertices.

Theorem 1* If G is an R(k,d) graph with $k\geq 3$ and $d\leq k-2$, then $G=K_{k+d}$.
Proof: Suppose G is not complete and that x and y are non-adjacent vertices. Let $v_1,\ldots,v_{k-2}$ be any other distinct vertices and let $f(x,y,v_1,\ldots,v_{k-2}) = \{g_1,\ldots,g_d\}$. Suppose that $f(x,y,g_1,\ldots,g_d,v_{d+1},\ldots,v_{k-2}) = \{h_1,\ldots,h_d\}$ and that $f(h_1,y,g_1,\ldots,g_d,v_{d+1},\ldots,v_{k-2}) = \{i_1,\ldots,i_d\}$. Since x is not adjacent to y, $x\neq i_j$ for $1\leq j\leq d$. Then

$$f(x,y,i_1,\ldots,i_d,v_{d+1},\ldots,v_{k-2}) \supset \{h_1,g_1,\ldots,g_d\}.$$

This contradicts the fact that G is an R(k,d) graph. Thus G is complete and by (ii) is K_{k+d}.

For $d\geq k-1$, R(k,d) graphs other than the complete graphs may possibly exist; however we show in the next theorem that all such graphs are strongly regular. A strongly regular graph is a regular graph with the property that the number of vertices adjacent to a pair of vertices depends only on whether or not the two vertices are adjacent. We say that a strongly regular graph has parameters (n,a,λ,μ) when it is regular of degree a on n vertices, two adjacent vertices have λ common neighbors, and two non-adjacent vertices have

*At the time of publication we have learned that J. Plesník has published some related results in Glasník Matematicki (1975).

μ neighbors in common. For example, a pentagon is a strongly regular graph with parameters (5,2,0,1).

Strongly regular graphs originated in the study of partial geometries [4] and led to the idea of association schemes [5]. They are significant in a variety of other areas as well: for example, in the study of rank three permutation groups [8], in coding theory [3], and in eigenvalue properties of two-graphs [12]. The interrelations of the parameters n, a, λ, and μ are well understood; we refer to [2, pp. 74-112] and to a recent survey article [9].

In [6] it is shown that R(2,d) graphs are regular and hence strongly regular. We obtain a similar result for $k>2$.

<u>Theorem 2</u> If G is an R(k,d) graph for $k \geq 3$, $d \geq k-1$ then G is strongly regular and is either K_{k+d} or has parameters (n,a,λ,μ) satisfying:

$$\binom{a-1}{k-2}\binom{d}{k-1} = \binom{d-1}{k-2}\binom{\lambda}{k-1} \tag{1}$$

$$\binom{a}{k-2}\binom{d}{k-1} = \binom{d}{k-2}\binom{\mu}{k-1} . \tag{2}$$

Proof: If $n=k+d$, then G is a K_{k+d} by (ii). Suppose now that $n>k+d$, so that G is not complete.

Let x and y be two vertices of G. Let A be the set of all k-1-sets whose elements are adjacent to both x and y. We count in two ways pairs (X,α) where X is a k-2-set, not containing y, whose elements are adjacent to x, and α is in A and is such that $f(X,x,y)$ contains α. Suppose x has degree a. If x is adjacent to y and if there are λ vertices adjacent to both x and y, then

$$\binom{a-1}{k-2}\binom{d}{k-1} = \binom{d-1}{k-2}\binom{\lambda}{k-1}.$$

If x is not adjacent to y and if there are μ vertices adjacent to both x and y, then

$$\binom{a}{k-2}\binom{d}{k-1} = \binom{d}{k-2}\binom{\mu}{k-1}.$$

The main observation to be made is that if we select a k-1-set α from A and if $f(\alpha,x) = W$, a d-set, then y is in W if and only if y is adjacent to x. Furthermore, a k-2-set X whose elements are adjacent to x but do not include y is a subset of W if and only if $f(X,x,y)$ contains α.

Suppose b is the degree of y. Since equations (1) and (2) hold with a replaced by b, a=b. Hence G is regular of degree a. Since λ and μ depend only on a, k and d, G is strongly regular with parameters (n,a,λ,μ) satisfying (1) and (2).

From equations (1) and (2) it can be deduced that $\lambda > \mu$ since

$$\frac{d(a-k+2)}{a(d-k+2)} = \frac{\lambda(\lambda-1)\cdots(\lambda-k+2)}{\mu(\mu-1)\cdots(\mu-k+2)}. \qquad (3)$$

Furthermore, from standard eigenvalue arguments for strongly regular graphs (as in [9]) we know that

$$(\lambda-\mu)^2 + 4(a-\mu) = s^2 \qquad (4)$$

for some integer s and

$$2a + (\lambda-\mu)(n-1) = rs \qquad (5)$$

for some integer r.

The conditions imposed by equations (1)-(5) are extremely restrictive, and at present we know of no example except for the complete graphs which meet all these. For example, the unsettled case with smallest parameters is that of R(3,2) graphs. We know that if there is such a strongly regular graph, other than K_5, it must have at least 639×10^9 vertices.

REFERENCES

1. R. W. Ahrens and G. Szekeres, On a combinatorial generalization of 27 lines associated with a cubic surface. _J. Austral. Math. Soc._ 10 (1969) 485-492.
2. N. Biggs, _Finite Groups of Automorphisms_, London Math. Soc. Lecture Note Series. Cambridge University Press, Cambridge (1971).
3. E. R. Berlekamp, J. H. van Lint, and J. J. Seidel, A strongly regular graph derived from the perfect ternary Golay code. _A Survey of Combinatorial Theory_ (J. N. Srivastava et al., eds.). North-Holland Publishing Company, Amsterdam (1973) 25-30.
4. R. C. Bose, Strongly regular graphs, partial geometries and partially balanced designs. _Pacific J. Math._ 13 (1963) 389-419.
5. R. C. Bose and Shimamoto, Classification and analysis of partially balanced incomplete block designs with two associate classes. _J. Am. Statist. Assoc._ 47 (1952) 151-184.
6. R. C. Bose and S. S. Shrikhande, Graphs in which each pair of vertices is adjacent to the same number d of other vertices. _Studia Sci. Math. Hungar._ 5 (1970) 181-195.
7. P. Erdös, A. Rényi and V. T. Sós, On a problem of graph theory. _Studia Sci. Math. Hungar._ 1 (1966) 215-235.
8. D. G. Higman, Finite permutation groups of rank 3. _Math. Z._ 86 (1964) 145-156.

9. X. L. Hubaut, Strongly regular graphs. Discrete Math. 13 (1975) 357-381.

10. J. Q. Longyear and T. D. Parsons, The friendship theorem. Nederl. Akad. Wetensch. Proc. Ser. A 75 = Indag Math. 34 (1972) 257-262.

11. T. D. Parsons, Ramsey graphs and block designs. I. Trans. Amer. Math. Soc. 209 (1975) 33-44.

12. J. J. Seidel, Graphs and two-graphs. Proceedings of the Fifth Southeastern Conference on Combinatorics, Graph Theory, and Computing. Utilitas Mathematica Publishing Inc., Winnipeg (1974) 125-143.

13. H. S. Wilf, The friendship theorem. Combinatorial Mathematics and Its Applications (D. J. A. Welsh ed.) Academic Press, New York (1971) 307-309.

LONG AND SHORT WALKS IN TREES

Paul C. Kainen

Case Western Reserve University

Cleveland, Ohio 44106

ABSTRACT

Given a graph G and a labeling λ of its vertices, say $v_1, \ldots, v_n$, define $h(\lambda)$ to be the sum $\sum_{1 \le i \le n} d(v_i, v_{i+1})$, where $v_{n+1} = v_1$ and $d(v_i, v_{i+1})$ is the distance from vertex v_i to v_{i+1}. Let $h(G)$ and $H(G)$ denote the minimum and maximum, respectively, of $h(\lambda)$ over all labelings λ of G. For G a tree, we give algorithms for computing $h(G)$ and $H(G)$ and for finding the corresponding labelings. An application is made to the Traveling Salesman Problem.

1. Introduction

Given a graph G and a specific listing or <u>labeling</u> λ of its vertices $v_1, \ldots, v_n$, we define $h(\lambda) = d(v_1, v_2) + d(v_2, v_3) + \ldots + d(v_{n-1}, v_n) + d(v_n, v_1)$, where $d(v,w)$ = distance from v to w in G. If G is a tree, there is a unique shortest-distance path λ_i from v_i to v_{i+1} $(i = 1, \ldots, n)$ and the concatenation of these paths is a walk $w(\lambda)$. An edge e appears in $w(\lambda)$ $\lambda(e)$ times so $h(\lambda) = \sum_e \lambda(e)$. Set $h(G) = \min_\lambda h(\lambda)$ and $H(G) = \max_\lambda h(\lambda)$.

In this paper we obtain $h(G)$ and $H(G)$ when G is a tree and derive algorithms for finding the corresponding minimal and maximal labelings. An outline of the paper is as follows: In Section 2, we set up the basic concept of usage numbers and use it to bound $\lambda(e)$. Section 3 defines non-consecutive labelings and uses a result in combinatorial number theory to find the maximal labeling. Finally, in Section 4, mminimal labelings are derived and related to the Traveling Salesman Problem.

Let us remark here that even if G is not a tree we still have a good bound on $h(G)$. For a connected graph G, G^3 is Hamiltonian. It follows that $h(G) \leq 3|V(G)|$ and, of course, $h(G) \geq |V(G)|$.

Much of the work presented in this paper has been done independently by R. N. Rao [2], who has also considered the more general case of edges with non-negative lengths. We should also like to thank Don Goldsmith and Roy Levow for their very helpful comments.

Finally, this author would like to thank the conference organizers and Western Michigan University for providing such a congenial atmosphere.

2. Usage numbers

For the remainder of this paper, G denotes a tree with n vertices. If e is an edge of G, then $G-e$ is a disconnected graph with two components G_1 and G_2. Let the usage number $u(e)$ of e denote the minimum of $|V(G_1)|$ and $|V(G_2)|$. The following result explains our terminology.

2.1 Lemma For any labeling λ of G, $2 \leq \lambda(e) \leq 2u(e)$.

Proof The walk $w(\lambda)$ induced by λ contains $\lambda(e)$ paths containing e of which at most 2 have any vertex as a common endpoint; hence, the second inequality holds. Since $w(\lambda)$ is closed, $\lambda(e) \geq 2$.

Summing over e yields the critical lemma.

2.2. Lemma If G is any tree with n vertices and if λ is any labeling of G, then

$$2(n-1) \leq h(\lambda) \leq 2 \sum_{e \in E(G)} u(e).$$

Given an edge $e = [v,w]$, let us orient e from v to w if the component G_1 of $G-e$ which contains v has exactly $u(e)$ vertices. It is easy to see that at most one edge of G has an ambiguous orientation. (This occurs when both components G_1 and G_2 contain the same number of vertices). Orienting the edges of G in this manner produces a directed graph D in which every endpoint of G is a source and in which there is a unique sink s.

The edges $e_1,\dots,e_k$ incident with s are oriented toward s and have usage numbers $u_i = u(e_i)$, $1 \le i \le k$.

2.3 Lemma With the previous notation, if G has n vertices, then $\sum_{i=1}^{k} u_i = n-1$ and $\max_{1\le i\le k} u_i = u \le \left[\frac{n}{2}\right]$.

We can use this lemma to develop an $O(n)$ algorithm for labeling the edges of G with their usage numbers. (The "obvious" method uses $O(n^2)$ steps.) Choose an external edge $e' = [v,w]$, where v is an endpoint of G. Label e' with $u(e') = 1$ and direct e' from v to w. An internal edge $e = [w,x]$ is available if all other edges $e_1',\dots,e_r'$ at w (or at x) are labeled and directed toward e. If an internal edge is available, label e with $u(e) = 1 + \sum_{i=1}^{r} u(e_i')$ unless $u(e) > \left[\frac{n}{2}\right]$. Otherwise, choose another external edge and continue.

3. Non-consectuive labelings

Call a labeling λ of G maximal if $h(\lambda) = H(G)$ (and minimal if $h(\lambda) = h(G)$). Thus, if λ is maximal, then $\lambda(e) = 2u(e)$ for every edge e.

We can now argue as in the proof of Lemma 2.1. For an edge e and associated vertex sets V_1, V_2 (formed by removing e from the tree G), define a graph G' depending on e and the labeling λ. $V(G') = V_1 \cup V_2$ and vertices $v_1 \in V_1$, $v_2 \in V_2$ are adjacent if and only if one of the $\lambda(e)$ paths induced by λ and passing through e has v_1 and v_2 as its endpoints. Since $\lambda(e) = 2u(e)$, each vertex of V_1 is an endpoint of 2 of these paths and hence no 2 consecutive vertices belong to V_1, that is, V_1 is a non-consecutive subset of V, and λ is called non-consecutive for e. We shall call a labeling λ which has this property for every edge non-consecutive.

The converse of the above remarks is also true and so we have shown:

3.1 Lemma λ is a maximal labeling if and only if λ is non-consecutive.

Since a subset of a non-consecutive set is also non-consecutive, we can sharpen the previous result slightly.

3.2 Lemma λ is a maximal labeling if and only if λ is non-consecutive for all edges e_i $(i = 1,\dots,k)$ incident with the sink s of the digraph D obtained from G as in the previous section.

But according to Lemma 2.3, we have now reduced finding a maximal labeling to solving the following problem in combinatorial number theory:

3.2 Problem Given $n-1 = u_1 + \dots + u_k$ with $u_i \le \left[\frac{n}{2}\right]$, find subsets S_i of $\{1,\dots,n-1\} = S$ such that (i) the S_i form a (disjoint) partition of S, (ii) $|S_i| = u_i$, and (iii) each S_i is a non-consecutive subset.

Since a non-consecutive subset of S is just an independent set of vertices in the cycle C_{n-1} of length $n-1$, we see that the problem as stated needs modification when $k = 2$ and n is even. Otherwise, we'd be trying to 2-color an odd-length cycle.

But in this case, $n = 2r$, $u_1 = r-1$ and $u_2 = r$. By including the sink vertex v in the set with $r-1$ vertices, we now have 2 sets of r vertices to choose from $2r$ vertices. That is, we only need 2-color C_{2r}.

For the other case when $k = 2$, namely $n = 2r + 1$, $u_1 = r = u_2$ and the problem is easily solved.

For $k \ge 3$, we still have an obstruction if $\max u_i > \left[\frac{n-1}{2}\right]$. But we can use the same trick as before, including the sink vertex s in one of the vertex sets with non-maximal cardinality, to replace Problem 3.2 by

3.3 Problem Given $n = u_1 + \dots + u_k$ with all $u_i \le \left[\frac{n}{2}\right]$ and $k \ge 3$, find subsets S_i of $S' = \{1,\dots,n\}$ such that (i) the S_i form a partition of S', (ii) $|S_i| = u_i$, and (iii) each S_i is a non-consecutive set.

Solving this problem would provide us with the required maximal labeling.

3.4 Theorem Problem 3.3 is solvable.

Proof We give an $O(n)$ algorithm due to Don Goldsmith for finding the subsets S_i. The first subset S_1 is given by $S_1 = \{1,3,\dots,2u_1 - 1\}$, where $u_1 = \max u_i$ (renumbering if necessary.) Since $u_1 \le \left[\frac{n}{2}\right]$, $2u_1 - 1 \le n-1$. Hence, S_1 is non-consecutive. To form S_2, S_3, etc, simply continue to enumerate the odd integers in S' until all are used up and then go on with the even integers.

4. Minimal labelings and the Traveling Salesman Problem

To find a minimal labeling λ of G is simply to find a labeling λ with $\lambda(e) = 2$ for every edge e of G. Thus, replacing each edge of G by 2 parallel edges converts the minimal-labeling problem into that of finding a closed eulerian trail in the new multigraph.

We can now give an interesting application to the Traveling Salesman Problem (TSP).

4.1 Heuristic Let K_n be the complete graph associated to a TSP in which distances obey the usual "triangle inequality" - ie., $d_{ij} + d_{jk} \leq d_{ik}$. Choose a minimal total length spanning tree G (by Kruskal's algorithm [1], for example) and find a minimal labeling λ of G. This labeling induces a Hamiltonian cccle in K_n whose length is a good approximation to the length of the shortest such cycle.

In subsequent work, we intend to apply these labeling results to a problem in mathematical biology.

REFERENCES

1. J. B. Kruskal, On the shortest spanning subtree of a graph and the travelling salesman problem, Proc. AMS 7 (1956), 48-50.

2. R. N. Rao, Optimal labelings on trees, Ph.D. Dissertation, Case Western Reserve University, 1976.

RANDOM CLUMPS, GRAPHS, AND POLYMER SOLUTIONS

J. W. Kennedy
University of Essex
Essex, ENGLAND

1. Introduction.

By extracting the underlying graph-theoretical nature of present day chemical and polymer theory in several areas [7 , 9] we have gained considerable insight into the proper character of the mathematical models that are used. One prime area still for research of this kind lies in the field of liquids and liquid mixtures, with polymer solutions as an example of especial technological importance. Here, as elsewhere in statistical mechanics, the key quantity associated with a system of molecules is its so called partition function Q from which all macroscopic (i.e. experimentally observable) properties of the system can, in principle at least, be derived [3]. It is in obtaining an expression for Q in terms of the microscopic structure of some underlying model, supposed to reflect a real physical system, that the importance of graph theory is manifest [cf. 14] .

2. Polymer solutions.

By virtue of rotation about its chemical bonds, a long polymer chain in solution can for many purposes be represented simply as a uniform density spherical cloud of polymer segments. Even for this simplified model it is difficult to find useful expressions for the partition function of a solution containing N polymer molecules in a volume V .

When the solution is sufficiently concentrated (i.e. N/V large enough) so that the polymer segments can be considered to be uniformly distributed throughout the entire volume V, one may use refinements of a theory due to Huggins and Flory [for references see 10.] to obtain semi-empirical expressions for the partition function (Q_* say) of the system.

When the polymer solution is dilute so that it consists of large regions of pure solvent separating polymer coils which interact with each other relatively rarely in pairs, triples etc., the assumption of uniformity clearly fails. This extreme also has been the subject of extensive theoretical study [15] based on treating the polymer coils as hard spheres subject to an interaction potential, analogous to theories of imperfect gases [3, ch 4].

A recent attempt to bridge the gap between dilute and concentrated polymer solution theories [10] used a so called 'soft sphere' model characterised by polymer coils distributed over the solution but free to interpenetrate each other. In spirit, this involved expanding the partition function for N polymer molecules in solution thus:

$$Q = Q_1^{n_1} Q_*^{(N-n_1)} \tag{1}$$

where Q_1 is the partition function for a single isolated polymer coil (cf. the hard sphere approach) and Q_* that for all other polymer molecules treated in the uniform density Flory-Huggins type approximation. The number n_1 of isolated polymer coils may be estimated from the known result for a Poisson distribution of spheres of equal size. Thus:

$$n_1 = N \exp(-8\rho) \tag{2}$$

where the density parameter ρ is just the total volume of the polymer coils divided by that of the solution.

We may regard eq. 1 as a first effort to use a soft-sphere expansion of the partition function Q thus:

$$Q = Q_1^{n_1} Q_2^{n_2} Q_3^{n_3} Q_4^{n_4} \cdots \tag{3}$$

where Q_i is the (mean) partition function for a cluster of i polymer coils and n_i the number of such clusters. Further progress requires the cluster size distribution.

Subsequently we shall need to incorporate distributions of coil sizes, shapes other than sperical and perturbations from random placement stemming from interactions between coils. At the present time, even the cluster size distribution for non-interacting spheres (circles) of equal size distributed randomly in 3-space (2-space) would furnish interesting developments in the theory of polymer solutions.

3.1. Random Clumps.

The aggregation of randomly placed objects appears in the literature in various disguises. Together with a number of one and two dimensional examples of practical interest, it is the subject of an interesting monograph by S. A. Roach entitled "The Theory of Random Clumping" [11]. Elsewhere it is called the clustering problem [4, 12] or unstructured percolation. Of course percolation theory itself has an extensive literature [see eg 2, 13] but this is principly concerned with either site or bond percolation on a lattice rather than the unstructured case which is more appropriate here.

The basic problem may be described as follows [cf. 4]. Suppose points are randomly distributed in q-space according to a Poisson point process. Let pairs of points be connected to each other whenever they are separated by a distance $x \le 2r$, (thus r is the radius of the q-space hypersphere The graph so formed is random in some sense but reflects constraints imposed by geometrical properties of the embedding space. I shall refer to it as a clumps graph and to its components as clumps.* The characteristic parameter in terms of which the properties of a clumps graph are expressed is the so called 'density parameter' $\hat{f}$, which is equivalently the average number of points contained in a hypersphere of radius $2r$, or the mean vertex degree of points in the clumps graph. For polymer solutions the more usual measure of density (ρ) is the total volume of the spheres divided by that of the solution. In q-dimensions the two density parameters are related by:

$$\hat{f} = 2^q \rho \quad (4)$$

On an infinite clumps graph there is a critical value for the density parameter $\hat{f}_c$ such that for $\hat{f} < \hat{f}_c$ the probability of finding an infinite clump is zero, while for $\hat{f} > \hat{f}_c$ the probability that a point selected at random belongs to an infinite clump is non-zero,

---

* Apparently a q-dimensional clumps graph corresponds to Berge's [loc. cit.] generalization of an interval graph to a family of convex sets in $\mathbb{R}^2$. Furthermore, by imposing torroidal boundary conditions (a device frequently employed in physical science) we obtain a generalization of circular arc graphs [loc. cit. Tucker].

and the probability that some infinite clump exists is unity. The existence proof for $\hat{f}_c$ yields weak upper and lower bounds [13], but these are widely separated and much of the literature is addressed to the problem of specifying $\hat{f}_c$ more closely. Critical density values are of considerable interest in physics in connection with phase transitions, e.g. ferromagnetism [2, 13]. Indeed one would expect polymer solutions to reflect ρ_c in their viscoelastic behavior. However, the main interest here is in the more general problem of clump size distribution, and this has been approached from a number of avenues recently at Essex University.

3.2. Numerical Results.

Using Monte Carlo methods Fremlin [4] developed a sophisticated computer program for growing clumps with sizes up to 25,000 on a mesh of grain size $2^{-15}r$ in two dimensions and $2^{-11}r$ in three dimensions. This has enabled him to give the probable critical densities in the 2- and 3-dimensional clumps probelm as $\hat{f}_{c,2} = 4.4 \pm 0.2$ and $\hat{f}_{c,3} = 2.7 \pm 0.1$ respectively.

Table 1. Simulated 2-dimensional clump size distribution

i	Fraction w_i of clumps with size i found in 1000 clumps, at density $\hat{f}$ =				
	0.5	1.0	1.5	2.0	2.5
1	.606	.389	.216	.130	.093
2	.248	.239	.194	.118	.076
3	.096	.146	.140	.102	.073
4	.030	.091	.107	.083	.058
5	.014	.059	.089	.072	.065
6	.005	.038	.069	.066	.051
7	.001	.018	.044	.058	.044
8		.008	.029	.058	.049
9		.008	.026	.037	.035
10		.001	.024	.037	.034
11		.001	.014	.036	.042
12		.001	.014	.023	.035
13		.000	.010	.023	.022
14		.000	.007	.019	.030
15		.001	.004	.020	.016
16			.004	.019	.020
17			.003	.017	.013
18			.001	.010	.015
19			.002	.010	.017
20			.002	.008	.016
Total Fraction $i \leq 20$	1.0	1.0	.999	.946	.804

Table 2. Simulated 3-dimensional clump size distribution

i	Fraction w_i of clumps with size i found in 1000 clumps, at density $\hat{f}$ =			
	0.5	1.0	1.5	2.0
1	0.611	0.351	0.227	0.149
2	0.233	0.204	0.138	0.104
3	0.090	0.132	0.126	0.065
4	0.033	0.087	0.081	0.068
5	0.016	0.072	0.076	0.058
6	0.009	0.051	0.045	0.043
7	0.003	0.032	0.047	0.031
8	0.004	0.020	0.038	0.033
9	0.001	0.017	0.042	0.026
10		0.011	0.026	0.030
11		0.010	0.027	0.021
12		0 006	0.016	0.026
13		0.001	0.017	0.021
14		0.004	0.017	0.012
15		0.000	0.011	0.018
16		0.000	0.015	0.017
17		0.000	0.003	0.013
18		0.001	0.009	0.018
19		0.000	0.004	0.013
20		0.001	0.001	0.010
Total fraction $i \leq 20$	1.0	1.0	.966	.776

Dr. Fremlin has kindly supplemented his results [4] by a more detailed analysis of small clumps at densities below $\hat{f}_c$. In each case 1000 clumps were generated to obtain the faction w_i of each size as recorded in tables 1 and 2. It is noted that w_i corresponds to the <u>weight</u> <u>fraction</u> of clumps with size i in the real clumps graph.

The scatter in these results (see also § 3.3) which increases with both i and $\hat{f}$, could certainly be reduced by using larger samples. The w_i values are useful in assessing the worth of approximate functions postulated to model the clump size distribution, however, the prospect of their direct use for expanding the partition function <u>numerically</u> according to eq. 3 is discouraging.

3.3 Analytic Results.

Analytic results pertaining to clump size distributions are sparse. Roach [11] gives a number of approximations, but these have limited scope.

The probability $w_{1,q}$ that a point selected at random in a q-dimensional clumps graph itself forms a clump of size 1 is well known from Poisson statistics (cf. eq. 2):

$$w_{1,q} = e^{-\hat{f}} \tag{5}$$

Dr. D. D. Freund obtained an exact result for $w_{2,2}$, the chance that a point chosen at random in a 2-dimensional clumps graph is part of a clump of size 2:

$$w_{2,2} = 2\,\hat{f}\,e^{-\hat{f}} \sum_{i=0}^{\infty} (-\hat{f}/\pi)^i \frac{1}{i!} \int_0^{\pi/3} \sin u\,(u + \sin u)^i \, du \tag{6}$$

His proof of this expression is given in an appendix to this paper. Computed values for the function appeared in table 3.

Table 3.

Fraction $w_{2,2}$ of 2-dimensional clumps of size 2 at density $\hat{f}$

$\hat{f}$	$w_{2,2}$	$\hat{f}$	$w_{2,2}$	$\hat{f}$	$w_{2,2}$
0	0	2.0	0.123548	4.0	0.016745
0.2	0.15081	2.2	0.103384	4.2	0.013507
0.4	0.22762	2.4	0.085878	4.4	0.010882
0.6	0.25790	2.6	0.070906	4.6	0.0087608
0.8	0.25994	2.8	0.058254	4.8	0.0070490
1.0	0.24584	3.0	0.047662	5.0	0.0056706
1.2	0.22340	3.2	0.038858	5.2	0.0045632
1.4	0.197544	3.4	0.031588		
1.6	0.171270	3.6	0.025616		
1.8	0.146304	3.8	0.020728		

Fremlin's simulated clump results (tables 1 and 2) are compared with equations 5 and 6 in figures 1 and 2.

To our knowledge there are no other exact results, of this kind available. Clearly the geometric integrals become increasingly more complex as the clump size increases and are unlikely to put w_i in a convenient form for use in the partition function.

LEGEND TO FIGURES

Figure 1. Weight fraction (w_1) of clumps of size 1 as a function of the density parameter $\hat{f}$ (see text). Symbols are Fremlin's Monte Carlo results (tables 1 and 2) for clump simulations in 2- and 3-dimensions (• and x respectively). Curve is the well known Poisson result (eq. 5) valid for any number of dimensions.

Figure 2. Weight fraction (w_2) of clumps of size 2 as a function of $\hat{f}$, cf. figure 1. The 2-dimensional clump simulations (•) follow closely the exact result (solid curve) for this case due to Freund (see appendix). The 3-dimensional clump simulations (x) are displaced downwards and to the left, lying intermediate between Freund's curve and that for the weight fraction of size 2 components in an infinite random graph (dashed curve), see § 4. It would appear that the random graph can be viewed as a q-dimensional clumps graph in the limit of large q.

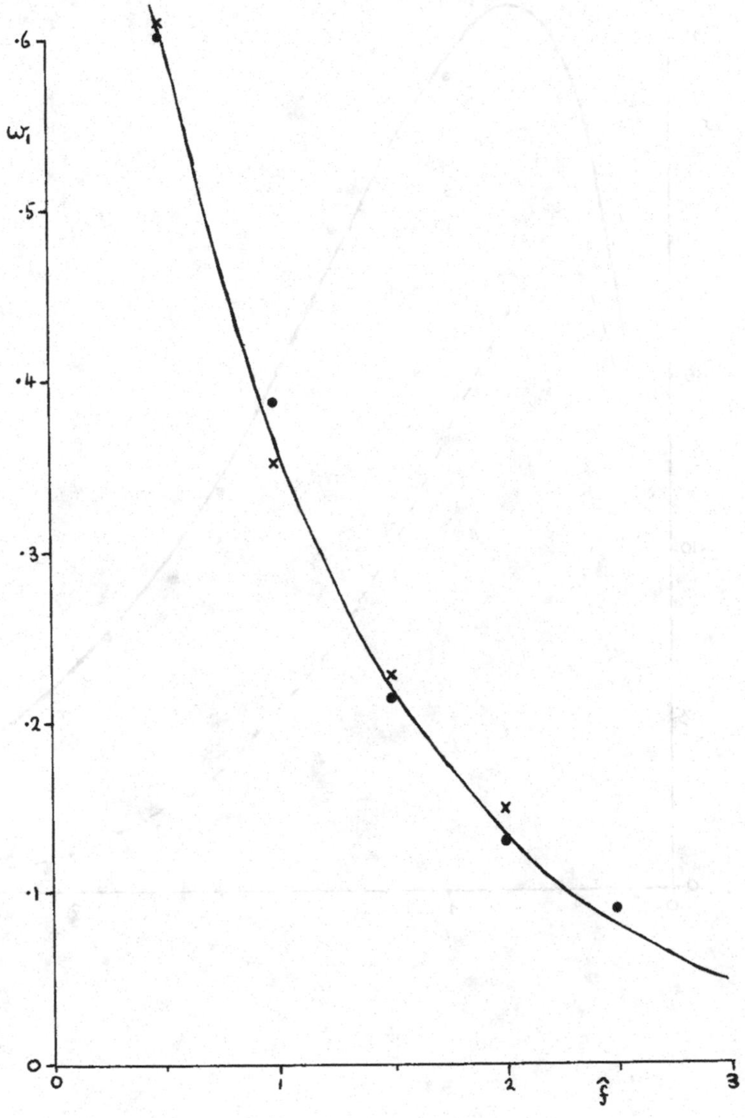

·6
·5
·4
·3
·2
·1
0
ω_1
0
1
2
3
$\hat{f}$

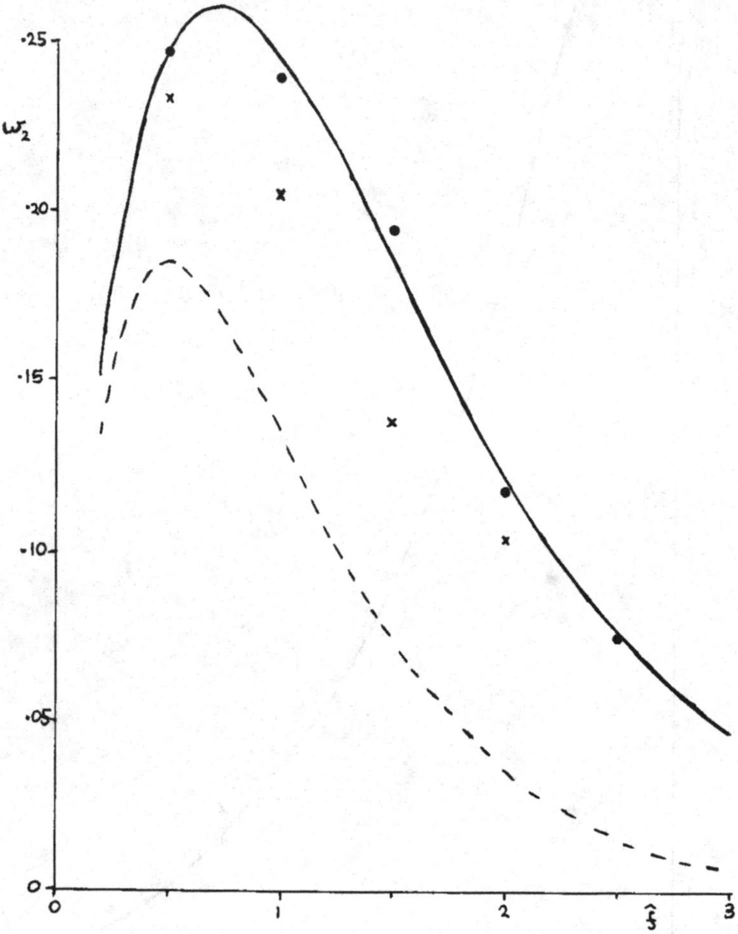
·25
·20
·15
·10
·05
0
w_2
0
1
2
3
$\hat{\xi}$

4. Random Graphs.

A graph R_N on N points is called <u>random</u> if for any pair of its points (i and j) there is a probability z that the bond $\langle i, j \rangle$ exists and probability $(1-z)$ that it does not exist. In some respects the statistical problem on random clumps is similar to that defined on random graphs, especially if we preserve the notion of $\hat{f}$ as the mean vertex degree and hence express the bond probability in R_N as:

$$z = \hat{f}/(N-1). \tag{7}$$

It should be emphasized that the two problems are <u>not</u> equivalent because of correlations, imposed by geometric constraints in the clumps graph, ignored when one considers random graphs. Thus in R_N all bonds are equally probable, whereas on a clumps graph if bonds $\langle i, j \rangle$ and $\langle j, k \rangle$ are known to exist then the probability that bond $\langle i, k \rangle$ exists also, is enhanced.

The random graph problem is of interest in its own right but in the present context it is hoped that methods for solving it can be adapted to give (at least approximate) answers to the more difficult, more physically interesting clumps problem. It is in this spirit that I present the following two results:

If R_N is a random graph on N points with bond probability z,

i. *The probability that R_N has s components =* Coefft[θ^s *in* $H_N(\theta)$] *where:*

$$H_N(\theta) = \theta h_{1,N-1}$$

$$h_{j,k} = \sum_{i=0}^{k} \binom{k}{i} (1-z)^{k-i} z^i h_{j-1+i,k-i} \; ; \quad j = 1, 2, \ldots, n>0$$

$$h_{j,0} = 1 \qquad h_{0,k} = H_k(\theta) \tag{8}$$

ii. *The probability that a point picked at random in R_N is contained in a component of size* i = Coefft[θ^i *in* $G_N(\theta)$] where:

$$G_N(\theta) = \theta\, g_{1,N-1} \equiv \sum_i w_i\, \theta^i$$

$$g_{j,k} = \sum_{i=0}^{k} \binom{k}{i} (1-z)^{k-i} z^i \theta^i g_{j-1+i,k-i} \;;\; j = 1, 2, \ldots, n > 0$$

$$g_{j,0} = 1 \qquad\qquad g_{0,k} = 1 \tag{9}$$

The first of these results contains, of course, the partitions of N into s components whose probabilities may be obtained from G_N (eq. 9). Although no use has been made of this result as yet, it is not difficult to envisage its implimentation to compute, for example, the probability that R_N is connected (viz. $s = 1$) as a function of the bond probability z.

It is the second result that is of more immediate interest in connection with the clumps problem and hence polymer solutions. For small values of N we could work with eq. 9 directly; in polymer solutions, however, we are interested in passage to the so called 'thermodynamic limit' which here corresponds to $N \to \infty$. Eq. 9 may be expanded to obtain the first few coefficients:

$$w_0 = 0$$

$$w_1 = (1-z)^{N-1}$$

$$w_2 = (N-1)(1-z)^{2N-4}\, z$$

$$w_3 = \left\{(N-1)(N-2)(1-z) + \binom{N-1}{2}\right\}(1-z)^{3N-9}\, z^2$$

$$w_4 = \left\{(N-1)(N-2)(N-3)(1-z)^3 + \left[\binom{N-1}{2}(N-3) + (N-1)\binom{N-2}{2}\right](1-z)^2 \right.$$
$$\left. \binom{N-1}{2}(N-3)(1-z) + \binom{N-1}{3}\right\}(1-z)^{4N-16}\, z^3$$

$$w_5 = \left\{(N-1)(N-2)(N-3)(N-4)(1-z)^6\right. \tag{10}$$
$$+\left[\binom{N-1}{2}(N-3)(N-4) + (N-1)\binom{N-2}{2}(N-4) + (N-1)(N-2)\binom{N-3}{2}\right](1-z)^5$$
$$+\left[\binom{N-1}{2}(N-3)(N-4) + \binom{N-1}{2}\binom{N-3}{2} + (N-1)(N-4)\binom{N-2}{2}\right](1-z)^4$$

$$+\left[\binom{N-1}{3}(N-4) + \binom{N-1}{2}(N-3)(N-4) + (N-1)\binom{N-2}{3}\right](1-z)^3$$

$$+\left[\binom{N-1}{3}(N-4) + \binom{N-1}{2}\binom{N-3}{2}\right](1-z)^2$$

$$+\binom{N-1}{3}(N-4)(1-z) + \binom{N-1}{4}\Big\}(1-z)^{5N-25}z^4$$

Using eq. 7 to replace z, preserving a fixed mean vertex degree $\hat{f}$, then in the limit of large N we find:

$$G_\infty(\theta) = e^{-\hat{f}}\theta + \hat{f}e^{-2\hat{f}}\theta^2 + \frac{3}{2}\hat{f}^2e^{-3\hat{f}}\theta^3 + \frac{8}{3}\hat{f}^3e^{-4\hat{f}}\theta^4 + \frac{125}{24}\hat{f}^4e^{-5\hat{f}}\theta^5 + \ldots \qquad (11)$$

as a generating function for the weight fraction of components in an infinite random graph. The terms in eq. 11 follow the general form $i^{i-2}\hat{f}^{i-1}e^{-i\hat{f}}/(i-1)!$ $(i = 1, 2, \ldots, 5)$. The unprove conjecture that this holds for $i > 5$ leads to

$$G_\infty(\theta) = \sum_{i=1}^{\infty} \frac{i^{i-2}}{(i-1)!}\hat{f}^{i-1}e^{-i\hat{f}}\theta^i \qquad (12)$$

which is the Borel-Tanner probability distribution [8]:

$$B(i;k,f) = \frac{k}{(i-k)!}i^{i-k-1}\hat{f}^{i-k}e^{-i\hat{f}} (i = k, k+1, \ldots) \qquad (13)$$

for the case $k = 1$. This distribution arises in connection with the theory of queues where $B(i;k,\hat{f})$ represents the probability that exactly i members of a queue will be served before the queue first vanishes, starting with k members, assuming Poisson arrivals and a constant service time t, and with $\hat{f}$ equal to the traffic intensity.

The same distribution is furthermore now recognized as a limiting case of a Galton-Watson cascade branching process [5] for generating $(1, p)$-trees which was adapted and extended so successfully by Gordon [6] and coworkers [see eq. 7] for treating many aspects of statistical polymer science. The weight fraction generating function for $(1, p)$-trees is [7]:

$$W(\theta) = \sum_i w_i\,\theta^i = \theta F_o(\mathbf{u})$$
$$u(\theta) = \theta F_1(u) \qquad (14)$$

where the link probability generating functions are:

$$F_o(\theta) = (1-\alpha+\alpha\theta)^p \quad F_1(\theta) = (1-\alpha+\alpha\,\theta)^{p-1} \tag{15}$$

and α is the probability that one point gives rise to another in the cascade process. From eq. 14 and 15, one finds by Lagrange expansion [7].

$$w_i = \frac{(pi-i)!}{(pi-2i+2)!(i-1)!}\; p\,\alpha^{i-1}\,(1-\alpha)^{pi-2i+2} \tag{16}$$

In the limiting case $(p \to \infty, \alpha \to 0,$ such that $\alpha\, p = f)$ eq. 16 reverts to the Borel-Tanner distribution (eq. 13) with $k = 1$, which is apparently generated by eq. 14 with (from eq. 15):

$$F_o(\theta) \approx e^{-\hat{f}(1-\theta)} = F_1(\theta) \tag{17}$$

For $\hat{f} < 1$ the first two moments of the Borel-Tanner distribution (eq. 13) are known [8] to be:

$$\sum_i i\; B(i;k,\hat{f}) = k/(1-\hat{f}) \tag{18}$$

and

$$\sum_i i^2\; B(i;k,\hat{f}) = \frac{k^2 + k(1-k)\hat{f}}{(1-\hat{f})^3} \tag{19}$$

so that if the conjectured eq. 12 is correct the weight average component size in R_∞, $\hat{w}$, is (eq. 18):

$$\hat{w} = 1/(1-\hat{f}) \tag{20}$$

This diverges at the critical values $\hat{f}_{c,r} = 1$ which is the random graph analogue of the critical density in the clumps problem, see § 3.1.

The first 5 terms of the conjectured eq. 12 are correct (eq. 11) and furnish immediate comparisons with the random clumps problem. The fraction $(e^{-\hat{f}})$ of isolated points is, as expected, the same for both situations (cf. eq. 5).

The weight fraction of components of size 2 in R_∞ may be compared with Freund's exact result (eq. 6) for 2-dimensional clumps and with Fremlin's Monte Carlo simulations (tables 1 and 2) for 2- and 3-dimensional clumps. This is in fact done in figure 2. That

Fremlin's 3-dimensional clump results interpose the two curves in figure 2 fosters our view of a random graph as an q-dimensional clump graph in the limit of large q; a view further supported by the three critical density values. It is in this limit that the aforementioned geometric constraints are expected to vanish. What a pity that we are unable to choose among E. A. Abbott's [1] worlds for our experiments with polymer solutions.

Acknowledgements

I express my thanks to Dr. D. H. Fremlin for tables 1 and 2, Dr. D. D. Freund for table 3 and the appendix, and to Professor M. Gordon for his ever useful discussions. Thanks are also due to the Science Research Council (UK) for a Fellowship.

APPENDIX Weight Fraction of 2-Dimensional Clumps with Size 2

by: D. D. Freund*, Department of Mathematics, University of Essex.

The event that a point p_o chosen at random on a two dimensional clump graph is contained in a clump of size 2, may be described thus: the nearest neighbour to p_o lies at a distance x, and no other points occur in a distance less than $2r$ from either of these two points and greater than x from p_o. Let the probability of this event be w_2.

Let $P(x)$ be the probability that the nearest neighbour lies in the interval $x, x + dx$. Then:

$$P(x) = 2\pi x\lambda \; e^{-\pi\lambda x^2} \, dx \tag{A1}$$

where $\lambda = \hat{f}/4\pi r^2$ (cf. §3.1) is the mean number of points per unit area. Hence:

$$w_2 = \int_o^{2r} P(x) \; e^{-\lambda A(x)} dx \tag{A2}$$

where $A(x)$ is the shaded region in the following sketch and $e^{-\lambda A(x)}$ the (Poisson) probability that no points fall in $A(x)$.

---

*Present address: 630 Mariposa Ave., Sierra Madre California 91024.

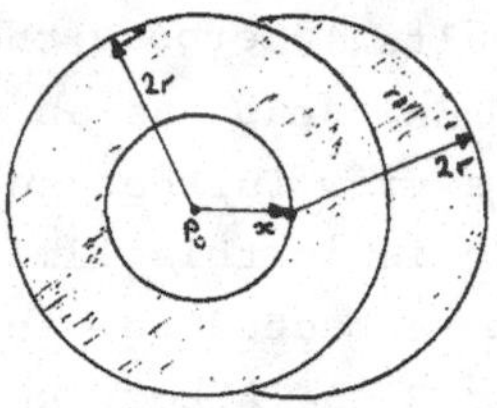

A simple calculation gives:

$$A(x) = 8\pi r^2 - \pi x^2 - 8r^2 \cos^{-1}\left(\frac{x}{4r}\right) + x\left(4r^2 - \frac{1}{4}x^2\right)^{\frac{1}{2}} \qquad \text{(A3)}$$

Hence:

$$w^2 = 2\pi\lambda e^{-8\pi\lambda r^2} \int_o^{2r} x \exp(8r^2 \lambda \cos^{-1}(\tfrac{x}{4r}) - \lambda x\sqrt{4r^2 - x^2/4})\,dx$$

Let $u = 2 \sin^{-1}(x/4r)$

$$\therefore \quad w_2 = 8\pi r^2 \lambda e^{-8\pi\lambda r^2} \int_o^{\pi/3} \sin u \; e^{4r^2\lambda(\pi - u - \sin u)} \, du$$

$$= 2\hat{f}\, e^{-\hat{f}} \sum_{i=0}^{\infty} (-\hat{f}/\pi)^i \frac{1}{i!} \int_o^{\pi/3} \sin u (u + \sin u)^i du \qquad \text{(A4)}$$

which is eq. 6 of the text.

References.

1. E. A. Abbott, Flatland - A Romance of Many Dimensions. Dover Publications, New York, (1952).

2. J. W. Essam, Percolation and Cluster Size. Chap. 6 in Phase Transitions and Critical Phenomena, Ed. C. Domb and M. S. Green, Volume 2, Academic Press, (1972).

3. R. P. Feynman, Statistical Mechanics: A Set of Lectures. Frontiers in physics series, W. A. Benjamin, Reading, (1972).

4. D. H. Fremlin, The clustering problem: some Monte Carlo results. J. de Physique, in press, (1976).

5. I. J. Good, The joint distribution for the size of generations in a cascade process. Proc. Cambridge Philosophical Society, 51, 240-242, (1955).

6. M. Gordon, Good's Theory of cascade processes applied to the statistics of polymer distributions. Proc. Royal Society A, 268, 240-259, (1962).

7. M. Gordon and W. B. Temple, The Graph-like State of Matter and Polymer Science. Chap. 10 in Chemical Applications of Graph Theory, Ed. A. T. Balaban, Academic Press, New York, (1976: in press).

8. F. A. Haight and M. A. Breuer, The Borel-Tanner distribution. Biometrika, 47, 143-150, (1960).

9. K. Kajiwara and M. Gordon, The graph-like state of matter V, J. Chemical Physics, 59, 3623-3632, (1973): and references to other parts of series contained therein.

10. R. Koningsveld, W. H. Stockmayer, J. W. Kennedy and L. A. Kleintjens, Liquid-liquid phase separation in multicomponent polymer solutions XI. Dilute and concentrated polymer solutions in equilibrium. Macromoleculues, 7, 73-79, (1974).

11. S. A. Roach, The Theory of Random Clumping. Methuen & Co. Ltd. London, (1968).

12. F. D. K. Roberts and S. H. Storey, A three dimensional cluster problem. Biometrika, 55, 258-260, (1968).

13. V. K. S. Shante and S. Kirkpatrick, An introduction to percolation theory. Advances in Physics, 20, 325-357, (1971).

14. F. Y. Wu, Graph theory in statistical physics. Adv. Math., submitted (1976).

15. H. Yamakawa, Modern Theory of Polymer Solutions. Harper & Row, New York, (1971).

MIXED RAMSEY NUMBERS: EDGE CHROMATIC NUMBERS VS. GRAPHS

Linda M. Lesniak
Louisiana State University
Baton Rouge, LA 70808

Albert D. Polimeni
SUNY, College at Fredonia
Fredonia, NY 14063

Donald W. VanderJagt
Grand Valley State Colleges
Allendale, MI 49401

Abstract

Let $\chi_1(G)$ denote the edge chromatic number of the graph G. For any positive integer m and any graph G, the mixed Ramsey number $\chi_1(m,G)$ is defined as the least positive integer p such that if H is any graph of order p, then either $\chi_1(H) \geq m$ or $\overline{H}$ contains a subgraph isomorphic to G. The number $\chi_1(m,G)$ is investigated for certain classes of graphs G, including star graphs, complete graphs, paths, and cycles.

MIXED RAMSEY NUMBERS: EDGE CHROMATIC NUMBERS VS. GRAPHS

Linda M. Lesniak
Louisiana State University
Baton Rouge, LA 70808

Albert D. Polimeni
SUNY, College at Fredonia
Fredonia, NY 14063

Donald W. VanderJagt
Grand Valley State Colleges
Allendale, MI 49401

For positive integers m and n the classical Ramsey number $r(m,n)$ is the least positive integer p such that if H is any graph of order p, then either H contains a subgraph isomorphic to the complete graph K_m or the complement $\overline{H}$ of H contains a subgraph isomorphic to K_n. In [2] the concept of mixed Ramsey number is introduced. In particular, if $\chi(G)$ denotes the chromatic number of the graph G and m is any positive integer, then the mixed Ramsey number $\chi(m,G)$ is defined in [2] as the least positive integer p such that if H is any graph of order p, then either $\chi(H) \geq m$ or $\overline{H}$ contains a subgraph isomorphic to G. The edge chromatic number $\chi_1(G)$ of the graph G is the least number of colors needed to color the edges of G so that adjacent edges are colored differently. For a positive integer m and a graph G, we define the mixed Ramsey number $\chi_1(m,G)$ to be the least positive integer p such that if H is any graph of order p, then either $\chi_1(H) \geq m$ or $\overline{H}$ contains a subgraph isomorphic to G. Alternatively, $\chi_1(m,G)$ is the least positive integer p such that for any factorization $K_p = F_1 \oplus F_2$ (i.e., F_1 and F_2 are edge-disjoint spanning subgraphs of K_p such that the union of their edge sets is the edge set of K_p), either $\chi_1(F_1) \geq m$ or F_2 contains a subgraph isomorphic to G. It is the object of this paper to investigate $\chi_1(m,G)$ for some special classes of graphs.

Henceforth, if a graph H contains a subgraph isomorphic to a graph G, then we shall simply say that G is a subgraph of H. If G_1 and G_2 are graphs with disjoint vertex sets V_1 and V_2 and edge sets E_1 and E_2, respectively, then their <u>union</u> $G_1 \cup G_2$ is the graph with vertex set $V_1 \cup V_2$ and edge set $E_1 \cup E_2$. Their <u>join</u> $G_1 + G_2$ is the graph with vertex set $V_1 \cup V_2$ and edge set $E_1 \cup E_2$, together with all edges joining V_1 with V_2. We denote by $\Delta(G)$ the maximum degree among the vertices of the graph G and by $\delta(G)$ the minimum degree among the vertices of G. Finally, we will make repeated use of the fact that

$$\chi_1(K_p) = \begin{cases} p & \text{, if } p \text{ odd} \\ p-1, & \text{if } p \text{ even.} \end{cases}$$

As an illustration of the techniques to be used we now show that

$$\chi_1(m,K(1,\ell)) = \begin{cases} m+\ell & \text{, if } m+\ell \text{ odd} \\ m+\ell-1, & \text{if } m+\ell \text{ even,} \end{cases}$$

where $K(1,\ell)$ denotes the star graph of order $\ell+1$. Suppose first that $m+\ell$ is odd. If $K_{m+\ell} = F_1 \oplus F_2$, with $\chi_1(F_1) \leq m-1$, then it follows that $\Delta(F_1) \leq m-1$. Since $\deg_{F_1} v + \deg_{F_2} v = m+\ell-1$ for each vertex v of $K_{m+\ell}$, $\delta(F_2) \geq (m+\ell-1)-(m-1) = \ell$. Thus $K(1,\ell)$ is a subgraph of F_2 and we conclude that $\chi_1(m,K(1,\ell)) \leq m+\ell$. It now suffices to exhibit a factorization $K_{m+\ell-1} = F_1 \oplus F_2$, where $\chi_1(F_1) \leq m-1$ and $K(1,\ell)$ is not a subgraph of F_2. Since $m+\ell-1$ is even, we have that $K_{m+\ell-1} = L_1 \oplus L_2 \oplus \ldots \oplus L_{m+\ell-2}$, where each L_i is a spanning 1-regular subgraph of $K_{m+\ell-1}$, $1 \leq i \leq m+\ell-2$. Then let $F_1 = L_1 \oplus L_2 \oplus \ldots \oplus L_{m-1}$ and $F_2 = L_m \oplus \ldots \oplus L_{m+\ell-2}$. It follows that $\chi_1(F_1) \leq$

$m - 1$ and, since $\deg_{F_2} u = \ell - 1$ for each vertex u of F_2, the graph $K(1,\ell)$ is not a subgraph of F_2. So in case $m + \ell$ is odd, $\chi_1(m,K(1,\ell)) = m + \ell$. The case $m + \ell$ even is handled in a similar manner.

We now proceed to the main theorems, the first of which yields a formula for $\chi_1(m,K_n)$. This result makes use of a theorem of P. Turán [6], which we now state.

Theorem. For positive integers p and n, with $3 \leq n \leq p$, let $T(p,n)$ be the smallest positive integer such that every graph of order p with $T(p,n)$ edges contains K_n as a subgraph. Then $T(p,n) = \binom{p}{2} + 1 - t(p - n + 1 + r)/2$, where $p = t(n - 1) + r$, $0 \leq r < n - 1$. Furthermore, the only graph of order p with $T(p,n) - 1$ edges which fails to contain K_n as a subgraph is the complete $(n - 1)$-partite graph $K(p_1,p_2,\ldots,p_{n-1})$, where $p_1 = p_2 = \ldots = p_r = t + 1$ and $p_{r+1} = p_{r+2} = \ldots = p_{n-1} = t$ (if $r = 0$, then $p_i = t$ for $1 \leq i \leq n$).

Theorem 1. If m and n are positive integers, then

$$\chi_1(m,K_n) = \begin{cases} n & \text{if } m = 1 \\ m(n - 1) + 1 & \text{if } m \text{ is even} \\ (m - 1)(n - 1) + 1 & \text{if } m \text{ is odd and } m \geq 3 . \end{cases}$$

Proof. It is readily seen that $\chi_1(1,K_n) = n$ and $\chi_1(m,K_1) = 1$. Also, that $\chi_1(m,K_2)$ is given by the above formula follows from the fact that $K_2 = K(1,1)$, a star graph. In what remains we assume $n \geq 3$ and distinguish between m odd and m even.

Case 1. Assume that m is even.

We first show that $\chi_1(m,K_n) > m(n - 1)$ by exhibiting a factorization $K_{m(n-1)} = F_1 \oplus F_2$, where $\chi_1(F_1) \leq m - 1$ and K_n is not a subgraph of F_2. Let $F_1 = (n - 1)K_m$; that is F_1 consists of $n - 1$ disjoint copies of K_m. Thus $\chi_1(F_1) = m - 1$ since m is even. Also, $F_2 = \overline{F}_1 = K(m,m,\ldots,m)$, the complete $(n-1)$-partite graph of order $m(n-1)$, which does not contain K_n as a subgraph. Next we show that $\chi_1(m,K_n) \leq m(n-1)+1$. Suppose $K_{m(n-1)+1} = F_1 \oplus F_2$, where $\chi_1(F_1) \leq m-1$. Then there is a coloring of the edges of F_1 with at most $m-1$ colors, each color class containing at most $m(n-1)/2$ edges since m is even. Thus there are at least

$$E = \binom{m(n-1)+1}{2} - \frac{(m-1)m(n-1)}{2}$$

edges in F_2 . By Turán's theorem, the number of edges which will insure that K_n be a subgraph of F_2 is

$$\begin{aligned} T &= T(m(n-1)+1,n) \\ &= \binom{m(n-1)+1}{2} + 1 - m(m(n-1)+1-n+1+1)/2 \end{aligned}$$

If $E - T \geq 0$, then we may conclude that K_n is a subgraph of F_2 . Since $E - T = m - 1 > 0$ we have $\chi_1(m,K_n) = m(n-1)+1$ if m is even.

<u>Case</u> 2. <u>Assume that</u> m <u>is odd, with</u> $m \geq 3$.

We proceed by induction on n , recalling that $n \geq 3$. We first show that $\chi_1(m,K_3) = 2m-1$. As in previous cases we show $\chi_1(m,K_3) > 2m-2$ by exhibiting a factorization $K_{2m-2} = F_1 \oplus F_2$, where $\chi_1(F_1) \leq m-1$ and K_3 is not a subgraph of F_2 . Let $F_1 = 2K_{m-1}$ and hence, $F_2 = K(m-1,m-1)$. Then $\chi_1(F_1) = m-2$ since $m-1$ is even, and K_3 is not a subgraph of $K(m-1,m-1)$. Next we claim that $\chi_1(m,K_3) \leq 2m-1$. Suppose $K_{2m-1} = F_1 \oplus F_2$ and $\chi_1(F_1) \leq m-1$. Then there are at most $(m-1)^2$ edges in F_1 and hence, at least

$$E = \binom{2m-1}{2} - (m-1)^2 = m(m-1)$$

edges in F_2 . By Turán's theorem the number of edges which will insure that K_3 be a subgraph of F_2 is $T = T(2m - 1,3) = 1 + m^2 - m$. Also, the only graph of order $2m - 1$ having $m^2 - m$ edges which does not contain K_3 as a subgraph is the complete bipartite graph $K(m - 1,m)$. However, if $F_2 = K(m - 1,m)$, then K_m will be a subgraph of F_1 , contradicting the assumption $\chi_1(F_1) \leq m - 1$. Thus K_3 is a subgraph of F_2 . So $\chi_1(m,K_3) = 2m - 1$.

Next we assume that $\chi_1(m,K_{n-1}) = (m - 1)(n - 2) + 1$ for some $n \geq 4$. Similar to previous cases, choosing $F_1 = (n - 1)K_{m-1}$ illustrates that $\chi_1(m,K_n) > (m - 1)(n - 1)$. It remains to show that $\chi_1(m,K_n) \leq (m - 1)(n - 1) + 1$. Suppose that $K_{(m-1)(n-1)+1} = F_1 \oplus F_2$ where $\chi_1(F_1) \leq m - 1$. Thus there is a coloring of the edges of F_1 resulting in at most $m - 1$ color classes, each class containing at most $(m - 1)(n - 1)/2$ edges. Thus there is at least one vertex v of F_1 such that $\deg_{F_1} v \leq m - 2$. If $\deg_{F_1} v = r$ let $v_1, v_2, \ldots, v_r$ denote those vertices adjacent to v in F_1 . The removal of $v, v_1, \ldots, v_r$ from both F_1 and F_2 results in a factorization of $K_{(m-1)(n-1)-r}$ and since $r \leq m - 2$, this factorization induces a factorization of $K_{(m-1)(n-2)+1}$, say $F_1' \oplus F_2'$. Since $\chi_1(F_1) \leq m - 1$ and F_1' is a subgraph of F_1 , it follows that $\chi_1(F_1') \leq m - 1$. Hence by the induction hypothesis K_{n-1} is a subgraph of F_2' and consequently, of F_2 . In F_2 the vertex v is adjacent to each vertex of K_{n-1} ; thus K_n is a subgraph of F_2 . So $\chi_1(m,K_n) = (m - 1)(n - 1) + 1$.

Since every graph G of order n is a subgraph of K_n , we have an immediate consequence of Theorem 1.

Corollary 1.1. For any graph G of order n and any positive integer m , the mixed Ramsey number $\chi_1(m,G)$ exists and $\chi_1(m,G) \leq \chi_1(m,K_n)$.

We now investigate $\chi_1(m,G)$ when G is a path; that is, we determine $\chi_1(m,P_n)$ where P_n is a path of order n.

In a graph G of order p, a <u>hamiltonian path</u> is a path in G which contains all the vertices of G. It is well known [1, p. 135] that if $\delta(G) \geq \frac{p-1}{2}$, then G contains a hamiltonian path. Note that if a graph contains P_k, then it contains P_j for $j < k$. Some values of $\chi_1(m,P_n)$ are easily determined.

<u>Theorem</u> 2. For positive integers m and n, $n \geq m$,

$$\chi_1(m,P_n) = \max\{n, 2m - 1\} .$$

<u>Proof</u>. The theorem is true for $m = 1$. Thus we assume $n \geq m \geq 2$. Let $t = \max\{n, 2m - 1\}$ and suppose $K_t = F_1 \oplus F_2$. If $\chi_1(F_1) \leq m - 1$, then $\delta(F_2) \geq t - m \geq \frac{t-1}{2}$. Thus F_2 contains a hamiltonian path, so contains P_t and hence P_n as a subgraph. Thus $\chi_1(m,P_n) \leq t$.

To show $\chi_1(m,P_n) \geq \max\{n, 2m - 1\}$, we exhibit the appropriate factorizations of K_{n-1} and K_{2m-2}. Since $\chi_1(\overline{K}_{n-1}) < m$, P_n is not a subgraph of K_{n-1}, and $K_{n-1} = \overline{K}_{n-1} \oplus K_{n-1}$, we know $\chi_1(m,P_n) \geq n$. Also, $\chi_1(K(m-1, m-1)) = m - 1$ and P_n is not a subgraph of $2K_{m-1}$ since $n \geq m$, so we must have $\chi_1(m,P_n) \geq 2m - 1$. Thus $\chi_1(m,P_n) \geq \max\{n, 2m - 1\}$, establishing the necessary equality.

The case for which $m > n$ is not as straightforward as that above. The cases for $n = 1$ and $n = 2$ are included in Theorem 1 since in these cases P_n is a complete graph. The following result gives a lower bound for $\chi_1(m,P_n)$ when $m > n \geq 3$.

<u>Theorem</u> 3. For $m > n \geq 3$, $\chi_1(m,P_n) \geq \max\{m + 1, m + n - r - 1\}$, where r is defined by:

if n is odd, $m - 1 = s(n - 1) + r$, where $0 \leq r < n - 1$;

if n is even, $m - 1 = (2s - 1)(n - 1) + r$, where $0 \leq r < 2n - 2$.

<u>Proof</u>. Let n be odd and r be defined by $m - 1 = s(n - 1) + r$,

where $0 \leq r < n - 1$. Then $m + n - r - 1 \geq m + 1$ and $m + n - r - 2 = (s + 1)(n - 1)$. Let F_1 be the complete $(s + 1)$-partite graph $K(n - 1,\ldots,n - 1)$ and $F_2 = \overline{F}_1 = (s + 1)K_{n-1}$. Since F_1 has even order, $\chi_1(F_1) = \Delta(F_1)$ (see [5]), so $\chi_1(F_1) = s(n - 1) \leq m - 1$. Furthermore, F_2 does not contain P_n as a subgraph. But $K_{m+n-r-2} = F_1 \oplus F_2$, so $\chi_1(m,P_n) \geq m + n - r - 1 = \max\{m + 1, m + n - r - 1\}$.

Now consider n even and r defined by $m - 1 = (2s - 1)(n - 1) + r$, where $0 \leq r < 2n - 2$; then $m + n - r - 2 = (2s)(n - 1)$. Let F_1 be the complete $(2s)$-partite graph $K(n - 1,\ldots,n - 1)$ and $F_2 = (2s)K_{n-1}$. Then F_1 has even order, so again $\chi_1(F_1) = \Delta(F_1) = (2s - 1)(n - 1) \leq m - 1$. Also F_2 does not contain P_n as a subgraph and $K_{n+m-r-2} = F_1 \oplus F_2$, so we have $\chi_1(m,P_n) > m + n - r - 2$. Since $n \geq 4$, the factorization $K_m = (K_1 \cup K_{m-1}) \oplus K(1,m - 1)$ illustrates $\chi_1(m,P_n) > m$. Thus $\chi_1(m,P_n) \geq \max\{m + 1, m + n - r - 1\}$.

It will be useful in what follows to consider the special cases for $r = 0$ and $r = 1$.

<u>Corollary</u> 3.1. Let $m > n \geq 3$. If n is even with $m \equiv n \pmod{2n - 2}$ or if n is odd with $m \equiv n \pmod{n - 1}$, then $\chi_1(m,P_n) \geq m + n - 1$.

<u>Corollary</u> 3.2. Let $m > n \geq 3$. If n is even with $m \equiv n + 1 \pmod{2n - 2}$ or if n is odd with $m \equiv n + 1 \pmod{n - 1}$, then $\chi_1(m,P_n) \geq m + n - 2$.

We now obtain an upper bound for $\chi_1(m,P_n)$, when $m > n \geq 3$. The following proof uses the fact that if G is a connected graph of order at least three, then either G is hamiltonian or G contains a path of order at least $1 + 2\delta(G)$, (see [1, p. 135]).

<u>Theorem</u> 4. Let $m > n \geq 3$. If n is even with $m \equiv n \pmod{2n - 2}$ or if n is odd with $m \equiv n \pmod{n - 1}$, then $\chi_1(m,P_n) =$

$m + n - 1$. Otherwise $\chi_1(m,P_n) \le m + n - 2$.

Proof. First assume that it is neither the case that n is even with $m \equiv n(\text{mod } 2n - 2)$ nor that n is odd with $m \equiv n(\text{mod } n - 1)$. Let $K_{m+n-2} = F_1 \oplus F_2$, with $\chi_1(F_1) \le m - 1$. Then $\delta(F_2) \ge n - 2$, so each component of F_2 contains at least $n - 1$ vertices. Suppose that each component contains exactly $n - 1$ vertices. Then $m + n - 2 = k(n - 1)$ for some positive integer k , so $m - n = (k - 2)(n - 1)$; that is, $m \equiv n(\text{mod } n - 1)$. By our assumptions, then, n is even. Furthermore, k is odd; for otherwise, $m - n = (k/2 - 1)(2n - 2)$, so that n is even with $m \equiv n(\text{mod } 2n - 2)$. Since each of the k components of F_2 contains $n - 1$ vertices, the complete k-partite graph $H = K(n - 1,\ldots,n - 1)$ is a subgraph of F_1 . But $k(n - 1)$ is odd, so $\chi_1(H) = 1 + \Delta(H) = 1 + (k-1)(n-1) = m$ However, this implies that $\chi_1(F_1) \ge m$, which contradicts the fact that $\chi_1(F_1) \le m - 1$. Thus there is a component C of F_2 which contains at least n vertices. Since $\delta(C) \ge n - 2$, either C is hamiltonian or C contains a path of order at least $1 + 2(n - 2)$. For $n \ge 3$, this implies that F_2 contains P_n as a subgraph, so that $\chi_1(m,P_n) \le m + n - 2$.

Assume now that either n is even with $m \equiv n(\text{mod } 2n - 2)$ or n is odd with $m \equiv n(\text{mod } n - 1)$. If $K_{m+n-1} = F_1 \oplus F_2$ with $\chi_1(F_1) \le m - 1$, then $\delta(F_2) \ge n - 1$, so each component of F_2 has at least n vertices. As above, each component of F_2 is either hamiltonian or has a path of order at least $1 + 2(n - 1)$. In any case F_2 contains P_n as a subgraph, so $\chi_1(m,P_n) \le m + n - 1$. By Corollary 3.1, $\chi_1(m,P_n) = m + n - 1$.

In Theorem 4, a class of graphs is identified for which $\chi_1(m,P_n)$ is determined when $m > n$. Another such class is obtained from Theorem 4 and Corollary 3.2.

Corollary 4.1. Let $m > n \ge 3$. If n is even with

$m \equiv n + 1 (\mathrm{mod}\ 2n - 2)$ or if n is odd with $m \equiv n + 1 (\mathrm{mod}\ n - 1)$, then $\chi_1(m,P_n) = m + n - 2$.

For integers $m > n \geq 3$, Theorem 3 provides a lower bound for $\chi_1(m,P_n)$ and Theorem 4 provides an upper bound. In general, these two bounds are not equal. Examples show that in some cases, the lower bound gives the value of $\chi_1(m,P_n)$ while in others, the upper bound is attained.

A graph G of order $p \geq 3$ is called <u>pancyclic</u> if G contains a cycle of length k for each integer k satisfying $3 \leq k \leq p$. It is well-known (see [4]) that if a graph G of order $p \geq 3$ has minimum degree at least $p/2$, then G is hamiltonian. In [3], it was shown that such a graph G is, in fact, either pancyclic or has even order p and is isomorphic to $K(p/2,p/2)$. In the case that $G = K(p/2,p/2)$, the graph G contains a cycle of length k if and only if k is even and $4 \leq k \leq p$.

Employing the results discussed in the previous paragraph, we next obtain the value of $\chi_1(m,C_n)$ for various values of m and n.

<u>Theorem</u> 5. For integers $m \geq 1$ and $n \geq 3$,

$$\chi_1(m,C_n) = \begin{cases} n & , \text{ if } n \geq 2m \\ 2m+1 & , \text{ if } n \leq 2m-1,\ m \text{ is even, and } n \text{ is odd} \\ 2m & , \text{ if } m+1 \leq n \leq 2m-1,\ m \text{ is even, and } n \text{ is even} \\ 2m & , \text{ if } m+1 \leq n \leq 2m-1 \text{ and } m \text{ is odd.} \end{cases}$$

<u>Proof</u>. We first show that if $n \geq 2m$, then $\chi_1(m,C_n) = n$. The factorization $K_{n-1} = \overline{K}_{n-1} \oplus K_{n-1}$ indicates that $\chi_1(m,C_n) > n - 1$. In order to show that $\chi_1(m,C_n) \leq n$, let $K_n = F_1 \oplus F_2$ be a factorization of K_n and suppose that $\chi_1(F_1) \leq m - 1$. Then $\delta(F_2) \geq n - m \geq n/2 = |V(F_2)|/2$. Therefore F_2 is hamiltonian, i.e. F_2 contains C_n as a subgraph.

Next, assume that $n \leq 2m-1$ and let $K_{2m+1} = F_1 \oplus F_2$ be a factorization of K_{2m+1}. If $\chi_1(F_1) \leq m-1$, then $\delta(F_2) \geq m+1 > |V(F_2)|/2$, implying that F_2 is pancyclic and so contains cycles of lengths k for $3 \leq k \leq 2m-1$. In particular, F_2 contains C_n as a subgraph. Thus $\chi_1(m,C_n) \leq 2m+1$. Now, if m is even and n is odd, consider the factorization $K_{2m} = F_1 \oplus F_2$, where $F_1 = 2K_m$ and $F_2 = K(m,m)$. Then $\chi_1(F_1) = m-1$ and F_2 does not contain C_n as a subgraph. Therefore $\chi_1(m,C_n) > 2m$ and we conclude that $\chi_1(m,C_n) = 2m+1$ if $n \leq 2m-1$, m is even, and n is odd.

Assume now that $n \leq 2m-1$ and that either m is odd or m and n are both even. In this case, we show that the inequality $\chi_1(m,C_n) \leq 2m$ holds. Let $K_{2m} = F_1 \oplus F_2$ be a factorization of K_{2m} and suppose $\chi_1(F_1) \leq m-1$. Then $\delta(F_2) \geq m = |V(F_2)|/2$ so that either F_2 is pancyclic or $F_2 = K(m,m)$. If F_2 is pancyclic, then C_n is a subgraph of F_2. If, on the other hand, $F_2 = K(m,m)$, then $F_1 = 2K_m$. Since $\chi_1(F_1) \leq m-1$, it cannot be the case that m is odd. Thus m and n are both even and so F_2 contains C_n as a subgraph. We conclude that if $n \leq 2m-1$ and either m is odd or m and n are both even, then $\chi_1(m,C_n) \leq 2m$.

The proof will be complete once we have shown that if $m+1 \leq n \leq 2m-1$, then $\chi_1(m,C_n) > 2m-1$. Consider the factorization $K_{2m-1} = F_1 \oplus F_2$, where $F_1 = K(m-1,m-1) \cup K_1$ and $F_2 = 2K_{m-1} + K_1$. Then $\chi_1(F_1) = m-1$. Furthermore, F_2 contains C_k as a subgraph if and only if $3 \leq k \leq m$ and so does not contain C_n.

Our final result provides bounds for $\chi_1(m,C_n)$ for those pairs of integers m and n which are not included in Theorem 5.

<u>Theorem</u> 6. For integers $m \geq n \geq 3$,

$$2m-1 \leq \chi_1(m,C_n) \leq 2m, \quad \text{if } m \text{ and } n \text{ are odd;}$$

$$\max\{m+1, m+n-r-1\} \leq \chi_1(m,C_n) \leq 2m, \quad \text{if } n \text{ is even and } m-1 = (2s-1)(n-1)+r \quad (0 \leq r < 2n-2)$$

Proof. The bound $\chi_1(m,C_n) \leq 2m$ was verified in the proof of the previous theorem. Since $\chi_1(m,C_n) \geq \chi_1(m,P_n)$, the inequality $\chi_1(m,C_n) \geq \max\{m+1, m+n-r-1\}$ is immediate from Theorem 3 when $m > n$ and from Theorem 2 when $m = n$. Finally, the fact that $\chi_1(m,C_n) > 2m-2$ when n is odd is shown by the factorization $K_{2m-2} = 2K_{m-1} \oplus K(m-1,m-1)$.

References

1. M. Behzad and G. Chartrand, Introduction to the Theory of Graphs. Allyn and Bacon, Boston (1971).
2. J. M. Benedict, G. Chartrand, and D. R. Lick, Mixed Ramsey Numbers. To appear.
3. J. A. Bondy, Pancyclic graphs I. J. Combinatorial Theory, Ser. B 11 (1971) 80-84.
4. G. A. Dirac, Some theorems on abstract graphs. Proc. London Math. Soc. 2 (1952) 69-81.
5. R. Laskar and W. Hare, Chromatic numbers for certain graphs. J. London Math. Soc. 4 (1972) 489-492.
6. P. Turán, Eine Extremalaufgabe aus der Graphentheorie. Mat. Fiz. Lapok. 48 (1941) 436-452.

PLANAR AND OUTERPLANAR CAYLEY GRAPHS OF FREE GROUPS

H. W. Levinson
Rutgers University
New Brunswick, N. J., 08903

Abstract

All finite planar graphs are subgraphs of certain planar Cayley graphs of free groups; outerplanar graphs are subgraphs of certain outerplanar Cayley graphs of free groups. It is shown for these particular planar presentations of free groups that special planar is equivalent with outerplanar.

INTRODUCTION

The purpose of this paper is to correct an earlier theorem about a class of presentations of free groups whose Cayley diagrams contained subgraphs isomorphic with all finite planar graphs and to explore the relations between the notions of point-symmetry, local finiteness, outerplanarity, and the embedding construction given in [3].

1. PLANAR EMBEDDINGS IN PLANAR CAYLEY GRAPHS

Let G be any finite connected graph on n vertices. Direct the edges of G arbitrarily and color each with a separate color. Let v_1 be a given vertex in G and consider n-1 copies $G_2', G_3', \ldots, G_n'$ of G. G_i' contains a vertex v_i' corresponding to v_i in G for each $2 \leq i \leq n$. Identifying v_1 with $v_2', v_3', \ldots, v_n'$ results in a graph in which each edge color enters and exits from the site in the former G occupied by v_1. Call this process "G-saturating v_1". According to Theorem 1 of [3], the continuation of this G-saturation at all vertices, old and new, not already G-saturated results in a Cayley graph Γ_G of a presentation of a free group. Γ_G contains G as a subgraph and Γ_G is planar if and only if G is planar.

It is a simple consequence of the embedding process described above that if G were composed of several blocks $B_1, \ldots, B_k$, then Γ_G would be the Cayley graph of the free product of free groups whose presentations had Cayley graphs $\Gamma_{B_1}, \ldots, \Gamma_{B_k}$. For this reason, the finite graphs G considered in the discussions below shall be assumed not to contain cut points, which means that all the blocks of Γ_G are isomorphic with G.

THEOREM 1. *Let* F *be a presentation of a free group on* e *generators and* k *defining relators in which each generator appears twice among all the generating symbols of all the defining relators, once to the exponent +1 and once to the exponent -1. Let each proper subset of the defining relators share at least one generating symbol with its compliment. Finally, let there be some juxtaposition of (cyclic permutations of) all the defining relators which freely equals* 1. *Then the graph of* F *is planar.*

Proof. Label segments of the boundaries of k disks with symbols representing the generators of F so that each disk is bounded by a defining relator of F. Fitting all the disks together according to the schema imposed by requiring a juxtaposition of cyclic permutations of them to be freely equal to 1 results in a 2-cell embedding of a graph in the plane in which each simple circuit bounding a finite face corresponds to a defining relator of F. The boundary of the infinite face freely equals 1, which means that if all adjacent similarly labeled oppositely directed edges are identified, one obtains a 2-cell embedding of a graph on the sphere. Each edge is differently labeled from any other, since it is the result of identifying the edges of two formerly separate disks. Call this graph G. Now Γ_G must be the Cayley graph of F, since all circuits in Γ_G may be described in terms of conjugates of juxtapositions of cyclic permutations of circuits of G and the defining relators of F form a directed cyclic basis for G by [5, p. 76]. Since G was planar, Γ_G must also be planar. Q.E.D.

If each relator of F were the consequence of juxtaposing (cyclic permutations of) the rest, then some juxtaposition of (cyclic permutations of) all must be freely equal to 1. Thus Theorem 3 of [3] is true.

Call the class of all presentations of free groups satisfying the premises of Theorem 1 $\mathfrak{F}$. As a simple consequence of Euler's formula, we have

COROLLARY. *A necessary condition for a presentation* F *on* e *generators and* k *defining relators to be in* $\mathfrak{F}$ *is that* $e - k + v = 1$, *where* v *is the rank of the free group* F.

Proof. The v generators of F which freely generate F account for the edges of a spanning tree for G. Thus G must have v+1 vertices. This, together with the assumed planarity of G and Euler's formula, gives the desired result. Q.E.D.

Following Levinson and Rapaport in [4], an embedding of a Cayley graph on an orientable surface shall be called *point-symmetric* if the counterclockwise cyclic succession of edges is the same at each vertex.

THEOREM 2. *Each planar graph may be embedded in a point-symmetric planar Cayley graph of a presentation of a free group.*

Proof. Let G be planar and embedded on a sphere with each edge differently colored and arbitrarily directed. Let $f_1, f_2, \ldots, f_n$ be a minimal set of faces of G such that each vertex is incident with at least one face. Assign each vertex to one and only one of the faces. Form copies G_i of G embedded in the plane by puncturing the face assigned vertex v_i. Thus v_i is on the exterior of G_i which is embedded on the plane. Now the counterclockwise cyclic succession of edges about any vertex v_k is the same in G_i for all i since puncturing a face does not alter any sucession of edges. (Note that since all the edges of G were colored differently and directed arbitrarily, no two distinct vertices of G have an edge incident to them identically colored and directed.) As a result, if we use the G_i

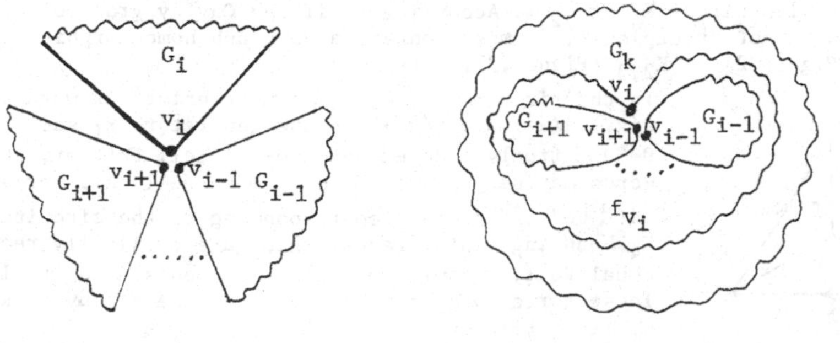

Figure 1

as the G_i' were used to construct Γ_G in the discussion prior to Theorem 1, the succession of edges about any given vertex is solely dependent on the succession of the G_i blocks about that vertex. We may, in fact, take any of the successions as our standard succession, say $G_1, G_2, \ldots, G_{v+1}$, and arrange this succession at every G-saturation of every vertex, both exterior and interior as shown in Figure 1. Since there is no impediment to this proceedure at any stage, a point-symmetric planar embedding of Γ_G results. Q.E.D.

Following Strasser and Levinson in [6], an embedding of a Cayley graph in an orientable surface is called *special* if it is point-symmetric and locally finite in the sense that no vertex is an accumulation point of vertices. If the special embedding is on a plane or on a sphere, it is called *special planar.*

THEOREM 3. *A finite graph* G *is outerplanar if and only if* Γ_G *is special planar.*

Proof. Let G be outerplanar. Then, by Theorem 2, Γ_G may be embedded point-symmetrically in the plane. Since only one face of G and its copies is used in the construction of Γ_G, all vertices of Γ_G lie on one face. Thus Γ_G is outer-planar. Since any outerplanar graph is trivially locally finite, Γ_G must be special planar.

If G were planar but not outerplanar, each planar embedding of G would contain a vertex v whose only adjacent faces were finite. G-saturating v would therefore necessarily place copies of G within some finite face f. Continuing the G-saturation of all new vertices being introduced within f quickly leads to the overcrowding which necessitates that vertices cluster. This precludes local finiteness for Γ_G which therefore cannot be special planar. Q.E.D.

Let R denote the set of roots of the defining relators of a presentation. (i.e. The actual defining relators are powers of the words in R.) Consider the words in R to be cyclic words in the generators. Let x be any generator. Form the word S as follows. Write x^{+1}. Follow x^{+1} by the symbol and its exponent appearing to the right of x^{-1} in R. Continue in this manner until one reaches x^{+1} again. Due to the condition on all presentations in $\mathcal{F}$ that each proper subset of the set of defining relators share some symbol with its compliment, one does eventually reach x^{+1} again. It was proven, under minor restrictions in [6] that if R gives rise to a word S containing all the generators, each to the exponents +1 and -1, then that presentation has a special planar Cayley graph and conversely. From this and Theorem 3, we have the following.

COROLLARY. *The subclass of* $\mathcal{F}$ *in which all outerplanar graphs are found as subgraphs of Cayley graphs is distinguished by the additional condition that their defining relators give rise to single words* S *containing all generators once to the exponent* +1 *and once to the exponent* -1.

2. ARBITRARY OUTERPLANAR CAYLEY GRAPHS

A graph is outerplanar if and only if it does not contain a subgraph which is homeomorphic with either K_4 or $K_{2,3}$. Accordingly, if the Cayley graph of a group presentation is not outerplanar, it must contain a subgraph homeomorphic with either K_4 (Figure 2) or $K_{2,3}$ (Figure 3).

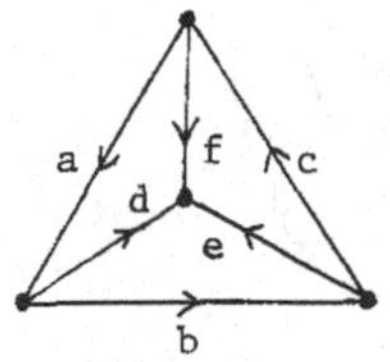

Figure 2.

In the first instance, there must exist six words, $a(x_i), b(x_i), \ldots, f(x_i)$, in the generators x_i such that each is freely reduced and non-trivial. Further, these words may be juxtaposed to form three longer words, adf^{-1}, bed^{-1}, and cfe^{-1}, corresponding to the circuits of K_4 bounding finite faces, which are cyclically reduced, equal to 1, but without proper segments equal to 1. These three words completely determine a copy of K_4 by [5, p. 76].

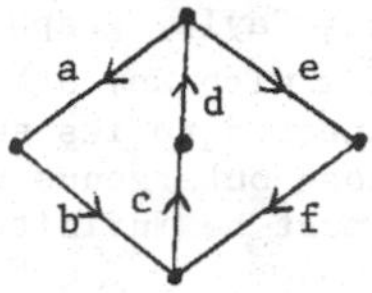

Figure 3.

In the second instance, there must exist three words, A=ab, B=cd, and C=ef, in the generators x_i, which are freely reduced, each of length two or greater in the generators, such that AB and BC are cyclically reduced, equal to 1, and without proper segments equal to 1. That these two words completely determine $K_{2,3}$ is clear from [5, p. 76] and the fact that they bound the only finite faces of $K_{2,3}$.

Thus we have

THEOREM 4. *The Cayley graph of a group presentation is not outerplanar if and only if either* i) *there exist six freely reduced, non-trivial words,* a,...,f *in the generators, such that* adf^{-1}, bed^{-1}, *and* cfe^{-1} *are each cyclically reduced, equal to* 1, *without proper segments equal to* 1 *or* ii) *there exist three freely reduced words,* A, B, *and* C, *each of length two or greater in the generators, such that* AB *and* BC *are both cyclically reduced, equal to* 1, *and without proper segments equal to* 1.

We now generalize the notion of outerplanarity.

DEFINITION. *The* <u>*outergenus*</u> *of a graph* G *is the minimum of the genera of all (orientable) surfaces on which* G *may be embedded such that all vertices of* G *are accessible from a single face.*

The (ordinary) genus of a graph is a lower bound for its outergenus, and a crude upper bound may be found as follows. Embed a triangulated n-gon on the sphere. This graph contains 2n-3 edges. Affix $\frac{1}{2}(n-2)(n-3)$ handles to the sphere in such a way that each carries an edge to complete the graph to K_n. Thus the outergenus of K_n is less than or equal to $\frac{1}{2}(n-2)(n-3)$. Since each graph on n vertices is a subgraph of K_n, the outergenus of a graph on n vertices is bounded from above by $\frac{1}{2}(n-2)(n-3)$.

To prove the outergenus of any infinite Cayley graph is 0 or infinite, the following lemma is needed.

LEMMA. *Let* K *be a graph consisting of* n *copies of either* K_4 *or* $K_{2,3}$, *each successively connected to the next by a bridge. Then the outergenus of* K *is* n.

Proof. It is easily verified that the outergenus of both K_4 and $K_{2,3}$ is 1. Thus if K consists of n blocks, each successively connected to the next by bridges and each of outergenus 1, K may be embedded on an n-torus by first embedding each of its blocks separately on n 1-tori, and forming their connected sum, one by one connecting each newcommer with a bridge from one of its vertices to a vertex in its predecessor. Were there an embedding of K on an m-torus, $m < n$, in which each of its vertices is accessible from a single face, all components of K could be separately accomodated on such a surface, each with all its vertices accessible from the same face as any other. Thus a new point in this region could be connected to each point in every component without improper crossings. This new point is a cut vertex for our new graph. The blocks of this new graph each contain a copy of either K_5 or $K_{3,3}$. Thus each is of (ordinary) genus 1 or greater. Since the genus of a graph is the sum of the genera of its blocks [1], we have an embedding of a graph of genus n or greater on a surface of lower genus. With this contradiction, the lemma is proven. Q.E.D.

It is probable that the outergenus is additive in the same manner as the ordinary genus.

THEOREM 5. *The outergenus of an infinite Cayley graph is either* 0 *or infinite.*

Proof. Suppose C is an infinite Cayley graph which is not outerplanar. Then there is a finite subgraph K within C which is either homeomorphic with K_4 or with $K_{2,3}$. By the result in [2] that any finite subgraph of an infinite Cayley graph C appears infinitely repeated in C such that no copy shares a vertex with any other, there are n distinct copies of K within C. Since each may be connected to its successor by a bridge within C, we have produced a subgraph of C whose outergenus is n or greater by the Lemma. Since n is arbitrary, the genus of C must be infinite.

Q.E.D.

We end with the following corollary which is immediate from Theorem 5.

COROLLARY. *If the outergenus of any Cayley graph of a group is finite and non-zero, then the group must be finite.*

The author wishes to acknowledge and thank Professor E. R. Strasser for her kind encouragement and suggestions with respect to several of the concepts used in this paper.

REFERENCES

1. J. Battle, F. Harary, Y. Kodama and J.W.T. Youngs, Additivity of the genus of a graph, Bull. Amer. Math. Soc. 68, (1962), 565-568.

2. H. Levinson, On the genera of graphs of group presentations, Ann. N. Y. Acad. Sci. 175, Art. 1, (1970), 277-284.

3. H. Levinson, On the genera of graphs of group presentations III, J. Comb. Theory, Series B 13, (1972), 298-302.

4. H. Levinson and E. S. Rapaport, Planarity of Cayley diagrams, Graph Theory and Applications. Springer-Verlag, Berlin (1972), 183-188.

5. C. W. Marshall, Applied Graph Theory. Wiley-Interscience, New York (1971).

6. E. R. Strasser and H. Levinson, Planarity of Cayley diagrams: planar presentations, Proc. 6th S-E Conf. Combinatorics, Graph Theory, and Computing. Utilitas Mathematica, Winnipeg (1975), 567-593.

COLORING RESTRICTIONS

Roy B. Levow
Florida Atlantic University
Boca Raton, FL 33432

Abstract

In this paper we introduce the notion of a coloring restriction, a generalization of the motion of a relative coloring. To each vertex of a subgraph of a graph G, a set of admissible colorings is assigned. The problem is to characterize those coloring restrictions which admit a consistent coloring with a specified number of colors.

The notion of a coloring restriction is a natural extension of the notion of relative coloring introduced by Kainen [2]. Rather than restricting the coloring on a vertex to a single color, a set of admissible colors is specified. Coloring restrictions appear to offer a useful tool for studying the extension of boundary colorings of planar graphs, studied by the author in [3]. They may also be helpful for the study of generalized colorings of the type studied by Meyer [4].

As to notation we shall generally adhere to that of Harary [1]. Our graphs will be connected, without loops or multiple edges. If u and w are adjacent, we shall write $u \sim w$.

Definition: Let G be a graph containing a subgraph H, and let C be a set of colors. A coloring restriction on G is a mapping from the vertices of H to the non-empty subsets of C, assigning to each vertex v of H a non-empty subset C_v of C.

We shall use the notation $\{C_v\}_{v \in H}$ to indicate a coloring restriction for $H \subset G$, or sometimes more simply $\{C_v\}$.

A coloring of G is consistent with a coloring restriction on G if, for each vertex v in the subgraph H, the color assigned to v is in C_v.

Let H and H' be subgraphs of G with $H \subset H'$. Let $\{C_v\}_{v \varepsilon H}$ and $\{C'_v\}_{v \varepsilon H'}$ be coloring restrictions on G. We say that the coloring restriction $\{C'_v\}$ is a refinement of the coloring restriction $\{C_v\}$ if $C'_v \subset C_v$ for every $v \varepsilon H$.

If G' is a subgraph of G and $\{C_v\}_{v \varepsilon H}$ is a coloring restriction on G, the induced coloring restriction on G' is given by $\{C_v\}_{v \varepsilon H\ G'}$.

We are principally interested in the application of coloring restrictions to planar graphs, and we therefore restrict our attention to such graphs. We assume our graphs to be drawn in the plane. If G is a planar graph and ∂G is a face bounding cycle, we say vertices u and w of ∂G are adjacent on ∂G if they are consecutive vertices in the cyclic order going around the face bounded by ∂G. If u and w are adjacent vertices of ∂G which are not adjacent on ∂G, we say the edge uw is a diagonal. A diagonal uw induces a splitting of G into two complementary subgraphs which contain only uw in common. The complementary subgraphs are the subgraphs of G bounded by the edge uw and a path along part of ∂G.

The interior of G consists of those vertices of G not on ∂G. We say G has triangulated interior if every face of G except possibly the face bounded by ∂G is a triangle.

The following theorem characterizes those coloring restrictions on outerplanar graphs which have consistent colorings.

<u>Theorem 1</u>: Let G be an outerplanar graph. Let C be a set of $k \geq 3$ colors. Let $\{C_v\}_{v \varepsilon G}$ be a coloring restriction with colors from C. Suppose there are adjacent vertices u and w of G and $a \varepsilon C_u$, such that $C_w \neq \{a\}$; and for every other vertex v of G, $|C_v| \geq 3$. Then there is a k-coloring of G consistent with $\{C_v\}$.

Note that this condition may always be refined to the special case $C_u = \{a\}$, $C_w = \{b\}$ with $b \neq a$; and $|C_v| = 3$ for all other $v \varepsilon G$. In our proof of the theorem we shall assume that we are in this special case.

<u>Proof</u>: We may assume that G is maximal outerplanar, adding edges to G if necessary. We work by induction on the number of vertices in G. The lemma is obviously true for $n \leq 3$. We assume it is true for all $n < n_0$ and consider the situation for $n = n_0 \geq 4$.

If uw is a diagonal for G, consider subgraphs induced by it. Each satisfies our condition under the induced coloring restriction. The consistent k-colorings guaranteed by the induction hypothesis agree on u and w, so they induce a consistent k-coloring of G.

Suppose therefore that u and v are adjacent along the boundary of G. As $u \sim w$, at most one of u,w has degree 2. Hence there is another vertex of G, say t, with $\deg t = 2$. Let G_t be the graph obtained from G by deleting t. By the induction hypothesis there is a k-coloring of G_t consistent with the induced coloring restriction on G_t. As $|C_t| \geq 3$, this coloring can be extended to a k-coloring of G consistent with the original coloring restriction.

The theorem is thus proved.

Before moving to our second theorem, we require the following technical lemma.

Lemma 1: Let G be a planar graph with face bounding cycle ∂G. Suppose G has at least one interior vertex. Then either

(i) There are at least 3 vertices of ∂G with degree at most 3, or

(ii) G has an interior vertex of degree at most 4, or

(iii) There are interior vertices x and y with $x \sim y$ and $\deg x = 5$.

Proof: It is a well known consequence of the Euler polyhedron formula that

$$4d_2 + 3d_3 + 2d_4 + d_5 = 12 + d_7 + 2d_8 + 3d_9 + \cdots$$

where d_i is the number of vertices of degree i in a planar triangulation. Consider the graph obtained from G by adding a vertex adjacent to each of the vertices of ∂G and adding edges, if necessary, to obtain a triangulation. It is easily verified that if G does not satisfy (i) then G has an interior vertex of degree at most 5. If this vertex has degree less than 5, then (ii) holds. Thus suppose the only internal vertices of G of degree 5 are adjacent only to vertices on ∂G.

Suppose v is such a vertex and is adjacent to vertices v_i, $i = 1,2,\cdots,5$. If G contains only these 6 vertices, we are clearly in case (i). Thus there is a subgraph of G bounded by the segment of ∂G between v_j and v_k, and the edges v_jv and vv_k which contains no other vertices on ∂G adjacent to v, but which does contain other vertices of G. Delete the edges v_jv and vv_k and replace them with the edge v_jv_k is it is not present. This is possible as v is not adjacent to any other vertices in this subgraph. Working inductively, we see that this graph satisfies the lemma. It must satisfy (i), and at least one of the vertices of ∂G of degree 3 or less is a vertex other than v_j or v_k.

If this is the only subgraph bounded by ∂G and consecutive pairs of edges from v, the remaining three vertices adjacent to v have degree 3. If there is only one other such subgraph containing other vertices, the two vertices of degree at most 3 together with the remaining vertex adjacent to v yield 3 vertices of degree at most 3 on ∂G. Finally, if there are at least 3 such subgraphs, then those 3 subgraphs yield 3 vertices of degree at most 3 on ∂G as required.

We are now prepared to prove the second theorem.

Theorem 2: Let G be a planar graph with a face bounding cycle ∂G. Let C be a set of $K \geq 5$ colors. Let $\{C_v\}_{v\varepsilon\partial G}$ be a coloring restriction with colors from C. If $|C_v| \geq 4$ for all but at most two vertices u and w of ∂G and if $u \sim w$ implies $|C_u \cup C_w| \geq 2$, then there is a k-coloring of G consistent with $\{C_v\}$.

Proof: By induction on the number of interior vertices of G, and for a fixed number of interior vertices on the number of vertices of ∂G.

Suppose G is maximal outerplanar. We may assume that $C_u = \{a\}$ and $C_w = \{b\}$ by taking a suitable refinement. If there is a vertex x distinct from u and w with $\deg x = 2$, the graph obtained by deleting w is consistently colorable and the coloring may be extended to all of G. If u and w are the only two vertices of degree 2 in G, form G' by deleting u and w. Refine the coloring restriction by eliminating a from the colorings available for the vertices adjacent to u and b from those adjacent to w. The resulting coloring restriction satisfies the preceding theorem. Thus there is a consistent coloring of the reduced graph which can be extended to a consistent coloring of G. Clearly the result also holds for arbitrary outerplanar graphs.

If G has an interior vertex, by the preceding lemma we consider three cases. If case (i) of the lemma holds, there is a vertex x distinct from u and w which has degree at most 3. The graph obtained from G by deleting x and adding an edge joining the vertices adjacent to x along ∂G satisfies the conditions of the theorem; and a consistent k-coloring of this graph can be extended to a consistent k-coloring of G.

If case (ii) of the lemma holds, a consistent k-coloring of the graph obtained from G by deleting the vertex of degree less than 5 can be clearly extended to a consistent k-coloring of G.

If case (iii) of the lemma holds, then by standard techniques y may be identified with another vertex adjacent to x in the graph obtained from G by deleting y, unless G contains a vertex bounding triangle. If there is a vertex bounding triangle, its interior vertices are deleted. In either situation the resulting graph may be consistently k-colored and that coloring extended to a consistent k-coloring of G.

The next two theorems deal with the case where six or more colors are available.

Theorem 3: Let G be a planar graph with face bounding cycle ∂G. Let C be a set of $K \geq 6$ colors. Let $\{C_v\}_{v \in \partial G}$ be a coloring restriction with colors from C. If $\{C_v\}$ satisfies the conditions of theorem 1, then there is a consistent k-coloring of G.

Proof: By induction on the number of vertices of G. If G has a diagonal xy, let G_1 be the subgraph induced by xy which contains u and w. G_1 with the induced coloring restriction satisfies the conditions of the theorem, and thus admits a k-coloring. Suppose such a k-coloring applies color c_x to vertex x and c_y to vertex y. Let G_2 be the complementary subgraph of G. The refinement of the induced coloring restriction obtained by replacing C_x by c_x and C_y by c_y satisfies the conditions of the theorem. Thus there is a consistent k-coloring for G_2. These coloring for G_1 and G_2 together yield a consistent k-coloring of G.

If G contains no diagonals, let x be a vertex of ∂G with $|C_x| \geq 3$. Suppose $\deg x \geq 3$ and let $D \subset C_x$ with $|D| = 3$. Let G' be the graph obtained from G by deleting x. A coloring restriction for G' is obtained as follows. If $v \in \partial G \{x\}$, use C_v. If v is in the interior of G and is adjacent to x in G, use $C \setminus D$. The induction hypothesis may be applied to G' with this coloring restriction to produce a consistent k-coloring of G'. This may be extended to a consistent k-coloring of G, as at least one of the colors in D does not appear on any vertex adjacent to x. If all vertices of ∂G except u and w are of degree 2, the theorem is easily verified.

All possibilities having been considered, the theorem is thus proved.

We next consider the situation in which our graph has been partially colored and we wish to extend this coloring to the entire graph. Let G be a planar graph with face bounding cycle ∂G. Let $P(G)$ be the subgraph of G constructed recursively as follows: Let $G_0 = \partial G$.

Having constructed G_i let G_{i+1} be the subgraph of G induced by the vertices of G which are adjacent to 4 or more vertices of G_i together with the vertices of G_i. Eventually we will have $G_k = G_{k+1} = \ldots,$ for some k. $P(G) = G_k$.

It is easily verified $|P(G)| \leq 2|\partial G| - 3$.

Theorem 4: Let G be a planar graph with face bounding cycle ∂G. Let C be a set of $k \geq 6$ colors. Let $\{C_v\}_{v \varepsilon \partial G}$ be a coloring restriction on ∂G. There is a k-coloring of G consistent with $\{C_v\}$ if and only if there is a k-coloring of $P(G)$ consistent with $\{C_v\}$.

Proof: The "only if" direction is trivial. Suppose there is a consistent k-coloring of $P(G)$. As the vertices of each component of $G\ P(G)$ adjacent to vertices of $P(G)$ are adjacent to at most three vertices of $P(G)$, it is possible to construct a coloring restriction on these vertices which assigns at least three colors to each such vertex in such a way that no vertex is assigned a color used by a vertex of $P(G)$ to which that vertex is adjacent. A simple generalization of theorem 3 then guarantees the existence of a consistent coloring of $G\ P(G)$. This coloring together with coloring of $P(G)$ provide the required consistent k-coloring of G.

Given a k-coloring of an n-cycle in a planar graph with $k \geq 6$, it is thus only necessary to examine at most $2n-6$ other vertices of the graph to determine whether the coloring can be extended to a k-coloring of the graph.

As a final note, we conjecture that if G is a planar graph with face bounding cycle ∂G, and $\{C_v\}_{v \varepsilon \partial G}$ is a coloring restriction in 5 colors with $|C_v| \geq 3$ for all $v \varepsilon \partial G$, then there is a k-coloring of G consistent with $\{C_v\}$.

References

1. F. Harary, Graph Theory, Addison-Wesley, Reading, Mass., 1969.

2. P. C. Kainen, Relative colorings of graphs, J. Combinatorial Theory, 6(1969), 259-262.

3. R. B. Levow, Coloring planar graphs with five or more colors, Proc. 6th S-E. Conf. Combinatorics, Graph Theory and Computing, Boca Raton, 1974, 549-561.

4. W. Meyer, Five coloring planar maps, J. Combinatorial Theory 13(1972), 72-82.

IF A HADAMARD MATRIX OF ORDER 24 HAS CHARACTER EXACTLY 2, ITS TRANSPOSE IS KNOWN

Judith Q. Longyear
Wayne State University
Detroit, Michigan 48202

Abstract

In [1], the order 24 Hadamard matrices of character at least 3 was classified. In this paper, we show that those with character exactly 2 have already been classified since the transpose character must be at least 3. It is still not known if there are inequivalent matrices of this order with both character and transpose character 1 .

Familiarity with [1] is assumed. Let H be of order 4 and character exactly 2, with A(row 1, row 2) = A(row 3, row 4) = columns 1 , ... , 6 . The remaining columns may be permuted so that H has the form:

	1 , ... , 6	7 , ... , 12	13 , ... , 18	19 , ... , 24
1	+	+	+	+
2	+	+	-	-
3	+	-	+	-
4	+	-	-	+
	Q_1	Q_2	Q_3	Q_4 .

Rows 5 , ... , 24 have 3+ and 3- in each of Q_1 , Q_2 , Q_3 , Q_4 . It is easy to see, since the columns are orthogonal, that every Q_k must have exactly two rows each of all the following forms:

$$\begin{Bmatrix} 111--- \\ ---111 \end{Bmatrix}, \begin{Bmatrix} 11-1-- \\ --1-11 \end{Bmatrix}, \begin{Bmatrix} 11--1- \\ --11-1 \end{Bmatrix}, \begin{Bmatrix} 11---1 \\ --111- \end{Bmatrix}, \begin{Bmatrix} 1-11-- \\ -1--11 \end{Bmatrix},$$

$$\begin{Bmatrix} 1-1-1- \\ -1-1-1 \end{Bmatrix}, \begin{Bmatrix} 1-1--1 \\ -1-11- \end{Bmatrix}, \begin{Bmatrix} -111-- \\ 1---11 \end{Bmatrix}, \begin{Bmatrix} -11-1- \\ 1--1-1 \end{Bmatrix}, \begin{Bmatrix} -11--1 \\ 1--11- \end{Bmatrix}.$$

That is, Q_i could have two rows 111--- , 111--- or 111--- , ---111 or ---111 , ---111 .

Any 2 rows of H (after the first 4) may "agree" or be of the same form in 0 , 1 or 2 of the Q_i . Three, and therefore four, agreements would cause char $H \geq 3$.

If two rows agree in one Q_i , call them Roman mates, if in two, call them Greek mates. Every row has two Greek mates, one Greek and two Roman mates, or four Roman mates.

Relative to any fixed but arbitrary row 5, with columns permuted within the Q_i to:

111--- , 111--- , 111--- , 111---

the remaining rows may be taken to have either 2 or 3+'s in the first 3 columns. In any Q where row j (with $j \geq 6$) does not mate with row 5 , there is a unique +1 or a unique -1 in the first 3 columns of Q and the other in the last three columns. Thus we may describe any row j by the number of +'s in the first halves of the Q's, and an indication of where the unique +'s and -'s occur. Thus 2ba 0 lac 3 represents 1-1 , 1-- , --- , 111 , 1-- , 11- , 111 , --- .

Relative row types.

Greek	3 , 0 , 1-- , 2--	in some order.
Roman	3 , 1-- , 1-- , 1--	or 0 , 2-- , 2-- , 2-- in some order
Chinese	2-- , 2-- , 1-- , 1--	in some order.

We consider compatibility of rows by types.

LEMMA. Only the following compatibilities exist.

1)	Greek Greek	0	3	1--	2--	
		2--	1--	0	3	
2)	Greek Roman	0	3	1xx	2xx	
		1--	1--	3	1yy	$x \neq y$
	or	1--	1--	1yy	3	$x \neq y$
3)	Greek Chinese	0	3	1xx	2xx	
	a)	2--	2--	1yy	1zz	
	b)	2--	1--	2yy	1zz	
	c)	2--	1--	1yy	2zz	

in a) the number of x's in {yy} = the number of x's in {zz}.

b) $y \neq x \neq z$

c) the number of x's in {yyzz} = 2 .

4)	Roman Roman	3	1--	1xx	1xx	exactly two x's are y's.
		1--	3	1yy	1yy	

(recall that row 5 could be represented 3333).

5)	Roman Chinese	3	1xx	1xx	1xx
		2--	2yy	1zz	1zz

The number of x's in {zzzz} = the number of x's in {yy} .

6)	Chinese Chinese				
		2xx	2xx	1xx	1xx
	a)	2yy	2yy	1yy	1yy
	b) or	2zz	1ww	2ww	1zz

in a) exactly 2y's are x's.

b) the number of x's in {zzzz} = the number of x's in {wwww} .

<u>First case</u>.

Suppose that every row has 2 Greek mates. Then further suppose that some rows 5, 6, 7, 8 completely mate each other, thus:

(6)	0	3	1aa	2aa
(7)	2aa	1aa	0	3
(8)	2aa	2aa	1aa	1aa .

By arguments similar to those in [1], H can have at most 8 more rows of type 2211, and at most 4 each of the remaining Chinese types, so up to isomorphism of the first 8 rows of H we may take (9) to be 2ab 2ab 1bb 1bb . This row has 4 possible Greek mates with one of Q being Q_1 . Three of these cannot be completed without going to higher character, but (1) 2ab 1ab 2bc 1bc can finish two ways.

(11)	2ba	1ba	2bb	1bb					
(12)	2ba	2ba	1cb	1cb					
(13)	2ca	2ca	1bc	1bc					
(14)	2ac	1ac	2cb	1cb					
(15)	2ca	1ca	2cc	1cc					
(16)	2ac	2ac	1cc	1cc					
(17)	2cc	2cc	1ab	1ab	or	2cc	1bb	1ab	2ac
(18)	2bc	1bc	1ab	2ab		2bc	2cb	1ab	1ac
(19)	2cc	1cc	1ca	2ca		2cc	2bb	1ca	1ba
(20)	2bc	2bc	1ba	1ba		2bc	1cb	1ba	2ca
(21)	2cb	2cb	1ca	1ca		2cb	1bc	1ca	2ba
(22)	2bb	1bb	1ba	2ba		2bb	2cc	1ba	1ca
(23)	2cb	1cb	1ac	2ac		2cb	2bc	1ac	1ab
(24)	2bb	2bb	1ac	1ac		2bb	1cc	1ac	2ab
		2GA					2GB		

If no four rows completely mate each but some six do, every completion has character at least 3 . If some 8 do, every completion requires 4 rows that mate completely. The last possible completion is:

(8) 2aa 2aa 1ab 1ab

(9) 2bb 2bb 1aa 1aa

(10) 2bc 1bc 1ab 2ab

(11) 2bb 1bb 1ba 2ba

(12) 2bc 2bc 1cb 1cb

(13) 2ac 2ac 1ba 1ba

(14) 2ac 1ac 2cb 1cb

(15) 2ab 2ab 1cc 1cc

(16) 2ab 1ab 2bc 1bc

(17) 2ca 1ca 2cc 1cc

(18) 2ba 2ba 1bc 1bc

(19) 2ca 2ca 1ca 1ca

(20) 2ba 1ba 2bb 1bb

(21) 2cc 1cc 1ca 2ca

(22) 2cb 2cb 1bb 1bb

(23) 2cc 2cc 1ac 1ac

(24) 2cb 1cb 1ac 2ac.

2GC

These three cases must now be examined in detail. Using row 5, divide Q_i into S_{2i-1} and S_{2i} (so that in 2ab for instance, the a refers to S_{2i-1} and the b to S_{2i}).

For each of the matrices obtained, there are potentially 8 inequivalent matrices to be obtained by leaving rows 1-5 alone but switching some of the S_{2i-1}, S_{2i} pairs. All 24 have high transpose character.

In 2GA and 2GC, every switch has transpose character 12 or 6^2, in the obvious way.

In 2GC, the transpose character is at least 4^3 for each switch since every switch preserves or completely inverts all of one of the following sets.

O_1. col i by col i + 6 for i = 1 , 4 , 13 , 16

O_2. col i by col i + 6 for i = 2 , 3 , 5 , 6

O_3. col 14 by col 21, col 15 by col 20, col 17 by col 24, col 18 by col 23.

The second case.

Suppose some row has 2 Roman mates and every row has at least one Greek mate. If some row has 2 Roman which are Roman mates, we may write:

(6) 0 3 laa 2aa

(7) laa laa 3 lbb

(8) laa lbb lbb 3

Each of (7) and (8) has 2 possible Greek mates, but no completion is possible.

Thus the Roman mates of any row having Roman mates are not mated and so (8) lab lab lbb 3 .

The least tiresome way to approach this case is by where the Greek mates of (7) and (8) lie.

$Q_{2\&4}$ and $Q_{1\&3}$. There are 4 possible Greek mate pairs, none of which admits both the second Q_1aa and Q_2aa .

$Q_{1\&4}$ and $Q_{2\&3}$. Two possible pairs. One has no second Q_1ab , the other no second Q_4bc .

$Q_{1\&4}$ and $Q_{1\&3}$. No Greek mate for (7) admits the second Q_2aa .

$Q_{1\&2}$ and $Q_{2\&3}$. No pair of Greek mates admits the second Q_{lab}.

$Q_{1\&2}$ and $Q_{1\&3}$. No Greek mate for (8) admits the second Q_2aa .

$Q_{1\&2}$ for both. Both ways having the second Q_3 & Q_4bb in different lines have character 3. Of the several ways having these in one line, only one may be completed with character 2. This is:

2ac	2ac	1bb	1bb
2ac	1ac	2cc	1cc
2aa	1aa	1cc	2aa
2ab	1ab	1aa	2cc
2bb	2ba	1bc	1bc
2cb	2cb	1cb	1cb
2ca	1cb	2bc	1bc
2ba	1bb	2cb	1cb
2ba	2cb	1ab	1ba
2bb	1ca	1ab	2ba
2ca	2bb	1ba	1ab
2cb	1ba	1ba	2ab
2cc	2bc	1ac	1ca
2bc	1cc	1ac	2ca
2bc	2cc	1ca	1ac
2cc	1bc	1ca	2ac

It is easy to see that each of the 8 matrices derived from this has transpose character 4 or 6 .

<u>Third case</u>.

Some row has four Roman mates. Up to isomorphism of the first four lines of H, there are five ways for a line to have 4 Roman mates. If three each meet each other as Roman mates, with one such mate occurring when the fourth Roman mates row 5, then looking for the remaining mates soon shows that no low character completion exists.

If three meet each other as Roman mates, no mate in the remaining Q, we may take (6), ..., (9) as

(6)	3	1aa	1aa	1aa
(7)	1aa	3	1aa	1bb
(8)	1aa	1aa	3	1cc
(9)	1ab	1ab	1ab	3

Using the automorphism group, there are essentially 5 ways to get lines with the second Q_4 aa , bb , and cc . In three of these, there are not enough Q_1 ba and ca. The remaining two have the same set of Q_1 ba and ca. Both admit only Q_2 ba and ca with a pair that raise the character to 3 .

If exactly two mate each other, then every set of mates for the Q_4 parts fails to complete.

If no two are mates, take (6) , ... , (9) as

(6)	3	1aa	1aa	1aa
(7)	1aa	3	1ab	1ab
(8)	1ab	1ab	3	1ac
(9)	1ac	1ac	1ac	3 .

There are 5 possible Q_2 aa. Using the automorphism group, all possible mates for (6)-(9) are easily obtained. It is then simple to see that there are only two essentially different ways to mate (6)-(9) completely.

	first mates							
(10)	2ac	1ab	1ab	2aa	2aa	2aa	1ac	1ac
(11)	2ba	2aa	1aa	1bc	2ab	1ac	2aa	1ab
(12)	2aa	2ba	1bc	1ab	2ac	1ab	1ab	2aa
(13)	2ab	1bc	2ba	1ac				
(14)	2bb	1ac	2ac	1bb				

The first mates can only be completed by

(15) 2ca 2cb 1bb 1cc

(16) 2ca 1ca 2ba 1bb

(17) 2ba 1bb 1cb 2bc

(18) 2bb 2bc 1cc 1cc

(19) 2bc 1ba 1bc 2ca

(20) 2cc 2ca 1cc 1bc

(21) 2cc 1cc 1ca 2ba

(22) 2bc 1cc 2ca 1ca

(23) 2cb 1bb 2cb 1cb

(24) 2cb 1cb 1bb 2cb

This matrix may be renormalized using (10), (15), or (18) for row (5), and since (10) is this with all 3 of Q_2, Q_3, Q_4 switched, (15) is just Q_2 switched and (18) is Q_3 and Q_4 switched, we only have two matrices derivable here. In either one col 1 x col 19 = col 7 x col 13 = col 6 x col 22, so the transpose characters are at least 3.

The second mates are extremely tedious although not at all difficult.

Looking only at the S_{2i}, the remaining Chinese lines compatible with (5) - (12) are:

	2211	2121	2112
1.	aa ab	ba ac	cc aa
2.	ba ac	aa ab	bb aa
3.	aa ba	bb aa	ca cc
4.	ca ca	cc aa	aa ba
5.	ab bc	bb bb	cb cc
6.	cb cc	cc bb	ab bc
7.	ac cb	bc cc	cc bb
8.	bc cc	ac cb	bb bb

At most four in any column are compatible, so there must be four in each. The possible 4-sets are A = (1368), B = (1458), C = (2367), D = (2457).

From the fact that #2 in the family 2112 does not admit both #4 and #5 in the family 2121 and the fact that #1 in 2112 does not admit both #3 and #6 in 2121, and applying the automorphism group of lines 1-12, we see that only B, C, C and A, B, D are possible and distinct. In B, C, C however, the four lines having -a in Q_1 are incompatible, so we are left with ABD. The b's and c's here may be filled in in two non-isomorphic ways.

(13)	2ba	2ba	1ba	1bb		
(14)	2ca	2ca	1cb	1ca		
(15)	2ca	1ca	2ba	1bb		
(16)	2bb	1bb	2bb	1cb		
(17)	2ba	1ba	1bb	2ba		
(18)	2bb	1cb	1cb	2cb	Q_1 and	Q_3 are the same
					Q_2	Q_4
(19)	2cb	2cc	1bc	1bc	2bc	1cc
(20)	2bc	2bb	1cc	1cc	2cb	1bc
(21)	2cb	1cc	2cc	1cc	1bc	1bc
(22)	2bc	1bc	2ca	1bc	1cc	1ca
(23)	2cc	1bc	1ca	2ca	1cc	2ba
(24)	2cc	1cb	1bc	2bc	1bb	2cc

Renormalizing either of these by rows (6) , ... , (12) gives all possible switches, so we have exactly 2 to consider. In each of these, col 1 x col 7 = col 4 x col 10 = col 13 x col 19, so the transpose character is at least 3.

Thus every H24 with character 2 has higher transpose character. In particular, every H24 with character 1 must have transpose character 1 or at least 3. It is not known how many of these there are.

References

[1] Longyear, J. Q. Order 24 Hadamard matrices of character at least 3 , to appear.

CYCLE LENGTHS IN POLYTOPAL GRAPHS

Joseph Malkevitch
York College (C.U.N.Y.)
Jamaica, New York 11451

Abstract

If G is a hamiltonian 3-polytopal graph with n vertices, G will be called <u>almost pancyclic of order m</u> ($m \geq 3$, $m < n$) if G has cycles of all lengths other than m. Some constructions are given for 3-valent 3-polytopal graphs which are almost pancyclic of orde: m, and some related results and problems are discussed.

CYCLE LENGTHS IN POLYTOPAL GRAPHS

Considerable attention has been given over a long period of time to determining when a graph has a hamilton circuit. Only relatively recently, in [1,2,5-8] has attention been paid to investigating more completely the range of cycle lengths that can occur in a graph. The purpose of this note is to continue these investigations and to raise some new problems. For standard terminology see [3].

A graph G with n vertices is called <u>pancyclic</u> if G has cycles of length k, $3 \leq k \leq n$. A graph G with n vertices, which is hamiltonian, will be called <u>almost pancyclic of order m</u> ($m \geq 3$, $m < n$) if G has cycles of all lengths (≥ 3) other than m.

Suppose T is a plane tree which has no two-valent vertices. Let G(T) denote the plane graph obtained from T by passing a simple closed curve C through the 1-valent vertices of T. We will call the graph G(T) obtained in this way a <u>skirted tree</u>. Such graphs are known [5] to have a hamilton cycle. Using the above terminology, a recent result [5] of J. A. Bondy and L. Lovász can be stated:

<u>Theorem 1</u>. (Bondy and Lovász) If G(T) is a skirted tree, then G(T) is either pancyclic or almost pancyclic of order m, where m is even.

The theorem below shows that this result is in a sense best possible.

<u>Theorem 2</u>. If m is any even integer (≥ 4), then there exists a tree T with only 1-valent and 3-valent vertices such that G(T) is almost pancyclic of order m.

<u>Proof</u>. We begin the construction with the graph in Figure 1, which has no 4-cycle. (Ignore the dotted and squiggled lines for the moment.) The numbers on the faces indicate the number of sides of these faces. Note the symmetry of the "M and M' configurations" about the 5-gon, shown within the squiggled lines, which are not part of the graph.

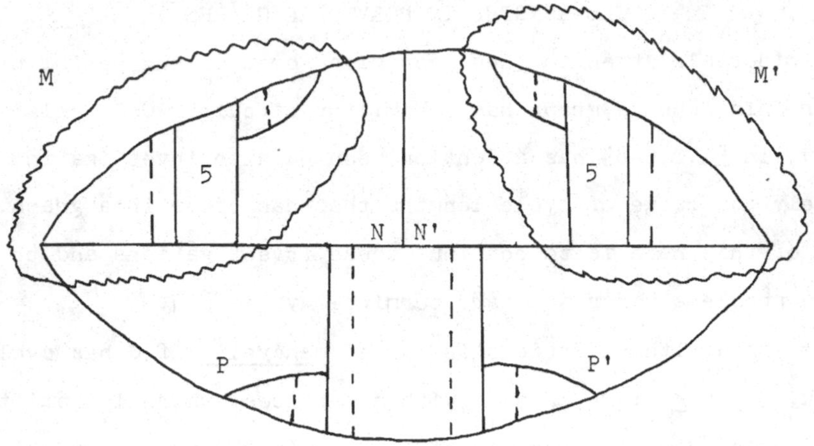

Figure 1

The cycle lengths inside the M and M' configurations do not include a 4-cycle, nor does a 4-cycle appear in the "lower half" of the graph. Note that Theorem 1 guarantees cycles of all other lengths. To get the next graph in the family, which lacks a 6-cycle, take the graph in Figure 1 and add the 8 dotted lines shown. Note that faces N, N', P, and P' now have two more sides each. This process can be repeated to obtain a graph lacking an m-cycle for each even m ($m \geq 4$).

Note that since the graphs constructed above are 3-connected, there exist 3-valent 3-polytopal almost pancyclic of order m graphs for every even m. Hence, it is natural to enquire if there also exist 3-valent 3-polytopal graphs which are almost pancyclic of order m, m odd ($m \geq 3$).

<u>Theorem 3</u>. For every odd integer ≥ 3, there exists a 3-valent 3-polytopal graph which is almost pancyclic of order m.

<u>Proof</u>. The graph G(3) in Figure 2(a) has no 3-cycle but has cycles of all other lengths. The graph G(5) shown in Figure 2(b) (ignore the dotted lines) has no 5-cycle. The dotted lines indicate how to

construct G(2S+3) from G(2S+1) ($S \geq 2$). G(2S+3) lacks only a (2S+3) cycle.

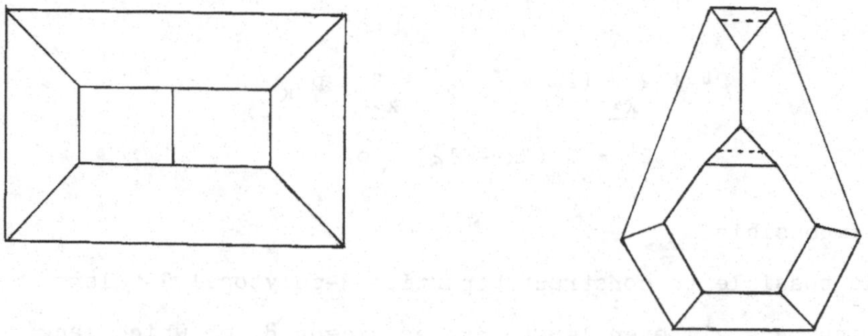

Figure 2

It can be shown that the cycles of other lengths are present.

In view of the ideas above and the special role bipartite graphs play in the theory of graphs with large numbers of cycle lengths, it is natural to consider the following problem: For each even integer $m \geq 4$, it is possible to construct a 3-valent 3-polytopal bipartite graph with n vertices which has no m cycle, but has cycles of length i, $4 \leq i\ (\neq m) \leq n$. For m = 4, the answer is no, since for any planar 3-valent graph, if p_k denotes the number of faces with k-sides:

$$2p_4 = 12 + \sum_{k>6} (k-6)\, p_k \qquad (*)$$

Rather more interesting is:

Theorem 4. If G is a plane 2-connected 3-valent bipartite graph then G has a 6-cycle.

Proof. If $p_6 \neq o$, then G has a 6-cycle. We will show that if $p_6 = o$, then G has two 4-gons which share an edge, and hence has a 6-cycle. Suppose, on the contrary, that no 4-gons in G share an edge. Hence, each 2k-gon in G ($k \geq 4$) is incident with $\leq k$ 4-gons.

Thus
$$4p_4 \leq \sum_{k \geq 4} k\, p_{2k} .$$

But (*) implies that:
$$4p_4 = 24 + 2 \sum_{k \geq 4} (2k - 6)\, p_{2k} .$$

Hence,
$$24 + 2 \sum_{k \geq 4} (2k - 6) p_{2k} \leq \sum_{k \geq 4} k\, p_{2k}$$
$$24 + \sum_{k \geq 4} (3k - 12) \leq 0$$

which is impossible.

It is possible to construct bipartite 3-polytopal 3-valent graphs which have all even length cycles except 8, or which lack only a single other even length cycle > 8, but no complete analysis has yet been obtained.

Motivated by the interesting properties of skirted trees, one can consider graphs obtained by amalgamating (any finite number) of skirted trees (or more arbitrary 3-connected graphs). Two such methods of amalgamation (S) and (T) are shown in Figure 3. (S) results in a 2-connected graph while (T) results in a 3-connected graph. The squiggled lines do not appear in the final graph (but do appear in the original graphs), while the dotted lines are part of the final graph.

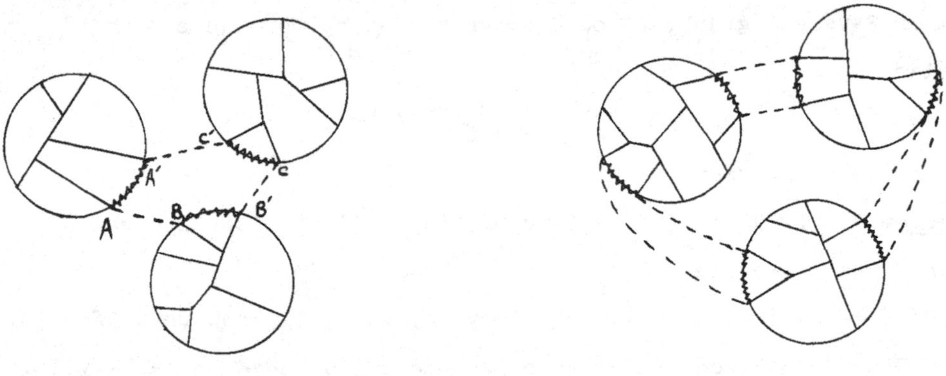

(S) Figure 3 (T)

Theorem 5. If G is the graph arising by amalgamations of type (S) or (T) from 3-valent skirted trees, then G has a hamilton circuit.

We first show the following lemma, which is of independent interest:

Lemma. If G is a 3-valent skirted tree, then every edge and every pair of edges in G lies on a hamilton cycle.

Proof. Any 3-valent skirted tree arises by a series of truncations from the tetrahedron. The tetrahedron has precisely three hamilton cycles, whose union covers each of its edges, each one twice. Each of these cycles can be extended uniquely to the new triangle created after a vertex is truncated in such a way that every new edge lies on precisely two of the extended cycles. Since each edge of a pair lies on precisely two hamilton cycles, the pair of edges lies on a hamilton cycle.

To prove the theorem, for an amalgamation of type (S), piece together the hamilton path from A to A', B to B' and C to C' in the skirted trees using the dotted lines. (The hamilton paths exist by the lemma.) In a similar way, a hamilton cycle can be pieced together for an amalgamated graph of type (T).

Figure 4 indicates a method of constructing 2-connected almost pancyclic graphs of order m (m odd), by using an amalgamation of type (S). Whether or not some analogue of the Bondy and Lovasz theorem can be proved for amalgamations is not yet known.

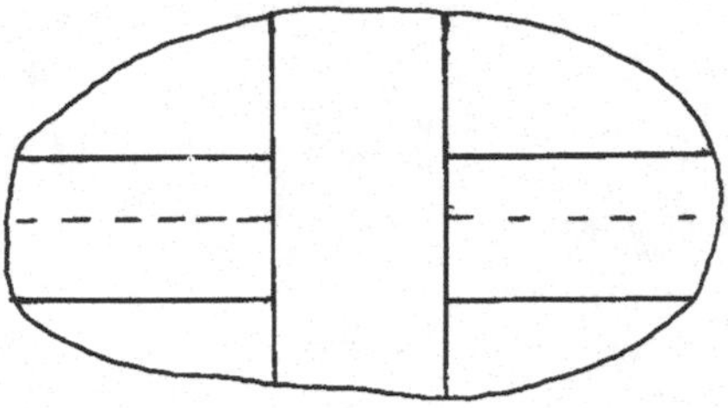

Figure 4

We close with the following conjecture and problems:

Conjecture 1. If G is a planar 4-valent 4-connected graph and G has a 4-cycle, then G is pancyclic.

Problem 1. If G is a hamilton 5-valent, 3-connected planar graph, must G be pancyclic?

Problem 2. If G(T) is a skirted tree, where T consists of only 4-valent and 1-valent vertices, or only 5-valent and 1-valent vertices, must G(T) be pancyclic?

REFERENCES

1. J. A. Bondy, Cycles in graphs. Combinatorial Structures and Their Applications. Gordon and Breach, New York (1970) 15-18.

2. J. A. Bondy, Pancyclic Graphs I, J. Combinatorial Theory 11 (1971) 80-84.

3. B. Grünbaum, Convex Polytopes. Interscience-Wiley, London, New York, Sydney (1967).

4. V. Jacos and S. Jendrol, A problem concerning j-pancyclic graphs. Matematicky Casopis. 24 (1974) 259-262.

5. L. Lovász and M. D. Plummer, On a family of planar bicritical graphs. Proc. London Math. Soc. (3) 30 (1975) 187-208.

6. J. Malkevitch, On the lengths of cycles in planar graphs. Recent Trends in Graph Theory. Springer-Verlag. Berlin, Heidelberg and New York (1970) 191-195.

7. G. H. J. Meredith, Cyclic and Colouring Properties of Graphs PhD Thesis. Southampton University (1973).

8. E. F. Schmeicher and S. L. Hakimi. Pancyclic graphs and a conjecture of Bondy and Chvatal. J. of Combinatorial Theory (B) 17 (1974) 22-34.

SUBGRAPH CONNECTIVITY NUMBERS OF A GRAPH

David W. Matula
Southern Methodist University
Dallas, Texas 75275

ABSTRACT

$\hat{\delta}$, $\hat{\lambda}$, $\hat{\kappa}$ and ω denote the maximum values of the minimum degree, edge-connectivity, vertex-connectivity and clique size, respectively, of the subgraphs of the graph G. These subgraph connectivity numbers are shown to be ordered by $\omega-1 \leq \hat{\kappa} \leq \hat{\lambda} \leq \hat{\delta}$ for any graph. Extremal variations between these subgraph connectivity numbers of a graph are investigated. The constraint $\hat{\delta} \leq 2\hat{\lambda} - 1$ is derived and equality is shown possible for certain graphs having $1 \leq \hat{\lambda} \leq 6$. Extremal relations between the chromatic number and the subgraph connectivity numbers are investigated. Computationally, the cut lemmas are described and shown to yield polynomial bounded algorithms for computing $\hat{\delta}$, $\hat{\lambda}$ and $\hat{\kappa}$.

I. INTRODUCTION AND SUMMARY

Four integer valued measures of the maximum intensity of connectivity within a graph G are provided by $\hat{\delta}(G)$, $\hat{\lambda}(G)$, $\hat{\kappa}(G)$ and $\omega(G)$ which denote, respectively, the maximum of the minimum degree, edge-connectivity, vertex-connectivity and clique size over the subgraphs of G. We have previously [3-6] separately investigated $\hat{\delta}$, $\hat{\lambda}$ and $\hat{\kappa}$. In this paper we seek to unify and extend results from the literature on the ordering and extremal relations between the four subgraph connectivity numbers $\hat{\delta}(G)$, $\hat{\lambda}(G)$, $\hat{\kappa}(G)$ and $\omega(G)$, and also to comment on the comparative algorithmic complexity of the determination of these four parameters.

In section II we derive the elementary inequality $\omega(G)-1 \leq \hat{\kappa}(G) \leq \hat{\lambda}(G) \leq \hat{\delta}(G)$ that parallels Whitney's result for the global connectivity numbers, and then give short proofs of the following cut lemmas. For G a graph with $E(G) \neq \phi$,

(i) if v is a minimum degree vertex in G, then

$$\hat{\delta}(G) = \max\{\delta(G), \hat{\delta}(G-v)\},$$

(ii) if $C \subset E(G)$ is a minimum cut of some component of G, then

$$\hat{\lambda}(G) = \max\{|C|, \hat{\lambda}(G-C)\},$$

(iii) If $S \subset V(G)$ is a minimum separating set of the connected graph G with $<A_1>, <A_2>, \ldots, <A_n>$ the components of G-S, then
$$\hat{\kappa}(G) = \max\{|S|, \max\{\hat{\kappa}(<A_i \cup S>) \mid 1 \le i \le n\}\}.$$

In section III we discuss algorithms for determining the subgraph connectivity numbers and note the complexity results that $\hat{\delta}(G)$ can be determined in $O(|E(G)|)$ steps, $\hat{\lambda}(G)$ in $O(|V(G)|\ |E(G)|^2)$ steps, and $\hat{\kappa}(G)$ in $O(|V(G)|^{3/2}\ |E(G)|^2)$ steps. Extremal relations between $\hat{\delta}$, $\hat{\lambda}$, $\hat{\kappa}$ and ω are investigated in section IV. We show that $\hat{\delta}(G) \le 2\hat{\lambda}(G) - 1$ for all G with equality obtained for certain G with $1 \le \hat{\lambda}(G) \le 6$. A class of graphs {G} are described for which $\hat{\kappa}(G)/\hat{\lambda}(G)$ approaches 3/7.

The chromatic number bound $\omega(G) \le \chi(G) \le \hat{\lambda}(G) + 1$ is noted in the final section, along with a class of graphs $\{G_n\}$ for which $\chi(G_n) = 2\hat{\kappa}(G_n) = 2n$ for all $n \ge 1$.

II. SUBGRAPH CONNECTIVITY AND THE CUT LEMMAS

We shall assume a graph to be finite without multiple edges. The minimum degree, $\delta(G)$, edge-connectivity, $\lambda(G)$, and vertex-connectivity, $\kappa(G)$, will be termed global connectivity numbers of the graph G. $H \subset G$ denotes that H is a subgraph of G, and the maximum values of $\delta(H)$, $\lambda(H)$ and $\kappa(H)$ over the subgraphs {H} of a graph G are defined as follows

(1)
bondage: $\hat{\delta}(G) = \max\{\delta(H) \mid H \subset G\}$,

edge-strength: $\hat{\lambda}(G) = \max\{\lambda(H) \mid H \subset G\}$,

vertex-strength: $\hat{\kappa}(G) = \max\{\kappa(H) \mid H \subset G\}$.

The variables $\hat{\delta}(G)$, $\hat{\lambda}(G)$, $\hat{\kappa}(G)$, along with the maximum clique size $\omega(G)$ will be termed the subgraph connectivity numbers of the graph G.

Whitney [8] established that the global connectivity numbers are ordered by

(2) $\kappa(G) \le \lambda(G) \le \delta(G)$ for any graph G.

Utilizing this result a corresponding ordering of the subgraph connectivity numbers is immediately obtained.

Theorem 1: For any graph G,

(3) $\omega(G) - 1 \le \hat{\kappa}(G) \le \hat{\lambda}(G) \le \hat{\delta}(G)$.

Proof: Since $\delta(K_n) = \lambda(K_n) = \kappa(K_n) = n-1$, the value $\omega(G)-1$ is a lower bound on the other subgraph connectivity numbers. For $H \subset G$ with $\kappa(H) = \hat{\kappa}(G)$, using (2),
$$\hat{\kappa}(G) = \kappa(H) \le \lambda(H) \le \hat{\lambda}(G),$$
and for $H' \subset G$ with $\lambda(H') = \hat{\lambda}(G)$,
$$\hat{\lambda}(G) = \lambda(H') \le \delta(H') \le \hat{\delta}(G),$$
thus establishing the ordering (3).▌

Any graph G having some $H \subset G$ with $\delta(H) = \hat{\delta}(G)$ where H is complete will result in equality throughout (3). Important instances of this case are any complete graph K_n, $n \geq 1$, where then

(4) $\omega(K_n)-1 = \hat{\kappa}(K_n) = \hat{\lambda}(K_n) = \hat{\delta}(K_n) = n-1$,

and any non-trivial tree T, where then

(5) $\omega(T)-1 = \hat{\kappa}(T) = \hat{\lambda}(T) = \hat{\delta}(T) = 1$.

For the complete bipartite graph $K_{n,m}$ for any $1 \leq n \leq m$,

(6) $\omega(K_{n,m}) = 2$, $\hat{\kappa}(K_{n,m}) = \hat{\lambda}(K_{n,m}) = \hat{\delta}(K_{n,m}) = n$.

For any planar graph P,

(7) $\hat{\kappa}(P) \leq \hat{\lambda}(P) \leq \hat{\delta}(P) \leq 5$,

with equality obtained throughout for the dual of the dodecahedron.

For a planar graph Q with girth at least four,

(8) $\hat{\kappa}(Q) \leq \hat{\lambda}(Q) \leq \hat{\delta}(Q) \leq 3$,

with equality obtained throughout for the cube.

For the preceding important classes of graphs the corresponding global and subgraph connectivity numbers and/or bounds coincided. It should be noted that the presence of a single vertex of degree one will force all global connectivity numbers to be no greater than unity, although the values of the subgraph connectivity numbers can be large and varied and may be determined by distinct subgraphs. In figure 1 the graph G has global connectivity numbers $\kappa(G) = 1$, $\lambda(G) = 2$ and $\delta(G) = 2$, and subgraph connectivity numbers $\omega(G) = 4$, $\hat{\kappa}(G) = 3$, $\hat{\lambda}(G) = 4$, and $\hat{\delta}(G) = 5$. Note in figure 1 that the graph on vertices 1-4 is a maximum clique, that the graph on vertices 1-10 is maximal with maximum vertex-connectivity three, that the graph on vertices 1-9 is maximal with maximum edge-connectivity four, and that the graph on vertices 1-19 is maximal with the maximum minimum degree five.

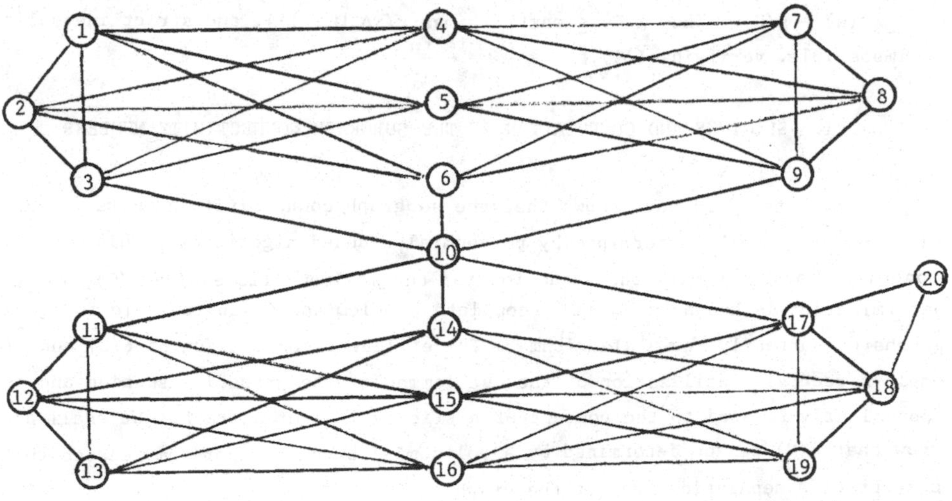

Figure 1. A graph with $\omega(G) = 4$, $\hat{\kappa}(G) = 3$, $\hat{\lambda}(G) = 4$, and $\hat{\delta}(G) = 5$.

Determination of the subgraph connectivity numbers is facilitated by the following three <u>cut lemmas</u> adapted from the results in [3-6].

<u>Lemma 1</u>: Let v be a vertex of minimum degree in the non-trivial graph G. Then

(9) $\hat{\delta}(G) = \max\{\delta(G), \hat{\delta}(G-v)\}$.

<u>Proof</u>: Choose $H \subset G$ so that $\delta(H) = \hat{\delta}(G)$. If $v \notin V(H)$, then $H \subset G-v$ and $\hat{\delta}(G) = \hat{\delta}(G-v)$. If $v \in H$, let $\deg(v|H)$ denote the degree of v in the graph H, so $\hat{\delta}(G) = \delta(H) \leq \deg(v|H) \leq \deg(v|G) = \delta(G)$. Thus $\hat{\delta}(G) \leq \max\{\delta(G), \hat{\delta}(G-v)\}$, and strict inequality is impossible by definition of $\hat{\delta}(G)$, verifying (9).∎

<u>Lemma 2</u>: Let G be a graph with non-void edge set and let $C \subset E(G)$ be a minimum cut set of some component of G. Then

(10) $\hat{\lambda}(G) = \max\{|C|, \hat{\lambda}(G-C)\}$.

<u>Proof</u>: Choose $H \subset G$ so that $\lambda(H) = \hat{\lambda}(G)$. If $C \cap E(H) = \phi$, then $H \subset G-C$ and $\hat{\lambda}(G) = \hat{\lambda}(G-C)$. If $C \cap E(H) \neq \phi$, then C contains a cut set of H and $|C| \geq \lambda(H) = \hat{\lambda}(G)$ Thus $\hat{\lambda}(G) \leq \max\{|C|, \lambda(G-C)\}$, and strict inequality is impossible by definition of $\hat{\lambda}(G)$, verifying (10).∎

<u>Lemma 3</u>: Let G be a connected graph which is not complete, and $S \subset V(G)$ a minimum separating vertex set of G. Let $A_1, A_2, \ldots, A_n$ be the vertex sets of the $n \geq 2$ components of G-S. Then

(11) $\hat{\kappa}(G) = \max\{|S|, \max_{1 \leq i \leq n}\{\hat{\kappa}(\langle A_i \cup S\rangle)\}\}$.

<u>Proof</u>: Choose $H \subset G$ so that $\kappa(H) = \hat{\kappa}(G)$. If $V(H) \subset S \cup A_i$ for some i, then $H \subset \langle A_i \cup S\rangle$ and $\hat{\kappa}(G) = \hat{\kappa}(\langle A_i \cup S\rangle)$. If $V(H) \cap A_i \neq \phi$ and $V(H) \cap A_j \neq \phi$ for $1 \leq i < j \leq n$, then S contains a separating vertex set of H. Thus $|S| \geq \kappa(H) = \hat{\kappa}(G)$. Thus $\hat{\kappa}(G) \leq \max\{|S|, \max_{1 \leq i \leq n}\{\hat{\kappa}(\langle A_i \cup S\rangle)\}\}$, and strict inequality is impossible, verifying (11).∎

III. SLICINGS AND COMPUTATION OF THE SUBGRAPH CONNECTIVITY NUMBERS

Our results [3-5] have shown that the subgraph connectivity numbers $\hat{\delta}(G)$ and $\hat{\lambda}(G)$ can be determined by polynomial bounded algorithms. This contrasts favorably with the situation for the maximum clique size $\omega(G)$, whose determination is known to be an NP-complete problem and is intractable for general graphs by currently known algorithms. The efficient computation of $\hat{\delta}(G)$ and $\hat{\lambda}(G)$ employs recursive utilization of the cut lemmas of the preceding section and is conveniently related to the concept of a slicing [4] of a graph. We shall also show that $\hat{\kappa}(G)$ can be determined by a polynomial bounded algorithm related to the concept of a separating tree of the graph.

For a graph G with non void edge set E(G), a <u>slicing</u> Z of G is an ordered partition $Z = (C_1, C_2, \ldots, C_n)$ of E(G) where C_i is a cut set of the subgraph $G - \bigcup_{j<i} C_j$ for $1 \leq i \leq n$. The slicing Z is a <u>min-degree star slicing</u> or <u>δ-slicing</u> if the cut C_i consists of the set of edges incident to a minimum (non-zero) degree vertex of $G - \bigcup_{j<i} C_j$ for $1 \leq i \leq n$, and Z is a <u>narrow</u> or <u>λ-slicing</u> if C_i is a minimum cut of some component of $G - \bigcup_{j<i} C_j$ for $1 \leq i \leq n$. Figure 2 illustrates the cuts of a δ-slicing and a λ-slicing of the same graph.

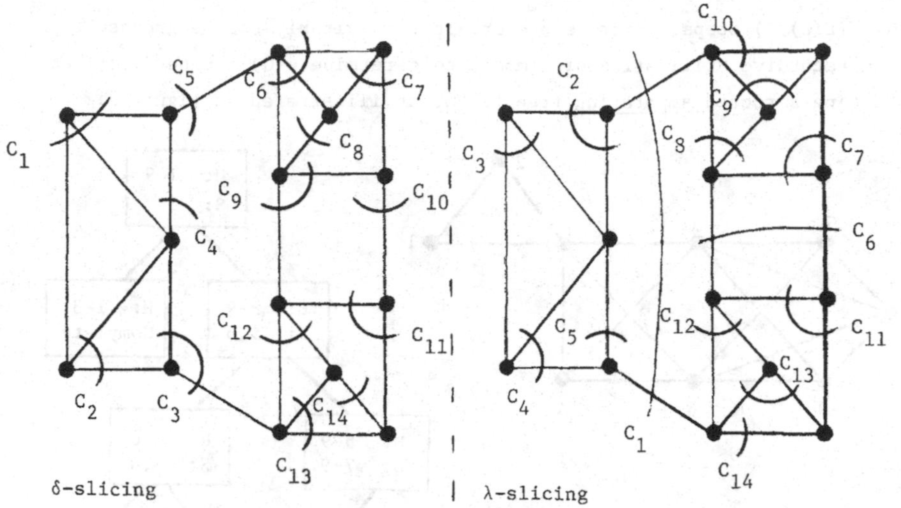

<u>Figure 2</u>: A δ-slicing and a λ-slicing of the graph G.

The following slicing theorem from [3, 4] is readily obtained from the cut lemmas, and thus provides verification that the graph G of figure 2 has $\hat{\delta}(G) = 3$, $\hat{\lambda}(G) = 2$.

<u>Theorem 2 [Slicing Theorem; 3,4]</u>: Let G be a graph with $E(G) \neq \phi$. Then

i) for any δ-slicing $Z = (C_1, C_2, \ldots, C_n)$ of G,

(12) $$\hat{\delta}(G) = \max_{1 \leq i \leq n}\{|C_i|\},$$

ii) and for any λ-slicing $Z' = (C_1', C_2', \ldots, C_m')$ of G,

(13) $$\hat{\lambda}(G) = \max_{1 \leq i \leq m}\{|C_i'|\}.$$

From the cut lemmas and slicing theorem, an $O(|V(G)|^2)$ algorithm for determining $\hat{\delta}(G)$ is obtained simply by recursively determining and deleting a vertex of minimum degree in the remaining graph. Employing more sophisticated data structures [6], a procedure yielding a δ-slicing and $\hat{\delta}(G)$ in $O(|E(G)|)$ steps can be obtained, which is of particular interest for sparse graphs.

Observe that for any slicing $Z = (C_1, C_2, \ldots, C_n)$ of G, the number of components of $G - \bigcup_{j<i} C_j$ is strictly increasing in i so that $n \leq |V(G)|-1$. Thus a λ-slicing and $\hat{\lambda}(G)$ can be obtained by sequentially solving $O(|V(G)|)$ edge-connectivity problems for particular subgraphs of G. Utilizing refinements of Dinac to the basic Ford-Fulkerson network flow theory, the edge-connectivity of any graph G can be determined [2] in $O(|V(G)|^{5/3} |E(G)|)$ steps, thus $\hat{\lambda}(G)$ can be determined in $O(|V(G)|^{8/3} |E(G)|)$ steps. Employing an alternative network flow approach, we have shown that a λ-slicing and $\hat{\lambda}(G)$ can be determined in $O(|V(G)| \; |E(G)|^2)$ steps, which is preferable for suitably sparse graphs.

The recursive utilization of lemma 3 to determine $\hat{\kappa}(G)$ can be described by constructing a rooted <u>separating tree</u> for G, as illustrated in figure 3.

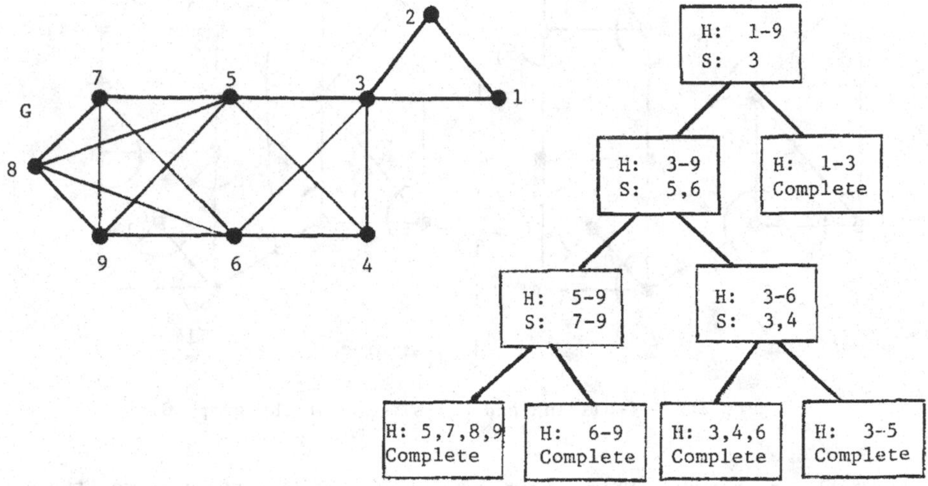

<u>Figure 3</u>: A graph G and a separating tree for G.

Each non-terminal node of the separating tree is labeled by a subgraph H along with a minimum separating vertex set S of H. The n nodes below this particular node are then labeled with the subgraphs $\langle A_1 \cup S\rangle, \langle A_2 \cup S\rangle, \ldots, \langle A_n \cup S\rangle$, where the $\langle A_i\rangle$ are the components of H-S. Every node labeled with a complete graph is then a terminal node of the separating tree. It follows readily from lemma 3 that for a separating tree T of the graph G, the maximum of (a) the largest separating set size on any node of T and (b) the largest vertex-connectivity of the complete subgraphs on the terminal nodes of T, must equal $\hat{\kappa}(G)$.

The construction of a full separating tree for the graph G provides an algorithm for determining $\hat{\kappa}(G)$, but this procedure is not efficient. Note that every clique of G must be the label of at least one terminal node of any separating tree, so the construction of any full separating tree has worst case complexity of an order exponential in $|V(G)|$. We now show that the following algorithm computes a portion of a separating tree sufficient to determine $\hat{\kappa}(G)$ where the number of non-terminal nodes computed is related linearly to $|V(G)|$.

Algorithm A: Given a graph G this algorithm determines $\hat{\kappa}(G)$.

1. $H \leftarrow G$.
2. Determine the components $H_1, H_2, \ldots, H_n$ of H. If any H_i is a complete graph then $k \leftarrow \max\{|V(H_i)-1| \,|\, H_i \text{ is complete}\}$, otherwise $k \leftarrow 0$. [k is a lower bound on $\hat{\kappa}(G)$].
3. Determine H' to be a component of H with the maximum number of vertices. If $|V(H')| \leq k+1$, stop [$k = \hat{\kappa}(G)$]; otherwise determine $S \subset V(H')$ to be a minimum separating vertex set of H', and determine $\langle A_1\rangle, \langle A_2\rangle, \ldots, \langle A_m\rangle$ to be the components of H'-S.
4. If any of the graphs $\langle A_i \cup S\rangle$, $1 \leq i \leq m$, are complete, then $m \leftarrow \max_i\{|A_i \cup S|-1 \,|\, \langle A_i \cup S\rangle \text{ is complete}\}$, otherwise $m \leftarrow 0$.
5. $k \leftarrow \max\{k, m, |S|\}$ [Improve lower bound].
6. Reconstruct H by deleting the component H' and adding as distinct components of H the graphs $\langle A_i \cup S\rangle$, $1 \leq i \leq m$, using distinct copies of the set S, and return to step 3.

Algorithm A may be related to the exploration of a separating tree T for G starting from the root node labeled with H=G. Steps 1 and 2 of Algorithm A determine if the root node is terminal or non-terminal, and if G is not connected the root is "processed" by determining the adjacent terminal and non-terminal nodes of the root which are then added to the explored portion of T. Step three determines the adjacent nodes of some non-terminal node of T to be added to the explored portion, and step 4 indicates which of the newly added nodes are terminal and non-terminal in T. The algorithm explores the tree guided by a branch-and-bound strategy successively updating the lower bound on $\hat{\kappa}(G)$ and branching on the tree node with the highest remaining upper bound on $\hat{\kappa}(G)$ as implicitly given by maximum $\{|V(H^*)|-2\}$ over the non-complete graphs H* remaining as non-terminal node labels to be processed. The efficiency of this search order is confirmed by the following theorem.

Theorem 3: Algorithm A determines $\hat{\kappa}(G)$ and requires no more than $O(|V(G)|^{3/2}|E(G)|^2)$ steps.

Proof: We first show that algorithm A terminates and determines $\hat{\kappa}(G)$. Let $G^* \subset G$ be a subgraph such that $\kappa(G^*) = \hat{\kappa}(G)$. Note that the value of k as assigned in steps 2 and 5 is always the vertex connectivity of some subgraph of G, so $k \leq \kappa(G^*)$ always holds. Furthermore, the successive values assigned to k are non-decreasing, and each modification of the graph H in steps 3-6 results in some j vertex component H' of H, where $j \geq k + 2$, being replaced by no more than j-1 components each having at most j-1 vertices. Note that H' as chosen in step 3 with $j \geq k+2$ vertices cannot be complete, so a minimum separating vertex set $S \subset V(H')$ always exists. Thus all components of H must eventually have at most k+1 vertices, yielding termination in step 3, and we need to show $k = \kappa(G^*)$ at termination. Initially in step 2, G* is in some component H* of H. If $G^* = H^*$ is complete, $k = \kappa(G^*)$ in step 2. Otherwise as long as there is a component H* containing G* and having $|V(H^*)| \geq \kappa(G^*) + 2 \geq k + 2$, such a component will eventually be split in step 3 until either $G^* = H^*$ with $|V(H^*)| \geq k + 1$, implying H* is complete with k set to $\kappa(G^*)$ by the preceding step 5, or a separating set $S \subset V(H^*)$ contains a separating set of G*, in which case $|S| = \kappa(G^*)$ and k is set to $\kappa(G^*)$ in the subsequent step 5, so k is set to $\hat{\kappa}(G)$ before termination of the algorithm.

The best known method [2] for finding the vertex connectivity of a graph G requires $O(|V(G)|^{1/2} |E(G)|^2)$ steps. Utilizing this vertex connectivity subprocedure the complexity of Algorithm A is dominated by the complexity expended in step 3 yielding $O(N|V(G)|^{1/2} |E(G)|^2)$ steps where N is the number of passes through step 3. We shall complete the theorem by showing $N \leq |V(G)|$.

In step 3 let $H_1, H_2, \ldots, H_j$ be the components of H for which $|V(H)| \geq \hat{\kappa}(G) + 2$, so then $|V(H)| \geq k + 2$ for the current value of k. At least one must exist for us to have to find a separating vertex set, and the choice of H' to be a component of H with the maximum number of vertices assures that $H' \in \{H_i | 1 \leq i \leq j\}$. Consider the <u>surplus</u> s defined by

$$s = \sum_{i=1}^{j} (|V(H_i)| - \hat{\kappa}(G) - 1)$$

where initially in the algorithm we note $s \leq |V(G)| - \hat{\kappa}(G) - 1$.

Suppose $S \subset V(H')$ is determined in step 3 and $\langle A_1 \cup S\rangle, \langle A_2 \cup S\rangle, \ldots, \langle A_p \cup S\rangle$ are the components replacing H' in step 6 for which $|A_i \cup S| \geq \hat{\kappa}(G) + 2$. If no more than one such component exists the surplus value s has been reduced by at least one unit in passing from step 3 to step 6. If $p \geq 2$, then since $|S| \leq \hat{\kappa}(G)$,

$$\sum_{i=1}^{p} (|A_i \cup S| - \hat{\kappa}(G) - 1) \leq (|S| + \sum_{i=1}^{p} |A_i|) + (p-1)(|S| - \hat{\kappa}(G)) - \hat{\kappa}(G) - p$$

$$\leq |V(H')| - \hat{\kappa}(G) - 2,$$

and in this case also the surplus s is reduced by at least one unit in passing through steps 3 to 6. Hence $N \leq |V(G)| - \hat{\kappa}(G)$, so Algorithm A requires at most $O(|V(G)|^{3/2} |E(G)|^2)$ steps. ▮

The complexity results for determination of the subgraph connectivity numbers are summarized in the following table.

Subgraph Connectivity Number	Complexity Upper Bound
$\hat{\delta}(G)$	$O(\lvert E(G)\rvert)$
$\hat{\lambda}(G)$	$O(\lvert V(G)\rvert \lvert E(G)\rvert^2)$
$\hat{\kappa}(G)$	$O(\lvert V(G)\rvert^{3/2} \lvert E(G)\rvert^2)$
$\omega(G)$	NP-complete problem

IV. BOUNDS AND EXTREMAL RELATIONS FOR SUBGRAPH CONNECTIVITY

Let $Z = (C_1, C_2, \ldots, C_n)$ be a λ-slicing of the graph G. The result $\hat{\lambda}(G) = \max_i\{|C_i|\}$ is obtained from the slicing theorem. We have noted that $n \leq |V(G)|-1$, thus as an immediate corollary

(14) $$\frac{|E(G)|}{|V(G)|-1} \leq \hat{\lambda}(G).$$

Let the <u>average degree</u> a(G) and <u>maximum subgraph average degree</u> $\hat{a}(G)$, be given by

(15) $$\begin{aligned} a(G) &= 2|E(G)|/|V(G)| \\ \hat{a}(G) &= \max_{H \subset G}\{a(H)\} \end{aligned} \qquad \text{for all G.}$$

$\hat{a}(G)$ may be considered another subgraph connectivity number, although it is not necessarily integer valued as are the others.

<u>Lemma 4</u>: For any graph G with $E(G) \neq \phi$,

(16) $$\frac{1}{2}\hat{a}(G) < \hat{\lambda}(G) \leq \hat{\delta}(G) \leq \hat{a}(G).$$

<u>Proof</u>: For $H \subset G$ with $a(H) = \hat{a}(G)$, using (14), (15),

$$\frac{1}{2}\hat{a}(G) = \frac{1}{2}a(H) = |E(H)|/|V(H)| < \hat{\lambda}(H) \leq \hat{\lambda}(G) \leq \hat{\delta}(G),$$

and for $H^* \subset G$ with $\delta(H^*) = \hat{\delta}(G)$,

$$\hat{\delta}(G) = \delta(H^*) \leq a(H^*) \leq \hat{a}(G).\blacksquare$$

It is possible for $\hat{\kappa}(G) < \frac{1}{2}\hat{a}(G)$. Let 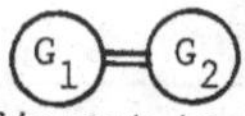 denote that the graph G_1 is <u>joined</u> to G_2, that is each vertex of G_1 is adjacent to each vertex of G_2. With I_n denoting a graph on n vertices with no edges, the graph $G_{n,m}$ having m copies of the complete graph K_n joined to I_n as shown

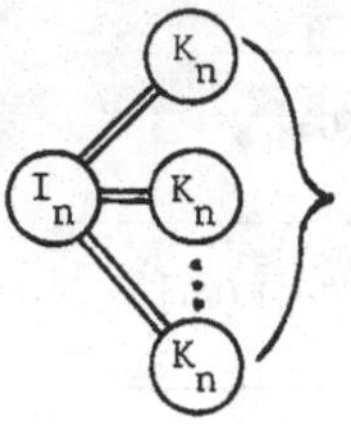

has

$$\hat{\kappa}(G_{n,m}) = n$$

(17) $\qquad\qquad\qquad$ for all n, $m \geq 1$,

$$\hat{a}(G_{n,m}) = \frac{m(3n-1)}{m+1}$$

so then $\hat{\kappa}(G_{n,m})/\hat{a}(G_{n,m}) \to 1/3$ as n, $m \to \infty$.

Whitney's ordering relation on the global connectivity numbers $\kappa(G) \leq \lambda(G) \leq \delta(G)$ can be quite slack in the sense that for any triple $1 \leq k \leq m \leq n$ it is easy [1] to construct a graph G with $\kappa(G) = k$, $\lambda(G) = m$ and $\delta(G) = n$. The preceding results indicate that the ordered subgraph connectivity numbers $\hat{\kappa}(G) \leq \hat{\lambda}(G) \leq \hat{\delta}(G)$ are more tightly related.

<u>Theorem 4</u>: For any graph G with $E(G) \neq \phi$,

(18) $\quad \hat{\delta}(G) \leq 2\hat{\lambda}(G) - 1$,

and equality is obtained for some G_i with $\hat{\lambda}(G_i) = i$ for $1 \leq i \leq 6$.

<u>Proof</u>: The inequality (18) is immediate from lemma 4. Graphs with $\hat{\lambda}(G_i) = i$, $\hat{\delta}(G_i) = 2i - 1$ for i = 1, 2, 3, are illustrated.

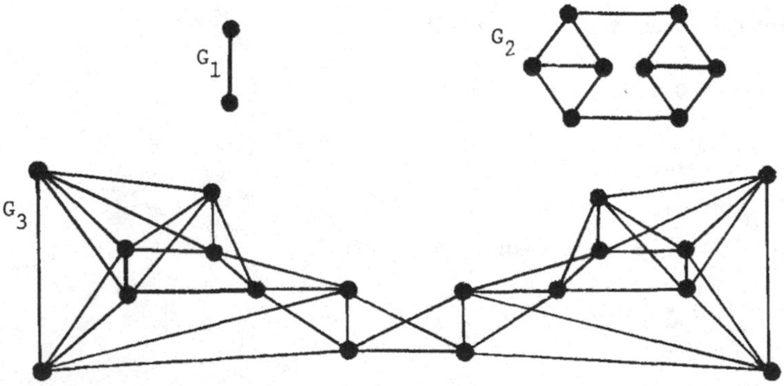

The procedure for constructing the graphs G_i, $3 \leq i \leq 6$, may be visualized by referring to the illustration of G_3. The construction starts with K_{i+1} where the degree of each of the i+1 vertices has a "residual need" of i-1 for a total of i^2-1. Adjoining a new vertex to i preceding vertices of non-zero residual need reduces the total residual need by unity as the added vertex has residual need of i-1. After properly adjoining $(i^2 + i - 2)/2$ vertices the vertex degree residual needs are 1, 2,..., i-1 yielding a total of i(i-1)/2 which cannot be further reduced in this manner. By linking two copies of this graph portion using i (or less) edges the residual need total becomes at most i^2-3i. Thus G_3 is completed at this stage. For G_4, the linking of the two 14 vertex graph portions yields a 28 vertex graph portion with residual need four, and the linking of two copies of this 28 vertex graph portion by four edges yields a 56 vertex graph G_4 with $\hat{\lambda}(G_4) = 4$, $\hat{\delta}(G_4) = 7$. For the construction of G_5 and G_6, the initial linking of two copies of graph portions does not reduce the total residual need but it can be assumed to be spread over 2(i-1) vertices, allowing the adjoining of more single vertices to reduce the residual need. Iterating these two procedures allows a desired graph G_5 on 336 vertices and G_6 on 2256 vertices to be constructed. ▌

The construction procedure of the preceding theorem does not converge to allow a graph G_7 with $\hat{\lambda}(G_7) = 7$ and $\hat{\delta}(G_7) = 13$. Further considerations suggest that the minimum ratio of $\hat{\lambda}(G)/\hat{\delta}(G)$ must become larger than 1/2 as $\hat{\lambda}(G) \to \infty$. A little more slack in the relation $\hat{\kappa}(G) \leq \hat{\lambda}(G)$ is possible in that the extremal ratio of $\hat{\kappa}(G)/\hat{\lambda}(G)$ must be bounded by a constant no larger than 3/7 as indicated by the class of graphs of figure 4 . Note that the graph G_n with independent sets on n and 3n points and complete subgraphs on 3n points joined as shown has $\hat{\kappa}(G_n) = 3n$, $\hat{\delta}(G_n) = \hat{\lambda}(G_n) = 7n-1$, and $|V(G_n)| = 30n$ for all $n \geq 1$, so that $\hat{\kappa}(G_n)/\hat{\lambda}(G_n) \to 3/7$ as $n \to \infty$.

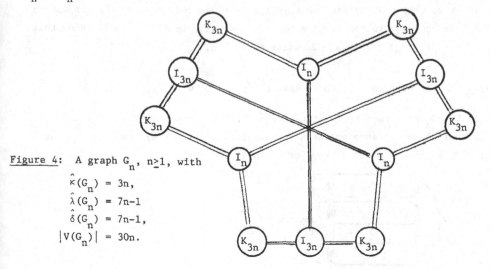

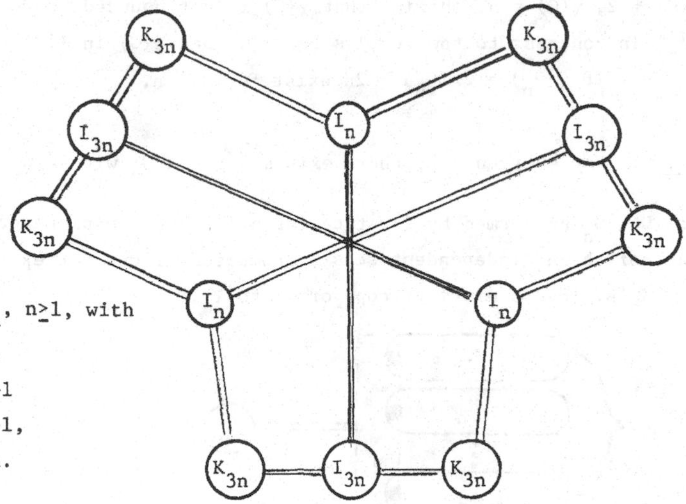

Figure 4: A graph G_n, $n \geq 1$, with
$\hat{\kappa}(G_n) = 3n$,
$\hat{\lambda}(G_n) = 7n-1$
$\hat{\delta}(G_n) = 7n-1$,
$|V(G_n)| = 30n$.

V. SUBGRAPH CONNECTIVITY AND CHROMATIC NUMBER

The chromatic number $\chi(G)$ of the graph G is bracketed by the subgraph connectivity numbers by the inequality

(19) $\omega(G) \leq \chi(G) \leq \hat{\lambda}(G) + 1 \leq \hat{\delta}(G) + 1.$

The upper bound on the chromatic number given by $\chi(G) \leq \hat{\delta}(G) + 1$ was derived in [7] and the stronger bound $\chi(G) \leq \hat{\lambda}(G) + 1$ was subsequently obtained in [3]. Note that $\omega(G_i) = \chi(G_i) = \hat{\lambda}(G_i) + 1 = i + 1 = \frac{1}{2}\hat{\delta}(G_i) + 1$ for the graphs G_i, $1 \leq i \leq 6$, constructed in the proof of theorem 4, so the edge-connectivity related bound can be a much sharper upper bound on $\chi(G)$. Computationally, it is shown in [6] that a $(\hat{\delta}(G)+1)$-coloring of G can be obtained in $O(|E(G)|)$ steps, and the methods for determining $\hat{\lambda}(G)$ are readily extended to provide an algorithm for determining a $(\hat{\lambda}(G) + 1)$-coloring of G in $O(|V(G)||E(G)|^2)$ steps.

The bracketing of $\chi(G)$ given by (19) can exhibit arbitrary slackness. Zykov [9] was the first of several authors to confirm the existence of graphs with $\omega = 2$, $\chi = n$ for all $n \geq 2$, so $\omega(G) = m$, $\chi(G) = n$ is attainable for some graph G for all $2 \leq m \leq n$. The graph $G = K_m \cup K_{n,n}$ demonstrates that $\chi(G) = m$, $\hat{\lambda}(G) = \hat{\delta}(G) = n$ is attainable for all $2 \leq m \leq n$.

The relation between $\hat{\kappa}(G)$ and $\chi(G)$ is more complex. For $\chi(G) \leq \hat{\kappa}(G)$, the graph $K_m \cup K_{n,n}$ also shows $\chi(G) = m$, $\hat{\kappa}(G) = n$ is attainable for all $2 \leq m \leq n$. For $\chi(G) > \hat{\kappa}(G)$, the following graph

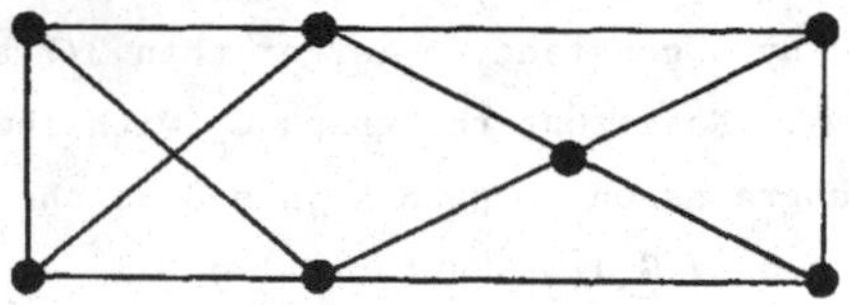

has $\hat{\kappa}(G) = 2$, $\chi(G) = 4$, showing that $\chi(G)$ is not bounded above by the expression $\hat{\kappa}(G) + 1$ in contrast to the results for $\hat{\lambda}(G)$ and $\hat{\delta}(G)$ in (19). We now show that graphs G_n with $\chi(G_n) = 2\hat{\kappa}(G_n) = 2n$ exist for all n.

Theorem 5: For every $n \geq 1$, there exists a graph G_n with $\hat{\kappa}(G_n) = n$, $\chi(G_n) = 2n$.

Proof: Let G_n be formed by first taking n disjoint complete graphs K_n, and then for each of the n^n independent sets I_n composed of one vertex from each of the initial K_n's, join a distinct copy of K_n to I_n.

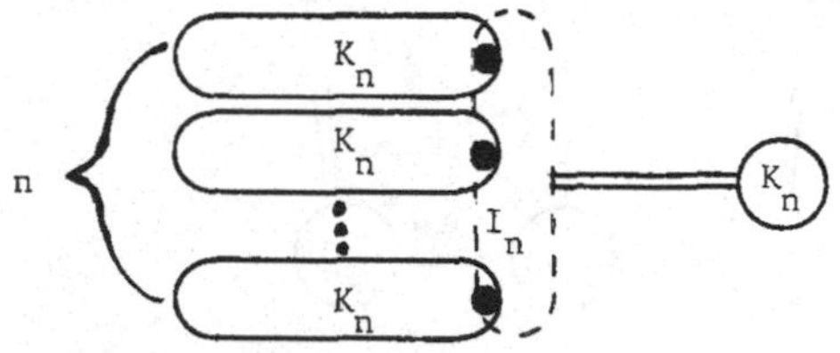

Clearly $\hat{\kappa}(G_n) = n$, so consider some minimum coloring of G_n. In this coloring some I_n must have all its vertices colored differently, and the K_n joined to this I_n must then utilize n more distinct colors, giving $\chi(G_n) \geq 2n$. A 2n-coloring of G_n is evident, so $\chi(G_n) = 2n$. ▮

The construction of further examples is aided by the following lemma, whose proof is straightforward from the cut lemmas.

Lemma 5: Let G be a graph with vertex v adjacent to all other vertices of G. Then

$$\begin{aligned} \omega(G) &= \omega(G-v) + 1, \\ \chi(G) &= \chi(G-v) + 1, \\ \hat{\kappa}(G) &= \hat{\kappa}(G-v) + 1, \\ \hat{\delta}(G) &= \hat{\delta}(G-v) + 1. \end{aligned} \tag{20}$$

Note that under the conditions of lemma 5, $\hat{\lambda}(G) > \hat{\lambda}(G-v) + 1$ is possible.

Now for $1 \leq n \leq m \leq 2n$, consider the graph with K_{2n-m} joined to the graph G_{m-n} described in the proof of theorem 5, and denote this graph by $K_{2n-m} + G_{m-n}$. Then from theorem 5 and lemma 5,

$$\hat{\kappa}(K_{2n-m} + G_{m-n}) = n, \ \chi(K_{2n-m} + G_{m-n}) = m \text{ for } 1 \leq n \leq m \leq 2n. \tag{21}$$

Our construction in theorem 5 yields an infinite class of graphs for which $\hat{\kappa}(G)/\chi(G) = 1/2$. This suggests and leaves open the following questions.

Question 1: Does there exist a graph G for which $\chi(G) \geq 2\hat{\kappa}(G) + 1$?

Question 2: What is the minimum value of α for which an infinite class of graphs $\mathcal{G}$ exists such that

$\hat{\kappa}(G) \leq \alpha\chi(G)$ for all $G \in \mathcal{G}$?

REFERENCES

1. G. Chartrand and F. Harary, Graphs with prescribed connectivities. Theory of Graphs. Academic Press, New York (1968) 61-63.
2. S. Even and R. E. Tarjan, Network flow and testing graph connectivity. SIAM J. on Comp. 4 (1975) 507-518.
3. D. W. Matula, The cohesive strength of graphs. The Many Facets of Graph Theory Springer-Verlag, Berlin (1969) 215-221.
4. D. W. Matula, k-components, clusters, and slicings in graphs. SIAM J. Appl. Math. 22 (1972) 459-480.
5. D. W. Matula, k-blocks and ultrablocks in graphs. J. Comb. Th. (to appear).
6. D. W. Matula, L. L. Beck, and J. F. Feld, Breadth-minimum search and clustering and graph coloring algorithms, submitted for publication.
7. G. Szekeres and H. S. Wilf, An inequality for the chromatic number of a graph. J. Comb. Th. 4 (1968) 1-3.
8. H. Whitney, Congruent graphs and the connectivity of graphs. Amer. J. Math. 54 (1932) 150-168.
9. A. A. Zykov On some properties of linear complexes. (Russian) Mat. Sbornik 24 (1949) 163-168. Amer. Math. Soc. Translation N. 79 (1952).

MATCHINGS IN GRAPHS III: INFINITE GRAPHS*

P. J. McCarthy
University of Kansas
Lawrence, KS 66045

Abstract

Let G be a graph, possibly infinite. Three successively larger sets of matchings in G are defined, the maximum, locally maximum, and quasi-locally maximum matchings. If G is finite the matchings of each type are maximum matchings in G. The maximum and locally maximum matchings are characterized in terms of alternating paths. Not every graph has a maximum matching, but Rado's selection theorem is used to show that every graph has a quasi-locally maximum matching. If G is locally finite, then G has a maximum matching, and every quasi-locally maximum matching in G is locally maximum.

*Research supported by University of Kansas General Research allocation 3239-5038.

1. INTRODUCTION

Let $G = (V,E)$ be a graph. All undefined terms have their meaning in [1]: thus G has no loops or multiple edges. The graph G is called locally finite if each vertex of G has finite degree. A matching in G is a subset M of E such that each vertex of G is incident with at most one edge in M. If M is a matching in G and v is a vertex of G incident with one of the edges in M, we say that M meets v. We set

$$\mu(M) = \{v: v \in V \text{ and } M \text{ meets } v\}.$$

If G is a finite graph, a matching M in G is called maximum if there is no matching in G containing more edges than M. Our long term aim is to study analogues of this concept for infinite graphs. In this paper we define several kinds of matchings in a graph such that if the graph happens to be finite, the matchings of each kind are maximum matchings in the graph. By looking at the relations between the kinds of matchings we obtain some insight into the problems that must be faced in the study of matchings in infinite graphs.

A matching M in G is called maximal if for every matching M' in G, if $M \subseteq M'$ then $M' = M$. By the definition of matching, the set of matchings in G has finite character. Hence, every matching in G is contained in a maximal matching. It is easy to find examples of finite graphs with maximal matchings that are not maximum.

2. MAXIMUM AND LOCALLY MAXIMUM MATCHINGS

A matching M in G is called maximum if for every matching M' in G, if $\mu(M) \subseteq \mu(M')$ then $\mu(M') = \mu(M)$, i.e., if $\mu(M)$ is maximal in the set $\{\mu(M'): M' \text{ is a matching in } G\}$. It can be shown that if G is a finite graph, then the two definitions of maximum matching are equivalent. The concept of maximum matching has a serious fault: not every graph has a maximum matching (see an example given below). We note that Brualdi showed that every locally finite graph has a maximum matching [3]. If M and M' are matchings in G, and if $M \subset M'$, then $\mu(M) \subset \mu(M')$: hence, a maximum matching in G is maximal.

A path in G can fail to terminate, can terminate at both ends, or can terminate at one end only. If it terminates at u and v it is called a uv-path: if it terminates only at u it is called a u-path.

Let M be a matching in G. A path P in G is M-alternating if the edges of P are alternately in M and not in M. If P is an M-alternating uv-path, and $u,v \notin \mu(M)$, then $P + M = E(P)\setminus M \cup M\setminus E(P)$ is a matching in G and $\mu(P + M) = \mu(M) \cup \{u,v\}$. If P is an M-alternating u-path, and $u \notin \mu(M)$, then $\mu(P + M) = \mu(M) \cup \{u\}$. Hence, if either of these kinds of paths exist, M is not maximum. We shall show that the converse is true, thus generalizing a result of Berge for finite graphs [2].

Theorem 1. A matching M in G is maximum if and only if there is no M-alternating u-path with $u \notin \mu(M)$ or M-alternating uv-path with $u,v \notin \mu(M)$.

Proof. Suppose M is not maximum. Then there is a vertex $u \notin \mu(M)$ and a matching M' in G such that $\mu(M) \cup \{u\} \subseteq \mu(M')$.

There is an edge $e_1 = uu_1 \in M'\setminus M$. Suppose $u_1 \in \mu(M)$: then there is an edge $e_2 = u_1u_2 \in M\setminus M'$. Note that $e_2 \neq e_1$ since $u \notin \mu(M)$. Since $u_2 \in \mu(M')$ there is an edge $e_3 = u_2u_3 \in M'\setminus M$. Note that $e_3 \neq e_2$ since $e_2 \notin M'$. Also, $u_3 \neq u$ since e_1 and e_3 are distinct edges in M'. Suppose $u_3 \in \mu(M)$: then there is an edge $e_4 = u_3u_4 \in M\setminus M'$. Continue in this way. If the suppositions continue to be valid, we obtain an M-alternating u-path. Otherwise, we obtain an M-alternating uv-path for some $v \notin \mu(M)$.

If H is a subgraph of G and if $F \subseteq E$, we write $H \cap F$ for the set of edges in F which are edges of H.

A matching M in G is <u>locally maximum</u> if for each finite subgraph H of G there is a finite subgraph H' of G such that H is a subgraph of H' and $H' \cap M$ is a maximum matching in H'. If G is finite and we take H = G we see that a locally maximum matching in a finite graph is maximum. This is not true for infinite graphs, even for infinite graphs with maximum matchings. For example, the graph

has the locally maximum matching $\{e_n : n \text{ even}\}$, which is not maximum. $\{e_n : n \text{ odd}\}$ is a maximum matching in G.

There is an analogue to Theorem 1 for locally maximum matchings.

Theorem 2. A matching M in G is locally maximum if and only if there is no M-alternating uv-path with $u,v \notin \mu(M)$.

Proof. Suppose there does exist such a path P. If H' is any finite subgraph of G having P as a subgraph, then $H' \cap M$ is not a maximum matching in H' by the theorem of Berge mentioned earlier. Thus, M is not locally maximum.

Conversely, suppose that no such path exists. Let H be a finite subgraph of G: we want to show that there is a finite subgraph H' of G such that H is a subgraph of H' and $H' \cap M$ is a maximum matching in H'. It is sufficient to do this under the assumption that H is induced. It is not, however, the fact that H is induced that is important, but rather the fact that if w and u are vertices of H and if $e = wu \in M$, then e is an edge of H.

Suppose $H \cap M$ is not maximum in H. By Berge's theorem, there is a pair of (distinct) vertices of H not met by $H \cap M$ and connected by an $(H \cap M)$-alternating path in H. Let $\nu(H)$ be the number of such pairs of vertices of H, and let u,v be one such pair. By assumption, at least one of u and v, say u, is met by M. Suppose $e = wu \in M$: then e is not an edge of H, and so by the property of H noted in the preceding paragraph, w is not a vertex of H. Let H_1 be the subgraph of G obtained by adding the vertex w and the edge e to H, and note that H_1 has the same property noted above for H. Let u_1,v_1 be a pair of vertices of H_1 not met by $H_1 \cap M$ and connected by an $(H_1 \cap M)$-alternating path Q in H_1. If e is an edge of Q, it is a terminating edge of Q, which means that either u_1 or v_1 is w. This is not true since w is met by $H_1 \cap M$. Also, $u_1 \neq u \neq v_1$. Hence, u_1,v_1 is a pair of vertices of H not met by $H \cap M$ and connected by an $(H \cap M)$-alternating path in H. Thus, $\nu(H_1) < \nu(H)$. We can repeat this argument, starting with H_1. After a finite number of repetitions we obtain a finite subgraph H' of G such that H is a subgraph of H' and $\nu(H') = 0$. By Berge's theorem, $H' \cap M$ is a maximum matching in H'.

Corollary 1. Every maximum matching in G is locally maximum.

Corollary 2. Every locally maximum matching in G is maximal.

A graph can have a locally maximum matching without having a maximum matching. In fact, for the graph in our example, every maximal matching is locally maximum.

Example. Let G be the complete bipartite graph with bipartition (V_1,V_2), where V_1 is countably infinite and V_2 is noncountably infinite. Let M be a maximal matching in G. Clearly, M meets each vertex in V_1 and not all vertices in V_2. If $v \in V_2 \setminus \mu(M)$, then $(\mu(M) \cap V_2) \cup \{v\}$ is countably infinite: hence there is a matching in G which meets each vertex in $\mu(M) \cup \{v\}$. Thus, M is not a maximum matching. However, M is locally maximum since every path in G connecting two vertices in V_2 has even length, and consequently cannot be M-alternating with both terminal edges not in M.

There are two natural questions to ask, and we pose them as problems.

Problem 1. Does every graph have a locally maximum matching?

Problem 2. Does there exist a countably infinite graph which has no maximum matching? If the answer is yes, does there exist a countably infinite graph which has a locally maximum matching but no maximum matching?

3. QUASI-LOCALLY MAXIMUM MATCHINGS AND LIMITS OF FINITE MAXIMUM MATCHINGS

A locally maximum matching M in G has the following property: if H is a finite subgraph of G, then there is a finite subgraph H' of G and a maximum matching M' in H' such that H is a subgraph of H' and $H \cap M = H \cap M'$. We call any matching in G with this property a <u>quasi-locally maximum matching</u>. If G is a finite graph, then every quasi-locally maximum matching in G is maximum. The most important property of quasi-locally maximum matchings is that they exist in every graph. This is a consequence of Theorem 4, proved below.

We use the following version of Rado's selection theorem, which is proved by modifying only slightly the proof of Rado's theorem in [4, p. 52].

Let $(A_e\colon e \in E)$ be a family of nonempty finite subsets of a set

S. Let $\{E_i : i \in I\}$ be a collection of finite subsets of E such that if $i,j \in I$ then there exists $k \in I$ such that $E_i \cup E_j \subseteq E_k$, and $E = \cup\{E_i : i \in I\}$. For each $i \in I$, let $f_i : E_i \to S$ be a mapping such that $f_i(e) \in A_e$ for each $e \in E_i$. Then there exists a mapping $f: E \to S$ such that

(a) $f(e) \in A_e$ for each $e \in E$, and

(b) for each $i \in I$ there exists $j \in I$ such that $E_i \subseteq E_j$ and $f|E_i = f_j|E_i$.

Theorem 3. Let $\{H_i : i \in I\}$ be a collection of finite subgraphs without isolated vertices of G such that if $i,j \in I$ then there exists $k \in I$ such that $E(H_i) \cup E(H_j) \subseteq E(H_k)$, and $E = \cup\{E(H_i) : i \in I\}$. For each $i \in I$, let M_i be a matching in H_i. Then there is a matching M in G such that for each $i \in I$ there exists $j \in I$ such that H_i is a subgraph of H_j and $H_i \cap M = H_i \cap M_j$.

Proof. We apply the selection theorem with $A_e = \{0,1\}$ for each $e \in E$, and $E_i = E(H_i)$ for each $i \in I$. Note that each $E(H_i)$ is nonempty since the vertex-set of a graph is always nonempty and H_i has no isolated vertices. For each $i \in I$ define $f_i : E(H_i) \to \{0,1\}$ by $f_i(e) = 1$ if $e \in M_i$ and $f_i(e) = 0$ if $e \in M_i$. By the selection theorem there is a mapping $f: E \to \{0,1\}$ such that for each $i \in I$ there exists $j \in I$ such that $E(H_i) \subseteq E(H_j)$ and $f|E(H_i) = f_j|E(H_i)$. Since H_i has no isolated vertices it is a subgraph of H_j. Let $M = \{e : e \in E \text{ and } f(e) = 1\}$. If $e_1, e_2 \in E$, $e_1 \neq e_2$, there is an $i \in I$ with $e_1, e_2 \in E(H_i)$. With $j \in I$ as above, we have $f(e_1) = f_j(e_1)$ and $f(e_2) = f_j(e_2)$. Thus, if e_1 and e_2 are adjacent, we cannot have both $f(e_1) = 1$ and $f(e_2) = 1$. Thus, M is a matching in G.

Let $i \in I$ and let $j \in I$ be as above. If $e \in H_i \cap M$ then $f_j(e) = f(e) = 1$, so $e \in H_i \cap M_j$. If $e \in H_i \cap M_j$, then $f(e) = f_j(e) = 1$, so $e \in H_i \cap M$. Therefore, $H_i \cap M = H_i \cap M_j$.

If, in Theorem 3, M_i is a maximum matching in H_i for each $i \in I$, then the matching M in G is said to be a <u>limit of finite maximum matchings</u>. If G is a finite graph, then every such matching is a maximum matching in G. Every graph G with at least one edge has a matching which is a limit of finite maximum matchings: we take simply $\{H_i : i \in I\}$ to be the collection of all finite subgraphs without isolated vertices of G. Suppose M is a limit of finite maximum matchings, and let H be a finite subgraph of G with at least one edge. Let H_0 be the subgraph of G obtained from H by deleting its isolated vertices. Then H_0 is a subgraph of H_i for some i, and if we let $j \in I$ be as in the statement of Theorem 3, then $H \cap M = H_0 \cap M = H_0 \cap H_i \cap M = H_0 \cap H_i \cap M_j = H_0 \cap M_j = H \cap M_j$. This proves the following result.

Theorem 4. Every matching in G that is a limit of finite maximum matchings is a quasi-locally maximum matching in G.

Corollary. Every graph has a quasi-locally maximum matching.

Theorem 5. Every locally maximum matching in G is a limit of finite maximum matchings.

Proof. Let M be a locally maximum matching in G. Let $\{H_i : i \in I\}$ be the collection of all finite subgraphs without isolated vertices of G. For each $i \in I$, let H_i' be a finite subgraph of G such that H_i is a subgraph of H_i' and $H_i' \cap M = M_i$ is a maximum matching in H_i': we can assume that each H_i' is without isolated vertices. Then, $\{H_i' : i \in I\}$ satisfies the requirements set forth in Theorem 3, and we let M' be the limit of the finite maximum matchings $\{M_i : i \in I\}$. For each $i \in I$ there exists $j \in I$ such that H_i' is a subgraph of H_j' and $H_i' \cap M' = H_i' \cap M_j = H_i' \cap H_j' \cap M = H_i' \cap M$. If $e \in E$, and if we choose $i \in I$ so that $e \in E(H_i')$, then we conclude that $e \in M$ if and only if $e \in M'$. Therefore, $M = M'$.

Thus, the concept of a limit of finite maximum matchings is intermediate to the concepts of locally maximum matching and quasi-locally maximum matching. In some infinite graphs, the three concepts coincide.

Theorem 6. If G is locally finite, then every quasi-locally maximum matching in G is locally maximum.

Proof. Let M be a quasi-locally maximum matching in G, and suppose M is not locally maximum. By Theorem 2 there is an M-alternating path P: $u = u_0, u_1, \ldots, u_n = v$, in G with $u, v \notin \mu(M)$. Let H be the subgraph of G induced by the set of all edges of G incident with at least one of $u_0, u_1, \ldots, u_n$. Then, H is a finite subgraph of G, and there is a finite subgraph H' of G such that H is a subgraph of H' and $H \cap M = H \cap M'$ for some maximum matching M' in H'. If e is an edge of H' incident with u_0, then e is an edge of H, and consequently $e \notin H \cap M$. Hence, $e \notin M'$. Thus, u is not met by M'. Likewise, v is not met by M'. Since P is an M'-alternating path in H', this implies that M' is not a maximum matching in H', a contradiction. Therefore, M is locally maximum.

Prompted by Theorems 5 and 6, we pose another problem.

Problem 3. Is there a quasi-locally maximum matching that is not a limit of finite maximum matchings? Is there a limit of finite maximum matchings that is not locally maximum?

References

1. M. Behzad and G. Chartrand, Introduction to the Theory of Graphs. Allyn and Bacon, Boston (1971).

2. C. Berge, Two theorems in graph theory. Proc. Nat. Acad. Sci. USA 43 (1957) 842-844.

3. R. A. Brualdi, Matchings in arbitrary graphs. Proc. Camb. Phil. Soc. 69 (1971) 401-407.

4. L. Mirsky, Transversal Theory. Academic Press, New York and London (1971).

STEINHAUS GRAPHS

John C. Molluzzo
St. John's University
Staten Island, N.Y. 10301

Abstract

A Steinhaus triangle is formed by starting with an arbitrary row of plus and minus signs. Each succeeding row is formed by placing a plus or minus sign under each pair of equal or opposite signs respectively. A Steinhaus graph is constructed by considering this triangle as the upper triangle of an adjacency matrix. The minimum number of lines possible in a Steinhaus graph is found and a theorem is obtained on the structure of these minimal graphs. Methods of constructing digraphs and valued graphs from Steinhaus triangles are discussed. Problems on the several types of graph are posed. Finally, a generalization of Steinhaus' original problem is made.

Introduction:

In [4, pp 47-48] Steinhaus posed a problem on a triangle of plus and minus signs formed as follows: begin with an arbitrary row S of n plus and minus signs. Each succeeding row is formed by placing a plus or a minus sign under each pair of equal or opposite signs respectively. We thus obtain a triangle T(S), uniquely determined by S, of n(n+1)/2 signs which we call the Steinhaus triangle generated by S. The triangle T(+ - - - +) appears in Figure 1.

```
+   -   -   -   +
  -   +   +   -
    -   +   -
      -   -
        +
```

Figure 1

Steinhaus' problem was to determine if it were possible to have a Steinhaus triangle with an equal number of plus and minus signs for every value of n for which n(n+1)/2 is even. An affirmative answer was obtained by Harborth in [3].

The purpose of this note is to introduce a method of constructing graphs and digraphs from Steinhaus triangles. A generalization of the Steinhaus triangle construction is given which gives rise to valued graphs and a generalization of Steinhaus' original problem.

Any terms used in this paper not herein defined may be found in [1] or [2].

Steinhaus Graphs:

To form the Steinhaus graph G(S) generated by S we consider T(S) as the upper triangle of a symmetric matrix A(S), minus signs being placed along the main diagonal. Thus the matrix

$$A(S) = \begin{bmatrix} - & + & - & - & - & + \\ + & - & - & + & + & - \\ - & - & - & - & + & - \\ - & + & - & - & - & - \\ - & + & + & - & - & + \\ + & - & - & - & + & - \end{bmatrix}$$

is the matrix formed from the triangle of Figure 1. The matrix A(S) can now be interpreted as the adjacency matrix of a graph G(S) by considering a plus sign as an adjacency and a minus sign as a non-adjacency. Figure 2 shows the Steinhaus graph G(+ - - - +) associated with the triangle of Figure 1.

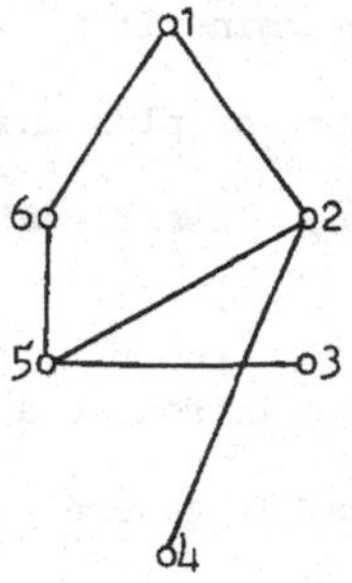

Figure 2

Note that the number of points p of G(S) is one greater than the number of signs in S.

A graph G with p points is a Steinhaus graph if there is a labeling of its points 1, 2,...,p such that the upper triangle of its adjacency matrix is a Steinhaus triangle. That not all graphs are Steinhaus graphs is easily seen by considering P_3, the path on three points. For any labeling of its points the upper triangle of its adjacency matrix consists of two plus and one minus sign which is impossible in a Steinhaus triangle.

The following two theorems combine to give a lower bound on the number of lines in a Steinhaus graph with p points.

Let S_0 be any sequence of $p-1 \geq 1$ consecutive signs taken from the periodic sequence + - - + - - + - -.... We then have the following theorem:

Theorem 1: If $G(S_0)$ is the Steinhaus graph generated by S_0, then the number of lines q of $G(S_0)$ is given by

$$q = \begin{cases} \left\{ p\,(p-1)/6 \right\}, & \text{if } p \equiv 2 \bmod 3 \text{ and } S_0 \text{ begins with } + \\ \left[p\,(p-1)/6 \right], & \text{otherwise} \end{cases}$$

where $[x]$ and $\{x\}$ are the greatest and least integer functions respectively.

Proof: Let $k = [p/3]$. Partition the adjacency matrix $A(S_0)$ of $G(S_0)$ as follows:

$$A(S_0) = \begin{bmatrix} A_{11} & \cdot & \cdot & \cdot & A_{1k} & C_1 \\ \cdot & & & & \cdot & \cdot \\ \cdot & & & & \cdot & \cdot \\ \cdot & & & & \cdot & \cdot \\ A_{k1} & \cdot & \cdot & \cdot & A_{kk} & C_k \\ R_1 & & \cdot & \cdot & R_k & D \end{bmatrix}$$

where each A_{ij} is a 3x3 matrix, and R_i and C_j are (p-3k)x3 and 3x(p-3k) matrices respectively. It is now easily seen that each A_{ij}, $i \neq j$, contains exactly three plus signs and each A_{ii} contains two. Each R_i and C_j contains p-3k plus signs. D has no plus signs except when $p \equiv 2 \bmod 3$ and S_0 begins with

+, in which case it contains two. Thus when D contains no plus signs $q = \frac{1}{2}(3k^2-k+2(p-3k)k) = \left[p \ (p-1)/6 \right]$. Likewise, when D has plus signs, $q = \left\{ p \ (p-1)/6 \right\}$.

In the paper by Harborth the following result is obtained, although in a different form.

Theorem 2: [Harborth] The minimum number of lines in a Steinhaus graph on p points is $\left[p \ (p-1)/6 \right]$.

Proof: Consider all possible Steinhaus triangles with two rows:

$$\begin{matrix} + & + \\ & + \end{matrix} \qquad \begin{matrix} + & - \\ & - \end{matrix} \qquad \begin{matrix} - & + \\ & - \end{matrix} \qquad \begin{matrix} - & - \\ & + \end{matrix}$$

Each of these triangles has at least one plus sign. Now each two row sub-triangle of an S_0 triangle contains exactly one plus sign. The result follows easily from this observation and Theorem 1.

We now investigate the structure of these Steinhaus graphs $G(S_0)$.

Theorem 3: Let $G(S_0)$ be the Steinhaus graph of S_0 with $p > 3$. Then $G(S_0)$ has two components. One component is the complete graph K_r on r points where r is $[p/3]$, $[(p+1)/3]$, or $[(p+2)/3]$ according as S_0 begins + - -, - + -, or - - +. If $p \geq 5$, the second component has diameter two.

Proof: We prove the theorem for the case where S_0 begins + - -, the others being similar. In the adjacency matrix $A(S_0)$ of $G(S_0)$ consider row i, $i \equiv 0 \bmod 3$. The plus signs in this row appear in columns numbered j, where $j \neq i$ and $j \equiv 0 \bmod 3$. Thus the subgraph of $G(S_0)$ induced by the points numbered 3, 6,..., 3r where $r = [p/3]$ is K_r.

To prove the second assertion, we consider those points not in the complete graph constructed above. The row numbers denoting the points of the graph, let i_1 and i_2 be two such points. There are three cases:

1. $i_1 \equiv 1$ and $i_2 \equiv 2 \bmod 3$. Then it is easily seen that i_1 and i_2 are adjacent.

2. i_1 and $i_2 \equiv 1 \bmod 3$. Then i_1, i_1+1, i_2 is a path between i_1 and i_2.

3. i_1 and $i_2 \equiv 2 \bmod 3$. Then i_1, i_1-1, i_2 is a path between i_1 and i_2.

There are many interesting questions one can ask about Steinhaus graphs. For example, given a graph G, when is it a Steinhaus graph? When is it not a Steinhaus graph? One can say for instance, as a simple corollary to Theorem 2, that no tree on $p>6$ points is a Steinhaus graph. There are trees with $p \leq 6$ which are Steinhaus graphs, namely K_1, K_2 and the two trees of Figure 3 labeled as shown.

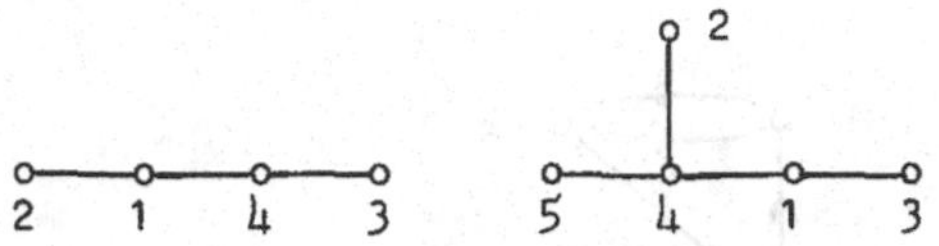

Figure 3

A complete characterization of Steinhaus graphs seems to be a difficult problem.

When one begins constructing Steinhaus graphs, it becomes immediately apparent that for a given number of points p several of the graphs are isomorphic. What are necessary and sufficient conditions on S_1 and S_2 that would insure $G(S_1) = G(S_2)$? Short of that, for a given p how many non-isomorphic Steinhaus graphs are there on p points. When $p = 4$ we have the following isomorphism equivalence classes:

$\{G(+\ +\ +)\}$, $\{G(+\ +\ -), G(-\ -\ -)\}$, $\{G(+\ -\ +), G(-\ +\ +)\}$, $\{G(+\ -\ -), G(-\ +\ -)\}$, and $\{G(-\ -\ +)\}$. Thus there are five non-isomorphic Steinhaus graphs on four points.

Finally, as noted in Theorem 3, the Steinhaus graphs $G(S_0)$ having the minimum number of lines are disconnected. What conditions on S will guarentee that G(S) be connected? Can one deduce the connectivity of G(S) from S?

<u>Generalizations</u>:

The construction of Steinhaus graphs can be generalized in at least two ways. First we consider two plus and minus sequences of the same length, S_1 and S_2. From S_1 we can work down and obtain the Steinhaus triangle $T(S_1)$. From S_2 we can work up and obtain an inverted Steinhaus triangle $T(S_2)$. These may be regarded as the upper and lower triangles respectively of the adjacency matrix $A(S_1,S_2)$ of a digraph $D(S_1,S_2)$ called the Steinhaus digraph generated by the pair (S_1, S_2). The diagraph D(+ - -, + - +) is depicted in Figure 4.

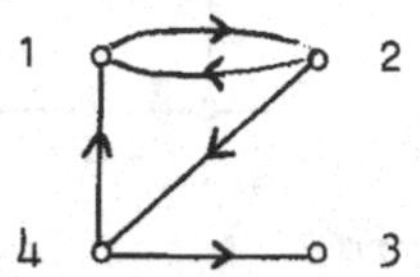

Figure 4

As in the case of Steinhaus graphs, many questions may be raised about Steinhaus digraphs. Aside from characterizing Steinhaus digraphs, other characterization problems are: describe simply the relationship between $D(S_1,S_2)$ and $D(S_2,S_1)$; characterize those digraphs of the form $D(S,S)$; when will $D(S_1,S_2)$ be symmetric?

With regard to connectedness, what conditions on S_1 and S_2 will determine the connectedness category of $D(S_1,S_2)$? Steinhaus digraphs may be in any connectedness category. For example, D(- +, + -), D(+ -, - +), D(- -, + -), D(+ +, + -) are in categories C_0, C_1, C_2, and C_3 respectively. It is even possible to obtain Steinhaus digraphs of the form D(S,S,) in each connectedness category.

To generalize in a second direction, note that if minus corresponds to 1 and plus to 0, then the rule for forming a Steinhaus triangle can be interpreted as addition mod2. This leads to the following: for a given integer

$m > 1$ consider a sequence S_m of n integers from 0 to m-1. Generate a sequence of n-1 integers below this by adding modm the two integers above in Steinhaus fashion. Continuing this way with succeeding rows, a triangle $T_m(S_m)$ is obtained. This is called an m-Steinhaus triangle. The triangle $T_4(321021)$ is depicted in Figure 5.

```
3   2   1   0   2   1
  1   3   1   2   3
    0   0   3   1
      0   3   0
        3   3
          2
```

Figure 5

Now, as in the previous cases, this triangle can be taken as the upper triangle of a symmetric matrix $A(S_m) = \left[a_{ij} \right]$, (m-1)'s being placed on the diagonal. This matrix can be considered as the adjacency matrix of a valued graph $G(S_m)$, $(m-1)-a_{ij}$ being the value of the edge (i,j).

The reader may enjoy asking questions concerning these valued graphs for himself.

We do, however, wish to mention the following which generalizes Steinhaus' original problem, thus returning us to our point of origin.

General Steinhaus Problem: Given an integer $m \geq 2$, for those values of n for which $n(n+1)/2$ is divisible by m, must there exist an m-Steinhause triangle in which there are an equal number of 0's, 1's, ..., (m-1)'s?

Acknowledgement: I wish to thank Dr. M. Capobianco for first bringing my attention to Steinhaus triangles and for the many stimulating discussions which followed. I also wish to thank Prof. E. Wang who brought my attention to Harborth's paper.

References

1. F. Harary, Graph Theor y, Addison-Wesley, Reading, Mass. (1969).

2. F. Harary, R.Z. Norman, and D. Cartwright, Structural Models: An Introduction to the Theor y of Directed Graphs, John Wiley and Sons, New York (1965).

3. H. Harborth, Solution of Steinhaus's Problem with Plus and Minus Signs, J. Combinatorial Theor y, Ser. A,12 (1972)

4. H. Steinhaus, One Hundred Problems in Elementary Mathematics, Pergamon, Elinsford, N.Y. (1963).

5. E. Wang, Problem E2541, Amer. Math. Monthly 82 (1975) 659-660.

PACKING AND COVERING CONSTANTS FOR RECURSIVE TREES

A. Meir and J.W. Moon
University of Alberta
Edmonton, Alberta, Canada

Abstract

Let $p_k(n)$ and $c_k(n)$ denote the average values of the k-packing and k-covering numbers of the $(n-1)!$ recursive trees with n nodes. We show that the limits of $p_k(n)/n$ and $c_k(n)/n$ as $n \to \infty$ exist and we discuss the problem of evaluating these limits.

§1. Introduction. Let $d(u,v)$ denote the distance between nodes u and v in a tree T. A subset P of nodes of T is a k-packing, where k denotes a fixed positive integer, if $d(u,v) > k$ for all u and v in P. A subset C of nodes is a k-covering if for every node u of T there exists a node v such that $d(u,v) \leq k$. The k-packing number of T is the number $P_k(T)$ of nodes in any largest k-packing in T and the k-covering number is the number of nodes in any smallest k-covering of T.

If F denotes a given family of trees, let $p_k(n)$ and $c_k(n)$ denote the average value of $P_k(T)$ and $C_k(T)$ over those trees T in F that have n nodes. In [4] (see also [1] and [2]) we showed that when the generating function $y = y(x)$ for trees in F satisfies a functional relation of the type $y = x\phi(y)$, subject to some mild conditions, then the packing and covering constants

$$p_k = \lim_{n \to \infty} p_k(n)/n \quad \text{and} \quad c_k = \lim_{n \to \infty} c_k(n)/n$$

exist and can be determined by solving a certain system of equations. Labelled trees, binary trees, and plane trees fall in this category. In [5] we considered the usually more difficult case when the generating function does not satisfy a functional relation of the type mentioned above. In particular, we showed that the constants p_1 and c_1 exist for both rooted and unrooted unlabelled trees and that each constant has the same value for these two families; we also gave estimates for these values.

Our object here is to consider a third family of trees, recursive trees, whose generating function does not satisfy a functional relation of the type $y = x\phi(y)$. In §2 we introduce some more notation and quote some general results. Our main results on recursive trees are in §3 and some numerical results are given in §4.

We showed in [3] that $P_{2k}(T) = C_k(T)$ for any tree T. Consequently, $p_{2k} = c_k$ for any family F and any k, so we shall henceforth restrict our attention to the packing constants.

§2. Preliminaries. Let y_n denote the number of trees with n nodes in a given family F of rooted trees. We assume that $y_1 = 1$ and that there exists a recurrence relation for y_n in terms of $y_1, \cdots, y_{n-1}$ for $n \geq 2$. We further assume that there exists an operator $\gamma\{h_1(x), h_2(x), \cdots\}$ defined for any (possibly infinite) sequence of power series $h_1(x), h_2(x), \cdots$, such that if

$$y = y(x) = \sum_1^\infty y_n \frac{x^n}{n!}$$

then

$$y(x) = \gamma\{y(x), y(x^2), \cdots\} .$$

For notational convenience we shall write this simply as

$$y(x) = \Gamma\{y(x)\} . \tag{2.1}$$

If $W(x,z)$ is any power series in the two variables x and z, we let

$$\Gamma\{W(x,z)\} = \gamma\{W(x,z), W(x^2,z^2), \cdots\} ,$$

where $W(x,z), W(x^2,z^2), \cdots$ are interpreted as functions of x for fixed values of z. (We are assuming that the nodes of trees in F are labelled; otherwise the factor $n!$ would not be present in the definition of $y(x)$.)

Let y_{ni} denote the number of trees T with n nodes in F for which $P_k(T) = i$, for $1 \leq i \leq n$. If

$$Y = Y(x,z) = \sum_{n,i} y_{ni} z^i \frac{x^n}{n!}$$

then $y(x) = Y(x,1)$. Furthermore, if $p_k(n)$ denotes the expected k-packing number of trees with n nodes in F and

$$M = M(x) = \sum_n p_k(n)\, y_n \frac{x^n}{n!} ,$$

then

$$(2.2) \qquad M = \frac{\partial}{\partial z} \big(Y(x,z)\big)\Big|_{z=1} .$$

We need to introduce some more terminology in order to state a defining relation for Y.

We say a rooted tree T with root r is a <u>type</u> (k,j) tree, where $0 \leq j \leq k$, if

$$\max\{d(r,P):\ P \text{ is a largest k-packing in } T\} = j.$$

Thus, for example, a type $(k,0)$ tree is one in which the root belongs to every largest k-packing. If f_{jni} denotes the number of type (k,j) trees T with n nodes in F such that $P_k(T) = i$, let

$$F_j = F_j(x,z) = \sum_{n,i} f_{jni} z^i \frac{x^n}{n!}$$

for $0 \leq j \leq k$. Then

$$(2.3) \qquad Y = F_0 + F_1 + \cdots + F_k$$

and

$$(2.4) \qquad y = f_0 + f_1 + \cdots + f_k$$

where $f_j(x) = F_j(x,1)$. Finally, let

$$S_j = F_0 + F_1 + \cdots + F_j$$

and

$$R_j = F_j + F_{j+1} + \cdots + F_k$$

for $0 \leq j \leq k$, and let $s_j = s_j(x) = S_j(x,1)$ and $r_j = r_j(x) = R_j(x,1)$.

Trees of type (k,j) were characterized in [4] and the characterization was used to derive defining relations for the F_j's when $y = x\phi(y)$ for some power series ϕ. The proof of the following more general result is essentially the same.

Theorem 1. If $y = \Gamma\{y\}$, then

$$F_0 = z\,\Gamma\{R_k\}\ ;$$

$$F_{j+1} = z\,\Gamma\{S_j/z + R_{k-j-1}\} - z\,\Gamma\{S_{j-1}/z + R_{k-j-1}\},\ \text{if } 0 \leq j < \frac{1}{2}(k-1)\ ;$$

and

$$F_{j+1} = \Gamma\{S_{k-j-2}/z + R_j\} - \Gamma\{S_{k-j-2}/z + R_{j+1}\},\ \text{if } \frac{1}{2}(k-1) \leq j \leq k-1.$$

Corollary 1.1. If $y = \Gamma\{y\}$ and $m = \left[\frac{1}{2}k\right]$, then

$$Y = z\,\Gamma\{S_{m-1}/z + R_m\} + (1-z)R_{m+1}\ .$$

Corollary 1.2. If $y = \Gamma\{y\}$ and $m = \left[\frac{1}{2}k\right]$, then

$$M = s_m + \frac{\partial}{\partial z}\left(\Gamma\{S_{m-1}/z + R_m\}\right)_{z=1}\ .$$

It is a straightforward but tedious exercise to show that Corollary 1.1 follows from Theorem 1 and equation (2.3). (Notice that the result reduces to $y = \Gamma\{y\}$ when $z = 1$.) Corollary 1.2 follows from Corollary 1.1 and equation (2.2).

§3. Main Results. A tree T with n labelled nodes, rooted at node 1, is a recursive tree if $n = 1$ or if T_n can be constructed by joining the n-th node to some node of a recursive tree with n-1 nodes. The branches of T with respect to the root node 1 are themselves recursive trees, or rather they would be if the nodes in each branch were relabelled according to the sizes of their original labels in increasing order. It is not difficult to show (see [2; p. 355]) that the generating function $y(x)$ of the family F of recursive trees satisfies the relation

$$y(x) = \int_0^x e^{y(t)}dt ;$$

hence, $y' = e^y$ and

$$y = \sum_1^\infty y_n \frac{x^n}{n!} = -\ln(1-x) = \sum_1^\infty \frac{x^n}{n},$$

since $y(0) = 0$. Thus the family F satisfies condition (2.1) with

$$\Gamma\{y(x)\} = \int_0^x e^{y(t)}dt . \tag{3.1}$$

Theorem 2. The generating functions f_j for the type (k,j) recursive trees satisfy the differential equations

$$f_0' = e^{r_k} ; \tag{3.2}$$

$$f_{j+1}' = e^{s_j + r_{k-j-1}} - e^{s_{j-1}+r_{k-j-1}}, \quad \text{if} \quad 0 \le j < \frac{1}{2}(k-1) ; \tag{3.3}$$

and

$$f_{j+1}' = e^{s_{k-j-2}+r_j} - e^{s_{k-j-2}+r_{j+1}}, \quad \text{if} \quad \frac{1}{2}(k-1) \le j \le k-1. \tag{3.4}$$

This follows from Theorem 1 and relation (3.1) upon setting $z = 1$ and differentiating with respect to x.

Theorem 3. If M denotes the generating function of the expected k-packing numbers for the family of recursive trees, then

$$M = s_m + \frac{1}{1-x}\int_0^x f_m(t)dt,$$

where $m = \left[\frac{1}{2}k\right]$.

Proof: We first observe that if $W(x,z)$ is any power series in x and z, then

$$\frac{\partial}{\partial z}\Gamma\{W(x,z)\} = \int_0^x e^{W(t,z)}\frac{\partial}{\partial z}W(t,z)dt ,$$

by (3.1). This and Corollary 1.2 imply that

$$M = s_m + \int_0^x e^{y(t)}\left(M(t)-s_{m-1}(t)\right)dt,$$

or that

$$M' = s_m' + e^y(M-s_{m-1}).$$

Since $e^y = y' = (1-x)^{-1}$ and $s_m = s_{m-1} + f_m$, this can be rewritten as

$$(1-x)M' - M = (1-x)s_m' - s_m + f_m.$$

This implies the required result since $M(0) = 0$.

Corollary 3.1. The k-packing constant p_k of the family of recursive trees exists and satisfies the relation

$$(3.5)\qquad p_k = \int_0^1 f_m(t)dt,$$

where $m = \left[\frac{1}{2}k\right]$; furthermore, if $k = 2m$ then

$$(3.6)\qquad p_k = 1 - \int_0^1 e^{-f_{m-1}(x)}dx.$$

The fact that p_k exists and satisfies relation (3.5) follows immediately from Theorem 3. Equation (3.3) and the fact that $y' = (1-x)^{-1}$ imply that

$$f_m' = \left(1 - e^{-f_{m-1}}\right)(1-x)^{-1}$$

when $k = 2m$; this and (3.5) yield relation (3.6). We remark that if $q_k(n)$ denotes the expected value of $P_k(T)\left(P_k(T)-1\right)$ over all recursive trees with n nodes, then it can be shown by a similar but more involved argument that $q_k(n)/n^2 \to p_k^2$ as $n \to \infty$. Hence the variance $\sigma^2(n)$ of $P_k(T)$ over all recursive trees with n nodes satisfies the relation $\sigma^2(n) = o(n^2)$. The following result is then implied by Chebyshev's inequality.

Corollary 3.2. If $P_k(T)$ denotes the k-packing number of a tree T chosen at random from the $(n-1)!$ recursive trees with n nodes, then

$$\Pr\{|P_k(T)/n - p_k| < \varepsilon\} \to 1$$

as $n \to \infty$ for any fixed positive ε.

§4. Numerical Results. If $k = 1$, then it follows from Theorem 2 and equation (2.4) that $f_0(t) = \log(1 - \log(1-t))$. Therefore,

$$p_1 = \int_0^1 \log(1 - \log(1-t))dt = \int_0^\infty \frac{e^{-u}}{1+u} du = .596347\cdots.$$

(This particular result was also given in [2, p. 355].)

It apparently is not easy to determine the explicit values of p_k when $k \geq 2$. We now show that $.3714 < p_2 < .3863$ by an ad hoc argument.

If we examine the small recursive trees when $k = 2$ we find that $f_0(x) > x + x^4/4!$ for $0 < x \leq 1$; hence,

$$p_2 > 1 - \int_0^1 \exp(-x-x^4/4!)dx > .3714,$$

by (3.6).

When $k = 2$ it follows from equations (3.2) and (3.3) that

$$f_1' e^{-f_1} = f_0'(e^{f_0} - 1) ,$$

or that

$$1 - e^{-f_1} = e^{f_0} - f_0 - 1,$$

upon integrating for 0 to x. It can be shown, using this and equations (3.3) and (3.4), that $f_1(x) > f_2(x)$ for $0 < x < 1$. Therefore,

$$(1-x)^{-1} = y' = e^{f_0+f_1+f_2} > e^{f_0+2f_2} = \left(f_0' e^{\frac{1}{2}f_0}\right)^2$$

for $0 < x < 1$, by (3.2). We deduce from this that

$$e^{-f_0(x)} > \left(2 - (1-x)^{\frac{1}{2}}\right)^{-2}$$

for $0 < x < 1$. Hence,

$$p_2 < 1 - \int_0^1 \left(2 - (1-x)^{\frac{1}{2}}\right)^{-2} dx = 2 \log 2 - 1 < .3863,$$

as required, by (3.6).

We remark that if $\theta(x) = \exp\{f_0(1 - e^{-x})\}$ for $0 \leq x < \infty$, then $\theta(x)$ is the solution of the equation

$$\theta'(x) = 2 + \log \theta(x) - \theta(x)$$

for which $\theta(0) = 1$; the formula for p_2 given by (3.6) can then be rewritten as

$$p_2 = 1 - \int_0^\infty \left(e^x \theta(x)\right)^{-1} dx.$$

The entries for recursive trees in the following table for $k \geq 2$ are estimates obtained from Theorem 3 and Corollary 3.1 by Mrs. Mary Willard with the aid of a computer. The entries for the other trees, given for comparison purposes, are taken from [4] and [5].

TABLE

	Recursive Trees	Labelled Trees	Binary Trees	Plane Trees	Rooted and Unrooted Unlabelled Trees
p_1	.596	.567	.586	.618	.606 ± .003
p_2	.375	.360	.378	.347	.333 ± .002
p_3	.263	.258	.250	.237	-
p_4	.185	.193	.174	.166	-
p_5	.138	.152	.130	.125	-

Packing Constants for Some Families of Trees

The preparation of this paper was facilitated by grants from the National Research Council of Canada.

REFERENCES

1. A. Meir and J.W. Moon, The expected node-independence number of random trees. Proc. Kon. Ned. Akad. v. Wetensch. 76 (1973) 335-341.

2. A. Meir and J.W. Moon, The expected node-independence number of various types of trees. Recent Advances in Graph Theory, Academia Praha (1975) 351-363.

3. A. Meir and J.W. Moon, Relations between packing and covering numbers of a tree. Pac. J. Math. 61 (1975) 225-233.

4. A. Meir and J.W. Moon, Packing and covering constants for certain families of trees, I. Journal of Graph Theory (to appear).

5. A. Meir and J.W. Moon, Packing and covering constants for certain families of trees, II. (submitted for publication).

THE MINIMUM DEGREE AND CONNECTIVITY OF A GRAPH

Ladislav Nebeský
Charles University
Prague, Czechoslovakia

Abstract

Relationships between the minimum degree of a connected graph and its connectivity (or edge-connectivity) are discussed. We give a sufficient condition for a connected graph G to contain at least two nonadjacent vertices of degree $\kappa(G)$, where $\kappa(G)$ denotes the connectivity of G.

1. Introduction. If G is a graph (in the sense of Behzad and Chartrand [1] or Harary [10]), then we denote by $V(G)$, $E(G)$, $\delta(G)$, $\lambda(G)$, and $\kappa(G)$ its vertex set, edge set, minimum degree, edge-connectivity, and connectivity, respectively.

It is well-known (see [10, p. 43-44] or [1, p. 119]) that

$$\delta(G) \geq \lambda(G) \geq \kappa(G),$$

for any graph G. Chartrand and Harary [4] proved that for every triple (i, j, k) of integers such that $i \geq j \geq k \geq 1$, there exists a connected graph G with $\delta(G) = i$, $\lambda(G) = j$, and $\kappa(G) = k$. This leads to natural questions, what graphs G contain vertices of degree $\lambda(G)$ or vertices of degree $\kappa(G)$ or, more generally, vertices of degree $< f(\kappa(G))$, where f is a function of $\kappa(G)$ such that $\kappa(G) < f(\kappa(G))$.

The following partial answer is due to Chartrand [3]: <u>If</u> G <u>is a nontrivial graph such that</u> $\delta(G) \geq [|V(G)|/2]$, <u>then</u> $\delta(G) = \lambda(G)$. This result was extended by Lesniak [13] as follows: <u>If</u> G <u>is a nontrivial graph such that</u> $\deg v + \deg w \geq |V(G)| - 1$ <u>for each pair</u> v, w <u>of nonadjacent vertices of</u> G, <u>then</u> $\delta(G) = \lambda(G)$. Here $\deg u$ denotes the degree of u.

Other partial answers to the aforementioned questions have a close relationship to the following concepts. Let G be a graph, and let $n \geq 1$ be an integer. We say that G is <u>minimally n-connected</u> if $\kappa(G) = n$ and for each $e \in E(G)$, $\kappa(G - e) = n - 1$. We say that G is <u>minimally n-edge-connected</u> if $\lambda(G) = n$ and for each $e \in E(G)$, $\lambda(G - e) = n - 1$. Finally, we say that G is <u>critically n-connected</u> if $\kappa(G) = n$ and for each $v \in V(G)$, $\kappa(G - v) = n - 1$.

Halin [8] proved that

(VM) <u>if</u> G <u>is a minimally n-connected graph</u> $(n \geq 1)$, <u>then</u> $\delta(G) = n$.

(Also see [9].) Mader [18] proved <u>that every minimally n-connected graph</u> $(n \geq 1)$ <u>contains at least</u> $n + 1$ <u>vertices of degree</u> n. (Minimally 2-connected graphs were studied by Dirac [6] and Plummer [24].)

Lick [15] proved (also, see [14]) that

(EM) <u>if</u> G <u>is a minimally n-edge-connected graph</u> $(n \geq 1)$, <u>then</u> $\delta(G) = n$.

Mader [17] proved that <u>every minimally n-edge-connected graph</u> $(n \geq 1)$ <u>contains at least</u> $n + 1$ <u>vertices of degree</u> n. Another extension of (EM) can be found in [21].

Chartrand, Kaugars, and Lick [5] proved (also see [14]) that

(VC) *if* G *is a critically n-connected graph, where* $n \geq 2$, *then* $\delta(G) < (3n-1)/2$ *and this inequality is the best possible.*

This implies that if G is a critically n-connected graph and $n = 2$ or 3, then $\delta(G) = n$. For $n = 2$, this result was proved by Kaugars [12]. In [23] it is proved that *every critically 2-connected graph with more than three vertices contains at least four vertices of degree two.* The complement of the graph which consists of two disjoint copies of the star $K(1, n-1)$, where $n \geq 3$, is an example of a critically n-connected graph which contains precisely two vertices of degree $< (3n-1)/2$.

Mader [16] proved a theorem which has (VM) and the inequalty in (VC) as its corollaries. Certain extensions of (VM) and (VC) were obtained in [22].

Finally, we add that some similar results are known for digraphs. See [2, pp. 31-32], [7], [11], and [19].

2. *Critical vertices*. Let G be a graph with $\kappa(G) = n \geq 1$. If $S \subseteq V(G)$, then we say that S is a *minimum cut-set* of G if $|S| = n$ and $G - S$ is disconnected. We say that a vertex v of G is *critical* if $\kappa(G - v) = n - 1$. We denote by Crit(G) the set of critical vertices of G. Obviously, the following three statements are equivalent: (1) $|V(G)| \geqslant n + 1$; (2) *there exists a minimum cut-set of* G; (3) *for each* $u \in \mathrm{Crit}(G)$, *there exists a minimum cut-set* S *of* G *such that* $u \in S \subseteq \mathrm{Crit}(G)$. It is clear that G is critically n-connected if and only if $\mathrm{Crit}(G) = V(G)$.

If G is a graph, $A \subseteq V(G)$, and each edge of G is incident with a vertex which belongs to A, then we say that A *covers* G.

Let G be a graph with $\kappa(G) = n \geq 2$ and $|V(G)| > n + 1$. In [20] the author proved that if $n = 2$ and Crit(G) covers G, then G contains at least two nonadjacent vertices of degree two. From (VC) it follows that a similar result does not hold for $n \geq 4$. Note that if G is the complement of the graph which consists of two disjoint copies of $K(1,2)$, then $n = 3$, and G contains precisely two vertices of degree three but they are adjacent.

In the next part of this paper it will be demonstrated that if Crit(G) contains a subset fulfilling certain properties, then G contains at least two nonadjacent vertices of degree $\kappa(G)$.

3. κ-Bases. Let G be a connected graph. We shall say that a nonempty subset B of $V(G)$ is a κ-basis of G if for each $u \in B$ there exist $v, w \in V(G) - B$ and a minimum cut-set S of G such that $u \in S \subseteq B$, and v and w belong to distinct components of $G - S$. Obviously, if B is a κ-basis of G, then $B \subseteq \text{Crit}(G)$, $\kappa(G) \leq |B| \leq |V(G)| - 2$, and $G - B$ is disconnected. It is easy to see that $|V(G)| > \kappa(G) + 1$ if and only if there exists a κ-basis of G.

PROPOSITION. *Let G be a graph with $\kappa(G) = 1$ and $|V(G)| \geq 3$. Then Crit(G) is a κ-basis of G.*

Proof. The result follows from the fact that for every cut-vertex v of G, each component of $G - v$ contains a vertex which is not a cut-vertex of G.

If F is an induced subgraph of a graph G, then we denote by $P_G(F)$ the set of vertices $u \in V(G) - V(F)$ adjacent with a vertex of F.

LEMMA. *Let G be a connected graph with $|V(G)| > \kappa(G) + 1$, and*

F_1 <u>and</u> F_2 <u>of</u> $G - B$ <u>such that</u> $|P_G(F_1)| = \kappa(G) = |P_G(F_2)|$.

<u>Proof</u>. Let $G_0 = G - B$. We say that an induced subgraph G' of G fulfils the property $\mathfrak{C}$ if

$$P_G(G') \subseteq B , |P_G(G')| = \kappa(G) , \text{ and } V(G') \cap V(G_0) \neq \emptyset$$
$$\neq V(G_0) - V(G') .$$

Obviously, $B \neq \emptyset$. Let $u \in B$. There exist a minimum cut-set S of G such that $u \in S \subseteq B$ and distinct components of $G - S$, say G_1 and G_2, with the property that
$V(G_1) \cap V(G_0) \neq \emptyset \neq V(G_2) \cap V(G_0)$.

Let $i \in \{1, 2\}$. Clearly, G_i fulfils $\mathfrak{C}$. There exists an induced subgraph F_i of G_i such that F_i fulfils property $\mathfrak{C}$ and any induced subgraph F_i' of F_i with $|V(F_i')| < |V(F_i)|$ does not fulfil $\mathfrak{C}$. Denote $R_i = P_G(F_i)$ and $H_i =$ $G \quad (V(F_i) \cup R_i)$. Obviously, $V(H_i) \cap V(G_0) \neq \emptyset$. It is easy to see that F_i is a component of $G - R_i$.

Assume that $V(G_i) \cap B \neq \emptyset$. Then there exists a minimum cut-set T_i of G such that $V(F_i) \cap T_i \neq \emptyset , T_i \subseteq B$, and at least two components of $G - T_i$ contain a vertex belonging to G_0 . Therefore, $G - T_i$ can be partitioned into vertex-disjoint graphs J_i' and J_i'' such that $V(J_i') \cap V(G_0) \neq \emptyset \neq V(J_i'') \cap V(G_0)$. It is not difficult to see that either

(1) $\quad V(F_i) \cap V(J_i') \cap V(G_0) \neq \emptyset \neq V(H_i) \cap V(J_i'') \cap V(G_0)$

or

(2) $\quad V(F_i) \cap V(J_i'') \cap V(G_0) \neq \emptyset \neq V(H_i) \cap V(J_i') \cap V(G_0)$.

Without loss of generality we consider (1). We denote by the subgraph of G induced by $V(F_i) \cap V(J_i')$. Similarly, we denote by L_i'' the subgraph of G induced by $V(H_i) \cap V(J_i'')$. Obviously, L_i' is an induced subgraph of F_i and $|V(L'_i)| < |V(F_i)|$.

We have

$$P_G(L_i') \subseteq (V(F_i) \cap T_i) \cup (T_i \cap R_i) \cup (R_i \cap V(J_i')) \subseteq B$$

and

$$P_G(L_i'') \subseteq (V(H_i) \cap T_i) \cup (T_i \cap R_i) \cup (R_i \cap V(J_i'')) \subseteq B .$$

Therefore

$$|P_G(L_i')| + |P_G(L_i'')| \leq |R_i| + |T_i| = 2\kappa(G) .$$

It is clear that $|P_G(L_i'')| \geq \kappa(G)$. Hence, $|P_G(L_i')| = \kappa(G)$. We get that L_i' fulfils $\mathfrak{C}$, which is a contradiction.

This means that $V(F_i) \cap B = \emptyset$. Then F_i is an induced subgraph of G_0. Since F_i is connected and $P_G(F_i) \subseteq B$, we have that F_i is a component of G_o, which completes the proof.

This leads us to the main result of this paper:

THEOREM. <u>Let</u> G <u>be a connected graph</u>. <u>If</u> G <u>has a κ-basis which covers</u> G, <u>Then</u> G <u>contains at least two nonadjacent vertices of degree</u> $\kappa(G)$.

COROLLARY. <u>Let</u> G <u>be a graph with</u> $\kappa(G) = 1$ <u>and</u> $|V(G)| \geq 3$. <u>If</u> Crit(G) <u>covers</u> G, <u>then</u> G <u>contains at least two nonadjacent vertices of degree one</u>.

<u>Remark</u>. Let m and n be integers such that $m \geq 2$ and $n \geq 1$. If $n = 1$, then we denote by G_{mn} the path P_{m+1}. If $n > 1$, then we denote by G_{mn} the join (see [10], p. 21) of P_{m+1} and K_{n-1}. We can see that (1) $|V(G_{mn})| = m + n$, (2) $\kappa(G_{mn}) = n$, (3) Crit(G) is a κ-basis of G, (4) Crit(G) covers G, and (5) G_{mn} contains precisely two vertices of degree n.

REFERENCES

1. M. Behzad and G. Chartrand, Introduction to the Theory of Graphs, Allyn and Bacon, Boston 1971.

2. C. Berge, Graphs and Hypergraphs, North-Holland, Amsterdam and London, American Elsevier, New York 1973.

3. G. Chartrand, A graph-theoretic approach to a communications problem, SIAM J. Appl. Math. 14 (1966), 778-781.

4. G. Chartrand and F. Harary, Graphs with prescribed connectivities, Theory of Graphs (P. Erdös and G. Katona, eds.), Akadémiai Kiadó, Budapest 1968 , 61-63.

5. G. Chartrand, A. Kaugars, and D. R. Lick, Critically n-connected graphs, Proc. Amer. Math. Soc., 32(1972), 63-68.

6. G. A. Dirac, Minimally 2-connected graphs, J. Reine Angew. Math., 228 (1967), 204-216.

7. D. P. Geller, Minimally strong digraphs, Proc. Edinburgh Math. Soc., 17 (Series II) (1970), 15-22.

8. R. Halin, A theorem on n-connected graphs, J. Combinatorial Theory, 7 (1969), 150-154.

9. R. Halin, Studies on minimally n-connected graphs, Combinatorial Mathematics and Its Applications (D. J. A. Welsh, ed.), Academic Press, London and New York 1971 , 129-136.

10. F. Harary, Graph Theory, Addison-Wesley, Reading, Massachusetts 1969.

11. T. Kameda, Note on Halin's theorem on minimally connected graphs, J. Combinatorial Theory,17 (Series B) (1974), 1-4.

12. A. Kaugars, A Theorem on the Removal of Vertices from Blocks, Senior Thesis, Kalamazoo College, Kalamazoo (Michigan), 1968.

13. L. Lesniak, Results on the edge-connectivity of graphs, Discrete Mathematics,8 (1974), 351-354.

14. D. R. Lick, Critically and minimally n-connected graphs, The Many Facets of Graph Theory (G. Chartrand and S. F. Kapoor, eds.), Lecture Notes in Mathematics 110, Springer-Verlag, Berlin, Heidelberg, New York 1969, 199-205.

15. D. R. Lick, Minimally n-line connected graphs, J. Reine Angew. Math., 252(1972), 178-182.

16. W. Mader, Eine Eigenschaft der Atome endlicher Graphen, Archiv der Mathematik, 22 (1971), 333-336.

17. W. Mader, Minimale n-fach kantenzusammenhängende Graphen, Math. Ann., 191(1971), 21-28.

18. W. Mader, Ecken vam Grad n in minimalen n-fach zusammenhängen den Graphen, Archiv Math., 23(1972), 219-224.

19. W. Mader, Ecken vom Innen- und Außengrad n in minimal n-fach kantenzusammenhängenden Digraphen, Archiv Math., 25(1974), 107-112.

20. L. Nebeský, A theorem on 2-connected graphs, Casopis pest. mat., 100(1975), 116-117.

21. L. Nebeský, On the minimum degree and edge-connectivity of a graph, to appear.

22. L. Nebeský, An upper bound for the minimum degree of a graph, submitted for publication.

23. L. Nebeský, On induced subgraphs of a block, submitted for publication.

24. M. D. Plummer, On minimal blocks, Trans. Amer. Math. Soc. 134 (1968), 85-94.

GENERALIZATIONS OF GRAPHICAL PARAMETERS

E.A. Nordhaus
Michigan State University
East Lansing, Michigan 48824

ABSTRACT

Graph theoretic parameters are often defined for a finite graph G with the aid of certain subsets of the vertex set $V(G)$ or of the edge set $E(G)$. One usually restricts attention to those subsets having maximum or minimum cardinality. Examples include the parameters associated with covering sets, matching sets, and separating sets. However, it may be desirable to consider a wider range of values, and we indicate how this can be accomplished for the vertex and edge versions of the parameters just mentioned.

Another generalization employs certain subsets of the union of $V(G)$ and $E(G)$, leading to such parameters as the total matching numbers and total connectivity numbers. The exact values of these parameters can be found for certain special classes of graphs and used to establish sharp upper and lower bounds for the corresponding parameters.

As a final example, one may generalize the concept of the genus of a graph by considering all compact orientable 2-manifolds in which a connected graph G has a 2-cell imbedding, and not only those for which the genus is a minimum.

1. INTRODUCTION

Let G be a connected undirected graph with no loops or multiple edges having vertex set $V(G)$ and edge set $E(G)$, with $|V(G)| = p \geq 2$ and $|E(G)| = q \geq 1$. The $p + q$ elements of the set $S = V(G) \cup E(G)$ are called elements of G, and suitable subsets of S are used to define various parameters for graphs.

The symbols $[x]$ and $\{x\}$ denote respectively the greatest integer and the least integer functions.

Our discussion indicates ways in which some of the classical parameters of graph theory can be generalized. In particular, we present a brief account of recent developments along these lines, focussing mainly on those parameters associated with covering sets, matching sets and separating sets. For additional details, see [1], [2], [6], and [7].

2. TOTAL COVERING SETS AND TOTAL MATCHINGS

If a vertex v of G is an end vertex of an edge e of G, the elements v and e are said to cover each other. In order to cover all elements of G by a subset of $S = V(G) \cup E(G)$, we use a more general definition of cover. Let a vertex v also cover itself and all adjacent vertices, and similarly let an edge e cover itself and all adjacent edges.

A subset C of S whose elements cover all the elements of G, and furthermore is minimal, is called a total cover for G. In the set of all total covers C of G, we define the total covering parameters $\alpha_2(G) = \inf |C|$ and $\alpha_2'(G) = \sup |C|$.

Two elements of G are called independent if neither one covers the other. A subset M of S is called a total matching if the elements of M are pairwise independent and M is maximal. In the set of all total matchings M of G, we define total matching (independence) numbers $\beta_2'(G) = \inf|M|$ and $\beta_2(G) = \sup|M|$.

If M is any total matching for a connected graph G, the maximal property implies that every element of G not in M is covered by some element of M. Since every element of M covers itself and is covered by no other element of M, then M is a total cover for G. The inequalities $\alpha_2(G) \leq \beta_2'(G) \leq \beta_2(G) \leq \alpha_2'(G)$ must then hold. By the examination of special classes of graphs, one may readily develop upper and lower bounds for each of these four parameters. The total covering parameters satisfy the inequalities $1 \leq \alpha_2(G) \leq [p/2] \leq \alpha_2'(G) \leq p - 1$, and the independence numbers satisfy the relations $1 \leq \beta_2'(G) \leq p - 1$ and $[p/2] \leq \beta_2(G) \leq p - 1$. The bounds given are sharp, with the possible exception of the upper bound for $\beta_2'(G)$.

We indicate briefly how the non-obvious inequalities can be obtained. The inequality $\alpha_2(G) \leq [p/2]$ is proved by induction on the set of connected graphs of order p, and $\alpha_2'(G) \leq p - 1$ is found by supposing that a total cover C for G has maximum cardinality at least p, and showing that a contradiction arises.

To prove that $\beta_2(G) \geq [p/2]$, one considers a total cover C with no vertices and having a minimum number of edges. One can show that no total cover C has fewer independent elements.

Vertex versions of the covering and independence numbers are obtained by restricting the total covers and the total matchings to subsets of $V(G)$. Then $\alpha_0(G) = \inf|C_0|$ and $\alpha_0'(G) = \sup|C_0|$, where C_0 is any total cover such that $C_0 \subset V(G)$. Similarly $\beta_0(G) = \sup|M_0|$ and $\beta_0'(G) = \inf|M_0|$, where M_0 is any total matching such that $M_0 \subset V(G)$. Analogous definitions hold for the edge versions of the covering and independence numbers, giving $\alpha_1(G) = \inf|C_1|$ and $\alpha_1'(G) = \sup|C_1|$, where C_1 is any total covers such that $C_1 \subset E(G)$. Similarly $\beta_1(G) = \sup|M_1|$ and

$\beta_1'(G) = \inf |M_1|$, where M_1 is any total matching for which $M_1 \subset E(G)$.

It is immediate that $\alpha_2(G) \leq \alpha_i(G) \leq \alpha_i'(G) \leq \alpha_2'(G)$ holds for $i = 0$ or 1, that $\beta_2(G) \geq \max(\beta_0(G), \beta_1(G))$ and that $\beta_2'(G) \geq \max(\beta_0'(G), \beta_1'(G))$.

A well known result due to Gallai [4] is that for a graph G of order p having no isolated vertices, $\alpha_0(G) + \beta_0(G) = p$ and $\alpha_1(G) + \beta_1(G) = p$. An extension of these results was made by Meng [5], who proved that $\alpha_0'(G) + \beta_0'(G) = p$ also holds. However, the sum $\alpha_2(G) + \beta_2(G)$ can exceed p, as shown by the complete graph K_p with p odd. In this case $\alpha_2(K_p) = \beta_2(K_p) = \{p/2\}$. One can show the general result that $p \leq \alpha_2(G) + \beta_2(G) \leq \{3p/2\} - 1$, where the upper bound may not be sharp.

3. TOTAL SEPARATING SETS AND TOTAL CONNECTIVITY

A connected graph G can be disconnected or reduced to a single vertex by removing an appropriate subset of the vertex set $V(G)$, and can always be disconnected by removing a suitable subset of the edge set $E(G)$. We restrict attention to separating sets which are minimal. An obvious generalization is to consider minimal subsets of $V(G) \cup E(G)$ whose removal from G yields a disconnected graph or reduces G to a single vertex. These are called total separating sets, and the minimum and the maximum cardinalities of such sets are denoted by $\kappa_2(G)$ and $\kappa_2'(G)$ respectively. A total separating set whose elements are vertices of G is called a minimal vertex separating set, and the minimum and maximum cardinalities of such sets are denoted by $\kappa_0(G)$ and $\kappa_0'(G)$. Similar remarks apply to minimal edge separating sets, and corresponding edge connectivity parameters $\kappa_1(G)$ and $\kappa_1'(G)$ arise.

From these definitions and a well know result of Whitney [9], one obtains the inequalities

$$1 \leq \varkappa_2(G) \leq \varkappa_0(G) \leq \varkappa_1(G) \leq \delta(G) \leq 2q/p \leq \Delta(G) \leq p-1,$$

where $\delta(G)$ and $\Delta(G)$ denote respectively the minimum and the maximum degrees of a connected graph G of order p having q edges. One can readily prove that $\varkappa_2(G) = \varkappa_0(G)$ and that $\varkappa_2'(G) = \max(\varkappa_0'(G), \varkappa_1'(G))$. The total connectivity parameters thus depend only on the parameters for vertex and edge connectivity. Nevertheless, one may investigate sharp upper bounds for $\varkappa_0'(G)$ and for $\varkappa_1'(G)$. These values measure in a sense the most inefficient way to disconnect a graph. No simple relation holds between $\varkappa_0'(G)$ and $\varkappa_1'(G)$, contrary to the situation for $\varkappa_0(G)$ and $\varkappa_1(G)$, for which $\varkappa_0(G) \leq \varkappa_1(G)$ always holds.

Using the fact that a minimal edge separating set separates G into two components, one can prove that $\varkappa_1'(G) \leq \min([p^2/4], q-p+2)$. Since $\varkappa_1'(K_p) = [p^2/4]$ and $\varkappa_1'(K_{m,n}) = 1 + (m-1)(n-1) = q - p + 2$, these upper bounds for $\varkappa_1'(G)$ are sharp.

The complete graph K_p cannot be disconnected by removing vertices, but can be reduced to a single vertex, so $\varkappa_0'(K_p) = p - 1$. In general one has $\varkappa_0'(G) \leq p - 1$, with equality if and only if G has at least one vertex of degree $p - 1$. If G is a connected graph of order $p \geq 4$ with q edges and maximum degree $\Delta(G) < p - 1$, then $\varkappa_0'(G) \leq \varkappa_1'(G) \leq q - p + 2$. The proof is similar to that made for $\varkappa_1'(G)$.

4. GENUS AND MAXIMUM GENUS

The genus $\gamma(G)$ of a connected graph G is defined as the minimum genus of a compact orientable 2-manifold in which g can be embedded. By adding the restriction that the faces are topologically homeomorphic to open 2-cells, one obtains a range of values for the genus of a graph, and can investigate the maximum genus of a graph. This is done by employing an algorithm due to Edmonds [3]. For the complete graph K_p one had the following sharp bounds on the genus:

$$\left\{\frac{(p-3)(p-4)}{12}\right\} \leq \gamma(K_p) \leq \left[\frac{(p-1)(p-2)}{4}\right] .$$

The lower bound has a long history, see Ringel and Youngs [8]. All values between the minimum and maximum values of $\gamma(K_p)$ are attainable.

REFERENCES

1. Alavi, Y., Behzad, M., Lesniak-Foster, L. M., and Nordhaus, E.A., "Total Matchings and Total Coverings of Graphs," Journal of Graph Theory, vol. 1 (1977), 135-140.

2. Alavi, Y., Behzad, M., and Nordhaus, E. A., "Minimal Separating Sets of Maximum Size," (in preparation).

3. Edmonds, J.R., "A Combinatorial Representation for Polyhedral Surfaces," Notices Amer. Math. Soc. 7 (1960), 646.

4. Gallai, T., "Über extreme Punct und Kantenmengen," Ann. Univ. Sci. Budapest, Eötvös Sect. Math. 2(1959), 133-138.

5. Meng, D., Matchings and Coverings for Graphs, Ph.D. thesis, Michigan State University, (1974).

6. Nordhaus, E. A., Stewart, B. M., and White, A. T., "On the Maximum Genus of a Graph," J. Combinatorial Theory, B 11, (1971), 258-267.

7. Nordhaus, E. A., Ringeisen, R. D., Stewart, B. M., and White, A. T., "A Kuratowski-Type Theorem for the Maximum Genus of a Graph", J. Combinatorial Theory, B 12 (1972), 250-267.

8. Ringel, G., and Youngs, J.W.T., "Solution of the Heawood Map-Coloring Problem," Proc. Nat. Acad. Sci. 60 (1968), 438-445.

9. Whitney, H., "Congruent Graphs and the Connectivity of Graphs," Amer. J. Math. 54 (1932), 150-168.

PURSUIT-EVASION IN A GRAPH

T.D. Parsons
The Pennsylvania State University

1. INTRODUCTION

Suppose a man is lost and wandering unpredictably in a dark cave. A party of searchers who know the structure of the cave is to be sent to find him. What is the minimum number of searchers needed to find the lost man regardless of how he behaves?

This question was raised by my spelunker friend Richard Breisch, who developed informal arguments for many plausible conjectures about the problem. There are many inequivalent mathematical formulations of this problem, depending on the nature of the cave and the possible behavior allowed the searchers and the lost man. Breisch did not make precise which formulation he intended, although he gave numerous examples. One example was that of a circular cave, which requires two searchers: the lost man could move so as to be always antipodal to a single searcher; however, two searchers could start from the same point and travel at constant speed in different directions around the circle, and by the time they met again they would have found the lost man with absolute certainty.

We shall assume that the cave can be regarded as a finite connected graph in which the searchers and the lost man must move continuously. The searchers must proceed according to a predetermined plan which will capture the lost man even if he were an arbitrarily fast, invisible evader who, clairvoyant, knows the searcher's every move.

2. THE SEARCH NUMBER OF A CONNECTED GRAPH

Let G be a finite connected graph without loops or multiple edges. We may assume that G is embedded in R^3 so that its vertices $v_1, v_2, \ldots, v_n$ are represented by distinct points, and its edges $\{v_i, v_j\}$ are represented by closed line segments $[v_i, v_j]$ in R^3 which intersect only at vertices of G. Regarded as a subset of R^3, G is a topological space with the relative topology. Then G is a compact locally connected metric space in which every connected set is arcwise connected.

A family $F = \{f_i \mid 1 \leq i \leq k\}$ of continuous functions $f_i : [0,\infty) \to G$ is a <u>search plan</u> for G if, for every continuous function $e : [0,\infty) \to G$ there exists $t_0 \in [0,\infty)$ and an $i \in \{1,2,\ldots,k\}$ such that $e(t_0) = f_i(t_0)$. Here we think of $e(t)$ and $f_i(t)$ as the positions at time t of an evader and the ith searcher. Then a search plan must catch any possible evader in a finite time.

The <u>search number</u> s(G) of G is the minimum cardinality of all search plans for G. We have $s(G) \leq n+1$ since we may let a searcher "sit" at each of the n vertices of G while an additional searcher follows any path which traverses every edge of G in a finite total time. Indeed, $s(G) \leq n$ follows by a slight modification of this argument; and if n' is the number of vertices of G of valence greater than 2, we have similarly $s(G) \leq n'+1$.

Our problem is to characterize the topological invariant s(G) in terms of the structure of G, and to give an algorithm which computes s(G) for an arbitrary finite connected graph G. This is a difficult problem. In general it is easy to get upper bounds for s(G), but lower bounds are much harder. We shall solve the problem completely for the case of trees, and give some indication of the results for arbitrary G.

To gain insight into the nature of s(G), we first consider two examples.

<u>Example 1</u>.

= G.

Here we have that G is a "chain" formed of finitely many "links"  each of which consists of finitely many (at least 3) paths of length 2 joining two vertices. If we start with 3 searchers at the leftmost vertex, have one walk to the next vertex of valence greater than 2, and have the third man go back and forth across all paths of length 2 joining the two other (temporarily fixed) searchers, we shall "eliminate" the first link. It is easily seen that we may thus successively eliminate each link until G has been effectively searched. Thus $s(G) \leqq 3$. In fact, $s(G) = 3$, but a rigorous proof requires some care.

Example 2.

G = 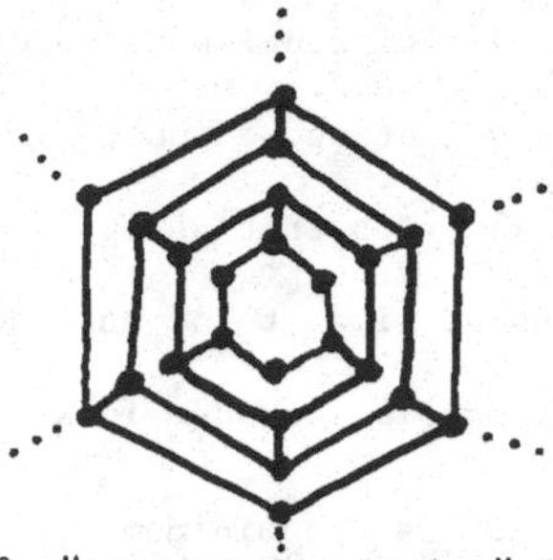

Here G is formed of $r \geqq 3$ "concentric circles" linked alternately by edges in the pattern shown. It can be shown by an argument similar to that used for Example 1 that $s(G) \leqq 5$, no matter how large the integer r is chosen. In fact, $s(G) = 5$ for all $r \geqq 3$, but this is rather difficult to prove even for the case $r = 3$.

These examples show that $s(G)$ is related to the ability of searchers to spread out through G, successively eliminating certain subgraphs in such a way that they become "searched out" and access to them is henceforth denied any possible evader still at large.

Examples such as these provide counterexamples to many "reasonable" conjectures attempting to characterize $s(G)$ in terms of other invariants of G.

3. THE SEARCH NUMBERS OF TREES

Our main result will be to characterize $s(T)$ for all trees T. In this section we just state our results. Rigorous proofs require considerable argument, outlined in the next section.

If v is a vertex of tree T, a branch of T at v is a maximal subtree T′ of T subject to the condition that v be of valence 1 in T′.

Example

T 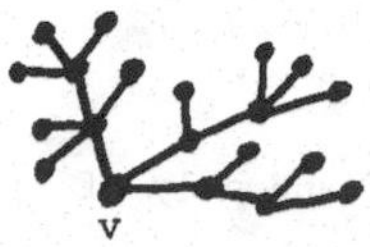 v branches of T at vertex v: 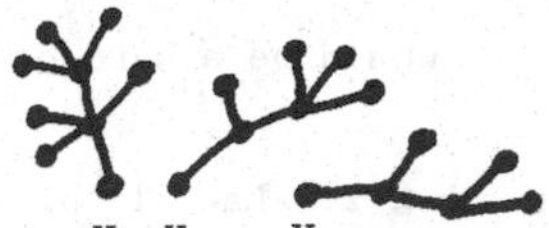 v v v

<u>Theorem</u>: Let $k \geqq 1$, and T be a tree. Then $s(T) \geqq k+1$ if and only if T has a vertex v at which there are at least three branches T_1, T_2, T_3 satisfying $s(T_j) \geqq k$ for $j = 1,2,3$.

This theorem can be used to characterize all trees with a given search number. Let us define recursively sets $\mathcal{T}_1, \mathcal{T}_2, \ldots$ of trees as follows:

$\mathcal{T}_1 =$ {}, and if $\mathcal{T}_k$ $(k \geqq 1)$ has been defined, let $\mathcal{T}_{k+1}$ contain one representative from each isomorphism class of the trees T resulting from the construction

(i) choose (with repetition of isomorphism types allowed) disjoint copies T_1, T_2, T_3 of trees in $\mathcal{T}_k$,

(ii) choose vertices $v_j \in T_j$ $(j = 1,2,3)$ such that for each j either v_j has valence 1 in T_j or else v_j is neither of valence 1 nor adjacent to a vertex of valence 1 in T_j,

(iii) let w be a new vertex not in T_1, T_2, T_3

(iv) form a tree T by identifying w with v_j if v_j has valence 1 in T_j, or by connecting w to v_j by an edge $[w, v_j]$ if v_j is not of valence 1 in T_j, (for $j = 1,2,3$). The vertex w is called the <u>root</u> of T.

It is clear from the above construction that $\mathcal{T}_1, \mathcal{T}_2, \ldots$ are all finite sets of trees having no vertices of valence 2, and whose roots have valence 3.

The first four sets are: $\mathcal{T}_1 =$ , and

$\mathcal{T}_4 =$ 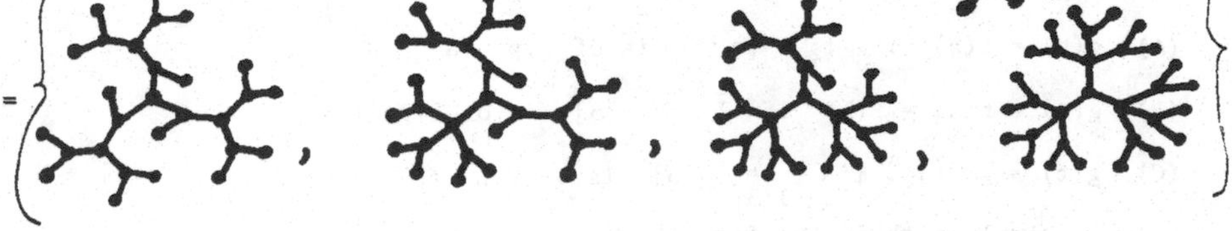

Thus $|T_k| = 1$ for $k = 1,2,3$ but $|T_4| = 4$. Determination of $|T_k|$ as a function of k would be a hard combinatorial problem.

Theorem: Let $k \geq 2$. Let T be a tree. Then $s(T) = k$ if and only if T has a subtree homeomorphic to a tree in T_k, but T has no subtree homeomorphic to a tree in T_{k+1}.

[Trivially, $s(T) = 1$ if and only if T is homeomorphic to a point or a single edge.]

4. SOME LEMMAS, THEOREMS, AND PROOFS

Were we to give full details of all proofs, this paper would become prohibitively long and tedious. Thus we will sketch our development usually with just enough details for a persistent reader to complete the proofs himself.

Lemma 1: Let H be a connected subgraph of G. Let $f : [0,\infty) \to G$ be continuous Then there exists a continuous $g : [0,\infty) \to H$ such that $g(t) = f(t)$ whenever $f(t) \in H$.

Proof: Let $S = \{t \in [0,\infty) \mid f(t) \notin H\} \neq \emptyset$. Since H is a closed subset of G, S is an open subset of $[0,\infty)$, thus S may be written as a finite or countably infinite union $S = \cup I_j$ of open disjoint intervals of $[0,\infty)$. Each I_j has its closure $\overline{I}_j$ in $[0,\infty)$ of one of the forms

(A) $\overline{I}_j = [a,\infty)$ where $f(a) \in V(H)$, the vertex set of H.

(B) $\overline{I}_j = [a,b]$ where $f(a) = f(b) \in V(H)$.

(C) $\overline{I}_j = [a,b]$ where $f(a) \neq f(b)$ and $f(a),f(b) \in V(H)$.

Define $g : [0,\infty) \to H$ by letting $g(t) = f(t)$ whenever $t \notin S$, and if $t \in I_j \subseteq S$ let

(A) $g(t) = f(a)$ if $\overline{I}_j = [a,\infty)$ is of type (A)

(B) $g(t) = f(a) = f(b)$ if $\overline{I}_j = [a,b]$ is of type (B)

(C) $g(t) = g_j(t)$ if $\overline{I}_j = [a,b]$ is of type (C), where $g_j : [a,b] \to H$ is a continuous simple path in H from f(a) to f(b).

Let $t_0 \in [0,\infty)$. Clearly g is continuous at t_0 if t_0 is not a boundary point of S. Let t_0 be a boundary point of S. Then $f(t_0) \in V(H)$, and t_0 has a neighborhood $\mathcal{U}$ in $[0,\infty)$ such that $t \in \mathcal{U}$ and $f(t) \in V(H)$ imply $f(t) = f(t_0)$. Thus at most two intervals I_j of type (C) intersect $\mathcal{U}$. It is now easy to show g is continuous at t_0, and the lemma follows.

Remark 1: If $f : [a,b] \to G$ is continuous and $0 \leqq a < b$, we may extend f by $f(t) = f(a)$ for $0 \leqq t \leqq a$ and $f(t) = f(b)$ for $b \leqq t < \infty$. Applying Lemma 1 to the extension of f gives a continuous g whose restriction to $[a,b]$ obeys $g : [a,b] \to H$ and $g(t) = f(t)$ whenever $f(t) \in H$.

Theorem 1: If H is a connected subgraph of G, then $s(H) \leqq s(G)$.

Proof: Let $\{f_i \mid 1 \leqq i \leqq k\}$ be a search plan for G. For $1 \leqq i \leqq k$, let $g_i : [0,\infty) \to H$ be continuous and obey $g_i(t) = f_i(t)$ whenever $f_i(t) \in H$. Then it is easy to see $\{g_i \mid 1 \leqq i \leqq k\}$ is a search plan for H, concluding the proof.

Let $M = [0,\infty) \times G$. If $S \subseteq [0,\infty)$ and $f : S \to G$, then $\text{graph}(f) = \{(t,f(t)) \mid t \in S\} \subseteq M$. Let $\mathcal{F} = \{f_i \mid 1 \leqq i \leqq k\}$ be a search plan for G. Let $F = \bigcup_{i=1}^{k} \text{graph}(f_i)$. Then F is a closed subset of M.

If (t,x) and $(t',x') \in M$, we write $(t,x) < (t',x') \bmod \mathcal{F}$ if $t < t'$ and there exists a continuous function $h : [t,t'] \to G$ such that $h(t) = x$, $h(t') = x'$, and $\text{graph}(h) \cap F = \emptyset$. We can interpret "$(t,x) < (t',x') \bmod \mathcal{F}$" as meaning than an evader at x at time t can move so as to reach x' at time t' without being caught from t to t'. The relation "$< \bmod \mathcal{F}$" is transitive.

For $t \in [0,\infty)$ let $G(t,\mathcal{F}) = G\setminus\{f_1(t),\ldots,f_k(t)\}$. Then $G(t,\mathcal{F})$ has finitely many connected components each of which is homeomorphic to some finite connected graph minus certain of its vertices of valence 1. If $x \in G(t,\mathcal{F})$, let $C(x,t,\mathcal{F})$ be the component of $G(t,\mathcal{F})$ which contains x.

Let $A(0,\mathcal{F}) = G(0,\mathcal{F})$, and for $0 < t < \infty$ let $A(t,\mathcal{F}) = \{x \in G \mid$ there exists $y \in G$ such that $(0,y) < (t,x) \bmod \mathcal{F}\}$. Then $A(t,\mathcal{F})$ is the set of points of G at which an uncaught evader may be located at time t, that is, the points "available" to an evader at time t. Let $U(t,F) = G\backslash A(t,F)$, the set of points "unavailable" to an evader at time t. When the search plan F is clear from context, we will abbreviate $G(t,F)$ as $G(t)$, and so forth. If $S \subseteq G$, then ∂S denotes the set of boundary points of S regarded as a subset of the topological space G.

The following statements are not difficult to prove:

(1) Suppose I is an interval of $[0,\infty)$, C is a connected subset of G, and $(I \times C) \cap F = \emptyset$. If (t_1,x_1) and $(t_2,x_2) \in I \times C$ and $t_1 < t_2$, then $(t_1,x_1) < (t_2,x_2) \bmod F$.

(2) Suppose $(t_1,x_1) < (t_2,x_2) \bmod F$, and that C_1,C_2 are compact connected subsets of G such that for $i = 1,2$, $x_i \in \text{interior}(C_i)$ and $C_i \subseteq C(x_i,t_i,F)$. Then there exist open intervals I_1, I_2 of $[0,\infty)$ such that $t_1 \in I_1$, $t_2 \in I_2$ and such that $(t',y_1) \in I_1 \times C_1$ and $(t'',y_2) \in I_2 \times C_2$ imply $(t',y_1) < (t'',y_2) \bmod F$.

(3) If $(t_1,x_1) < (t_2,x_2) \bmod F$ and for $i = 1,2$, $x_i' \in C(x_i,t_i,F)$, then $(t_1,x_1') < (t_2,x_2') \bmod F$.

(4) For all $t \in [0,\infty)$, $\partial A(t,F) = \partial U(t,F) \subseteq \{f_1(t),\ldots,f_k(t)\}$.

(5) If $x \in A(t,F)$, then there exist neighborhoods $N(x)$ of x in G and $N(t)$ of t in $[0,\infty)$ such that $(t',y) \in N(t) \times N(x)$ implies $y \in A(t',F)$.

<u>Lemma 2 (Compactness lemma)</u>: Let $F = \{f_i \mid 1 \leq i \leq k\}$ be a search plan for G. For each continuous function $e : [0,\infty) \to G$ let $t(e) = \min\{t \in [0,\infty) \mid$ for some i, $f_i(t) = e(t)\}$. Then $\sup_e t(e) < \infty$.

<u>Proof</u>: If $e : [0,\infty) \to G$ is continuous, then $\text{graph}(e) \cap F \neq \emptyset$ so the set $\{t \mid (t,e(t)) \in F\}$ is nonempty and has an infimum $t^* \geq 0$. Since F is closed, $(t^*,e(t^*)) \in F$. Thus $t(e) = t^*$ is well-defined.

Suppose $\sup_e t(e) = +\infty$. Let $S = \{(t,x) \in M \mid$ for every $t' > t$ there exists $x' \in G$ such that $(t,x) < (t',x') \bmod F\}$.

For every positive integer m there exists a continuous $e_m : [0,\infty) \to G$ with $t(e_m) > m$. Let $y_m = e_m(0)$ and $z_m = e_m(m)$. Then $(0,y_m) < (m,z_m) \bmod F$. Since every $y_m \in G(0,F)$, which has only finitely many components, we may let $C(0,F)$ be any component of $G(0,F)$ such that $\{m \mid y_m \in C(0,F)\}$ is infinite. Let $x_0 \in C(0,F)$. From (3) we get $(0,x_0) < (m,z_m) \bmod F$ for all m. This easily implies $(0,x_0) \in S$.

A similar argument, using (3) and the fact that $G(t',F)$ has finitely many components, gives

(6) If $(t,x) \in S$ and $t' > t$, then there exists $x' \in G$ such that $(t',x') \in S$ and $(t,x) < (t',x') \bmod F$.

Now $(0,x_0) \in S$. By (6) and induction, for $m = 0,1,2,\ldots$ there exists $(m,x_m) \in S$ such that $(m,x_m) < (m+1,x_{m+1}) \bmod F$ for all m. Let $e_m^* : [m,m+1] \to G$ be continuous such that $e_m^*(m) = x_m$, $e_m^*(m+1) = x_{m+1}$, and $\text{graph}(e_m^*) \cap F = \emptyset$. Define $e^* : [0,\infty) \to G$ be $e^*(t) = e_m^*(t)$ for $m \leq t \leq m+1$, $m = 0,1,2,\ldots$. Obviously e^* is continuous and $\text{graph}(e^*) \cap F = \emptyset$. This contradicts the hypothesis that F is a search plan for G.

It follows that $\sup_e t(e) < \infty$.

Remark 2: The finite number $\tau(F) = \sup_e t(e)$ is the capture time of F. We have that for every continuous $e : [0,\infty) \to G$ there is an $i \in \{1,2,\ldots,k\}$ and a t_0, $0 \leq t_0 \leq \tau(F)$ such that $e(t_0) = f_i(t_0)$. Thus the defining property "for every evader e there is a time $t(e)$ by which F catches e" has been strengthened to "there exists a time $\tau(F)$ by which F catches every evader e". The significance of $\tau(F) < \infty$ is that it allows us to replace the noncompact set $[0,\infty)$ by the compact set $[0,\tau(F)]$ in proofs. [This is its only significance; its size is irrelevant since letting $f_i^*(t) = f_i(\lambda t)$, where $\lambda > 0$ is a constant, gives another search plan F^* with $\tau(F^*) = \lambda^{-1}\tau(F)$.]

As an application of Lemma 2, we prove that if T is a tree with a vertex v such that every branch of T at v has search number $\leq k$, then $s(T) \leq k+1$. First, it is easy to show that for any G

(7) If $F = \{f_i \mid 1 \leq i \leq k\}$ is a search plan for G and $g_i : [0,\lambda] \to G$ are continuous such that $g_i(\lambda) = f_i(0)$ for $1 \leq i \leq k$ (where $\lambda \geq 0$ is constant), then if $f_i^* : [0,\infty) \to G$ are defined by $f_i^*(t) = g_i(t)$ for $0 \leq t \leq \lambda$ and $f_i^*(t) = f_i(t-\lambda)$ for $\lambda \leq t < \infty$, we have $\{f_i^* \mid 1 \leq i \leq k\}$ is also a search plan for G.

Now suppose $T_1, T_2, \ldots, T_r$ are the branches of T at v. By hypothesis, $s(T_j) \leq k$ for $j = 1,2,\ldots,r$. Let $f_{k+1}(t) = v$ for all $t \in [0,\infty)$. Using Lemma 2 and (7), we may easily define continuous $f_1,\ldots,f_k : [0,\infty) \to G$ such that these "searchers" first rid T_1 of any possible evader then proceed to v during $0 \leq t \leq \tau_1$, then they rid T_2 of any evader and return to v during $\tau_1 \leq t \leq \tau_2$, and so forth. Then $\{f_i \mid 1 \leq i \leq k+1\}$ is a search plan for T.

<u>Lemma 3</u>: Let H be a connected subgraph of G. Suppose $f_1,\ldots,f_k : [0,\infty) \to G$ are continuous, and for each $t \in [0,\infty)$, $|\{i \mid f_i(t) \in H\}| < s(H)$. Then there exists a continuous $e : [0,\infty) \to H$ such that for all $i \in \{1,2,\ldots,k\}$ and all $t \in [0,\infty)$, $e(t) \neq f_i(t)$.

<u>Proof</u>: Suppose that the lemma fails. Then for each continuous $e : [0,\infty) \to H$ we may define $t(e) = \min\{t \in [0,\infty) \mid \text{for some } i,\ e(t) = f_i(t)\}$. Now $F_1 = \bigcup_{i=1}^{k}$ graph (f_i) is a closed subset of M, so $F = F_1 \cap ([0,\infty) \times H)$ is a closed subset of $M(H) = [0,\infty) \times H$, and F has the properties that $H(t,F) = \{x \in H \mid (t,x) \notin F\} = H \backslash \{f_1(t),\ldots,f_k(t)\}$ has finitely many components for each $t \in [0,\infty)$, and $t(e) = \min\{t \mid (t,e(t)) \in F\}$. It is now easy to modify the proof of Lemma 2 to show that there exists $\tau \in [0,\infty)$ such that $\sup t(e) = \tau$. Let $h = s(H)$.

If $g_1,\ldots,g_{h-1}$ are continuous functions mapping a subset of $[0,\infty)$ into H, and if $(t,x) \in M(H)$, then (*) is the condition: $|\{i \mid f_i(t) = x\}| \leq |\{j \mid g_j(t) = x\}|$. Let $t_0 = \sup\{t' \in [0,\infty) \mid$ there exist continuous $g_1,\ldots,g_{h-1} : [0,t'] \to H$ such that (*) holds for all $(t,x) \in [0,t'] \times H\}$. It is easy to show that $t_0 > 0$. Suppose $t_0 < \infty$. Let $I = \{i \mid f_i(t_0) \notin H\}$. There exists $\delta > 0$ such that $\delta < t_0$ and $t \in [t_0-\delta, t_0+\delta]$ implies $f_i(t) \notin H$ for all $i \in I$. There exist continuous $g_1,\ldots,g_{h-1} : [0,t_0-\delta] \to H$ such that (*) holds for all $(t,x) \in [0,t_0-\delta] \times H$. It is straightforward, but rather tedious, to show how to extend $g_1,\ldots,g_{h-1}$ continuously so as to map $[0,t_0+\delta] \to H$ and to obey (*) for all $(t,x) \in [0,t_0+\delta] \times H$. This contradicts the definition of t_0, so we conclude $t_0 = +\infty$.

We conclude that there exist continuous $g_1,\ldots,g_{h-1} : [0,\tau] \to H$ satisfying (*) on $[0,\tau] \times H$. Letting $g_i(t) = g_i(\tau)$ for $\tau \leq t < \infty$ and $i = 1,2,\ldots,h-1$, we extend the $g_1,\ldots,g_{h-1} : [0,\infty) \to H$. Since $s(H) = h$, $\{g_1,\ldots,g_{h-1}\}$ is not a search plan for H, so there exists a continuous function $e : [0,\infty) \to H$ such that for all $t \in [0,\infty)$ and all $i = 1,2,\ldots,h-1$, $e(t) \neq g_i(t)$. By definition of τ, there exists $r \in \{1,2,\ldots,k\}$ and $t^* \leq \tau$ such that $f_r(t^*) = e(t^*)$. But then $e(t^*) = x \in H$ so $0 < |\{i \mid f_i(t^*) = x\}| \leq |\{j \mid g_j(t^*) = x\}|$ by (*), so there exists $j \in \{1,2,\ldots,h-1\}$ with $g_j(t^*) = e(t^*)$, a contradiction.

We conclude that the lemma holds.

Remark 3: The interval $[0,\infty)$ may be replaced in Lemma 3 by the interval $[a,b]$, where $0 \leq a < b$, by the same argument used in Remark 1.

Lemma 4: Let T_1,T_2,T_3 be disjoint trees each having at least one edge. For $j = 1,2,3$ let v_j be a vertex of valence 1 in T_j. Let T be the tree obtained by identifying v_1,v_2,v_3 into a single vertex v. If $s(T_j) = k$ for $j = 1,2,3$, then $s(T) = k+1$.

Proof: By the statement preceding (7), $s(T) \leq k+1$. By Theorem 1, $k = s(T_1) \leq s(T)$. Suppose $s(T) = k$. Let $F = \{f_i \mid 1 \leq i \leq k\}$ be a search plan for T. Let u_j be the vertex of T_j adjacent to v in T, and let w_j be the midpoint of the edge $[u_j,v]$ in T, for $j = 1,2,3$. Introduce w_1,w_2,w_3 as new vertices, giving a tree T^* which is the same set of points in R^3 as T but with a different combinatorial structure. Of course, F is still a search plan for T^*. Let T_j^* be the branch of T^* at w_j which does not contain v. Since T_j^* is homeomorphic to T_j, $s(T_j^*) = k$. Let $S_j = \{t \in [0,\infty) \mid \{f_1(t),\ldots,f_k(t)\} \subseteq T_j^*\}$. By Lemma 3, $S_j \neq \emptyset$ for $j = 1,2,3$. Note that S_1,S_2,S_3 are disjoint closed subsets of $[0,\infty)$.

Let $t_1 = \inf(S_1 \cup S_2 \cup S_3)$. Then $0 \leq t_1 < \infty$ and $t_1 \in S_1 \cup S_2 \cup S_3$. Let i_1 be the index in $\{1,2,3\}$ such that $t_1 \in S_{i_1}$. Suppose we have defined $t_m \in [0,\infty)$ and $i_m \in \{1,2,3\}$, where $m \geq 1$. Define i'_m, i''_m by $i'_m < i''_m$ and $\{i_m, i'_m, i''_m\} = \{1,2,3\}$. Let $t_{m+1} = \inf\{t \in [0,\infty) \mid t_m < t$ and $t \in S_{i'_m} \cup S_{i''_m}\}$. If $t_{m+1} = \infty$, then i_{m+1} and t_{m+2} are left undefined. If $t_{m+1} < \infty$, let i_{m+1} be the element of $\{i'_m, i''_m\}$ such that $t_{m+1} \in S_{i_{m+1}}$. By induction, this defines a finite or infinite sequence $t_1 < t_2 < \ldots$ of points in $[0,\infty)$ and $S_{i_1}, S_{i_2}, \ldots$ of subsets of $[0,\infty)$ such that

(a) whenever i_{m+1} is defined, $i_{m+1} \neq i_m$

(b) whenever t_m is defined, $\{f_1(t_m),\ldots,f_k(t_m)\} \subseteq T^*_{i_m}$

(c) whenever t_{m+1} is defined and $t_m \leq t < t_{m+1}$, $|\{f_1(t),\ldots,f_k(t)\} \cap (T^*_{i'_m} \cup T^*_{i''_m})| < k$.

By (a), we define j_m to be the unique element in $\{1,2,3\} \setminus \{i_m, i_{m+1}\}$ if $t_{m+1} < \infty$, or $j_m = i'_m$ if $t_{m+1} = \infty$. Then we have

(d) if $t_{m+1} < \infty$ and $t_m \leq t < t_{m+2}$, or if $t_{m+1} = \infty$ and $t_m \leq t < \infty$, then $|\{f_1(t),\ldots,f_k(t)\} \cap T^*_{j_m}| < k$.

By Remark 3, there exists $e_m : [t_m, t_{m+1}] \to T^*_{j_m}$ such that e_m is continuous and $f_i(t) \neq e_m(t)$ for $i = 1,2,\ldots,k$ and $t \in [t_m, t_{m+1}]$. Since $j_1 \neq i_1$ and $\{f_1(t_1),\ldots,f_k(t_1)\} \subseteq T^*_{i_1}$ and $t \notin S_1 \cup S_2 \cup S_3$ for $t < t_1$, if $t_1 > 0$ there exists $e_0 : [0,t_1] \to T^*_{j_1}$ such that e_0 is continuous and $f_i(t) \neq e_0(t)$ for all i and $t \in [0,t_1]$.

Let $\varepsilon > 0$ be less than the distance of any of w_1, w_2, w_3 from v, and let $N(v)$ be the neighborhood of v of all points less than distance ε from v. If t_m is defined, there exists $\delta_m > 0$ such that if $t \in [0,\infty) \cap [t_m - \delta_m, t_m + \delta_m]$, then $f_i(t) \notin N(v)$ for $i = 1,2,\ldots,k$. We may assume $\delta_1 < t_1$ if $t_1 > 0$, and that $t_{m-1} < t_m - \delta_m$ whenever t_m and t_{m-1} are defined. For each m such that t_m is defined, let $e^* : [t_m - \delta_m, t_m] \to T$ be a continuous "shortest path" from $e_{m-1}(t_m - \delta_m)$ to $e_m(t_m)$. Let $e^*(t) = e_0(t)$ for $0 \leq t \leq t_1 - \delta_1$, if $t_1 > 0$, and

$e^*(t) = e_m(t)$ for $t_m \leqq t \leqq t_{m+1}-\delta_{m+1}$ when $t_{m+1} < \infty$ (or for $t_m \leqq t < \infty$ if $t_{m+1} = \infty$).

If $t_1 < t_2 < \ldots$ is an infinite sequence, we easily see $\lim_{m\to\infty} t_m = \infty$. Thus we have $e^* : [0,\infty) \to T$ is continuous. It is easy to show $e^*(t) \neq f_i(t)$ for all $t \in [0,\infty)$ and all $i \in \{1,2,\ldots,k\}$. This contradicts our assumption that F is a search plan for T.

It follows that $s(T) = k+1$, proving the lemma.

A _twig_ of the (connected) graph G is an edge one of whose ends has valence 1 in G. If $[v_i,v_j]$ is a twig of G, with v_j of valence 1 in G, then $G' = G\backslash(v_i,v_j]$ is connected, and it is easy to see that $s(G) \leqq s(G')+1$. Using this fact, it is easy to show that if T is a tree with $s(G) = k$, then for each integer r with $1 \leqq r \leqq k$, T has a subtree T′ such that $s(T') = r$ but $s(T'') < r$ for every proper subtree T″ of T′. [That is, T′ is r-minimal.] We may now prove

Theorem 2: Let $k \geqq 1$, and T be a tree. Then $s(T) \geqq k+1$ if and only if T has a vertex v at which there are at least three branches T_1,T_2,T_3 satisfying $s(T_j) \geqq k$ for $j = 1,2,3$.

Proof: First, suppose T has such a vertex v with branches T_1,T_2,T_3 of T at v having $s(T_j) \geqq k$. Then v is of valence 1 in each T_j. By the preceding paragraph, T_j has a subtree T_j' with $s(T_j') = k$ in which v has valence 1. The subtree T′ spanned by T_1',T_2',T_3' has $s(T') = k+1$ by Lemma 4. Then $s(T) \geqq k+1$ by Theorem 1.

Next, suppose $s(T) \geqq k+1$. Letting T′ be a (k+1)-minimal subtree of T, we see that if T′ has a vertex v with branches T_1',T_2',T_3' of T′ at v satisfying $s(T_j') \geqq k$, then the branches T_1,T_2,T_3 of T at v containing T_1',T_2', T_3' will satisfy $s(T_j) \geqq k$. Thus we may assume without loss of generality that T itself is (k+1)-minimal. Then if w is any vertex of valence greater than 1 in T, each branch of T at w has search number at most k. Since $s(T) = k+1$, w must have at least one branch with search number equal to k.

Let w_1 be a vertex of valence greater than 1 in T which has a branch T_1 with the minimum number of edges such that $s(T_1) = k$. Let u_1 be the vertex of T_1 adjacent to w_1. All branches of T at u_1, except the branch containing w_1, have search number at most $k-1$. Start with $f_1, f_2, \ldots, f_k$ at u_1 at time $t = 0$, and let f_k stay at u_1 while $f_1, \ldots, f_{k-1}$ "clear" all these branches not containing w_1, and then return to u_1. Next, let $f_1, \ldots, f_k$ move along edge $[u_1, w_1]$ from u_1 to w_1. If w_1 has two other branches T_2, T_3 with $s(T_2) = k = s(T_3)$, then we are done. If not, w_1 must have exactly one other branch T_2 with search number k. Let w_2 be the vertex of T_2 adjacent to w_1. Now let f_k remain at w_1 while $f_1, \ldots, f_{k-1}$ clear all branches of T at w_1 not containing u_1 or w_2, and then return to w_1. Then have $f_1, \ldots, f_k$ move along $[w_1, w_2]$ from w_1 to w_2. The branch $T_1(w_2)$ of T at w_2 which contains w_1 obeys $s(T_1(w_2)) \geqq k$ since $T_1(w_2) \supset T_1$. If w_2 has two other branches $T_2(w_2)$, $T_3(w_2)$ with search number k, we are done. Otherwise it has exactly one branch $T_2(w_2) \neq T_1(w_2)$ with $s(T_2(w_2)) = k$. Let w_3 be the vertex of $T_2(w_2)$ adjacent to $w_2 \ldots$ Proceeding as before, we construct a sequence $w_1, w_2, w_3, \ldots$ and trees $T_1 \subsetneqq T_1(w_2) \subsetneqq T_1(w_3) \ldots$ such that at each stage, we have all searchers at some w_m, having "cleared" $T_1(w_m)$. Since $s(T) \geqq k+1$, we cannot have $T_1(w_m) = T$, so the above sequence is finite, and terminates with a w_m having at least three branches $T_j(w_m)$ with $s(T_j(w_m)) \geqq k$.

We note that the proof of Theorem 2 provides an algorithm for computing $s(T)$ for any tree T. It is easy to use Theorem 2 and induction on k to prove

(8) Let $k \geqq 2$. If T is a tree and $s(T) = k$, then T has a subtree homeomorphic to a tree in $\mathcal{T}_k$

where the family $\mathcal{T}_k$ was defined in Section 3.

For $k \geqq 1$, let $\alpha(k)$, $\beta(k)$ be the minimum, (respectively) maximum number of edges possible for trees in $\mathcal{T}_k$. It is easy to see that $\alpha(1) = \beta(1) = 1$, $\alpha(2) = \beta(2) = 3$, $\alpha(3) = \beta(3) = 9$, and for $k \geqq 4$, $\beta(k) = 3^{k-1} + 1/2(3^{k-2}-3)$ while $\alpha(k) = 3^{k-1}$. Since $\beta(k) < 3^k = \alpha(k+1)$ for all $k \geqq 1$, we have

(9) If $T \in \mathcal{T}_k$, then T has no subtree homeomorphic to a tree in $\mathcal{T}_{k+1}$.

Using (8), (9), and the construction for $\mathcal{T}_k$, it is easy to prove by induction on k that

(10) If $T \in \mathcal{T}_k$, then $s(T) = k$.

An immediate consequence of (8), (10) and Theorem 1 is

Theorem 3: Let $k \geq 2$. Let T be a tree. Then $s(T) = k$ if and only if T has a subtree homeomorphic to a tree in $\mathcal{T}_k$, but T has no subtree homeomorphic to a tree in $\mathcal{T}_{k+1}$.

This completes proof of the results stated in Section 3. We remark that if $\mathcal{T}_k^*$ is the set of trees in $\mathcal{T}_k$ having no proper subtree homeomorphic to a tree in $\mathcal{T}_k$, then a tree T is k-minimal if and only if it is homeomorphic to a tree in $\mathcal{T}_k^*$. Apparently $\mathcal{T}_k^* = \mathcal{T}_k$ for all k, but we have no proof.

5. FURTHER DEVELOPMENTS

We shall mention only a few of the many other results we have about $s(G)$, then we shall discuss the problem of a general algorithm for it.

A subdivision of G results from inserting some new vertices of valence 2. A contraction of G results from successive contractions of edges of G followed by insertion of some vertices of valence 2 (to avoid loops and multiedges).

It is easy to see that if H is a connected subgraph of some subdivision or contraction of G, then $s(H) \leq s(G)$. This can be used to compute (not necessarily best possible) lower bounds for $s(G)$. For example, the tree in $\mathcal{T}_3$ is a subgraph of a subdivision of K_4, so $s(K_4) \geq 3$. Actually $s(K_4) = 4$, and in fact $s(K_n) = n$ for $n \geq 4$, where K_n is the complete graph on n vertices.

It can be shown that if G' arises from G by adding an edge with at least one of its endpoints in G, then $s(G') \leq s(G) + 1$. [It would be difficult to characterize just when $s(G') = s(G) + 1$.]

Suppose $F = \{f_i \mid 1 \leq i \leq k\}$ is a search plan for G. The functions f_i are possibly very badly behaved (infinitely "wiggly", etc.) A first step in trying to get a general algorithm for $s(G)$ would be to try to replace F by an $F^* = \{f_i^* \mid 1 \leq i \leq k\}$ with nicely behaved "combinatorial" searchers. To do this, consider the sets $U(t,F)$ defined after Theorem 1. We have $U(0,F) = \{f_1(0),\ldots, f_k(0)\}$, and as t varies, $U(t,F)$ varies over certain subsets of G until we have $U(\tau(F),F) = G$, and G is "cleared" at time $\tau(F)$. Because of possible pathological behavior of the f_i in F, the behavior of $U(t,F)$ as a function of t could be complicated. However, it is clear that the homeomorphism type of $U(t,F)$ assumes only finitely many values, because by (4), $|\partial U(t,F)| \leq k$. The type of $U(t,F)$ can change only when some searcher f_i passes through a vertex of G. Careful study of the behavior possible for $U(t,F)$ as t varies is reminiscent of the study of critical points in Morse Theory, and leads to an algorithm for $s(G)$ involving a certain type of sequence of subgraphs of a subdivision of G. Discussion of this is better left to a sequel paper, for this paper is already too long.

6. OTHER MODELS

Many variations on our search problem come to mind. What if the searchers can "see" down edges, and "capture" means the evader is sighted? Or in any case, if the evader is "seen" but not captured, what if the searchers can radio each other to modify their "search plan" to incorporate the new information? Perhaps the evader can only move at speeds up to some bound, and the searchers are similarly constrained.

Variations even exist in higher dimensions: the evader can be a "point" on a surface and a "searcher" can be a homeomorphic image of the interval [0,1] (a "dragnet"). If so, the sphere needs two searchers.

The variations are endless, and some are interesting. Our version seemed the most "graph theoretic" in flavor, and that is why we pursued it here. (Hopefully, we caught it.)

7. ACKNOWLEDGMENTS

I am indebted to Richard Breisch and to Robert Wells for many hours of conversation about the "search problem". The "Compactness Lemma" (Lemma 2), was independently proved by Steve Simpson, Robert Wells, and by me, after I conjectured it was true.

REGULARITY IN TOURNAMENTS

Marcel Herzog
The Australian National University
Canberra, ACT 2600
Australia

and

K. B. Reid
Louisiana State University
Baton Rouge, LA 70803

Abstract

A nearly triply regular (abbreviated NTR) tournament is a doubly regular tournament in which every triple of distinct vertices dominates exactly j_1 or j_2 vertices, where $0 \leq j_1 < j_2$. We show that if T is NTR and if the outset of a pair of distinct vertices of T either is a singleton or spans a subtournament which is not strongly connected, then $j_1 = 0$ and $j_2 = 1$ and T is one of the quadratic residue tournaments of orders 7 or 11.

1. Introduction. A tournament of order n is regular of degree ℓ if all of its outdegrees equal ℓ and $\ell > 0$. For example, a directed 3-cycle is the only regular tournament of degree 1. It is easy to see that $n = 2\ell + 1$, there is only one regular tournament of degree 2, and there exist many regular tournaments of degree ℓ, for $\ell \geq 3$. In fact, the enumeration of regular tournaments is an open problem [5, p. 219]. One useful construction for a regular tournament of degree ℓ, proceeds as follows: Let the vertices be the integers modulo $2\ell + 1$ and let the ℓ-set S consist of exactly one integer from each of the ℓ sets $\{i, 2\ell+1-i\}$, $1 \leq i \leq \ell$. Let vertex p dominate vertex q if $q - p \equiv s \pmod{2\ell+1}$, for some s in S. The set S is called the symbol of the regular tournament so constructed. Such a tournament is vertex-symmetric, i.e. its group of automorphisms is transitive on its vertex set. Alspach [1] and Astie [2] have enumerated nonisomorphic vertex-symmetric tournaments of prime order.

A regular tournament of degree ℓ and order n is doubly regular if each outset spans a regular tournament of degree k. For example, the tournament of order seven with symbol $\{1,2,4\}$, called the quadratic residue tournament of order seven since 1,2, and 4 are the squares in the integers modulo 7, is doubly regular. It is easy to see that $\ell = 2k + 1$ and $n = 4k + 3$. It is known that the existence of a doubly regular tournament of order n, is equivalent to the existence of a skew Hadamard matrix of order $n + 1$ (e.g. see [3], [6], [10] and also [4]). Considerable effort has been expended in the last ten years in the search for new Hadamard and skew Hadamard matrices (see [9]). In fact, Szekeres [10] utilized this connection with tournaments to construct new skew Hadamard matrices. Other results on doubly regular tournaments have been given by Müller and Pelant [7] under the name of strongly homogeneous tournaments.

A <u>triply regular tournament</u> is a doubly regular tournament in which the outsets of all triples of distinct vertices contain the same number of vertices. There are no triply regular tournaments (see [3], p. 337] or [7, p. 379].). We give a new proof of this in Section 2. In an attempt to sharpen regularity in tournaments, let us say that a tournament T of order n, is <u>nearly triply regular</u> (abbreviated NTR) if there exist integers j_1, j_2, k, ℓ such that
$0 \le j_1 < j_2 \le k < \ell$ and

1) $o(x) = \ell$, for all vertices x in T,
2) $o(x,y) = k$, for all vertices $x \ne y$ in T, and
3) $o(x,y,z) = \begin{cases} j_1, & \text{or} \\ j_2 \end{cases}$, for all triple of distinct vertices $x, y, z,$ in T,

where $o(x_1, x_2, \ldots, x_m)$ denotes the number of vertices dominated by all the vertices $x_1, x_2, \ldots, x_m$. The three parts of this definition will be referred to as conditions 1), 2), and 3). Conditions 1) and 2) require T to be doubly regular, $\ell = 2k + 1$, and $n = 2\ell + 1 = 4k + 3$. For example, it is straightforward to check that NTR tournaments include the quadratic residue tournament of order 7 when $j_1 = 0$, $j_1 = 1 = k$, $\ell = 3$, and the <u>quadratic residue tournament of order 11</u> when $j_1 = 0$, $j_1 = 1$, $k = 2$, $\ell = 5$. (this tournament has symbol $\{1,3,4,5,9\}$). However, the quadratic residue tournaments of orders 19, 23, and 31 are not NTR. In Section 3, we prove that the values j_1 and j_2 in condition 3) are not assumed solely according to the structure of the tournament spanned by the triple of vertices. We also prove that the order 11 example given above is the only instance of a NTR tournament in which the outset of some two vertices spans a subtournament which is not strongly connected.

2. <u>Preliminary Notation and Results</u>. In the following, the outset (inset) of vertices $x_1, \ldots, x_m$ in a tournament will be denoted

$\mathcal{O}(x_1,\ldots,x_m)$ $(I(x_1,\ldots,x_m))$. That is, $\mathcal{O}(x_1,\ldots,x_m) = \mathcal{O}(x_1) \cap \ldots \cap \mathcal{O}(x_m)$. The cardinality of $\mathcal{O}(x_1,\ldots,x_m)$ $(I(x_1,\ldots,x_m))$ will be denoted $o(x_1,\ldots,x_m)$ $(i(x_1,\ldots,x_m))$. The notation $x \to y$ will be used both to denote the arc with initial vertex x and terminal vertex y and to denote the fact that vertex x dominates vertex y . Also, if A and B are disjoint sets of vertices in a tournament, then $A \Rightarrow B$ denotes $a \to b$ for all a in A and $b \in B$. If $x \to y$, then $B(x,y)$, the bypass set of $x \to y$, denotes $\mathcal{O}(x) \cap I(y)$, and $C(x,y)$, the cycle set of $x \to y$, denotes $\mathcal{O}(y) \cap I(x)$. If T is a NTR tournament, then the integers j_1 and j_2 in condition 3) of the definition in Section 1 are called the <u>parameters</u> of T . For a set of vertices A in a tournament, $\langle A\rangle$ denotes the subtournament with vertices A .

The nonexistence of triply regular tournaments follows immediately from the following slightly stronger theorem:

<u>Theorem 2.1</u>. There exists no tournament containing an arc $x \to y$ with $C(x,y) \neq \emptyset$ and $\mathcal{O}(x,y) \neq \emptyset$ such that for some integers ℓ , j_1 , and j_2

i) $o(y)$ is odd,

ii) $o(y,z) = \ell$, for every z in $C(x,y)$,

iii) $o(x,y,z) = j_1$, for every z in $\mathcal{O}(x,y)$, and

iv) $o(x,y,z) = j_2$, for every z in $C(x,y)$.

<u>Proof</u>. Suppose T is such a tournament. Let $\mathcal{O} = \mathcal{O}(x,y)$ and $C = C(x,y)$. Both $\mathcal{O}$ and C are nonempty. By iii), $\langle\mathcal{O}\rangle$ is regular and $|\mathcal{O}| = 2j_1 + 1$. Since $o(y) = |\mathcal{O}| + |C|$ is odd (by i)), $|C|$ is even. On the other hand, for z in C , $\mathcal{O}(y,z)$ is the disjoint union of $\mathcal{O}(z) \cap C$ and $\mathcal{O}(z) \cap \mathcal{O}$, so that, by iii) and iv), each vertex in C has outdegree, in $\langle C\rangle$, $\ell - j_2$. That is, $\langle C\rangle$ is regular and $|C| = 2(\ell-j_2) + 1$, a contradiction. The theorem follows.

Remark. The requirement that $\mathcal{O}$ and C be nonempty cannot be deleted from the statement of the theorem. For, consider the tournaments obtained from the 3-cycle $x \to y \to z \to x$ or the transitive triple $x \to y \to z$ and $x \to z$ by replacing z with a regular tournament.

It can also be seen that the theorem still holds if condition i) is replaced with i)′ $o(x)$ and $o(y)$ have the same parity, i)″ $o(x,z)$ is fixed for every z in $B(x,y)$, and i)‴ $o(x,y,z)$ is fixed for every z in $B(x,y)$.

Corollary 2.2. There exists no tournament T so that for some integers j,k,ℓ, $0 \leq j < k < \ell$,

a) $o(x) = \ell$, for every vertex x in T

b) $o(x,y) = k$, for every pair of distinct vertices x and y in T

c) $o(x,y,z) = j$, for every triple of distinct vertices x,y,z in T

To prepare for Section 3, we include a necessary lemma without proof.

Lemma 2.3. Let T be a NTR tournament with parameters j_1 and j_2, and let $x \to y$ in T. Then (using symbols in conditions 1), 2), 3) of Section 1) $|I(x,y)| = |B(x,y)| = |\mathcal{O}(x,y)| = k$ and $|C(x,y)| = k+1$.

A technical lemma is also included here.

Lemma 2.4. Let T be as in Lemma 2.3. Let α_1 denote the number of vertices in $C(x,y)$ each of which dominates exactly j_1 vertices in $\mathcal{O}(x,y)$. Then

$$\alpha_1(j_2-j_1) = \frac{(k+1)(2j_2-k)}{2}$$

Proof. Set $C = C(x,y)$ and $\mathcal{O} = \mathcal{O}(x,y)$. Let $\alpha_2 = k+1-\alpha_1$, so that α_2 vertices in C dominate exactly j_2 vertices in $\mathcal{O}$. Then, by condition 2) and Lemma 2.3,

$$\Sigma\ o(y,z) = k(k+1)\ ,$$

where the sum is over all z in C. Also, summing over all z in C,

$$\Sigma\ o(y,z) = \Sigma\ (|\mathcal{O}(z,y) \cap C| + |\mathcal{O}(z,y) \cap \mathcal{O}|) = \binom{k+1}{2} + (\alpha_1 j_1 + \alpha_2 j_2)\ ,$$

since the first term in the last line counts the sum of outdegrees in <C> and the second term counts the number of arcs from C to $\mathcal{O}$. Equating the two expressions for $\Sigma\ o(y,z)$ yields $\alpha_1 j_1 + \alpha_2 j_2 = \binom{k+1}{2}$, which when solved with $\alpha_1 + \alpha_2 = k + 1$ yields the result.

3. Main Results. First we show that the value of $o(x,y,z)$ in a NTR tournament is not determined solely by whether $\langle x,y,z\rangle$ is cyclic or transitive.

Theorem 3.1. Let T be a NTR tournament with parameters j_1 and j_2.

(a) For every arc $x \to y$, there exists z in $\mathcal{O}(x,y)$ such that $o(x,y,z) = j_1$.

(b) For every arc $x \to y$, there exists z in $C(x,y)$ such that $o(x,y,z) = j_2$.

(c) If $x \to y$ and for every z in $\mathcal{O}(x,y)$, $o(x,y,z) = j_1$, then there exist $z_1 \neq z_2$ in $C(x,y)$ such that $o(x,y,z_i) = j_i$, $i = 1,2$.

(d) If $x \to y$ and for every z in $C(x,y)$, $o(x,y,z) = j_2$, then there exist $z_1 \neq z_2$ in $\mathcal{O}(x,y)$ such that $o(x,y,z_i) = j_i$, $i = 1,2$.

Proof. (a) Suppose that $o(x,y,z) = j_2$ whenever z is in $\mathcal{O}(x,y)$. Then $\mathcal{O}(x,y)$ is regular of degree j_2 so that $o(x,y) = k = 2j_2 + 1$.

Then using the notation of Lemma 2.4, we note that $\alpha_1(j_2-j_1) = (j_2+1)(-1)$, contrary to $\alpha_1 \geq 0$ and $j_1 < j_2$. Part (a) follows.

(b) Suppose that $o(x,y,z) = j_1$ whenever z is in $C(x,y)$. Again, by Lemma 2.4, since $\alpha_1 = k+1$, we obtain $k = 2j_1$. Let β_1 denote the number of vertices in $\mathcal{O}(x,y)$ each of which dominates exactly j_1 vertices in $\mathcal{O}(x,y)$. Then $\Sigma\, o(x,y,z) = \beta_1 j_1 + (k-\beta_1)j_2$, where the sum is over all z in $\mathcal{O}(x,y)$. Hence,

$$\beta_1 j_1 + (k-\beta_1)j_2 = \binom{k}{2}, \quad \text{or}$$

$$\beta_1 = k + \frac{j_1}{j_2-j_1} .$$

But $\beta_1 \leq o(x,y) = k$, so that $j_1 = 0$ and, hence, $k = 0$, contrary to the definition of a NTR tournament. Part (b) follows.

(c) Suppose that $o(x,y,z) = j_1$ for every z in $\mathcal{O}(x,y)$. Then $\mathcal{O}(x,y)$ is regular of degree j_1 and $o(x,y) = k = 2j_1 + 1$. By Lemma 2.4, $\alpha_1 > 0$ since $k \neq 2j_2$. But, by part (b), $\alpha_1 < |C| = k+1$. Part (c) follows.

(d) Suppose that $o(x,y,z) = j_2$ for every z in $C(x,y)$. Thus $\alpha_1 = 0$, so that, by Lemma 2.4, $k = 2j_2$. Thus, not every w in $\mathcal{O}(x,y)$ satisfies $o(x,y,w) = j_1$ (as that would imply $o(x,y) = 2j_1 + 1$). But there is such a w by part (a). Part (d) follows.

This complete the proof of the theorem.

As an immediate corollary we obtain the following:

Corollary 3.2. There exists no NTR tournament with parameters j_1 and j_2 such that

$$o(x,y,z) = \begin{cases} r, & \text{if } \langle x,y,z\rangle \text{ is transitive} \\ s, & \text{if } \langle x,y,z\rangle \text{ is cyclic}, \end{cases}$$

where $\{r,s\} = \{j_1, j_2\}$.

As seen in Section 1, there exists a NTR tournament of order 11 in which $\langle \mathcal{O}(x,y) \rangle$ is a single arc for every $x \neq y$. We now prove that this is the only such example of a NTR tournament in which $\langle \mathcal{O}(x,y) \rangle$ is not strong for some pair of vertices $x \neq y$. We can then conclude that if there exists a NTR tournament of order n, $n \neq 7,11$, then $\langle \mathcal{O}(x,y) \rangle$ must be strongly connected.

<u>Theorem 3.3</u>. If T is a NTR tournament of order n with parameters j_1 and j_2, and if $\langle \mathcal{O}(x,y) \rangle$ is a singleton or not strongly connected for some $x \to y$ in T, then $j_1 = 0$ and $j_2 = 1$.

<u>Proof</u>. Suppose that $\langle \mathcal{O}(x,y) \rangle$ is not strongly connected for some $x \to y$ in T. Set $\mathcal{O} = \mathcal{O}(x,y)$ and $C = C(x,y)$. By condition 3) (of Section 1), vertices in $\langle \mathcal{O} \rangle$ have outdegree j_1 or j_2, and both values are assumed if $o(x,y) = k \geq 2$ as $\langle \mathcal{O} \rangle$ is not regular (regular tournaments are strongly connected). If $k = 1$, then $\ell = 3$ and $n = 7$. But this implies that T is a tournament of order 7 which contains no transitive subtournament of order 4. There is only one such tournament, the quadratic residue tournament of order 7 described in Section 1 (see [8, p. 226]). Thus, $j_1 = 0$ and $j_2 = 1$. In the following we suppose that $k \geq 2$. Consequently, $\langle \mathcal{O} \rangle$ has the form $A \Rightarrow D$, where A is nonempty and regular of degree $j_2 - 2j_1 - 1$ and D is nonempty and regular of degree j_1. Thus, $k = o(x,y) = |A| + |D| = 2j_2 - 2j_1$.

Assume, first, that $j_1 > 0$. Let z be a vertex in A. Since $o(y,z) = k$ (condition 2)), $|\mathcal{O}(z) \cap C| = k - o(x,y,z) = k - j_2 = j_2 - 2j_1$. Thus, $\mathcal{O}(y,z)$ consists of $j_2 - 2j_1 - 1$ vertices in A, all of D, and $j_2 - 2j_1$ vertices in C. By condition 3), each vertex in $\langle \mathcal{O}(y,z) \rangle$ has outdegree j_1 or j_2. In order for a vertex d in D to have outdegree j_2 in $\langle \mathcal{O}(y,z) \rangle$, d must dominate $j_2 - j_1$ vertices among the $j_2 - 2j_1$ vertices of $\mathcal{O}(y,z)$ in C (recall $A \Rightarrow D$). But $j_2 - 2j_1 < j_2 - j_1$, since $j_1 > 0$, so vertices in D have outdegree j_1 in $\langle \mathcal{O}(y,z) \rangle$. In summary, if vertex w in C is dominated by some z in A (so that w is in $\mathcal{O}(y,z)$), then $w \Rightarrow D$. Partition C as follows:

$$C_1 = \{w \mid w \text{ a vertex in } C \text{ and } w \Rightarrow A\}$$
$$C_2 = \{w \mid w \text{ a vertex in } C \text{ and } z \rightarrow w \text{ for some vertex } z \text{ in } A\} .$$

C_2 is nonempty. For otherwise, $C \Rightarrow D$ so that for vertex z in A, $k = o(y,z) = o(x,y,z) = j_2$, contrary to the fact that j_2 is an outdegree in $\langle\mathcal{O}\rangle$ (which has order k). Similarly, C_1 is nonempty.

Next, we claim that $|A| = |C_2|$. To see this we count the number (nonzero) of arcs between A and C_2. On one hand, there are $|A|\,|C_2|$ such arcs. On the other hand, let p (q) be the number of arcs from C_2 to A (from A to C_2). Since, for c_2 in C_2, $o(x,y,c_2) \geq |D| = 2j_1 + 1 > j_1$, we see that (by condition 3)) $o(x,y,c_2) = j_2$. Then $p = \Sigma(o(x,y,c_2) - |D|) = \Sigma\,(j_2-2j_1-1) = |C_2|(j_2-2j_1-1)$, where the sum is over all c_2 in C_2. Also,

$$q = \Sigma(o(a,y) - o(x,a,y)) = \Sigma(k - j_2) = |A|(k-j_2) = |A|(j_2-2j_1) ,$$

where the sum is over all a in A. Thus,

$$|A|\,|C_2| = p + q = |C_2|(j_2-2j_1-1) + |A|(j_2-2j_1) , \quad \text{or}$$
$$|A|(j_2-2j_1) = |C_2|(j_2-2j_1) ,$$

since $|A| = 2(j_2-2j_1-1) + 1$ (i.e. $\langle A\rangle$ is regular of degree $j_2 - 2j_1 - 1$). If $j_2 - 2j_1 = 0$, then $k = 2j_2 - 2j_1 = j_2$, contrary to the fact that j_2 is an outdegree in a $\langle\mathcal{O}\rangle$, a tournament of order k. So $|A| = |C_2|$. As $|C_1| + |C_2| = k + 1 = 2j_2 - 2j_1 + 1$ and $|A| + |D| = k = 2j_2 - 2j_1$ (Lemma 2.3 is used here), we also deduce that $|C_1| = |B| + 1 = 2j_1 + 2$ and $|A| = |C_2| = 2j_2 - 4j_1 - 1$.

Next we claim that $C_1 \Rightarrow C_2$. Let c_2 be a vertex in C_2. Since $c_2 \Rightarrow D$ and $|D| = 2j_1 + 1 > j_1$, $o(x,y,c_2) = j_2$. Thus, $\mathcal{O}(c_2,y)$ includes j_2 vertices in $\mathcal{O}$, $(j_2 - 2j_1 - 1)$ in A, all

$2j_1 + 1$ in D, and $o(y,c_2) - j_2 = k - j_2$ vertices in C. By condition 3), $o(d,y,c_2)$ is j_1 or j_2 for d in D. But $o(d,y,c_2) \leq j_1 + (k-j_2) = j_2 - j_1 < j_2$, since $j_1 > 0$ and d dominates at most $k - j_2$ vertices in $\mathcal{O}(c_2,y) - D$ (recall that $A \Rightarrow D$). Thus, $o(d,y,c_2) = j_1$ and $(\mathcal{O}(c_2,y) \cap C) \Rightarrow D$. Consequently, any vertex c_1 in C_1 which is in $\mathcal{O}(c_2,y)$ dominates all of $\mathcal{O}$, i.e. $o(x,y,c_1) = k$, again contrary to the fact that $k > j_2 > j_1$. Thus, no c_2 dominates any vertex in C_1, i.e. $C_1 \Rightarrow C_2$.

Finally, $k = o(y,c_2) = o(x,y,c_2) + |\mathcal{O}(c_2) \cap C_1| + |\mathcal{O}(c_2) \cap C_2| = j_2 + |\mathcal{O}(c_2) \cap C_2|$, since $C_1 \Rightarrow C_2$. So $|\mathcal{O}(c_2) \cap C_2| = k - j_2$, i.e. $\langle C_2\rangle$ is regular of degree $k - j_2$ and $|C_2| = 2(k-j_2) + 1 = 2j_2 - 4j_1 + 1$. This contradicts the value of $|C_2|$ obtained above.

So $j_1 = 0$ and $|B| = 1$, $k = 2j_2$, $|A| = k - 1 = 2j_2 - 1$. Assume in the following that $j_2 > 1$. Set $B = \{b\}$. As $|C| = k + 1$, $o(b,y) = k$, and $A \Rightarrow b$, there is a unique vertex w in C such that $w \rightarrow b$ and $b \Rightarrow (C-w)$. So outdegrees of vertices in tournament $\langle\mathcal{O}(y,b)\rangle = \langle C-w\rangle$ are 0 or j_2, by condition 3). All such outdegrees are not j_2, as otherwise $\langle C-w\rangle$ is regular, which implies that $|C-w| = 2j_2 + 1$. But, by Lemma 2.3, $|C| = k + 1 = 2j_2 + 1$. So $\langle C-w\rangle$ contains a receiver, say u (i.e. $\mathcal{O}(u) \cap (C-w)$ is empty). Since $w \rightarrow b$, $o(x,y,w) \geq 1$, so $o(x,y,w) = j_2$ and w dominates exactly $j_2 - 1$ vertices of A. Note that $b \rightarrow u$ and $C - w \Rightarrow u$, so $u \rightarrow w$ and $u \Rightarrow A$ since $o(u,y) = k$. But then $\mathcal{O}(u,w,y)$ consists of exactly the $j_2 - 1$ vertices of A dominated by w, i.e. $j_2 = o(u,w,y) = j_2 - 1$, a contradiction.

Thus, $j_1 = 0$ and $j_2 = 1$. This in turn implies that $k = 2$, $\ell = 5$, and $n = 11$. The theorem follows.

<u>Corollary 3.4</u>. (a) The unique NTR tournament such that $k = 1$ is the quadratic residue tournament of order 7.

(b) The unique NTR tournament such that $\langle\mathcal{O}(x,y)\rangle$ is not strongly connected, for some vertices $x \neq y$ in T, is the quadratic residue tournament of order 11.

(c) The only NTR tournaments with parameters $j_1 = 0$ and $j_2 > 0$ are the quadratic residue tournaments of orders 7 and 11. In both cases $j_2 = 1$.

Proof. The proof of (a) is contained in the proof of the theorem. The proof of (b) is tedious but straightforward using the ideas of the proof of the theorem and the result that $j_1 = 0$, $j_2 = 1$, $k = 2$, $\ell = 5$, and $n = 11$ (which follows from the theorem). Thus, we omit the proof of (b). Part (c) follows from (a) and (b).

In conclusion, we see that any new NTR tournament, T , must be such that $\langle\mathcal{O}(x,y)\rangle$ is strongly connected for all vertices $x \neq y$ in T . However, we conjecture that there exist only a (small) finite number of NTR tournaments.

References

1. B. Alspach, On point-symmetric tournaments. Canad. Math. Bull. 13 (1970) 317-323.
2. A. Astie, Groupes d'automorphismes des tournois sommet-symétriques d'ordre premier et dénombrement de cas tournois. C. R. Acad. Sc. Paris 275 (1972) 167-169.
3. E. Brown and K. B. Reid, Doubly regular tournaments are equivalent to skew Hadamard matrices. J. Combinatorial Theory 12 (1972) 332-338.
4. P. Delsarte, J. M. Goethals, and J. J. Seidel, Orthogonal matrices with zero diagonal. II. Can. J. Math. 23 (1971) 816-832.
5. F. Harary and E. M. Palmer, Graphical Enumeration. Academic Press, New York and London (1973).
6. E. C. Johnsen, Integral solutions to the incidence equation for a finite projective plane, cases of orders $n \equiv 2 \pmod 4$.

Pacific J. Math. 17 (1966) 97-120.

7. V. Müller and J. Pelant, On strongly homogeneous tournaments. *Czechoslovak Math. J.* 24 (1974) 378-391.

8. E. T. Parker and K. B. Reid, Disproof of a conjecture of Erdős and Moser on tournaments. *J. Combinatorial Theory* 9 (1970) 225-238.

9. A. P. Street, J. S. Wallis, and W. D. Wallis, *Combinatorics: Room Squares, Sum-free Sets, Hadamard Matrices*. Lecture Notes in Mathematics, 292, Springer-Verlag, Berlin-Heidleberg-New York (1972).

10. G. Szekeres, Tournaments and Hadamard matrices. *L'Enseignement. Math.* 15 (1969) 269-278.

ON GRAPHS OF EMBEDDING RANGE ONE

R.D. Ringeisen
Indiana University-Purdue University
at Fort Wayne
Fort Wayne, Indiana, 46805

In this paper we characterize those connected planar and toroidal graphs of embedding range one.

I. Preliminary definitions and results.

Throughout this paper G will be a connected graph. The genus of G, $\gamma(G)$, and maximum genus, $\gamma_M(G)$, are the smallest and largest of the numbers $\gamma(N)$, respectively, where N is a compact orientable 2-manifold in which G has a two cell embedding. The embedding range of a connected graph G, denoted R(G), is given by $R(G) = \gamma_M(G) - \gamma(G)$. The Betti number $\beta(G)$ is determined by $\beta(G) = E - V + 1$ where E and V are the number of edges and vertices of G, respectively. In [3], the following theorems are given:

THEOREM A. $\gamma_M(G) \leq \left[\frac{\beta(G)}{2}\right]$. Equality holds if and only if the embedding has one or two faces, according as $\beta(G)$ is even or odd, respectively.

THEOREM B. Let H be a connected subgraph of the connected graph G. Then $\gamma_M(H) \leq \gamma_M(G)$.

THEOREM C. Let G be a connected graph with blocks B_i, $i = 1, \ldots, n$. Then $\gamma_M(G) \geq \sum_{i=1}^{n} \gamma_M(B_i)$.

COROLLARY. If G consists of two connected graphs G_1 and G_2 sharing exactly one vertex, then $\gamma_M(G) \geq \gamma_M(G_1) + \gamma_M(G_2)$.

In [2], equality is shown for theorem C in the case where no two cyclic blocks of G share a cutpoint. We will refer to this property as the additivity result. A graph is called upper embeddable if equality holds in theorem A. The following theorem appears in [5] and is known as the edge adding technique

(henceforth denoted E.A.T.)

THEOREM D. Let i and j denote nonadjacent vertices of a connected graph G, which has a two cell embedding T with i in the boundary of face F_i and j in the boundary of face F_j. Let G' be the graph formed when the edge ij is added to G. Then:

(a) If $F_i \neq F_j$, G' has a 2-cell embedding with one less face than T.

(b) If $F_i = F_j$, then G' has a 2-cell embedding with one more face than T. Furthermore, the directed edges ij and ji appear in different faces of this embedding for G'.

When a path P joins two nonadjacent vertices of G so that only the endpoints of P are vertices in G we will call P a path which is open disjoint from G. Clearly, if one adds a path rather than an edge, the vertices u and v in the E.A.T. may be adjacent. The E.A.T. implies the following two remarks:

REMARK 1. If an open disjoint path is added between two nonadjacent vertices appearing in the same face boundary for an embedding of a graph G, then the embedding implied by the E.A.T. has each vertex of P in at least two faces.

REMARK 2. Suppose a vertex v appears in two faces for some embedding T, of G. If an open disjoint path is added to G with one endpoint v, then the graph so formed has an embedding with one less face than T.

We now describe some graphs for easy reference later. Let N_1 be a cycle of any length. Form N_2 by joining any two not necessarily distinct vertices of N_1 by an open disjoint path so that a graph (not a multi- or pseudo-graph) is formed. For $k \geq 3$, form N_k from N_{k-1} by choosing two not necessarily distinct vertices of N_{k-1} which were not vertices of N_{k-2}, and adding a path between them. N_k is called a chain graph of length k. A butterfly graph with n wings, denoted B_n, is formed by beginning with a nontrivial path of any length and adding n pairwise disjoint paths of arbitrary length between the endpoints of P. (To avoid multigraphs, we insist that no more than one of the n + 1 paths be of length one.) A rose graph with n petals, denoted R_n, is obtained by attaching n cycles to a vertex b in such a way that no two of the cycles share more than b. Figure 1 shows a graph of each type.

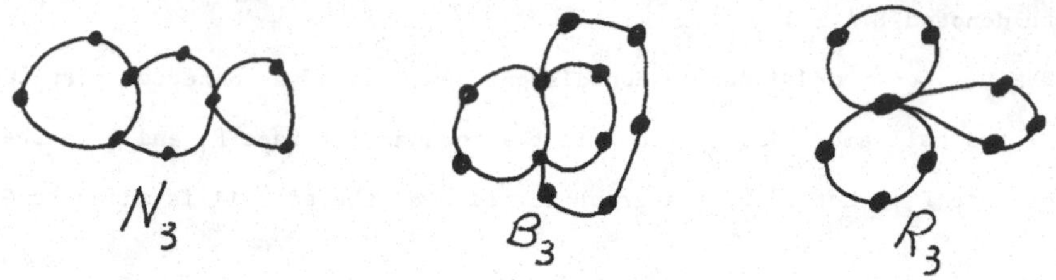

Figure 1

Because R_2 and B_2 are embeddable on the torus, mathematical induction may be combined with (judicious use of) the E.A.T. to yield the following remark.

REMARK 3. All chain, butterfly, and rose graphs are upper embeddable.

II. Planar graphs of Embedding Range One.

Throughout this section and the next, we will be adding paths between non-adjacent vertices of a graph. In all cases the path P will be open disjoint from the graph G. The graph obtained will be denoted by $G \cup P$. Also, note that a planar graph has $R(G) = 1$ if and only if $\gamma_M(G) = 1$.

Let T_1 be K_4 with an edge deleted. For $n \geq 2$, let T_n be the graph obtained by replacing the vertices of an n cycle by disjoint copies of the complete graph K_3 so that exactly two vertices of each K_3 have degree three. For each $i \geq 1$, define $H_i = \{G|G$ is a graph homeomorphic with $T_i\}$. Let $H = \{G|G$ is homeomorphic with $M \cup P$, where $M \in H_1$, and P is a path open disjoint from M$\}$.

THEOREM 1: G is a planar block with $R(G) = 1$ if and only if G is in H or G is in H_n for some $n \geq 1$.

Proof. Suppose G is in H. When a path is added to a graph in H_1, the resultant graph, G, has $\beta(G) = 3$. Thus $\gamma_M(G) \leq 1$ and since G contains a homeomorph of T_1, we also have $\gamma_M(G) \geq 1$ by Remark 3 and Theorem B. If G is in H_n, then $\gamma_M(G) = \gamma_M(T_n)$. Remark 3 gives $\gamma_M(T_1) = 1$. We thus assume $n \geq 2$. Because T_n, $n \geq 2$ contains a homeomorph of T_1, $\gamma_M(T_n) \geq 1$. Now let T be the graph obtained from T_n by removing one edge of the cycle from which T_n was created. By the additivity result and $\gamma_M(K_3) = 0$, we have $\gamma_M(T) = 0$. Thus $\gamma_M(T_n) \leq 1$, since the addition of an edge can increase maximum genus by at most one.

Clearly, graphs of either type are planar blocks.

Now let G be any planar block with $\gamma_M(G) = 1$. Then G contains a subgraph M in H_1. Furthermore, if G has vertices not in M, then there is path in G, open disjoint from M, joining vertices of M. We show that either G is in H_n for some n or that G has at most one such path. When an open disjoint path is added to M, a graph homeomorphic with one of the four graphs in Figure 2 results.

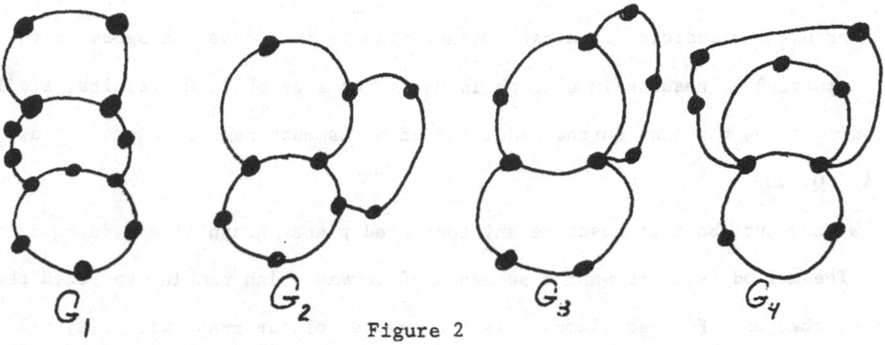

Figure 2

Because for each G_i, $i = 1, \ldots, 4$, $\beta(G_i \cup P) = 4$, where P is an open disjoint path between non-adjacent vertices of G_i, we need only show that $G_i \cup P$ is upper embeddable.

Claim: Let K be a homeomorph of G_2 and P open disjoint from K. Then $K \cup P$ is upper embeddable.

Proof of claim. Either at least one endpoint of P has degree three in K or neither does. By constructing an appropriate embedding for K like the one in Figure 3, any degree three vertex lies in both faces of a two face embedding for K on the torus; Remark 2 thus completes the proof if either endpoint has degree three.

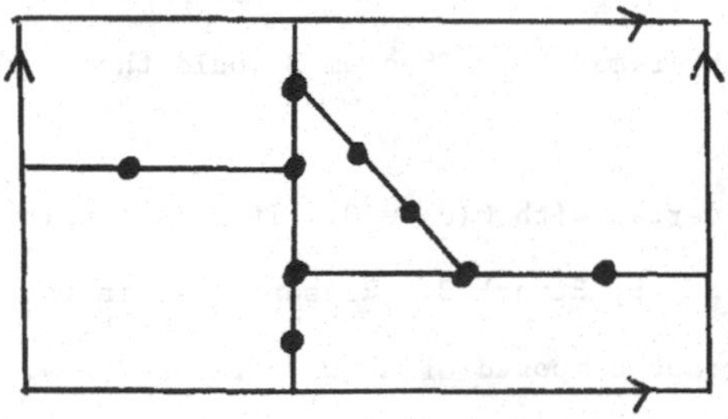

Figure 3

If both endpoints of P have degree two, then $K \cup P$ is homeomorphic to a prism graph as described in [4] and is thus upper embeddable [also 4]. One now notices that when any path P' is added to G_i, i = 3, 4, a graph having a subgraph homeomorphic with the $K \cup P$ of the claim results. Furthermore, $G_i \cup P'$ has a path between nonadjacent vertices of $K \cup P$. Thus, for G_i, i = 3, 4, no other path may be added.

For G_1, the addition of a path either results in a case analogous to the last paragraph or results in a graph in H_3. When a graph in H_3 results, similar reasoning shows that any further addition of paths must result in graphs in H_k, $k \geq 4$. Q.E.D.

We now set about to describe any connected planar graph of embedding range one. The method is to present a sequence of lemmas which combine to yield the desired theorem. For each lemma G is a connected planar graph with $R(G) = 1$.

LEMMA 1. G has at most one block which is not a cycle or K_2.

Proof. Any block not of the described type has a two winged butterfly B_2, having maximum genus one, as a subgraph. Theorems B and C then combine to complete the proof.

We denote by M(c) the number of cyclic blocks in which a vertex c is contained.

LEMMA 2. a) For each vertex c of G, $M(c) \leq 3$, b) If $M(c) = 3$, then each block containing c is a cycle. Furthermore $M(v) = 1$ for every other vertex of G.

Proof. For each cyclic block in which a cutpoint c is contained, c lies in a cycle. Hence, if $M(c) \geq 4$, then G has a subgraph which is a four petalled rose R_4 and $\gamma_M(R_4) = 2$, by remark 3. Theorem B would then force G to have maximum genus at least two.

Now suppose c is a vertex with $M(c) = 3$. If B is cyclic block which is not a cycle, then $\gamma_M(B) \geq 1$ by Remark 3. Reasoning as in the last paragraph, G would then have a subgraph composed of a two petalled rose, R_2, sharing the vertex c with B. By the corollary to theorem C we have that $\gamma_M(G) \geq \gamma_M(B) + \gamma_M(R_2) \geq 2$. Hence all blocks at c must be cycles.

If G has two vertices with multiplicity two or more, then G has a connected subgraph which contains two rose graphs, R_2 and R_2', sharing at most one cutpoint. Then by the corollary to theorem C, $\gamma_M(G) \geq 2\gamma_M(R_2) = 2$. Q.E.D.

LEMMA 3. Suppose G is not a block and has no cutpoint of nontrivial multiplicity three. If G has more than one cutpoint of multiplicity two, then every block is a cycle and there exists a cycle which contains every cutpoint of G. If G has exactly one cutpoint of multiplicity two then at most one of its blocks is not a cycle; this non-cycle block is in H_n, for some n.

Proof. Suppose G has at least two cutpoints of multiplicity two. If they do not share a block, then G contains two R_2 graphs joined by a path. The additivity result combines with theorem B to force $\gamma_M(G) \geq 2$. Hence any two such cutpoints share a block. If more than one block has two or more cutpoints, this last conclusion creates a violation of the definition of cutpoint. Should any block not be a cycle then it must contain a butterfly subgraph B_2. G would then have a chain subgraph, N_4 and hence $\gamma_M(G) \geq 2$ by Remark 3 and theorem B

If a non-cycle block B is not in H_n, for some n, then theorem 1 guarantees that it is in H. Then it is easy to see that B and any cycle with which it shares a vertex create a subgraph with maximum genus two. Hence under these circumstances B must be in H_n, for some n.

THEOREM 2. G is a planar connected graph with R(G) = 1 if and only if $G \setminus \bigcup x$, where the union is taken over all bridges x of G, can be written as the disjoint union of connected graphs, $G \setminus \bigcup x = \bigcup_{i=1}^{n} G_i$, $(G_i \cap G_j = \phi,\ i \neq j)$ so that exactly one G_i is not a cycle. This exceptional graph is either a block of maximum genus one, of form R_3, or a graph G as described in Lemma 3.

Proof. Suppose G is any connected planar graph with R(G) = 1 . The additivity result gives $1 = \gamma_M(G) = \sum_{i=1}^{n} \gamma_M(G_1)$. Hence, there is exactly one G_i of maximum genus one. Since each G_i is bridgeless, any cutpoint must have multiplicity larger than one. Lemmas 1-3 combine to give the result.

Clearly, the graphs described have maximum genus one and hence embedding range one.

III. Toroidal graphs of embedding range one.

For the embedding range of a toroidal graph to be one it must have maximum genus two. Since $\gamma_M(K_5) = 3$ [see 3] and $\gamma_M(K_{3,3}) = 2$ [see 5], Kuratowski's theorem implies that any toriodal graph of maximum genus two must contain a subgraph homeomorphic with $K_{3,3}$. Given edges x, y and edges z, w, all of $K_{3,3}$ so that either both pairs are incident or neither pair is, we note that there is a graph isomorphism of $K_{3,3}$ to itself so that $\{x,y\}$ is mapped to $\{z,w\}$. This fact will allow some freedom when selecting arbitrary edges in the proof of the next theorem. For the proof of theorem 3 it is convenient to have $K_{3,3}$ embedded with labelled vertices. Let one of the bipartile sets be {A,B,C} and the other be {1,2,3}. When one defines an Edmond's embedding scheme by $p_A = p_B = p_C$ (123), $p_2 = p_3 =$ (ACB), $p_1 =$ (ABC), an embedding with one face on a surface of genus two results. The face is schematically given in Figure 4, where we imagine the edges directed counterclockwise.

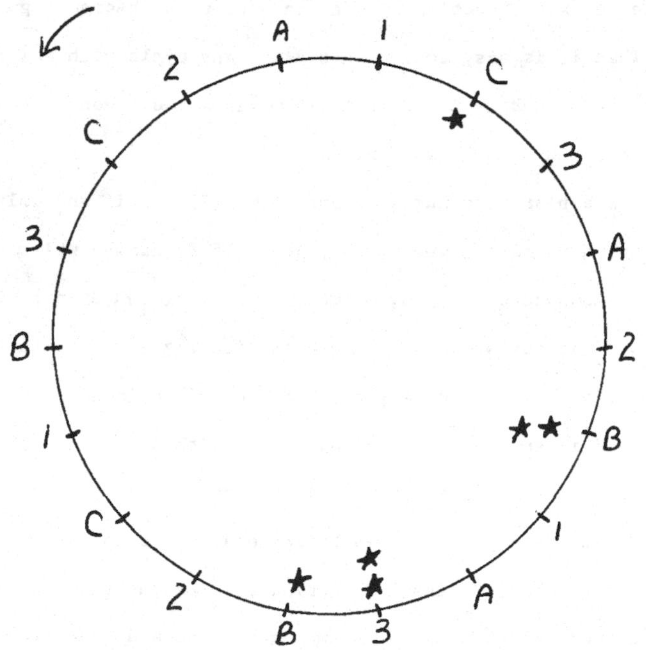

Figure 4

THEOREM 3. G is a toroidal block with $R(G) = 1$ if and only if G is homeomorphic with K or with $K \cup P$, where K is a homeomorph of $K_{3,3}$ and P is path between vertices of K, open disjoint from K.

Proof. Clearly, the E.A.T. applied to the given embedding gives $\gamma_M(K \cup P) = 2$.

Let G be any toroidal block of embedding range one. Let K be the subgraph of G homeomorphic with $K_{3,3}$. If K is a proper subgraph, then because G is a block it must contain a subgraph homeomorphic with $K \cup P$, for some path P, open disjoint from K. If this subgraph is proper, then there are two vertices of $K \cup P$ which may be joined by an open disjoint path in such a manner that a graph of maximum genus two results. We show that the addition of any open disjoint path to $K \cup P$ produces a graph of maximum genus three. Henceforth, we assume that P was added to K as in Remark 1. Remark 2 implies that no other path with an endpoint on P may be added. We thus consider a graph formed from K by the addition of two paths, P_1 and P_2, so that each is disjoint from the other and open disjoint from K.

We discuss $K_{3,3}$ rather than K. If either P_1 or P_2 has an endpoint of degree two in K, we will often consider the path to terminate at an interior point on an edge in $K_{3,3}$. We consider cases based on the degrees in K of the endpoints of P_1 and P_2. In all cases the order in which the endpoints are stated will refer to the directed edges in Figure 4. In cases 1 and 2 we assume that no edge of $K_{3,3}$ has more than one endpoint of $P_1 \cup P_2$ on it.

Case 1: Neither path has both endpoints of degree two.

a) P_1 has both degree three endpoints.

Let the endpoints be B and C if they are nonadjacent, 1 and 3 if they are. (The proof is the same in either case.) If one constructs P_1 from B to C as starred in the face of Figure 4, one then has an embedding of $K_{3,3} \cup P_1$ with each vertex of $K_{3,3}$ appearing in both faces. Since P_2 must have an endvertex in $K_{3,3}$, remark 2 completes this case.

b) Each path has exactly one endpoint of degree two.

If the edge endpoint for P_1 is incident with its vertex endpoint, choose them to be 1A and 1, respectively. If not choose 1A and B or 1A and 3, depending on the adjacency relationship between the vertices involved. In each case, when an embedding is made by joining the appropriate edge and the vertex identified by ** in Figure 4, a two face embedding in which we have each $K_{3,3}$ vertex in both faces is formed. Because P_2 must have one endpoint in $K_{3,3}$, remark 2 completes the case.

Case 2: P_1 has both endpoints as interior points of different edges of $K_{3,3}$.

a) P_2 has at least one vertex in $K_{3,3}$.

If the endpoints of P_1 appear on incident edges of $K_{3,3}$ choose them as 1A and 1B in Figure 4, otherwise choose 1A and 2B. In either case, each vertex of $K_{3,3}$ is then in each face of the embedding so formed. Thus, remark 2 completes the proof.

b) P_2 has neither endpoint a vertex of $K_{3,3}$.

We assume the endpoints of P_2 lie on different edges of $K_{3,3}$. Otherwise we go to case 3. Choose the endpoints of P_1 to be 1A and 1B if the edges on which they occur are incident, 1A and B3 otherwise. The details are similar to those of the other cases. For each of the eight possible ways in which P_2's endpoints might be incidence related to those of P_1 (and to each other), there is an appropriate choice of edges of $K_{3,3}$ so that a path with endpoints on these edges can join points in different faces of the embedding for $K \cup P_1$. For example, if the endpoints of P_1 are on incident edges as are those of P_2 while exactly one endpoint of P_2 lies on an edge incident with exactly one edge containing an endpoint of P_1, then choose C2 and B2 as the endpoint edges for P_2.

Case 3. Two vertices of degree two on the same edge of $K_{3,3}$ are endpoints for P_1 or P_2.

a) The paths have an endpoint on a common edge.

Choose the edge to be 1A. Notice that A1 occurs in Figure 4 between 2B and B2 and between B1 and 1B. Suppose the degree two endpoint on 1A of P_i is u_i, i = 1, 2. Let x be the edge in $K_{3,3}$ from 2 to B (or 1 to B, appropriately).

One may then choose the direction (location) of x in the face of the embedding so that a path from u_1 to x places u_2 (on 1A) in one face and A1 in the other. Hence u_2 is in both faces of the embedding formed for $K \cup P_1$.

b) P_1 has both endpoints on the same edge of $K_{3,3}$, while P_2 has neither endpoint on that edge.

If an edge for an endpoint of P_2 is incident with the common edge, choose the common edge to be 1A for P_1 and one end edge as 1B for P_2. If not choose 1A for P_1 and 2B for an end edge for P_2. Choose one endpoint of P_1 on 1A and the other on A1. Then the endpoints for P_2 may be regarded as in different faces in the embedding formed for $K \cup P_1$, since B1 and 1B, as well as 2B and B2, lie in different faces of the embedding for $K \cup P_1$

In all possible cases, then, $K \cup P_1 \cup P_2$ has a one face embedding and hence has maximum genus three, which completes the proof. Q.E.D.

THEOREM 4. G is a toroidal connected graph with R(G) = 1 if and only if $M(c) \geq 1$ for at most one vertex of G. Furthermore exactly one block of G is not a cycle or K_2. This unusual block is as described in theorem 3. If M(c) = 2, then c has one block homeomorphic with $K_{3,3}$ and the other a cycle.

<u>Proof</u>: Clearly, the graphs as described are toroidal with embedding range one.

Since G is toroidal it must have exactly one toroidal block, B, by genus additivity [1]. If B is not homeomorphic with $K_{3,3}$, then should some other cyclic block share a vertex with B, the graph so formed would have maximum genus three, because the proof of theorem 3 shows us that any vertex of B may be regarded as in both faces of the maximum embedding for B. If B is homeomorphic with $K_{3,3}$, then one can show that a path from a vertex to itself added to $K_{3,3}$ results in a graph with the property just described. Should two cyclic blocks share a vertex, the graph so formed would have maximum genus one and, by theorem B and the corollary to theorem C, G would have maximum genus at least three. Q.E.D.

It is an open question as to whether there are graphs of higher genus and embedding range one.

References

1. J. Battle, F. Harary, Y. Kodama, and J.W.T. Youngs, Additivity of the genus of a graph, Bull. Amer. Math. Soc. 68(1962), 565-568.

2. E. Nordhaus, R. Ringeisen, B. Stewart, and A. White, A Kuratowski-Type Theorem for the Maximum Genus of a Graph, J. Combinatorial Theory B, 12, No. 3, (1972), 260-267.

3. E. Nordhaus, B. Stewart, A. White, On the Maximum Genus of a Graph, J. Combinatorial Theory, 11, No. 3 (1971), 258-267.

4. R. Ringeisen, Graphs of Given Genus and Arbitrarily large Maximum Genus Discrete Math. 6 (1973), 169-174.

5. R. Ringeisen, Determining All Compact Orientable 2-Manifolds upon which $K_{m,n}$ has 2-cell Imbeddings, J. Combinatorial Theory, 12, No. 2 (1972), 101-104.

NON-EXISTENCE OF GRAPH EMBEDDINGS*

Gerhard Ringel
University of California, Santa Cruz
Santa Cruz, California

Twenty-five years ago I prepared my first paper [11]. It showed the existence of a triangular embedding of the complete graph K_n into a non-orientable surface for each $n \equiv 3 \pmod 6$. In other words, I proved the formula,

$$(1) \quad \bar{\gamma}(K_n) = \left\{\frac{(n-3)(n-4)}{6}\right\}, \quad (n \neq 7)$$

for the non-orientable genus of the complete graph K_n, only in the above mentioned case. At this time my thesis supervisor, Sperner, recommended that I search for a general proof, as he predicted another mathematician would do so as soon as my proof was published. Although today this formula is proven for all $n \neq 7$, even the most up-to-date proof is divided into 12 cases, i.e. for each $n \pmod{12}$, and each particular case requires its own ad hoc proof.

There is also a well-known formula for the orientable genus of K_n:

$$(2) \quad \gamma(K_n) = \left\{\frac{(n-3)(n-4)}{12}\right\}, \quad \text{for } n \geq 3.$$

The proofs for both formulas are presented in my book Map Color Theorem [13], which has been available for more than a year. One may wonder if this book is still up-to-date, and I believe that basically it is.

---

* Research supported by the National Science Foundation.

Again, as for formula (1), the proof for formula (2) is divided into 12 cases. In fact, if one uses the very elegant method of current graphs to prove the formula, the need for 12 cases becomes compulsory and ineluctable. Even though there are interesting explanations and interpretations (Alpert and Gross [4]) of the method of current graphs and quotient graphs (branched coverings) as of yet there have been no improvements in the various ad hoc proofs. Especially when one reviews the present proof of the orientable case for $n \equiv 6 \pmod{12}$, the urgency for a nicer, shorter proof becomes desirable. However, whenever I mention this to young graph theorists, they reply that they are not interested in dealing with one specific case, but would rather discover a general theorem which can solve the problem in one step, thereby dispensing with the necessity of ad hoc proofs. Of course I tried this also 25 years ago, and after a certain time I felt that such a theorem probably did not exist. At that time it was just a feeling, and today I will justify this feeling somewhat.

First, I would like to make a remark concerning the non-orientable genus of K_n. The worst and most awkward case in my book is the case $n \equiv 1 \pmod{12}$. Immediately after the book was printed, my friend, Mark Jungerman [8], found a neat and brief proof for this case.

Triangular Embeddings. We shall now consider graphs which are connected and where each vertex has valence ≥ 3. Let G be a graph and S a closed surface. If there exists an embedding of G into S we write $G \subset S$. Moreover, if this is a triangular embedding (all faces are triangles) then we write $G \lhd S$. Let $E(S)$ be the Euler characteristic of S. We denote the number of vertices and edges of G by α_0 and α_1 respectively.

Theorem 1. If an embedding $G \subset S$ exists then

(3) $\alpha_1 \leq 3\,\alpha_0 - 3E(S)$.

Theorem 2. If an triangular embedding $G \lhd S$ exists then

(4) $\alpha_1 = 3\,\alpha_0 - 3E(S)$.

These are well known theorems [13, p. 55-59]. Condition (4) is practically the Euler formula. Heawood [5] in 1890 had the strong feeling that (4) already guarantees the existance of the embedding $G \lhd S$ at least in the case where G is the complete graph K_n.

However there are many counter examples. For instance take S = sphere and $G = K_6 - K_3$. This graph is obtained from K_6 by removing 3 edges forming a 3-cycle (closed way of length 3). It is easy to see that $G \lhd S$ does not exist although (4) is true. Let W_n be a graph with $n+1$ vertices $(n \geq 3)$ where one vertex P has valence n and all the others have valence 3 with the property that $W_n - P$ is a n-cycle. We call W_n a wheel of order n and P the center of W_n. (If $n = 3$ each vertex of W_3 can play the center).

Theorem 3. If a triangular embedding of a graph G into a closed surface exists then G has the property

(5) If P is a vertex of valence h in G then there exists a wheel-subgraph in G of order h with center P.

Proof. Consider the embedding $G \lhd S$. Since P has valence h, P is incident with h triangles. Their vertices and edges form a wheel of order h and with center P.

The graph $K_6 - K_3$ does not have property (5) and is therefore not triangularly embeddable into any surface.

It is now the question: are the two properties (4) and (5) together sufficient for the existence of a triangular embedding $G \triangleleft S$?

We denote the orientable or non-orientable surface of genus p by S_p or N_p respectively. So S_1 is the torus and N_2 the Klein's bottle. For a long time there were only two examples known where (4) and (5) are not sufficient:

1) $K_7 \ntriangleleft N_2$ (Franklin 1934, [3])

2) $K_8 - K_2 \ntriangleleft N_3$ (Ringel 1955, [12]).

Surprisingly within the last couple of years there were many more counter examples discovered. By $K(n,n,n,n)$ we mean the quadripartite graph with 4 times n vertices.

3) $K(2,2,2,2) \ntriangleleft N_2$ (Jungerman 1976 [10])

4-8) K_9 - 3 edges $\ntriangleleft S_2$ (Huneke 1975 [6]).

Of course there are five different ways to subtract 3 edges from K_9. For all five possibilities (4) and (5) are true and the embedding still does not exist. This was discovered by Jungerman and Ringel and was proved using a computer. Independently Huneke published a written proof.

9) $K_{11} - K_5 \ntriangleleft S_3$ (Jungerman 1974 [7])

10) $K(3,3,3,3) \ntriangleleft S_4$ (Jungerman 1975 [9])

11) $K_{13} - K_6 \ntriangleleft S_5$ (Jungerman 1974 [7])

If we only consider orientable surfaces S_p of genus p then theorem 2 can be written in the form:

Theorem 2a. *If a triangular embedding* $G \triangleleft S_p$ *exists then*

(6) $p = \gamma(G) = \frac{1}{6}(\alpha_1 - 3\alpha_0 + 6)$.

Herein $\gamma(G)$ denotes the genus of G, i.e. the smallest genus p with $G \subset S_p$.

We consider the following graph F_n with $5n$ vertices and $20n - 10$ edges (see Figure 1 for $n = 8$).

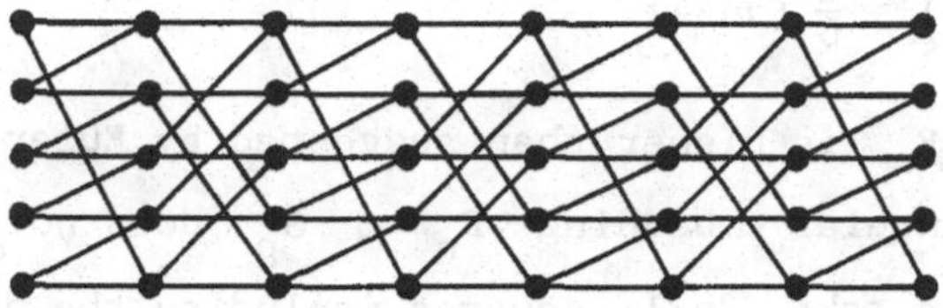

Figure 1

The vertices are denoted by the pairs

$$(r, i)$$

with $r = 0, 1, 2, 3, 4$, and $i = 1, 2, 3, \ldots, n$. The first component r we consider as residue class (mod 5). Two different vertices (r,i) and (s,j) are joined by an edge if

$i = j$ or if

$s = r$ and $j = i+1$ $(i = 1,2,\cdots,n-1)$ or if

$s \equiv r-1 \pmod 5$ and $j = i+1$ $(i = 1,3,5\cdots)$ or if

$s \equiv r - 2 \pmod 5$ and $j = i+1$ $(i=2,4,\cdots)$.

In Figure 1 we see 8 sets of 5 vertices arranged in vertical columns. If we join each pair of vertices by an edge if they are in the same column we obtain an illustration of F_8.

It can easily be checked that the graph F_n satisfies conditions (5).

The graph F_n contains a subgraph H consisting of n disjoint copies of K_5. Using the additivity of the genus of a graph found by Battle, Harary, Kodama, Youngs [1], we obtain

$$\gamma(F_n) \geq \gamma(H) = n\, \gamma(K_5) = n.$$

Now assume $F_n \lhd S_p$ then theorem 2a would say

(7) $p = \frac{1}{6}(\alpha_1 - 3\alpha_0 + 6) = \frac{1}{6}(5n-4).$

So the actual genus of F_n is higher than suggested by Euler's Formula (7). And a triangular embedding $F_n \lhd S_p$ does not exist.

As a second example take $2n-1$ squares including the diagonals arranged as in Figure 2 ($n = 6$). Equivalently take a sequence of $2n-1$ copies of K_4 where each copy has one edge (including the two vertices) in common with the next copy. Then take one extra vertex P and join it with each of the other $4n$ vertices (these $4n$ edges are not drawn in the Figure 2).

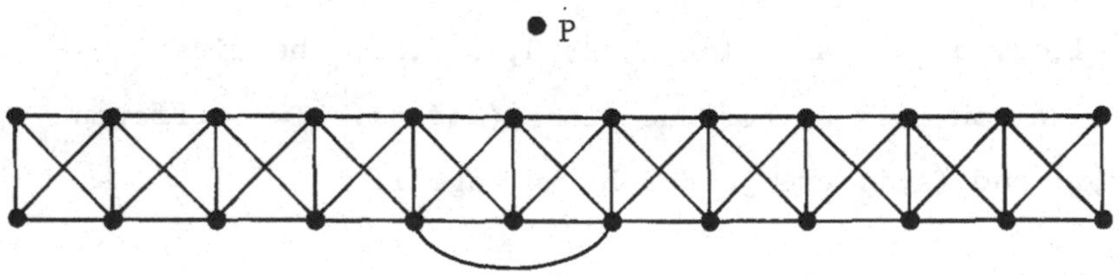

Figure 2

Then add one extra edge as in Figure 2 bottom. The resulting graph G_n has

$$\alpha_0 = 4n+1 \quad \text{vertices and}$$

$$\alpha_1 = 14n-3 \text{ edges.}$$

It can be checked that G_n satisfies property (5). G_n has a subgraph H consisting of n copies of K_5 where any two of them have only the vertex P in common. Therefore

$$\gamma(G_n) \geq \gamma(H) = n\,\gamma(K_5) = n.$$

Here we have used the fact that the genus of a graph equals the sum of the genus of its blocks. In this case each block is a copy of K_5 (see again [1]). However Euler's Formula (6) suggests $p = \frac{n}{3}$ as the genus of G_n _if_ there exists a triangular embedding of G_n into S_p. But in reality $\gamma(G_n) \geq n$, hence $G_n \not\triangleleft S_p$. These two examples dispose of some conjectures I occasionally have heard.

Bouchet gave me another infinite series of counter-examples which work for orientable and non-orientable surfaces simultaneously. (Personal communication).

From these examples we see that the properties (4) and (5) together do not guarantee the existence of an embedding $G \triangleleft S$. However, in all the counter examples, S was different from the sphere. In fact if we assume S = sphere, (4) and (5) is sufficient. In other words the following theorem holds [14]:

<u>Theorem of Skupien</u>: <u>Let</u> G <u>be a graph with</u> α_0 <u>vertices and</u> α_1 <u>edges</u>. <u>Assume</u> $\alpha_1 = 3\alpha_0 - 6$ <u>and</u> G <u>satisfies property</u> (5). <u>Then</u> G <u>is a planar graph</u>.

<u>Torus and Klein's Bottle</u>. It is known that both graphs K_7 and $K(2,2,2,2)$ are embeddable into S_1 but not into N_2, as I previously mentioned. There are more of this kind. Let Q_4 the 4-dimensional cube graph and T the graph of Figure 3.

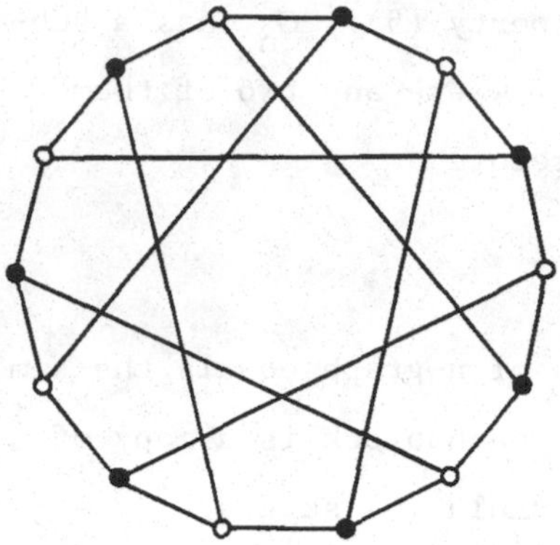

Figure 3

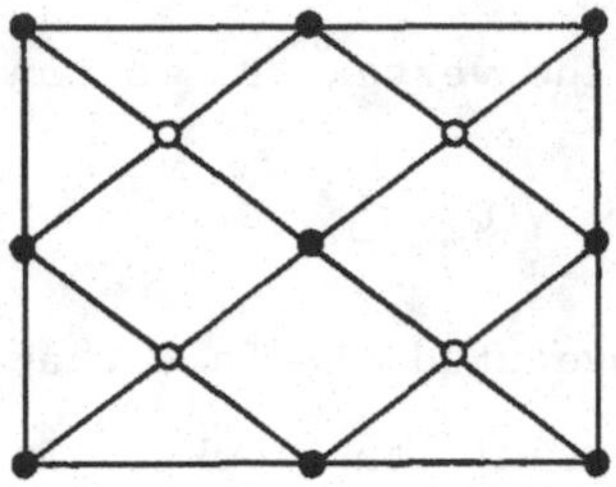

Figure 4

It is one of Tutte's [16] cages. Here we have the following:

$Q_4 \subset S_1$, $Q_4 \not\subset N_2$ (Stahl 1975 [15])

$T \subset S_1$, $T \not\subset N_2$.

We omit the proof here. There is also a nice graph which is embeddable in both surfaces, namely the bipartite graph K(4,4):

$K(4,4) \subset S_1$, $K(4,4) \subset N_2$.

It is remarkable that Figure 4 can illustrate both embeddings. It only depends whether the four sides of the rectangle are identified as $a\ b\ a^{-1}\ b^{-1}$ generating a torus or as $a\ c\ a^{-1}\ c$ generating Klein's Bottle.

The graphs Q_4, T and K(4,4) don't have 3-cycles. So the faces in the embeddings are not triangles, in fact, they are quadrangles in case of Q_4 and (K(4,4) and hexagons in case of T. The properties (4) and (5) have to be modified accordingly.

Now we consider again triangular embeddings only. By C_4 we denote a 4-cycle.

$K_8 - C_4 \lhd N_2$ (Ringel 1955 [12]).

Suprisingly this graph is not embeddable into the torus:

$$K_8 - C_4 \not\triangleleft S_1 \quad \text{(Duke and Haggard 1972 [2])}.$$

The proof is a discussion of all possibilities. It took me quite a while to find a graph which is not embeddable in S_1 and N_2. The graph R is pictured in Figure 5.

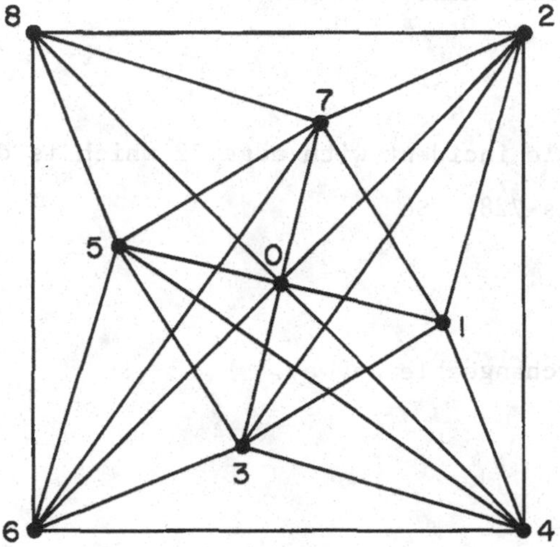

Figure 5

It looks very regular, however notice that there is no edge connecting 1 and 8. The graph R has $\alpha_0 = 9$ vertices and $\alpha_1 = 27$ edges. So (4) is satisfied with $E(S) = 0$. Property (5) can also be checked easily. Despite all of this:

$$R \not\triangleleft S_1 \text{ and } R \not\triangleleft N_2.$$

Here I like to give a proof. Assume an embedding $R \triangleleft S$ exists with $E(S) = 0$. We describe the embedding as in [13] by a combinatorial scheme satisfying Rule R [13, p. 70]. Since there are only two 3-cycles incident with edge 71, the line

1. 0 7 2 ··

is part of the scheme. There are only two more vertices 3 and 4 adjacent to 1. So the line reads:

1. 0 7 2 3 4

(or 3 and 4 are exchanged, which is not really a different case because of symmetry). From Rule R we get

7. 0 1 2 · · ·

There is only one 3-cycle incident with edge 72 which is different from 721 and 270. It is 728. So

7. 0 1 2 8 5 6

Again 5 and 6 are interchangeable anyway.
From Rule R we get

8. 5 7 2 ···

and the two blanks are 06 or 60. But there is no 3-cycle 826 in the graph R. So we have

8. 5 7 2 0 6

Now we construct row 2 by comparing with rows 8, 7, 1 in that order using Rule R. We obtain

2. 0 8 7 1 3 4

where last number has to be 4 in order to complete row 2. Now construct row 0 by comparing with rows 1, 7, 8, 2 in that order. We obtain

0. 4 1 7 6 8 2

as a complete row 0. However 0 is also adjacent to 5 and 6. This is a contradiction. So the two embeddings of R do not exist.

Finally I like to mention that there are graphs which triangulate both the torus and Klein's Bottle, for instance $K_8 - D$ where D is one of the three graphs .

REFERENCES

1) Battle, J., Harary, F., Kodama, Y., Youngs, J.W.T., Additivity of the genus of a graph. Byll. Amer. Math. Soc. 68, 565-568 (1962).

2) Duke, R.A., and Haggard, G., The genus of subgraphs of K_8, Israel J. of Math., Vol. 11, No. 4, 1972, pp 452-455.

3) Franklin, P., A Six Colour Problem, J. Math. Phys. 13, 363-369 (1934).

4) Gross, J. and Alpert, S., The topological theory of current graphs, to appear.

5) Heawood, P.J., Map Colour Theorem, Quart, J. Math. 24, 332-338 (1890).

6) Huneke, J.P., A Minimum Vertex Triangulation. Submitted to J. Combinatorial Theory.

7) Jungerman, M., Orientable Triangular Embeddings of K_{18}-K_3 and K_{13}-K_3, J. Combinatorial Theory, Series B, Vol. 16, 293-294 (1974).

8) Jungerman, M., A New Solution for the Nonorientable Case 1 of the Heawood Map Color Theorem, J. Combinatorial Theory, Vol. 19, 69-71 (1975).

9) Jungerman, M., The Genus of the Symmetric Quadripartite Graph. J. Combinatorial Theory, Series B, Vol. 19, 181-197 (1975).

10) Jungerman, M., The Non-Orientable Genus of the Symmetric Quadripartite Graph, Submitted to the J. Combinatorial Theory, Series B, 1976.

11) Ringel, G., Farbensatz für nichtorientierbare Flächen beliebigen Geschlechts. Journal für die reine und angewandte Mathematik, Band 190, Seite 129-147 (1952).

12) Ringel, G., Wie man die geschlossenen nichtorientierbaren Flächen in möglichst wenig Dreiecke zerlegen kann, Math. Annalen, Bd. 130, S. 317-326 (1955).

13) Ringel, G., Map Color Theorem, Springer-Verlag, Die Grundlehren der mathematischen Wissenschaften in Einzeldarstellungen Band 209, 1974.

14) Skupien, Z., Locally Hamiltonian and Planar Graphs, Fundamenta Mathematica, Vol. 58, (193-200) 1966.

15) Stahl, S., Self-dual Embeddings of Graphs, Doctoral dissertation, Western Michigan University, August 1975.

16) Tutte, W.T., A family of cubical graphs, Proc. Cambridge Philos. Soc. 43, 459-474 (1947).

FOOD WEBS, COMPETITION GRAPHS, AND THE BOXICITY OF ECOLOGICAL PHASE SPACE

Fred S. Roberts
Rutgers University
New Brunswick, NJ 08903

Abstract

Two species in an ecosystem compete if and only if they have intersecting ecological niches. Competition can be defined independently by using a food web for the ecosystem, and this notion of competition gives rise to a competition graph. This paper briefly describes the problem of representing the competition graph as an intersection graph of boxes (k-dimensional rectangles representing ecological niches) in Euclidean k-space and then discusses the class of graphs which arise as competition graphs of (acyclic) food webs.

1. Ecological Phase Space

The normal, healthy environment of a species of animal or plant can be characterized by considering lower and upper bounds on various dimensions such as temperature, moisture, pH, etc. If k different dimensions are used, and lower and upper bounds are set on each dimension, the region in k-dimensional Euclidean space of all points lying within the bounds on every dimension is a k-dimensional rectangle with sides parallel to the coordinate axes. Following Danzer and Grünbaum [1], we call such a region a box. This region corresponds to what is often called the species' ecological niche. The k-dimensional Euclidean space is called ecological phase space. A basic ecological principle is that two species compete if and only if their ecological niches overlap. Joel Cohen, in some unpublished work, has suggested the following problem: if we start with an independent notion of competition, what is the minimum number of dimensions which must be used to describe an ecological phase space in which at least the notion of competition is captured, i.e., in which exactly the competing species have overlapping niches?

To be more precise about this problem, suppose we start with an independent notion of competition and define a <u>competition graph</u> $G = (V, E)$ (an undirected graph) by taking as vertices the species in an ecosystem and taking an edge between two species if and only if they compete. (We follow the graph-theoretical notation and terminology of Roberts [6].) We would like to find a number k and an assignment to each species u of a box $B(u)$ in Euclidean k-space such that for all $u \neq v$ in $V(G)$,

$$\{u, v\} \in E(G) \quad \text{iff} \quad B(u) \cap B(v) \neq \emptyset. \tag{*}$$

There is an assignment satisfying (*) if and only if G is the intersection graph of a family of boxes in k-space. (For the definition of intersection graph, see Roberts [6, Sec. 3.4].) The smallest k such that there is such an assignment is called in Roberts [7] the <u>boxicity</u> of G. Recent results on the computation of boxicity are contained in the paper by Gabai [3].

Joel Cohen suggests defining competition in an ecosystem as follows. Start with a <u>food web</u> for the ecosystem, a directed graph whose vertices are the species of the system and which has an arc from vertex u to vertex v if and only if u preys on v. Food webs throughout this paper will be assumed, as is commonly done, to be acyclic digraphs. Given a food web F, we say that species u and v <u>compete</u> if and only if they have some common prey. Figure 1 shows a simple food web and the competition graph defined using this notion of competition. For example, the Rabbit and the Insect compete because they both prey on Grass.

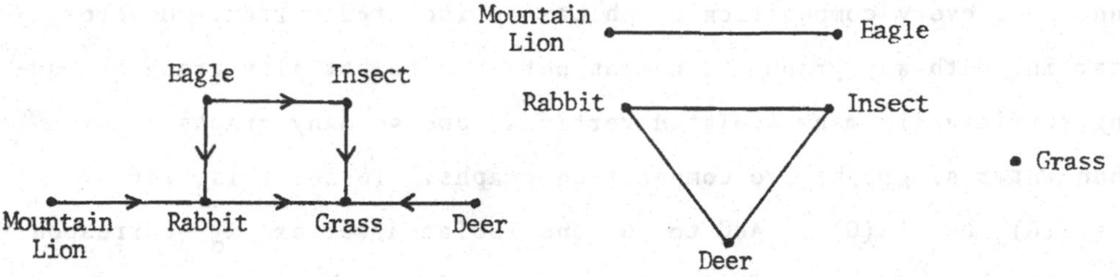

Figure 1.

Cohen and others have analyzed the number of dimensions needed to account for competition as defined from various empirically-determined food webs. Remarkably, each such food web has led to a competition graph with boxicity at most 1. (Graphs with boxicity 0 are the complete graphs.) Graphs with boxicity at most 1 are exactly the ones which can be represented as the intersection graphs of intervals on the line, the so-called <u>interval graphs</u>. Thus, every food web which has been investigated as of this writing gives rise to a competition graph which is an interval graph. (See Roberts [6, Sec. 3.5] for a larger example than the one in Fig. 1.) No one is sure whether this is an accident, perhaps resulting from a not very thorough analysis of food webs, or whether it reflects an important ecological discovery. No one is sure of the interpretation of the single dimension needed to describe competition.

Mathematically, one is led to ask: can graphs other than interval graphs arise as competition graphs of (acyclic) food webs? In fact, what graphs arise this way? We shall investigate these questions in what follows.

2. The Competition Number $k(G)$

We shall say that a graph G is a <u>competition graph</u> if it arises as the competition graph from some acyclic food web. Not every graph is a competition graph. For every acyclic digraph has a vertex with no outgoing arcs (Harary, Norman, and Cartwright [4, p. 64]), and hence every competition graph has an isolated vertex. However, starting with any graph G, we can obtain a competition graph by adding sufficiently many isolated vertices, and so many graphs other than interval graphs are competition graphs. To see this, let $e = e(G)$ be $|E(G)|$. Add to G one isolated vertex x_α corresponding to each edge α of G. Construct a food web F with vertex set

$$V(F) = V(G) \cup \{x_\alpha: \ \alpha \in E(G)\}$$

and with an arc from the end points a and b of edge α to vertex x_α. Then F is a food web for the graph $G \cup I_e$, the disjoint union of G and a graph I_e consisting of e isolated vertices.

It now makes sense to define $k(G)$, the <u>competition number</u> of G, as the smallest k such that $G \cup I_k$ is a competition graph. (The number $k(G)$ is 0 if G is already a competition graph.) Characterization of the class of competition graphs reduces to the problem of computing $k(G)$ for all graphs G. To begin our investigation of $k(G)$, we shall use the notation $n(G)$ for $|V(G)|$.

<u>Proposition 1</u>. If G is a graph without triangles (cliques of size 3), then $k(G) \geqq e(G) - n(G) + 2$.

<u>Proof</u>. Let $n = n(G)$, $e = e(G)$, and $k = k(G)$. Suppose that $G \cup I_k$ is a competition graph, with corresponding food web F. According to Harary, Norman, and Cartwright [4, p. 269], we can assign the integers $1, 2, \ldots, n+k$ to the $n+k$ vertices of F so that every arc goes from a lower number to a higher number. In particular, it follows that the vertex labelled 1 has no incoming arcs and the vertex labelled 2 has at most one incoming arc. For each edge $\alpha = \{u, v\}$ of G, there is a vertex a_α in F such that u and v both prey on a_α in F. Moreover, since G has no triangles, the a_α are distinct. It follows that $G \cup I_k$ has at least e vertices a_α. Moreover, since these e vertices all have at least two incoming

arcs in F, at least two of the vertices of $G \cup I_k$ are not a_α's, the vertices labelled 1 and 2. It follows that $n+k-2 \geq e$. Q.E.D.

That the bound in Proposition 1 is not sharp is easily seen by taking G to be I_3. For $k(G) = 0$ while $e(G)-n(G)+2 = -1$. We shall see below (Theorem 2) that the bound is sharp if G is connected, has no triangles, and has at least two vertices. Proposition 1 also shows that if Z_n is the circuit of length n, $n > 3$, then $k(Z_n) \geq 2$. We shall see below that $k(Z_n) = 2$. (It is easy to show that $k(Z_3) = 1$.)

Proposition 2. For every positive integer k, there is a graph G such that $k(G) > k$.

Proof. Consider two copies of Z_k, $k \geq 4$. Join corresponding vertices by an edge, as shown in Fig. 2 for $k = 4$. The resulting graph G has no triangles, it has $e = 3k$, and $n = 2k$. Thus, by Proposition 1, $k(G) \geq 3k - 2k + 2 = k + 2$. Q.E.D.

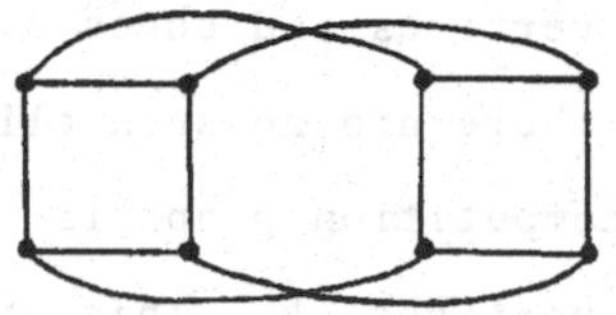

Figure 2.

As an aside, we note that $k(G)$ can be much larger than $n(G)$. Let $f(n)$ be the largest $k(G)$ for all graphs G with n vertices. Since $k(G) \leq e(G)$, it follows that $f(n) \leq n(n-1)/2$. However, for n even, for example, it is easy to show that $f(n) > n(n-4)/4$. To see this, let $p = n/2$. Let G be the complete bipartite graph $K(p, p)$. Then G has no triangles, $e(G) = p^2$, and $n(G) = 2p$. Hence, by Proposition 1, $k(G) \geq p^2 - 2p + 2 = (n/2)^2 - n + 2$, which is greater than $n(n-4)/4$.

3. The Elimination Number m(G)

If a is a vertex of a graph G, the (open) neighborhood of a, $N(a)$, is defined as follows:

$$N(a) = \{b \in V(G): \{a, b\} \in E(G)\}.$$

We will also use $N(a)$ for the subgraph generated (induced) by this set of vertices. The number $\gamma(a)$ is defined to be 0 if $N(a) = \emptyset$ and otherwise it is the least number of cliques which cover the vertices and edges of $N(a)$. Finally, $G\Delta a$ will denote the subgraph generated by vertices of G other than a, less edges of $N(a)$.

Suppose $a_1, a_2, \ldots, a_n$ is any ordering of the vertices of the graph G. This induces a sequence of graphs $G_1, G_2, \ldots, G_n$ as follows: G_1 is G, and G_i is $G_{i-1}\Delta a_{i-1}$. This sequence and the following concepts are illustrated in Fig. 3. Let $N_i(a_i)$ be $N(a_i)$ in G_i and let γ_i be $\gamma(a_i)$ in G_i, i.e., 0 if $N_i(a_i) = \emptyset$ and otherwise the least number of cliques of G_i which cover the vertices and edges of $N_i(a_i)$. Finally, let $K_{i1}, K_{i2}, \ldots, K_{i\gamma_i}$ be a set of cliques covering the vertices and edges of $N_i(a_i)$. Of course, if $N_i(a_i) = \emptyset$, $\gamma_i = 0$ and there are no such cliques. We shall build a food web F whose competition graph is $G \cup I_k$ for some k by adding sufficiently many vertices k. This will give us an upper bound for $k(G)$.

The food web F is constructed in steps. In brief, we construct a sequence of food webs $F_1, F_2, \ldots, F_{n-1}$, with $F_{n-1} = F$. The food web F_i will be obtained from the food web F_{i-1} by adding a common prey for and hence competition between species u and v if and only if u and v are species in G, they compete in G_i, and they do not compete in G_{i+1}. Since G_n has no competing vertices, $F=F_{n-1}$ will be a food web with competition between two vertices if and only if those vertices compete in G. The steps of the construction are as follows. Each step is illustrated in Figure 3.

Step 1. Build a food web F_1 by starting with the vertices of G. Introduce new vertices $x_{11}, x_{12}, \ldots, x_{1\gamma_1}$. (If $\gamma_1 = 0$, introduce no new vertices.) Add arcs from a_1 and vertices of K_{1j} to x_{1j}. In F_1, vertex a_1 competes with vertices of $N(a_1)$ (since the K_{1j} cover the vertices of $N(a_1)$) and vertices within $N(a_1)$ compete if and only if they did in $G = G_1$ (since the K_{1j} cover the edges of $N(a_1)$). There are no other competing vertices in F_1. It now remains to introduce the competitions in $G\Delta a_1 = G_1\Delta a_1 = G_2$. For species u and v compete in F_1 if and only if u and v are in G and edge {u, v} is in G but not in G_2. Let $h_1 = \gamma_1$ count the number of new vertices introduced.

Step 2. Let $A_1 = \{a_1\}$. Let $g_1 = |A_1|$. Note that F_1 has no arcs leading into vertices of A_1 or $V(G_1)$.

Step 3. Extend food web F_1 to a food web F_2 by introducing vertices $x_{21}, x_{22}, \ldots, x_{2\gamma_2}$. (If $\gamma_2 = 0$, introduce no vertices.) The vertex a_1 can be used for x_{21}. Otherwise, new vertices from outside are used. Add arcs from a_2 and vertices of K_{2j} to x_{2j}. In F_2, the following new competitions have been added: a_2 with vertices of $N_2(a_2)$, and between vertices within $N_2(a_2)$ if these vertices competed in G_2. It now remains to introduce the competitions in $G_2\Delta A_2 = G_3$. For species u and v compete in F_2 if and only if u and v are in G and edge {u, v} is in G but not in G_3. Altogether, h_2 vertices were added from the outside, where h_2 is 0 if $\gamma_2 = 0$ or if $\gamma_2 \leqq g_1 = 1$, and h_2 is $\gamma_2 - g_1$ otherwise.

Step 4. Let A_2 be obtained from A_1 by deleting vertex a_1 if it was used in Step 3 and adding vertex a_2. Let $g_2 = |A_2|$. Note that F_2 has no arcs leading into vertices of A_2 or $V(G_2)$.

Assume that a food web F_{i-1} has been constructed with no arcs from any vertex into vertices of the set A_{i-1} or into vertices of $V(G_{i-1})$. Assume that F_{i-1} has an edge between vertices u and v if and only if u and v are in G and edge {u, v} is in G but is not in G_i.

Step 2i-1. Extend food web F_{i-1} to a food web F_i by introducing vertices $x_{i1}, x_{i2}, \ldots, x_{i\gamma_i}$. (If $\gamma_i = 0$, introduce no vertices.) The vertices of A_{i-1} should be used for as many of these vertices as possible. Add arcs from a_i and vertices of K_{ij} to x_{ij}. In F_i, the following new competitions have been added: a_i with vertices of $N_i(a_i)$, and between vertices within $N_i(a_i)$ if these vertices competed in G_i. No other competitions have been added since there were no arcs in F_{i-1} to vertices in A_{i-1} or, of course, to any of the new vertices introduced from outside. Thus, species u and v compete in F_i if and only if u and v are in G and edge {u, v} appears in G but not in G_{i+1}. Altogether, h_i vertices have been added from the outside, where h_i is 0 if $\gamma_i = 0$ or if $\gamma_i \leq g_{i-1}$, and h_i is $\gamma_i - g_{i-1}$ otherwise.

Step 2i. Let A_i be obtained from A_{i-1} by deleting those vertices of A_{i-1} used for x_{ij}'s in Step 2i-1, and adding vertex a_i. Let $g_i = |A_i|$. Observe that F_i has no arcs into vertices of the set A_i or into vertices of $V(G_i)$.

The food web $F = F_{n-1}$ has been constructed so that the sub-web F_i introduces exactly the competitions removed from G_i to obtain G_{i+1}. Thus, F introduces exactly the competitions in G. More precisely, F has an edge between vertices u and v if and only if u and v are in V(G) and edge {u, v} is in E(G). It is left to observe that F is acyclic. This follows since each arc of F goes from a vertex of G_i to a vertex not in G_i. The number of vertices added to F from G is given by $h_1 + h_2 + \ldots + h_{n-1} = k$.

Figure 3.

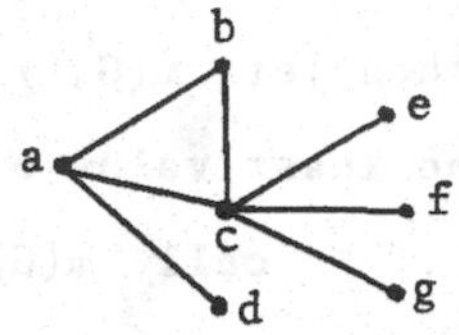 $G_1 = G$	**Steps 1 and 2** $a_1 = a$ $N_1(a_1) = \{b,c,d\}$ $\gamma_1 = 2$ $K_{11} = \{b,c\}$ $K_{12} = \{d\}$	Arcs added to obtain F_1 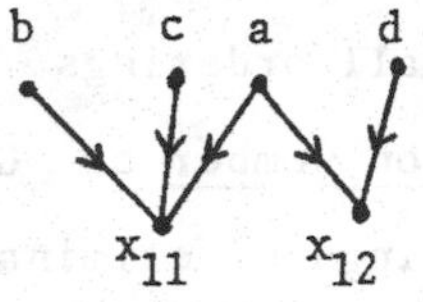	$h_1 = \gamma_1 = 2$ $A_1 = \{a\}$ $g_1 = 1$
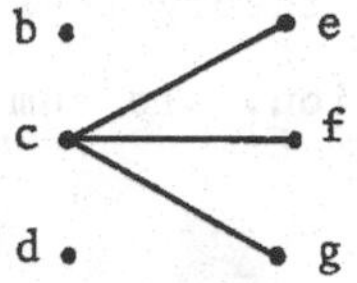 $G_2 = G_1 \Delta a_1$	**Steps 3 and 4** $a_2 = b$ $N_2(a_2) = \emptyset$ $\gamma_2 = 0$	Arcs added to obtain F_2 none	$h_2 = 0$ $A_2 = \{a, b\}$ $g_2 = 2$
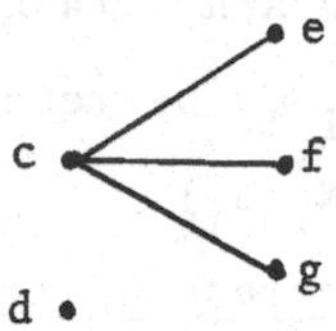 $G_3 = G_2 \Delta a_2$	**Steps 5 and 6** $a_3 = c$ $N_3(a_3) = \{e,f,g\}$ $\gamma_3 = 3$ $K_{31} = \{e\}$ $K_{32} = \{f\}$ $K_{33} = \{g\}$	Arcs added to obtain F_3 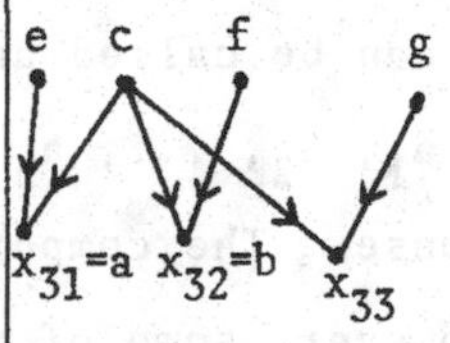	$h_3 = \gamma_3 - g_2 = 1$ $A_3 = \{c\}$ $g_3 = 1$
 $G_4 = G_3 \Delta a_3$	**Steps 7 and 8** $a_4 = d$ $N_4(a_4) = \emptyset$ $\gamma_4 = 0$	Arcs added to obtain F_4 none	$h_4 = 0$ $A_4 = \{c, d\}$ $g_4 = 2$

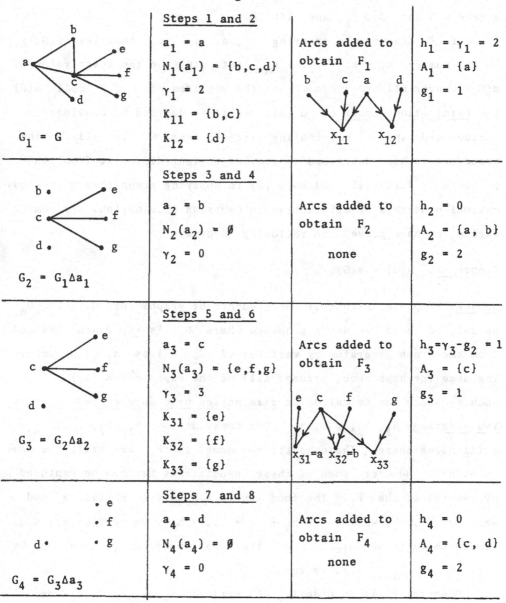

Continue with $a_5 = e$, $a_6 = f$, $a_7 = g$. Add no more arcs to obtain F_5 and F_6. Obtain $h_5 = h_6 = 0$.

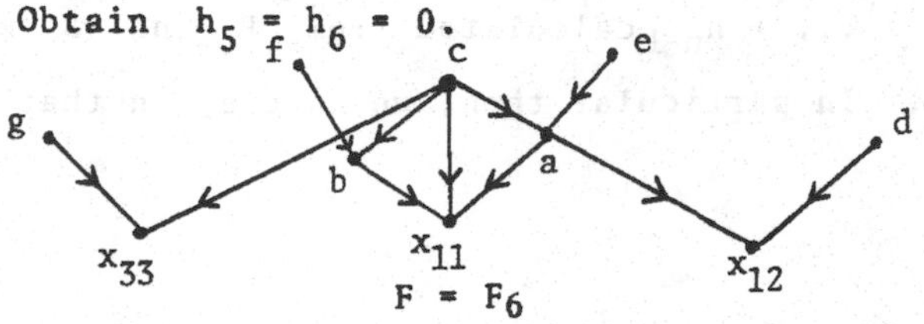

$F = F_6$

There are no competitions between these vertices in F. Thus, F is a food web for $G \cup I_k$, and $k(G) \leqq k$.

If P denotes the ordering $a_1, a_2, \ldots, a_n$, then let $m(G,P)$ be the number $h_1 + h_2 + \ldots + h_{n-1}$. Let $m(G)$ be the least value of $m(G,P)$ over all orderings P of the vertices of G. We call $m(G)$ the <u>elimination number</u> of G because it is defined by considering various orderings of eliminating vertices from G. The elimination procedure we have described is similar to (but different from) the one used by Parter [5] and Rose [8] in applying graph theory to study optimal orderings of elimination in Gaussian elimination. To summarize, we have proved the following theorem.

<u>Theorem 1</u>. $k(G) \leqq m(G)$.

<u>Remark</u>. Suppose instead of the sequence of graphs $G_1, G_2, \ldots, G_n$ as defined above, we use a sequence where G_i is any graph obtained from the graph generated by vertices of G_{i-1} less a_{i-1} by removing some (perhaps none, perhaps all) of the edges of $N_i(a_i)$. Any such sequence can be called an <u>elimination procedure corresponding to the ordering</u> $a_1, a_2, \ldots, a_n$. Our construction $F_1, F_2, \ldots, F_{n-1}$ still makes sense. The competitions added in F_i are still the same as before. However, some of these competitions may now be captured by several of the F_i. The food web $F = F_{n-1}$ is thus still a food web for $G \cup I_k$, where $k = h_1 + h_2 + \ldots + h_{n-1}$. In particular, G_i can be taken to be $G_{i-1} - a_{i-1}$, the subgraph of G_{i-1} generated by vertices of G_{i-1} less vertex a_{i-1}.

Suppose P is an ordering of vertices $a_1, a_2, \ldots, a_n$ and $S = G_1, G_2, \ldots, G_n$ is a corresponding elimination procedure. The number $h_1 + h_2 + \ldots + h_{n-1}$ calculated from P and S will be denoted $m(G,P,S)$. In particular then, we have shown that $k(G) \leq m(G,P,S)$.

<u>Corollary 1</u>. If $n > 3$, $k(Z_n) = 2$.

<u>Proof</u>. We know by Proposition 1 that $k(Z_n) \geq 2$. But removing vertices around the circumference of Z_n in order leads to an ordering P such that $m(G,P) = 2$. Q.E.D.

A vertex a in a graph G is called <u>simplicial</u> if $N(a)$ is a clique or is empty.

<u>Corollary 2</u>. Suppose it is possible to order the vertices of G as $a_1, a_2, \ldots, a_n$ so that a_1 is a simplifical vertex of $G = G_1$ and for all i, a_i is a simplicial vertex of $G_i = G_{i-1}\Delta a_{i-1}$. Then $k(G) \leq 1$.

<u>Proof</u>. With this ordering P, $\gamma_i = 1$ for all i except those i where a_i is isolated in G_i. Moreover, $h_i = 1$ for the first i such that $\gamma_i = 1$, and $h_i = 0$ otherwise. Thus, $m(G,P)$ is 0 or 1. Q.E.D.

A graph G is called a <u>rigid circuit graph</u> if G does not have Z_n, $n > 3$, as a generated subgraph.

<u>Corollary 3</u>. Every rigid circuit graph G has $k(G) \leq 1$.

<u>Proof</u>. Dirac [2, Theorem 4] proves that every rigid circuit graph G has a simplicial vertex a_1. Let $G_1 = G$ and let $G_2 = G_1 - a_1$. Since every generated subgraph of a rigid circuit graph is rigid circuit, G_2 has a simplicial vertex a_2. Let $G_3 = G_2 - a_2$. Let a_3 be simplicial in G_3. And so on. By the Remark after Theorem 1, if $a_1, a_2, \ldots, a_n$ is the ordering P and $G_1, G_2, \ldots, G_n$ is the elimination procedure S, then $k(G) \leq m(G,P,S)$. But as in the proof of Corollary 2, it is easy to see that $m(G,P,S) \leq 1$. Q.E.D.

<u>Corollary 4</u>. Every interval graph G has $k(G) \leq 1$.

<u>Proof</u>. Every interval graph is a rigid circuit graph.

Remark. Figure 4 shows a graph G which has $k(G) \leqq 1$ but which is not a rigid circuit graph. This can be proved using the ordering of vertices shown and Corollary 2. Figure 4 also shows a graph H which has $k(H) \leqq 1$ but which has no ordering of vertices of the type described in Corollary 2. The proof that $k(H) \leqq 1$ uses the ordering of vertices shown and Theorem 1.

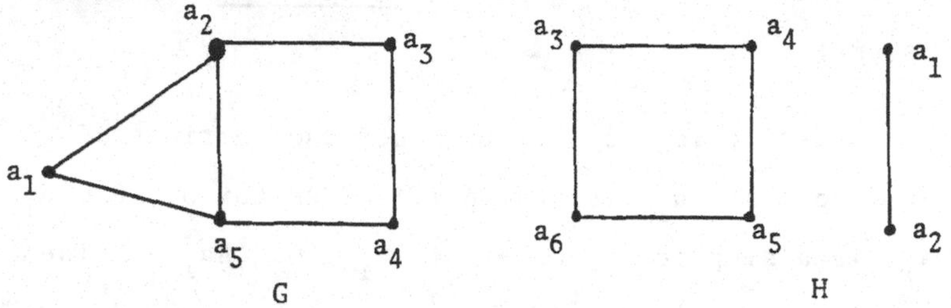

Figure 4.

Theorem 2. If G is connected, $|V(G)| > 1$, and G has no triangles, then $k(G) = m(G) = e(G) - n(G) + 2$.

To prove Theorem 2, we first note the following lemmas.

Lemma 1. If G is connected, $|V(G)| > 1$, and G has no triangles, then there is a vertex a in G such that $G\Delta a$ is connected.

Proof. Since G has no triangles, $G\Delta a = G - a$ for all a. Now any end vertex a of a spanning tree of G has the property that $G\Delta a = G - a$ is connected.

Q.E.D.

Lemma 2. Suppose $P = a_1, a_2, \ldots, a_n$ is an ordering of vertices of G and $P_2 = a_2, a_3, \ldots, a_n$ is an ordering of vertices of $G_2 = G\Delta a_1$. Then if $m(G_2,P_2) > 0$,

$$m(G,P) = \gamma_1 + m(G_2,P_2) - 1.$$

Proof. In the ordering P, $h_1 = h_1(P) = \gamma_1$. The first nonzero term $h_i(P_2)$ is reduced by 1 in calculating $m(G,P)$, since vertex a_1 is available. Every subsequent $h_i(P_2)$ stays the same. Q.E.D.

Theorem 2 is now proved by induction on $|V(G)|$. If $|V(G)| = 2$, then G must consist of a single edge. It is easily seen that $k(G) = m(G) = e(G) - n(G) + 2 = 1$.

Assume the theorem holds for graphs with $n-1$ vertices and let $n(G) = n > 2$. Since G is connected, has no triangles, and $|V(G)| > 1$, Lemma 1 says that there must be a vertex a such that $G\Delta a$ is connected. Let $a_1 = a$ and let $G_2 = G\Delta a$. Since $n(G) > 2$, $|V(G_2)| > 1$. Let $P_2 = a_2, a_3, \ldots, a_n$ be an ordering of vertices of G_2 such that $m(G_2,P_2) = m(G_2)$. Since G_2 has more than one vertex and is connected, $m(G_2,P_2) > 0$. Moreover, G_2 is connected and has no triangles. Consider the ordering $P = a_1, a_2, \ldots, a_n$ for G. By Lemma 2 and by the inductive assumption,

$$\begin{aligned} m(G,P) &= \gamma_1 + m(G_2,P_2) - 1 \\ &= \gamma_1 + m(G_2) - 1 \\ &= \gamma_1 + [e(G_2) - n(G_2) + 2] - 1 \\ &= \gamma_1 + e(G_2) - n(G_2) + 1. \end{aligned}$$

Since G has no triangles, $N(a_1)$ has no edges, and so G_2 is obtained from G by removing the edges leading from a_1. There must be γ_1 such edges. Thus, $e(G_2) = e(G) - \gamma_1$. Also, $n(G_2) = n(G)-1$. Thus,

$$\begin{aligned} m(G,P) &= \gamma_1 + [e(G)-\gamma_1] - [n(G)-1] + 1 \\ &= e(G) - n(G) + 2. \end{aligned}$$

It follows, using Theorem 1 and Proposition 1, that

$$k(G) \leq m(G) \leq m(G,P) = e(G) - n(G) + 2 \leq k(G).$$

Hence, Theorem 2 is proved.

We mention one corollary of Theorem 2.

Corollary. If G is connected, $|V(G)| > 1$, and G has no triangles, then $k(G) = 1$ if and only if G is a tree.

Proof. We have

$$e(G) - n(G) + 2 = 1 \text{ iff } e(G) = n(G) - 1.$$

Q.E.D.

4. Further Questions

Since we have seen that so many graphs are competition graphs, it is surprising that only interval graphs seem to arise from food webs in the real world. It might be useful to determine the properties of the class of food webs whose competition graphs are interval graphs. Other questions specifically about competition graphs are also of interest. For example, is $k(G)$ always equal to $m(G)$ and what graphs have $k(G) = 1$?

References

1. L. Danzer and B. Grünbaum, Intersection properties of boxes in R^d. Mimeographed, Department of Mathematics, Univ. of Washington, Seattle, August 1967.

2. G. A. Dirac, On rigid circuit graphs. Abhandlungen Mathematischen Seminar Universität Hamburg 25 (1961) 71-76.

3. H. Gabai, Bounds for the boxicity of a graph. Mimeographed, York College (CUNY), April 1976, submitted to SIAM J. Appl. Math.

4. F. Harary, R. Z. Norman, and D. Cartwright, Structural Models: An Introduction to the Theory of Directed Graphs. John Wiley & Sons, Inc., New York (1965).

5. S. Parter, The use of linear graphs in Gauss elimination. SIAM Review 3 (1961) 119-130.

6. F. S. Roberts, Discrete Mathematical Models, with Applications to Social, Biological, and Environmental Problems. Prentice-Hall, Inc., Englewood Cliffs, New Jersey (1976).

7. F. S. Roberts, On the boxicity and cubicity of a graph. Recent Progress in Combinatorics. Academic Press, New York (1969) 301-310.

8. D. J. Rose, Triangulated graphs and the elimination process. J. Math. Anal. & Appl. 32 (1970) 597-609.

HAMILTONIAN PERSISTENCY IS PERIODIC AMONG ITERATED LINE DIGRAPHS

John Roberts
University of Louisville
Louisville, Kentucky 40208

§1 LINE DIGRAPHS

While in general we adopt the common conventions of Behzad and Chartrand [1] and Harary [3], it is convenient to make the following exception. The directed graphs (briefly, digraphs) are actually pseudodigraphs; i.e., the digraphs have finitely many vertices and arcs but may contain loops as well as parallel arcs. The vertex set of a digraph D is denoted by V(D) and the arc set by A(D). A digraph is <u>nonempty</u> if it has arcs.

The <u>line digraph</u> L(D) of a nonempty digraph D has vertex set $V(L(D)) = \{\bar{e}: e \in A(D)\}$ where $\bar{e}$ is adjacent to $\bar{f}$ in L(D) if and only if

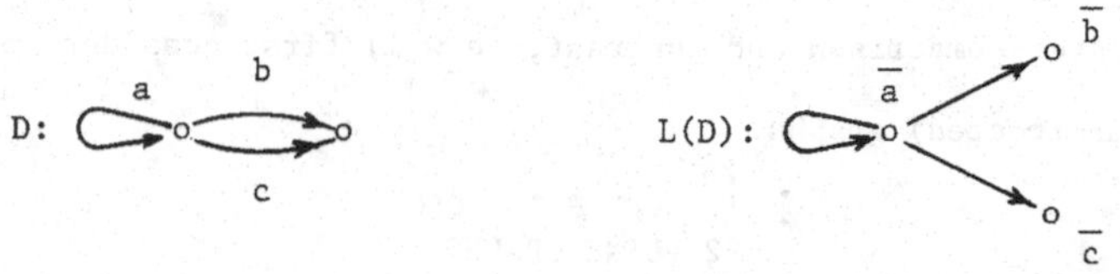

Figure 1

the arc e is adjacent to the arc f in D. For example, the line digraph of a cycle is a cycle of the same length while the line digraph of a path is a path of length one less. The digraphs in Figure 1 illustrate that a line digraph never has parallel arcs even though its "predecessor" may have both loops and parallel arcs. Also, a line digraph L(D) has loops if and only if D has loops. In general, a line digraph has an n-cycle (n distinct vertices) if and only if its predecessor has an n-circuit (n distinct arcs).

If a line digraph L(D) is nonempty, then $L^2(D) = L(L(D))$ exists. Moreover, if $n \geq 1$ and $L^n(D)$ is nonempty, then the iterated line digraph $L^{n+1}(D) = L(L^n(D))$ exists. Thus, each nonempty digraph D gives rise to a (possibly terminal) sequence

$$\ell(D)\colon L(D),\ L^2(D),\ L^3(D),\ \ldots$$

of iterated line digraphs whose n^{th} term is $L^n(D)$ provided that $L^{n-1}(D)$ exists and is nonempty. Because of the relationship between circuits in a digraph and the cycles in its line digraph and since the line digraph of a path is a shorter path, the sequence of iterated line digraphs is infinite if and only if the original digraph has a cycle.

In the following, we investigate the occurrence of infinitely many hamiltonian digraphs in $\ell(D)$. One can easily show that if $\ell(D)$ contains a hamiltonian digraph, then D must be the disjoint union of a strongly connected nonempty digraph D_1 and an acyclic digraph D_2. Since the acyclic digraph D_2 will eventually cease to contribute to the sequence of iterated line digraphs, it will suffice if we consider only strongly connected nonempty digraphs, each having infinitely many hamiltonian digraphs in its sequence of iterated line digraphs. However, for later comparison and contrast, we will first consider the analogous problem for (undirected) graphs.

§2 LINE GRAPHS

The <u>line graph</u> $L(G)$ of a nonempty graph G is obtained by associating a new vertex with each edge of G and joining vertices in $L(G)$ if the corresponding edges are adjacent in G. As with the line digraph, the line graph of a cycle is a cycle of the same length while the line graph of a path is a path of length one less. However, as illustrated by the graphs in Figure 2, the line graph of an acyclic graph is not necessarily acyclic. In fact, $L(G)$ is acyclic if and only if

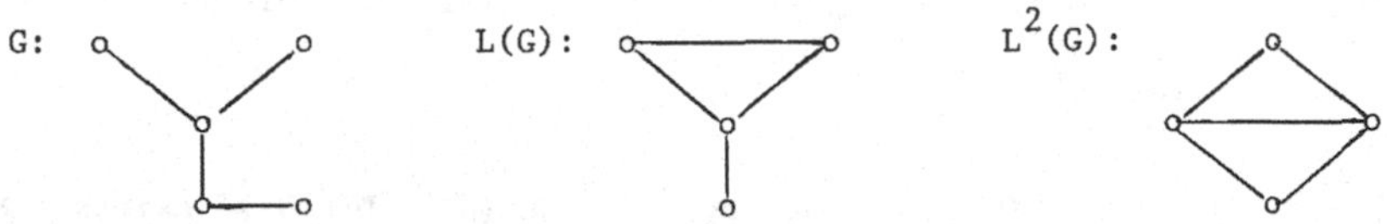

Figure 2

G is the disjoint union of paths (not all of which are trivial).

Clearly, each nonempty graph G gives rise to a (possibly terminal) sequence

$$\ell(G): L(G),\ L^2(G),\ L^3(G),\ \ldots$$

of iterated line graphs. Moreover, $\ell(G)$ is infinite if and only if G is not the disjoint union of paths. One may easily verify that a line graph L(H) is hamiltonian if H has a circuit incident with all edges of H . Consequently, if any graph in $\ell(G)$ is hamiltonian, then all subsequent line graphs exist and are hamiltonian. Thus, either $\ell(G)$ contains infinitely many hamiltonian graphs or none at all. The question of whether or not $\ell(G)$ contains infinitely many hamiltonian graphs (for a given graph G) was answered by

THEOREM 1 (Chartrand [2]): *If G is a graph with p vertices and exactly one component different from a path, then $L^{p-3}(G)$ is hamiltonian.*

The preceding result, whose sharpness is illustrated by the graphs in Figure 2, yields the following characterization of those graphs having a sequence of iterated line graphs containing infinitely many hamiltonian graphs.

COROLLARY 2: *The sequence $\ell(G)$ contains infinitely many hamiltonian line graphs if and only if G has exactly one component different from a path.*

§3 HAMILTONIAN PERSISTENCY AND PERIODICITY

Unlike line graphs, the presence of a hamiltonian cycle in a digraph does not imply that its line digraph is hamiltonian. This is illustrated by the digraphs in Figure 3. The digraphs E and $L^2(E)$ are hamiltonian while L(E) is not. In

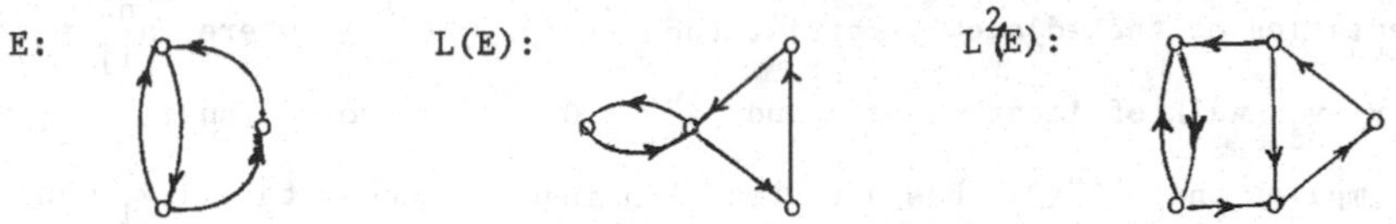

Figure 3

fact, as we shall see later, $L^n(E)$ is not hamiltonian if $n \geq 3$. The digraphs in Figure 3 also illustrate that a strongly connected digraph has a hamiltonian line digraph if and only if it is eulerian (i.e., it has an arc-spanning circuit).

Thus, $\ell(D)$ contains infinitely many hamiltonian digraphs if and only if it contains infinitely many eulerian digraphs.

Since eulerian digraphs are characterized by being strongly connected, nonempty and locally regular (i.e., the indegree $\mathrm{id}\, v$ equals the outdegree $\mathrm{od}\, v$ at each vertex v), and since we have restricted our attention to strongly connected nonempty digraphs, it suffices to determine how often locally regular digraphs occur in an iterated sequence of line digraphs.

To examine the degree structure of a typical iterated line digraph $L^n(D)$, it may be helpful to note the following. If an arc e joins vertex u to vertex v in a digraph D , then $\bar{e}$ is a vertex in $L(D)$ and represents a unique u-v walk of length one in D . Also, the arcs in $L(D)$ correspond to unique vertices in $L^2(D)$ and to unique walks of length two in D . Similarly, there is a "natural" one-to-one correspondence between the walks of length n in D and the vertices in $L^n(D)$. Also, if the arc e joins the vertex u to the vertex v , then $\mathrm{id}\,\bar{e} = \mathrm{id}\, u$ and $\mathrm{od}\,\bar{e} = \mathrm{od}\, v$ since e is adjacent from the $\mathrm{id}\, u$ arcs entering u and adjacent to the $\mathrm{od}\, v$ arcs leaving v . Similarly, if $\overline{e_1 e_2 \ldots e_n}$ is a vertex of $L(D)$ "naturally" corresponding to a u-v walk of length n in D , then $\overline{e_1 e_2 \ldots e_n}$ has indegree $\mathrm{id}\, u$ and outdegree $\mathrm{od}\, v$. Consequently, the degree structure of $L^n(D)$ is uniquely determined by the degree structure of D in conjunction with the walks of length n in D . As such, if $L^n(D)$ is eulerian, then every walk of length n in D must terminate with a vertex having outdegree equal to the indegree of the initial vertex.

Let D be a digraph with $V(D) = \{v_1, v_2, \ldots, v_p\}$. If $B(D)$ is the boolean representation of the adjacency matrix, then $B^n(D) = (a^n_{ij})$ where $a^n_{ij} = 1$ if D has a v_i-v_j walk of length $n \geq 1$ and $a^n_{ij} = 0$ otherwise. Thus, $a^n_{ij} = 1$ in $B^n(D)$ implies that $L^n(D)$ has a vertex with indegree equal to $\mathrm{id}\, v_i$ and outdegree equal to $\mathrm{od}\, v_j$. Consequently, $B^n(D)$ determines the entire degree structure of $L^n(D)$ if $L^n(D)$ exists. Since there are only finitely many p-by-p 0,1-matrices, the sequence

$$B(D),\ B^2(D),\ B^3(D),\ \ldots$$

must eventually become periodic. Thus, one may use a finite subsequence of it to determine if $\ell(D)$ contains infinitely many eulerian (and therefore, infinitely many hamiltonian) digraphs.

For example, returning to the digraph E in Figure 3, let v_1 be the vertex with indegree two, v_2 be the vertex with outdegree two and v_3 be the remaining vertex. Then,

$$B(E) = \begin{pmatrix} 0 & 1 & 0 \\ 1 & 0 & 1 \\ 1 & 0 & 0 \end{pmatrix} \qquad B^2(E) = \begin{pmatrix} 1 & 0 & 1 \\ 1 & 1 & 0 \\ 0 & 1 & 0 \end{pmatrix}$$

$$B^3(E) = \begin{pmatrix} 1 & 1 & 0 \\ 1 & 1 & 1 \\ 1 & 0 & 1 \end{pmatrix} \qquad B^4(E) = \begin{pmatrix} 1 & 1 & 1 \\ 1 & 1 & 1 \\ 1 & 1 & 0 \end{pmatrix}$$

and $B^n(E)$ is a p-by-p matrix of ones for $n \geq 5$. Since $a_{11}^n = 1$ for all $n \geq 2$, the digraph E has a v_1-v_1 walk of length n and, therefore, $L^n(E)$ has a vertex with indegree equal to $\text{id } v_1 = 2$ and outdegree equal to $\text{od } v_1 = 1$. Consequently, $L^n(E)$ is not eulerian if $n \geq 2$ and, therefore, $L^n(E)$ is not hamiltonian if $n \geq 3$. Thus, $L^2(E)$ is the only hamiltonian digraph in the sequence $\ell(E)$ of iterated line digraphs.

Moreover, because of the periodicity associated with the sequence $B(D), B^2(D), B^3(D), \ldots$, there is a periodicity associated with the occurrence of hamiltonian line digraphs in $\ell(D)$ if infinitely many occur. A digraph D is said to have a <u>hamiltonian period</u> if hamiltonian digraphs occur periodically in $\ell(D)$. We may now present the following result.

<u>THEOREM</u> 3: <u>The sequence</u> $\ell(D)$ <u>contains infinitely many hamiltonian digraphs if and only if the digraph</u> D <u>has a hamiltonian period</u>.

Recall that the sequence $\ell(G)$ of iterated line graphs contains infinitely many hamiltonian graphs if it contains one hamiltonian graph. While we cannot say the same of the sequence $\ell(D)$ of iterated line digraphs, the following provides an analogous result.

THEOREM 4: <u>If the sequence</u> $\ell(D)$ <u>for a nonempty digraph</u> D <u>contains two hamiltonian digraphs, then it contains infinitely many hamiltonian digraphs.</u>

§4 FIXED SCORE CYCLIC DIGRAPHS

In this section we provide two characterizations of strongly connected nonempty digraphs having a hamiltonian period. To facilitate this, let $S: s_1,s_2,\ldots,s_k$ denote a sequence of k positive integers and let $d_F(u,v)$ be the distance from vertex u to vertex v in the digraph F.

An S-<u>scoring</u> of a cycle C in a digraph D is a 4-tuple $(C,u,a,b)_S$ such that u is a vertex of C, the length k of S divides the length of C, and $d_C(u,v) = j$ implies that $\text{id } v = s_{a+j}$ and $\text{od } v = s_{b+j}$ (where the subscripts are modulo k). Clearly, if $d_C(u,v) = j$ and if $(C,u,a,b)_S$ is an S-scoring of C, then so is $(C,v,a+j,b+j)_S$. Moreover, any two such S-scorings will be considered as equivalent. A strongly connected nonempty digraph D is called S-<u>score cyclic</u> if each cycle in D has an S-scoring. Also, a digraph is said to be <u>score cyclic</u> if it is S-score cyclic for some sequence $S: s_1,s_2,\ldots,s_k$ of positive integers.

For example, cycles are score cyclic as is the digraph $C_{m,\ell}$ consisting of m cycles of length ℓ which are disjoint except for a vertex which is common to every cycle. Likewise, the digraph D_n (in Figure 4) consisting of two disjoint cycles $C: u_1,u_2,\ldots,u_{2n},u_1$ and $C': v_1,v_2,\ldots,v_{2n},v_1$ joined by the arcs $v_{2i-1}u_{2i-1}$ and $u_{2i}v_{2i}$ (for $1 \le i \le n$) is score cyclic; in particular, D_n is S-score cyclic where $S: 1,2$.

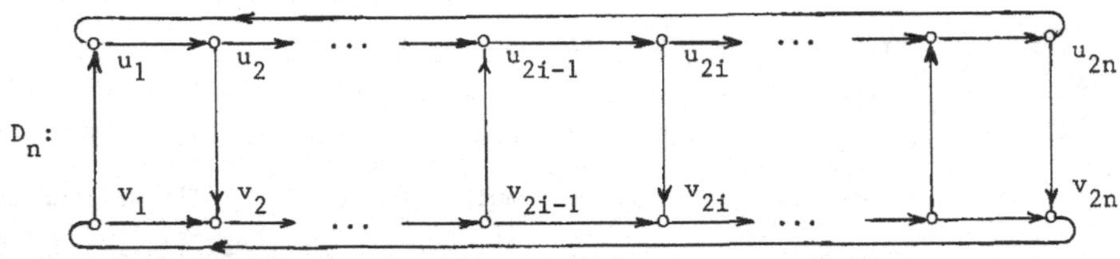

Figure 4

An S-score cyclic digraph is _fixed_ _S-score_ _cyclic_ if for any S-scoring $(C,u,a,b)_S$ and any cycle C' also containing u, then $(C',u,a,b)_S$ is an S-scoring of C'. Also, a digraph is _fixed_ _score_ _cyclic_ if it is fixed S-score cyclic for some sequence $S: s_1,s_2,\ldots,s_k$. For example, one may easily verify that each of the three preceding digraphs (i.e., cycles, $C_{m,\ell}$ and D_n) are fixed score cyclic. However, not all score cyclic digraphs are fixed score cyclic. In particular, consider the digraph D in Figure 5 which is S-score cyclic only for S: 1,1,2,2 (or a cyclic permutation thereof). Let C be the 4-cycle containing

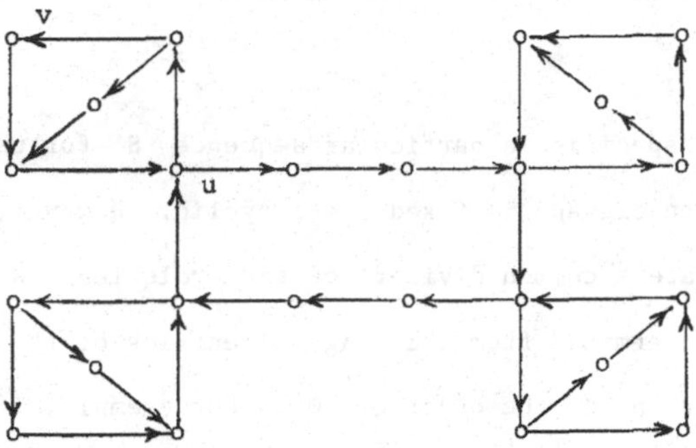

Figure 5

both u and v and let C' be the 8-cycle in D. Then $(C,u,4,3)_S$ is an S-scoring of C but $(C',u,4,3)_S$ is not an S-scoring of C'. With this in mind, we offer the following.

THEOREM 5: _A strongly connected digraph_ D _has a hamiltonian period if and only if it is fixed score cyclic._

The following result provides a characterization of fixed score cyclic digraphs and, therefore, yields a second characterization of strongly connected digraphs having a hamiltonian period.

THEOREM 6: _A strongly connected nonempty digraph_ D _is fixed score cyclic if and only if there is a sequence_ $S: s_1,s_2,\ldots,s_k$ _and a mapping_

$$\alpha: V(D) \rightarrow \{ (i,j): 1 \le i,j \le k \}$$

such that $\alpha(u) = (a,b)$ _implies that:_

(1) id $u = s_a$ *and* od $u = s_b$; *and*

(2) $\alpha(v) = (a+1, b+1)$ *whenever* $uv \in A(D)$.

In spite of the fact that the preceding result does not specify a particular sequence $S: s_1, s_2, \ldots, s_k$ it is not necessary to test every "possible" sequence of positive integers in hope of finding one that "works". In particular, (the following result asserts that) any sequence for which the digraph is score cyclic will suffice.

THEOREM 7: *If a digraph* D *is fixed score cyclic and* S-*score cyclic, then* D *is fixed* S-*score cyclic*.

The next result specifies a particular sequence S for use in determining whether or not a given digraph is fixed score cyclic. However, we first observe that the g.c.d. (greatest common divisor) of the cycle lengths associated with a digraph D can be determined from the diagonal entries of the matrices $B^i(D)$ where $1 \le i \le p$ and p is the order of D . For example, if $1 \le i \le p$ and $B^i(D)$ has a one on the diagonal, then either D has a cycle of length i or it has a closed walk (which is equivalent to a linear combination of cycles) of length i . Consequently, we can compute the g.c.d. of the cycle lengths by computing the g.c.d. of the integers i between 1 and p such that $B^i(D)$ has a one on the diagonal.

THEOREM 8: *Let* g *be the g.c.d. of the cycle lengths associated with a strongly connected nonempty digraph* D *and let* $v_1, v_2, \ldots, v_g$ *be any path in* D *having* g *vertices. Then,* D *is fixed score cyclic if and only if it is fixed* S-*score* cyclic where S: od v_1, od v_2, ..., od v_g .

We now illustrate the use of Theorems 6 and 8 by showing that the digraph D_n in Figure 4 is in fact fixed score cyclic as asserted. First of all, D_n is bipartite and, therefore, every cycle in D_n has even length. Also, the cycle $C': v_1, v_2, \ldots, v_n, v_1$ has length 2n while the cycle obtained from C' by replacing the subpath v_1, v_2 with the path v_1, u_1, u_2, v_2 has length 2n+2 . Thus, the g.c.d. of the cycle lengths associated with D_n is two. If we choose the path v_1, v_2 in D_n, then Theorem 8 yields the sequence $S: s_1, s_2$ where $s_1 = \text{od } v_1 = 2$ and $s_2 = \text{od } v_2 = 1$. If we let α be the mapping defined by

$$\alpha(u_{2i-1}) = \alpha(v_{2i}) = (2,1) \quad \text{and}$$

$$\alpha(u_{2i}) = \alpha(v_{2i-1}) = (1,2)$$

for $1 \le i \le n$, then $S: 2,1$ and α satisfy the sufficiency portion of Theorem 6 Thus, D_n is fixed score cyclic.

We also note that if a digraph D is fixed T-score cyclic, then $T: t_1, t_2, \ldots, t_k$ must be a subsequence of S (as in Theorem 8) which can be used to generate S (or a cyclic permutation thereof). Thus, Theorem 8 yields the longest sequence for which a digraph can be fixed score cyclic. Also, it $M: m_1, m_2, \ldots, m_k$ is a shortest sequence "generating" S, then the length k of M is the period between hamiltonian line digraphs in $\ell(D)$ if D has a hamiltonian period.

REFERENCES

1. M. Behzad and G. Chartrand, Introduction to the Theory of Graphs (Allyn and Bacon, Boston, 1971).

2. G. Chartrand, The existence of complete cycles in repeated line-graphs, Bulletin of the American Mathematical Society LXXI, 668-670 (1965).

3. F. Harary, Graph Theory (Addison-Wesley, Reading, Mass., 1969).

SOME PROBLEMS IN RAMSEY THEORY

R.J. Faudree and R.H. Schelp
Memphis State University
Memphis, TN 38152

Abstract

Some exact results in generalized Ramsey theory are given - others are conjectured. For example, all known results for the Ramsey number of the cycle - complete graphs are presented with attention focused on the unsolved parts. Non-standard results include introduction of "edge minimal" graphs that "arrow" a pair of graphs.

INTRODUCTION

Generalized Ramsey theory has received a great deal of attention over the last several years. The work has branched into several directions, making any survey article at best difficult.

The intent of this paper is not to survey the recent work done, but rather to focus on several specific problems, sacrificing breadth for some depth. Specifically, we will focus on several exact results giving all known results for the cycle-complete graph, a conjecture for the path-tree problem, and some results and a canonical coloring for a family of cycles. Other non-standard results introduced will involve the minimal number of edges needed by a graph to "arrow" a pair of graphs.

Many of the results presented have been worked on jointly by the authors with Professors P. Erdös and C.C. Rousseau, and appropriate recognition of their contribution should be noted by the reader.

NOTATION

For the most part our notation will conform with that used in [1] and [10]. All graphs considered will be finite, undirected, and without loops or multiple edges. Usually a single symbol will be used to denote a graph. If it is necessary for clarity to distinguish its vertex set V and its edge set E, then the graph will be denoted by $G(V,E)$. The <u>order</u> of the graph G will be indicated by $|G|$.

For $X \subseteq V$ or $Y \subseteq E$, the subgraph of G induced by X or induced by Y will be denoted correspondingly by $\langle X \rangle$ or $\langle Y \rangle$. A <u>cycle</u> and <u>path</u> with n vertices $x_1, x_2, \ldots, x_n$ will be denoted by $C_n = (x_1, x_2, \ldots, x_n, x_1)$ and $P_n = (x_1, x_2, \ldots, x_n)$ respectively, where for the cycle the subscript i in x_i will be taken modulo the cycle length n.

If $G_1, G_2, \ldots, G_t$ are graphs, define the <u>Ramsey number</u> $r(G_1, G_2, \ldots, G_t)$ to be the least positive integer n such that if the edges of K_n are colored in any fashion by t colors, then for some i the ith colored subgraph contains G_i as a subgraph. These numbers are generalized Ramsey numbers in the sense that the G_i need not be complete graphs. In the special case when $t = 2$ we refer to the coloring of the edges of K_n as a two coloring. <u>Throughout the paper all two colorings will be assumed to be in the colors red and blue</u>. Thus when determining that $r(G_1, G_2)$ is less than n, we show that a two colored K_n contains a red G_1 or a blue G_2. Also there will be times when we wish to two color the edges of a non-complete graph G. In particular if G is a graph, complete or non-complete, which when two colored always contains a red graph H or a blue graph L, then G will be said to <u>arrow</u> the pair (H,L) and will be written $G \rightarrow (H,L)$.

In the sequel we will also discuss a slight variation of the generalized Ramsey number just defined. For $\mathcal{H}$ and $\mathcal{K}$ both nonempty collections of graphs $r(\mathcal{H},\mathcal{K})$ will be the smallest positive integer n such that a two colored K_n contains a red H or a blue K for some $H \in \mathcal{H}$, $K \in \mathcal{K}$. In the special case when $\mathcal{H}$ is one of the collections $\{C_\ell \mid \ell \leq t\}$, $\{C_\ell \mid \ell \geq t\}$ we correspondingly write $r(\leq C_t,\mathcal{K})$, $r(C_t\geq,\mathcal{K})$ for $r(\mathcal{H},\mathcal{K})$. If $\mathcal{C}$ is a list of cycles, $\mathcal{C} = (C_{t_1},C_{t_2},\ldots,C_{t_k})$, then $r(\leq\mathcal{C})$ will be a shortened way of denoting $r(\leq C_{t_1},\leq C_{t_2},\ldots,\leq C_{t_k})$.

For X a subset of the vertex set, the set of all vertices adjacent to at least one vertex of X will be denoted $\Gamma(X)$. If the graph has been two colored and we wish to account only for those vertices adjacent by a red edge (blue edge) to at least one vertex of X, then we will denote this set by $\Gamma_R(X)$, $(\Gamma_B(X))$. The degree of vertex x in the graph will be denoted by $d(x)$, while if two colored its red degree (blue degree) will be denoted by $d_R(x)$ $(d_B(x))$.

The symbols $[\cdot]$ and $\{\cdot\}$ occur several times in this paper. Whenever x represents a real number, the symbols $[x]$ and $\{x\}$ will signify the greatest integer $\leq x$ and the least integer $\geq x$ respectively. Otherwise $\{\cdot\}$ will serve its usual purpose of denoting a set.

EXACT RESULTS

Many exact results are known in generalized Ramsey theory, others are partially solved. An interesting problem of those partially solved is finding $r(C_n,K_m)$ for arbitrary n and m. Although the complete solution is highly unlikely, since $r(K_3,K_m)$ for arbitrary m may never be solved, there are unsolved portions which should be tractable. We first state the known results, referencing the solver.

THEOREM 1.

(i) $r(C_n,K_m) = (n-1)(m-1)+1$ <u>for</u> $n \geq m^2-2$, [2].

(ii) $r(C_n,K_m) \leq \{(n-2)(m^{1/k}+2)+1\}(m-1)$ for $n \geq 3$, $m \geq 2$ where $k = [(n-1)/2]$, [7].

(iii) $r(C_n,K_m) \geq r(\leq C_n,K_m) \geq c(m/\log m)^{(n-1)/(n-2)}$ for n fixed, $n \geq 3$, c a constant dependent on n, and m sufficiently large, [11].

Surely the equality shown in Theorem 1(i) should hold for a larger range of n and m, in particular for $n \geq m$. For the moment we focus our attention on this problem. First we give an alternate proof for (i) which may lead to an improvement of the bound $n \geq m^2-2$. To this end we prove the following theorem.

THEOREM 2. For $m \geq 2$, $n \geq 3$

$$r(C_{n}^{\geq},K_m) = (n-1)(m-1)+1 .$$

Proof. The usual example establishes that $r(C_{n}^{\geq},K_m) \geq (n-1)(m-1)+1$. The proof is by induction on m, being trivial for $m = 2$.

Two color a complete graph G on $(n-1)(m-1)+1$ vertices with $m \geq 3$, $n \geq 3$, assuming the result holds for $m-1$ and $n \geq 3$. We consider two possible cases. First suppose there exists a vertex v in G with $d_R(v) \leq n-2$. Then by deleting $\Gamma_R\{v\} \cup \{v\}$ we obtain a two colored complete graph H with at least $(n-1)(m-2)+1$ vertices. By induction H contains a red C_ℓ for some $\ell \geq n$ or a blue K_{m-1}. Thus we may assume H contains a blue K_{m-1}. But since v is adjacent to every vertex of H with blue edges, we have that $\langle H \cup \{v\}\rangle$ contains a blue K_m. Finally we consider the remaining case where $d_R(x) \geq n-1$ for each $x \in G$. Thus it follows that G has a red path $P_t = (x_1,x_2,\ldots,x_t)$ with $t \geq n$. Assume this path is of maximal length. Then all red adjacencies of x_1 are with elements of the path. But $d_R(x_1) \geq n-1$, so that there exists a red edge (x_1,x_i) for some $i \geq n$. Hence G contains a red C_ℓ for some $\ell \geq n$.

THEOREM 3. *For* n *sufficiently large*

$$r(C_n,K_m) = (n-1)(m-1)+1 .$$

Proof. We only show that $r(C_n,K_m) \leq (n-1)(m-1)+1$. Thus two color the graph G, a complete graph with $(n-1)(m-1)+1$ vertices. Observe that the theorem follows from Theorem 2 unless G contains a red C_ℓ, for some $\ell > n$, and no red C_n or blue K_m. Assuming this to be the case, choose ℓ such that G contains a red C_ℓ and no red C_k for $\ell > k \geq n$. We see by the next theorem that, if n is sufficiently large, G contains a blue K_m, contrary to the earlier assumption.

THEOREM 4. *Let* m *be a fixed positive integer. If* G *is a Hamiltonian graph on* p *vertices such that* G *contains no* C_{p-1}, *then for* p *sufficiently large* G *contains an independent set with* m *vertices*.

Proof. Since G is Hamiltonian, let G be the cycle $(x_1,x_2,\ldots,x_p,x_1)$. Assume that there exists a t such that each path with t vertices on the Hamiltonian cycle G contains an independent set with $m - 1$ vertices, i.e. for each i the path $(x_{i+1},x_{i+2},\ldots,x_{i+t})$ contains $m - 1$ independent vertices. This is surely the case for $m-1 = 1$ or 2.

The theorem is proved by induction showing, for p sufficiently large, that there exists a t', dependent on t, such that each path with t' vertices on G contains an independent set with m elements. First observe if no pair of the set of the vertices $\{x_1,x_3,\ldots,x_{2t-1}\}$ are adjacent the theorem holds. Hence for convenience assume x_1 is adjacent to x_ℓ, ℓ odd, $3 \leq \ell \leq 2t-1$. Taking p sufficiently large consider the $r = m(\ell-3)$ edge disjoint paths $Q_j = (x_{\ell+(j-1)t},x_{\ell+(j-1)t+1},\ldots,x_{\ell+jt})$, $j = 1,2,\ldots,r$, each of which by assumption contains $m - 1$ independent vertices. In fact p is assumed larger than $\ell + tr$.

Denote the path $(x_1, x_2, \ldots, x_\ell)$ by Q_0. For the remainder of the proof we assume $\langle \bigcup_{j=0}^{r} Q_j \rangle$ contains no independent set with m vertices. We will show, since $\langle \bigcup_{j=0}^{r} Q_j \rangle$ contains no independent set with m vertices, that the cycle $Q = (x_1, x_\ell, x_{\ell+1}, \ldots, x_p, x_1)$ can be enlarged to a C_{p-1}, in contradiction to the hypothesis of the theorem. To this end consider the vertex x_2 which is not in Q. Since the edge disjoint subgraphs $\{Q_j\}_{j=1}^{m}$ of Q each contain $m-1$ independent vertices and $\langle \bigcup_{j=0}^{r} Q_j \rangle$ contains no independent set with m vertices, for each j, $1 \leq j \leq m$, there exists a $x_{i_j} \varepsilon Q_j$ such that (x_2, x_{i_j}) is an edge of G. But the set $\{x_{i_1-1}, x_{i_2-1}, \ldots, x_{i_m-1}\}$ is not independent, so that there exists a v and w, such that (x_{i_v-1}, x_{i_w-1}) is an edge of G. But then $(x_{i_v-1}, x_{i_w-1}, x_{i_w-2}, \ldots, x_{i_v}, x_2, x_{i_w}, x_{i_w+1}, \ldots, x_{i_v-2}, x_{i_v-1})$ is cycle of length $|Q| + 1$ containing x_2. Likewise using x_i, $3 \leq i \leq \ell-2$, and the paths $\{Q_j\}_{j=m(i-2)+1}^{m(i-2)+m}$ we can enlarge the cycle to a cycle containing $|Q|+\ell-3 = p-1$ vertices, contradicting the hypothesis. Hence the assumption that $\langle \bigcup_{j=0}^{n} Q_j \rangle$ contains no independent set with m vertices is false, and the theorem is proved.

Observe that a stronger result than that given in Theorem 4 could lead to the improvement of bound on n with respect to m given in Theorem 1(i). In particular in Theorem 4, if the independence number $\beta(G)$ is equal to cp for some constant c, then the bound would be improved to $n \geq \frac{1}{c} m$. Thus of general interest is the following question.

QUESTION 1. _Given that_ G _is a Hamiltonian graph with_ p _vertices,_ containing no C_{p-1}, find $\beta(G)$ or at least its order of magnitude.

An additional fact, which we know to be true for n sufficiently large and m fixed, would immediately prove Theorem 1(i) for all $n \geq m$. This is given in Question 2.

QUESTION 2. For $n \geq m$, show that $(n-1)(m-1) \to (C_n \text{ or } K_{n-1}, K_m)$. In fact probably the following better result holds, namely that

$$r(C_n \text{ or } K_{n-1}, K_m) = (n-2)(m-1)+1 \quad \text{when} \quad n \geq m.$$

To see that $(n-1)(m-1) \to (C_n \text{ or } K_{n-1}, K_m)$ implies $r(C_n, K_m) = (n-1)(m-1)+1$ for $n > m$, simply two color the complete graph G on $(n-1)(m-1)+1$ vertices. If this graph fails to contain a red C_n or a blue K_m, it must contain a red K_{n-1}. Remove this red K_{n-1} from G giving a two colored graph H. But H contains $(n-1)(m-2)+1$ vertices. Hence by inducting on m, we can assume $r(C_n, K_{m-1}) = (n-1)(m-2)+1$, giving that H contains a red C_n or a blue K_{m-1}. We are finished unless H contains a blue K_{m-1}. Thus we have a red K_{n-1} and a blue K_{m-1} disjoint from it. If G contains no blue K_m, each vertex of the red K_{n-1} is adjacent in red to some vertex of the blue K_{m-1}. But $n-1 > m-1$, so two distinct vertices of the red K_{n-1} are adjacent in red to the same vertex of the blue K_{m-1}. Hence we obtain the red C_n desired. A more careful argument also shows that $r(C_n, K_n) = (n-1)^2+1$ when the statement given in Question 2 holds.

We formulate our preceding discussion in terms of a question.

QUESTION 3. Find the range of integers n and m such that $r(C_n, K_m) = (n-1)(m-1)+1$. In particular show that the equality holds for $n \geq m$.

Closer inspection of the lower bound on $r(C_n, K_m)$ given in Theorem 1(iii) shows that for n fixed and m sufficiently large, there exists an $\alpha > 0$ such that $r(C_n,K_m) \geq m^{1+\alpha}$. Thus Theorem 1 cannot possibly hold for m much larger than n. Any improvement of the bounds given in (ii) and (iii) of Theorem 1 would also be useful. We thus pose the next question.

QUESTION 4. <u>How rapidly must</u> n <u>increase with</u> m <u>in order that</u> $r(C_n,K_m)$ <u>approaches</u> nm <u>asymptotically?</u>

Since in Theorem 1(iii) a lower bound is given for $r(\leq C_n,K_m)$, we give what is known about this Ramsey number. These results appear in [7].

THEOREM 5.

(i) <u>For all</u> $m \geq 2$, $r(\leq C_n,K_m) = \begin{cases} 2m-1 & \text{when } n \geq 2m-1 \\ 2m & \text{when } 2m-1 > n > m. \end{cases}$

(ii) <u>For</u> ε <u>fixed</u>, $\varepsilon > 0$, <u>there exists a constant</u> A_ε <u>such that</u> $r(\leq C_n,K_m) \leq \{(2+\varepsilon)m\}$ <u>when</u> $n \geq A_\varepsilon \log m$.

QUESTION 5. <u>For</u> k <u>a small positive integer determine the range of</u> n <u>with respect to</u> m <u>such that</u> $r(\leq C_n,K_m) = 2m+k$.

We next discuss $r(H,L)$ when H and L are trees.

One might be tempted to suspect that finding $r(T_n,T_m)$ should be easy. Here of course T_i denotes a tree with i vertices. This appears to be far from the truth. In fact it would be nice to know if $r(T_n,T_m) \leq n+m-2$, which we understand has not been proved. Since the Ramsey numbers for trees should be investigated further, we will give a conjecture concerning them. In [5] Cockayne has computed $r(K_{1,n}T_m)$ for certain classes of trees, but nothing has been done with the path-tree problem. Thus our conjecture concerns the path-tree. Let T_j^k, $j \leq k$, be a tree with $j + k$ vertices such that T_j^k is a subgraph of the complete bipartite graph $K_{j,k}$. We pose our conjecture in the next question.

QUESTION 6. Show

$$r(P_n,T_j^k) = \begin{cases} n+j-1 & \text{for } n \geq 2k-1 \\ \max\{\left[\frac{n}{2}\right]+k+j-1, 2k-1\} & \text{for } 2k-1 \geq n \geq k. \end{cases}$$

Observe that the conjecture says that the path-tree Ramsey number depends heavily on the smallest complete bipartite graph in which the tree is embeddable.

The graphs $K_{n-1} \cup K_{j-1}$, $\overline{K}_{k+j-1} + \overline{K}_{\left[\frac{n}{2}\right]-1}$, and $K_{k-1} \cup K_{k-1}$ taken as red graphs with their complements as the blue graphs, show that $r(P_n,T_j^k)$ is at least as large as indicated in the conjecture. We can in fact prove the following, which verifies the conjecture for n somewhat larger than k.

THEOREM 6. $r(P_n,K_{j,k}) = n+j-1$ for $n \geq (k-1)2^j+2$, and $k \geq j$.

COROLLARY 7. $r(P_n,T_j^k) = n+j-1$ for $n \geq (k-1)2^j+2$.

Proof of theorem. Two color the edges of a complete graph G on $n+j-1$ vertices. We assume throughout the proof that G contains no red P_n and no blue $K_{j,k}$. Let $Q_1 = (v_1,x_1,x_2,\ldots,x_\ell)$ be a maximal length red path in G.

We first show that $d_R(v_1) \leq k-1$. To see this suppose the opposite, that $t = d_R(v_1) \geq k$. Since Q_1 is maximal length, a red edge (v_1,y) in G implies $y = x_j$ for some $1 \leq j \leq \ell$. Hence $\Gamma_R(v_1) = \{x_{i_1},x_{i_2},\ldots,x_{i_t}\}$. But then for each i_j, $1 \leq j \leq t$, there exists a maximal length red path in G with x_{i_j} as an endpoint. Hence the t vertices $\{x_{i_1-1},x_{i_2-1},\ldots,x_{i_t-1}\}$ are adjacent in blue to each of the vertices $G \setminus Q_1$. Since $|G \setminus Q_1| \geq j$ and $t \geq k$, this gives a blue $K_{j,k}$, a contradiction to our assumption. Thus $d_R(v_1) \leq k-1$.

Next let $Q_2 = (v_2,x_1,x_2,\ldots,x_\ell)$ be a maximal length red path in $G \setminus \{v_1\}$. We claim $d_R(v_2) \leq 2k-2$. By the same argument as given above if $t = d_R(v_2) \geq 2k-1$, then there exist t vertices $\{x_{i_1-1}, x_{i_2-1}, \ldots, x_{i_t-1}\}$, each of which is adjacent in blue to each of the vertices of $G \setminus (Q_2 \cup \{v_1\})$. But $d_R(v_1) \leq k-1$, so at least $t-(k-1) \geq k$ of the vertices $\{x_{i_1-1}, x_{i_2-1}, \ldots, x_{i_t-1}\}$ are also adjacent in blue to v_1. Hence these k vertices and the vertices $(G \setminus Q_2) \cup \{v_1\}$ give a blue $K_{j,k}$, again a contradiction. Thus $d_R(v_2) \leq 2k-2$. Continuing in this fashion we find vertices $v_1, v_2, \ldots, v_j$ such that $d_R(v_i) \leq 2^{i-1}(k-1)$ for $1 \leq i \leq j$. But $\sum_{i=1}^{j} d_R(v_i) \leq (k-1)(2^j-1)$. Thus if $n+j-1-((k-1)(2^j-1)+j) \geq k$, then G contains a blue $K_{i,j}$ with the j vertices $\{v_1, v_2, \ldots, v_j\}$ as one of its parts, a contradiction. Since $n+j-1-((k-1)(2^j-1)+j) \geq k$ is equivalent to $n \geq (k-1)2^j+2$ the result follows.

Finally concerning exact results we investigate rather briefly the generalized Ramsey theory for multiple colors. Let C_o and C_e denote lists of cycles defined as

$$C_o = (C_{2t_1+1}, C_{2t_2+1}, \ldots, C_{2t_r+1}), \quad C_e = (C_{2\ell_1}, C_{2\ell_2}, \ldots, C_{2\ell_s}).$$

Then $r(C_n, C_e, C_o)$ will denote the Ramsey number for the graphs $(C_n, C_{2\ell_1}, C_{2\ell_2}, \ldots, C_{2\ell_s}, C_{2t_1+1}, C_{2t_2+1}, \ldots, C_{2t_r+1})$. The following results appear in a more general form and in greater detail in [6].

THEOREM 8. <u>For</u> n <u>sufficiently large each of the following results hold</u>.

(i) $r(C_n, C_e, C_o) = 2^r(n+\sum_{j=1}^{s} \ell_j - s - 1)+1$, <u>when</u> $t_i \geq 2^{r-i}$, $1 \leq i \leq r$.

(ii) $r(C_n, C_e, C_o) = (r(\leq C_o)-1)(n+\sum_{j=1}^{s} \ell_j - s-1)+1$.

(iii) $r(C_n, C_{2\ell+1}, C_{2k+1}) = \begin{cases} 4n-3, & \ell \geq 2,\ k \geq 1 \\ 5n-4, & \ell = 1,\ k = 1. \end{cases}$

Although no proof of the theorem will be given here, we do wish to give a canonical coloring which will establish that $r(C_n,C_e,C_o) > 2^r(n+\sum_{j=1}^{s} \ell_j-s-1)$ in the special case when $s = r = 2$. This coloring is given as a demonstration of the similarities of canonical multiple colorings and canonical two colorings. We need to five color the complete graph $K_{4(n+\ell_1+\ell_2-3)}$, which we denote by G. Partition the vertices of G into four sets of equal cardinality and denote these sets by V_1,V_2,V_3, and V_4. We color each of edges of the graphs $\langle V_i\rangle$ with the first three colors in an identical fashion. To do this decompose V_i (for i fixed) into disjoint sets V_1^*,V_2^*,V_3^* where $|V_1^*| = n-1$, $|V_2^*| = \ell_1-1$, $|V_3^*| = \ell_2-1$. Now color each edge joining a pair of vertices of V_1^* with the first color, each edge joining a vertex of V_2^* to a vertex of $V_1^* \cup V_2^*$ with the second color, and each edge joining a vertex of V_3^* to a vertex of $V_1^* \cup V_2^* \cup V_3^*$ with the third color. Next color edges joining vertices of V_1 to vertices of V_2 or those of V_3 to those of V_4 in the fourth color, and color edges joining vertices of $V_1 \cup V_2$ to vertices of $V_3 \cup V_4$ in the fifth color. This coloring clearly demonstrates the above inequality for $s = r = 2$.

Several other comments are in order. First observe that Theorem 8(ii) clearly indicates the usefulness of the Ramsey numbers $r(\leq C_o)$. In particular since $r(\leq C_3,\leq C_3) = r(C_3,C_3) = 6$ and $r(\leq C_3,\leq C_3,\leq C_3) = r(C_3,C_3,C_3) = 17$, (ii) implies for n large that $r(C_n,C_3,C_3) = 5n-4$ and $r(C_n,C_3,C_3,C_3) = 16n-15$. It can be shown that these two equalities hold for n quite small. This is done by using the Ramsey number given for the cycle-complete graph in Theorem 1(i). Specifically recall that $r(C_n,K_6) = 5n-4$ for $n \geq 34$ and $r(C_n,K_{17}) = 16n-15$ for $n \geq 17^2-2 = 287$. Thus for every three coloring (four coloring) of the edges of a K_{5n-4} (K_{16n-15}) we obtain, by identifying all colors other than the first, that the graph contains a C_n in the first color or a K_6 (K_{17}) in the identified colors. But since every two coloring (three coloring) of a K_6 (K_{17}) has a C_3 in one of the colors, it follows that $r(C_n,C_3,C_3) = 5n-4$ for $n \geq 34$ and $r(C_n,C_3,C_3,C_3) = 16n-15$ for $n \geq 287$. This result alone shows the value of knowing that $r(C_n,K_m) = (n-1)(m-1)+1$ holds when $n \geq m$. In particular this would improve the above lower bounds on n from 34 and 287 to 6 and 17 respectively.

We conclude the section on exact results with two additional questions which have been asked previously by P. Erdös.

QUESTION 7. *Determine if* $r(C_{2k+1},C_{2k+1},C_{2k+1}) = 4(2k+1)-3$ *when* $k \geq 2$.

QUESTION 8. *Find* $r(C_n,C_n,C_e,C_o)$ *when* n *is large*.

It is apparent that a more general strategy is needed to make significant progress with multiple coloring problems.

NON-STANDARD RESULTS

For the remainder of the paper we will look at some non-standard results in generalized Ramsey theory. The first of these results involves what is known as the size Ramsey number. Let G and H be graphs and let $C = \{L \mid L$ is a graph such that $L \to (G,H)\}$. The size Ramsey number of G and H, denoted $\hat{r}(G,H)$, is defined as $\hat{r}(G,H) = \min_{L \in C} |E(L)|$. If $G = H$, we denote $\hat{r}(G,H)$ by $\hat{r}(G)$.

First it is clear that $\hat{r}(G,H) \leq \binom{r(G,H)}{2}$. It has been shown (originally by V. Chvatal) when G and H are complete graphs, that this inequality is an equality. Also C.C. Rousseau observed that $\hat{r}(C_4) = \binom{r(C_4)}{2} = 15$, so that there are non-complete graphs such that $\hat{r}(G,H) = \binom{r(G,H)}{2}$.

There are of course many graphs where $\hat{r}(G,H)$ is considerably

smaller than $\binom{r(G,H)}{2}$. The simplest example involves stars. It is straightforward to show that $\hat{r}(K_{1,n},K_{1,m}) = m+n-1$ while
$$r(K_{1,n},K_{1,m}) = \begin{cases} m+n-1 & n,\ m \text{ both even} \\ m+n & \text{otherwise} \end{cases}.$$
Not so surprising is the fact that finding the exact value of $\hat{r}(G,H)$ is quite difficult. In fact letting nG denote n disjoint copies of graph G, it can be shown when $k = \ell$ that $\hat{r}(mK_{1,k},nK_{1,\ell}) = (k+\ell-1)(m+n-1)$. This result probably holds for $k \neq \ell$, although the method of proof used when $k = \ell$ fails to work when $k \neq \ell$.

QUESTION 9. <u>Show</u> $\hat{r}(mK_{1,k},nK_{1,\ell}) = (k+\ell-1)(m+n-1)$ <u>for</u> $k \neq \ell$.

Most of the results known involve in some respects a large star (or stars) attached to another graph. Thus we define three "star" operations. The join of two graphs is familar. Thus given a graph G, the graph $G + \bar{K}_n$ is obtained from G by introducing n new vertices and by joining each vertex of G to each of these n additional vertices. If G is of order n then $G + \bar{K}_n$ is of order $m + n$. In a similar way, we define $G \oplus \bar{K}_n$. This is the graph obtained from G by introducing, for each vertex v in G, n additional vertices and by joining v to these n vertices. Thus if G is of order m, then $G \oplus \bar{K}_n$ is of order $m(n+1)$. Finally we let v be a particular vertex of G and define $G * \bar{K}_n(v)$ to be the graph obtained by adding n vertices and joining just v to the n additional vertices. Thus for G of order m, $G * \bar{K}_n(v)$ is of order $m + n$. If the choice of v is immaterial, we shall write $G * \bar{K}_n$.

With this notation we list several results. Details concerning the proofs of these results and additional information on the size Ramsey number can be found in [8].

THEOREM 9. <u>For appropriate constants</u> $\{a_i\}_{i=1}^{10}$ <u>and</u> n <u>sufficiently large each of the following hold</u>:

(i) $a_1m^2n^2 \leq \hat{r}(K_m * \bar{K}_n) \leq a_2m^2n^2$.

(ii) $a_3m2^{m-1}n \leq \hat{r}(K_{m,n}) \leq a_4m^22^{m-1}n$.

(iii) $a_5m^2n^2 \leq \hat{r}(K_m + \bar{K}_n) \leq a_64^{2m}n^2$.

(iv) $a_7m^3n^2 \leq \hat{r}(K_m \oplus \bar{K}_n) \leq a_8m^4n^2$.

(v) $a_9n^22^{n/2} \leq \hat{r}(K_{n,n}) \leq a_{10}n^32^{n-1}$.

(vi) $mn/2 \leq \hat{r}(K_{m,k} * \bar{K}_n(v)) < 4m(2(n+k)-1)$.
where $m \leq k$ *and* v *is a vertex in the smaller part of* $K_{m,k}$.

(vii) G *non-bipartite implies* $\hat{r}(G * \bar{K}_n) > \frac{n^2}{2}$.

The reader will observe that there are essentially no results for the size Ramsey number which do not involve the star. In fact determination of the exact size Ramsey number for a simple graph like a path, P_n, appears to be quite difficult.

QUESTION 10. *Does* $\lim_{n\to\infty} \frac{\hat{r}(P_n,P_n)}{n^2} = 0$?

QUESTION 11. *Does* $\lim_{n\to\infty} \frac{\hat{r}(P_n,P_n)}{n}$ *exist?*

It is clear that much work remains to be done for the size Ramsey number.

Along a similar vein, suppose G, H, L are graphs such that $L \rightarrow (G,H)$ with L edge minimal in the sense that $L' \not\rightarrow (G,H)$ for each proper subgraph L' of L. Such a graph L is called (G,H)-irreducible. A natural problem arises. For fixed graphs G and H characterize the (G,H)-irreducible graphs or at least determine "how many" such non-isomorphic graphs exist. In [3] it is shown for $r, s \geq 3$ that there are infinitely many non-isomorphic (K_r,K_s)-irreducible graphs, but only finitely many (mK_2,nK_2)-irreducible ones. Also they show that $K_{1,2n-1}$ is the unique $(K_{1,n},K_{1,n})$-irreducible graph when n is odd. It is shown in [4] that for $n \geq 3$ there are infinitely many non-isomorphic (P_n,P_n)-graphs. This is trivial for $n = 3$ (simply consider the family of odd cycles) but surprisingly more difficult for $n \geq 4$. This, of course, raises many questions.

QUESTION 12. _Are there infinitely many non-isomorphic_ (P_n,P_m)-_irreducible graphs when_ $n \neq m$?

QUESTION 13. _For what graphs_ G _and_ H _are there only finitely many non-isomorphic_ (G,H)-_irreducible graphs_?

The final non-standard result which will be mentioned is the connected Ramsey number. The connected Ramsey number has been defined by Sumner in [12].

Let G and H be fixed graphs. Then the _connected Ramsey number_, $r_c(G,H)$, is defined as the smallest integer n such that each two coloring of a K_n, with both colored graphs (red and blue) _connected_ of order n, contains a red G or a blue H. Sumner shows that $r_c(G,H) = r(G,H)$ when G and H are blocks and determines the value of $r_c(P_n,P_m)$. In addition the values of r_c for the path-cycle, path-complete graph, path-star, star-complete graph are given in [9].

Obviously there are many similar "connected Ramsey type numbers" that could be considered. As an example, we could specify that both the red graph and the blue graph be two connected, instead of one connected. Of course, this would require some care in making the definition. For example a K_5 when two colored such that both colored graphs are two connected must contain a C_5 in the first color, while there exist two colorings of K_6 with both colored graphs two connected such that there is no C_5 in the first color and no C_n, $n \geq 7$, in the second color. This sort of problem can not occur in the one connected case discussed above.

REFERENCES

[1] M. Behzad and G. Chartrand, _Introduction to the Theory of Graphs_. Allyn and Bacon, Boston (1971).

[2] J.A. Bondy and P. Erdös, Ramsey numbers for cycles in graphs. _J. Combinatorial Theory Ser. B._ 14 (1973) 46-54.

[3] S.A. Burr, P. Erdös, and L. Lovász, On graphs of Ramsey type. (To appear).

[4] S.A. Burr, P. Erdös, R.J. Faudree, and R.H. Schelp, On Ramsey-minimal graphs. (To appear).

[5] E.J. Cockayne, Some tree-star Ramsey numbers. _J. Combinatorial Theory Ser. B_ 17 (1974) 183-187.

[6] P. Erdös, R.J. Faudree, C.C. Rousseau, and R.H. Schelp, Generalized Ramsey theory for multiple colors. (To appear).

[7] __________, On cycle-complete graph Ramsey numbers. (To appear).

[8] __________, The size Ramsey number. (To appear).

[9] R.J. Faudree and R.H. Schelp, Some connected Ramsey numbers. (To appear).

[10] F. Harary, _Graph Theory_. Addison-Wesley, Reading (1969).

[11] Joel Spencer, Asymptotic lower bounds for Ramsey functions. (To appear).

[12] David P. Sumner, The connected Ramsey number. (To appear).

EXACTLY THIRTEEN CONNECTED CUBIC GRAPHS HAVE INTEGRAL SPECTRA

Allen J. Schwenk*
U. S. Naval Academy
Annapolis, MD 21402

Abstract

The problem of identifying those graphs whose spectra consist entirely of integers was first posed by F. Harary. We examined some elementary procedures for constructing integral graphs in [6]. Although the general problem seems intractible, it is easy to find the seven connected graphs with integral spectra, maximum degree at most three, and minimum degree less than three. This article was inspired by Cvetković's attempt [4] to find the connected cubic integral graphs. He had displayed twelve such graphs, and had restricted the remaining possibilities to ninety-five potential spectral.

In this article we construct the sole graph omitted from Cvetković's list and prove that no other exist. We have just learned that Cvetković has recently collaborated with Bussemaker [1] to obtain the same result. Unlike their effort, the present article avoids the use of computer search to examine all the possibilities.

The thirteenth graph happens to have the same spectrum as one of the others. This cospectral pair confirms a conjecture of Balaban by being indistinguishable under a certain proposed chemical classification scheme.

* Research supported in part by a grant from the Naval Academy Research Council.

1. Introduction

Our goal is to identify all the connected cubic graphs with integral spectra. We begin with several simple observations which greatly reduce the number of candidates that need be examined. In section 2, we find the eight connected cubic integral bigraphs. The five remaining integral graphs are constructed in section 3. We close with some speculation on the possible application of our technique to the construction of Moore graphs.

The conjenction $G_1 \wedge G_2$ was defined by Miller [10] and appears in Harary [5, p. 25]. It has $V_1 \times V_2$ as its vertex set with (u_1, u_2) adjacent to (w_1, w_2) whenever both u_1 adj w_1 in G_1 and u_2 adj w_2 in G_2.

Lemma 1. If G is cubic with integral spectrum, then $B = G \wedge K_2$ is a cubic integral bigraph.

Proof. From the definition of conjunction we note that B is certainly cubic. Moreover, Cvetković and Lučić demonstrated in [3] just as we did [11] that the spectrum $G_1 \wedge G_2$ consists of all products $\lambda_1 \cdot \lambda_2$ where λ_i is an eigenvalue of G_i. Since G is integral and K_2 has the spectrum $\{+1, -1\}$, it is clear that B must be integral. Furthermore, the eigenvalues of B are paired (that is, λ is an eigenvalue if and only if $-\lambda$ is an eigenvalue), and so, by the pairing theorem (e.g., see Coulson and Rushbrooke [2]) B must also be a bigraph.

We observe that this lemma can be used repeatedly to construct arbitrarily many cubic integral graphs, however, when G is already a bigraph, $G \wedge K_2$ is merely two disjoint copies of G, and so not a new connected cubic integral graph. Thus, we need only consider conjunction for those G which contain an odd cycle.

This observation sets the general strategy we shall employ. We first identify all cubic integral connected bigraphs. Then we ask which of these B can be decomposed as $B = G \wedge K_2$.

2. Bipartite Cubic Integral Graphs

Let B be a bipartite cubic integral graph with $p = 2n$ vertices. Using superscripts to represent multiplicities, we may write its spectrum in the form

$$3\ ,\ 2^x\ ,\ 1^y\ ,\ 0^{2z}\ ,\ -1^y\ ,\ -2^x\ ,\ -3\ .$$

Let us let q and h stand for the numbers of quadrilaterals and hexagons in B . In [7] we noted that the sum of the kth powers of the eigenvalues is just the numbers of closed walks of length k. This result plus a bit of computation directly verifies the next lemma.

Lemma 2. The parameters n, x, y, z, q, h must satisfy the equations:

(1) $\frac{1}{2}\ \Sigma\ \lambda_i^o = 1 + x + y + z = n$

(2) $\frac{1}{2}\ \Sigma\ \lambda_i^2 = 9 + 4x + y = 3n$

(3) $\frac{1}{2}\ \Sigma\ \lambda_i^4 = 81 + 16x + y = 15n + 4q$

(4) $\frac{1}{2}\ \Sigma\ \lambda_i^6 = 729 + 64x + y = 87n + 48q + 6h$

A final useful lemma is due to Hoffman [8]. Let a regular graph G have distinct eigenvalues $r, \mu_2, \mu_3, \ldots, \mu_k$, and let A be the adjacency matrix of G while J is the matrix whose entires are all 1's .

Lemma 3. The adjacency matrix satisfies the equation

(5) $\prod_{i=2}^{k} (r - \mu_i)\ J = p \prod_{i=2}^{k} (A - \mu_i)\ .$

We are now ready to construct all cubic integral connected bigraphs.

<u>Theorem 1</u>. There are exactly eight cubic connected bigraphs with integral spectra. (see Figure 1.)

<u>Proof</u>. We consider cases depending upon which of the parameters x, y, and z may be zero.

<u>Case 1</u>. If $x = y = z = 0$, then by (1), $n = 1$ which is impossible for a cubic graph.

<u>Case 2</u>. If $x = y = 0$, but $z > 0$, we note that by (2) $n = 3$, whence $B = K_{3,3}$ is the first integral bigraph.

<u>Case 3</u>. If $x = z = 0$, but $y > 0$, equations (1) and (2) are solved simultaneously to find $n = 4$ and $y = 3$. The cube $K_2 \times K_2 \times K_2$ is the only eight point cubic bigraph, and provides the second integral bigraph.

<u>Case 4</u>. If $y = z = 0$, bux $x > 0$, solution of (1) and (2) provides $n = -5$ which is impossible.

<u>Case 5</u>. Only $x = 0$. Equation (2) requires $y = 3k$ and $n = k + 3$ with $k > 0$. But then (1) forces $z = 2 - 2k \leq 0$, so this case is impossible.

<u>Case 6</u>. Only $y = 0$. Hoffman's polynomial reduces to

$$(6) \quad 90\,J = 2n[A^4 - 4A^2] + 6n[A^3 - 4A]\,.$$

Let $w_k(i, j)$ denote the number of walks of length k from v_i to v_j. Now because B is a bigraph, for v_i adjacent to v_j every $v_i - v_j$ walk must have odd length, so the (i, j) entry of (6) is

$$(7) \quad 90 = 2n[0 - 4 \cdot 0] + 6n[w_3(i, j) - 4]\,.$$

In particular, $6n \mid 90$, and so $n \mid 15$. Thus $n = 3, 5$, or 15. Equation (2) requires $3 \mid x$, so setting $x = 3k \geq 3$, we find

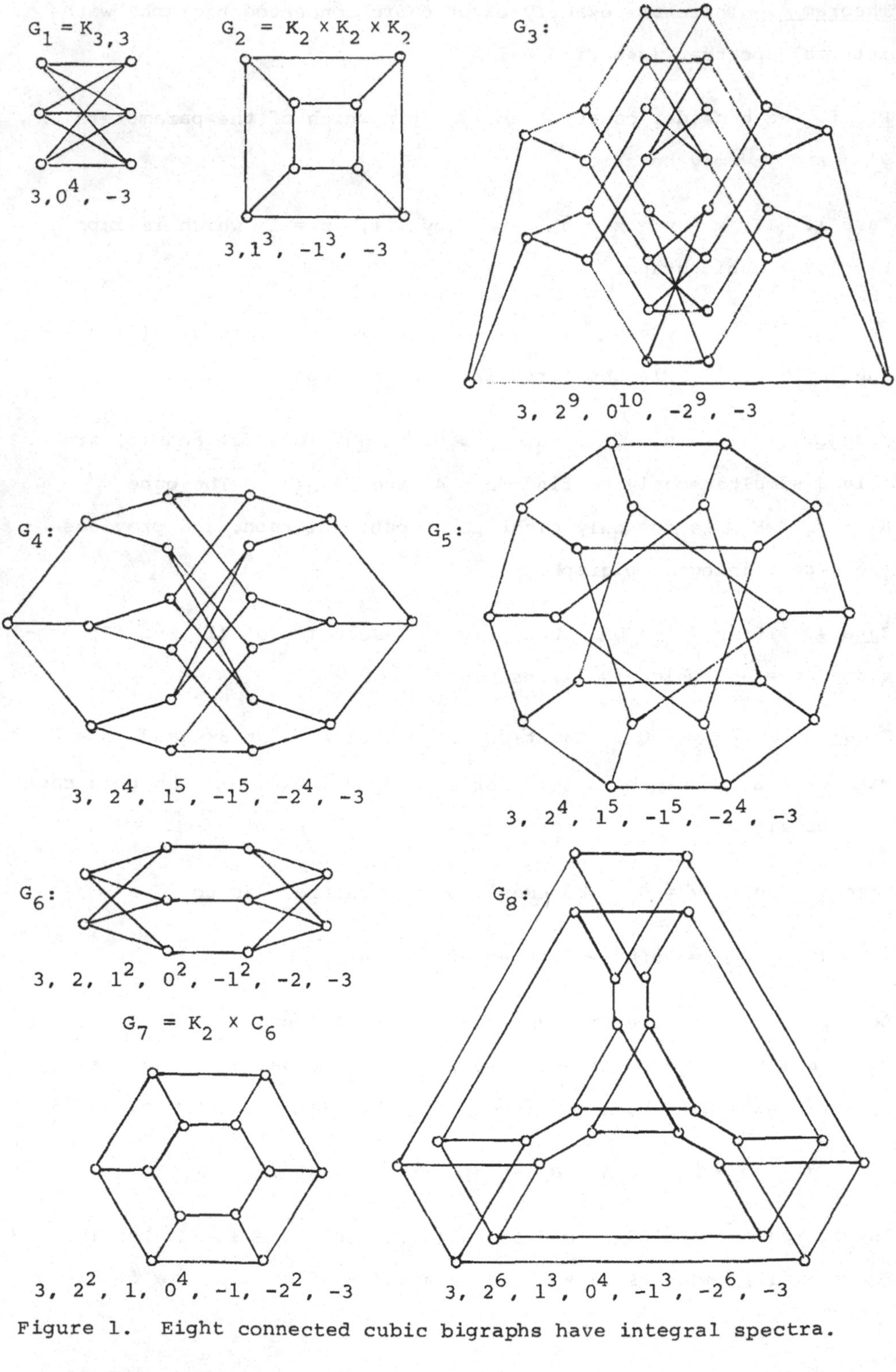

Figure 1. Eight connected cubic bigraphs have integral spectra.

$n = 4k + 3$, which can only occur if $n = 15$, $x = 9$, $z = 5$. Equation (6) becomes

$$(8) \quad 3J = [A^4 - 4A^2] + 3[A^3 - 4A] .$$

This forces B to have diameter 4, for otherwise a pair of points v_i, v_j at distance 5 produce in the (i, j) position of (8) the impossibility $3 = 0$. Finally, (3) and (4) require that $q = h = 0$, so the shortest cycle in B has length 8. Now focus attention on a particular edge $u_o w_0$. Starting from either endpoint and taking three steps (see Figure 2), we reach $4 + 8 + 16 = 28$ vertices, and these must be distinct to avoid cycles shorter than the girth of 8. Thus, every vertex of B is reached in this way, and the structure of B will be specified if we can decide how to join the 16 end-points via a regular subgraph H of degree 2. Considering the girth, H must be either a 16-cycle or two 8-cycles.

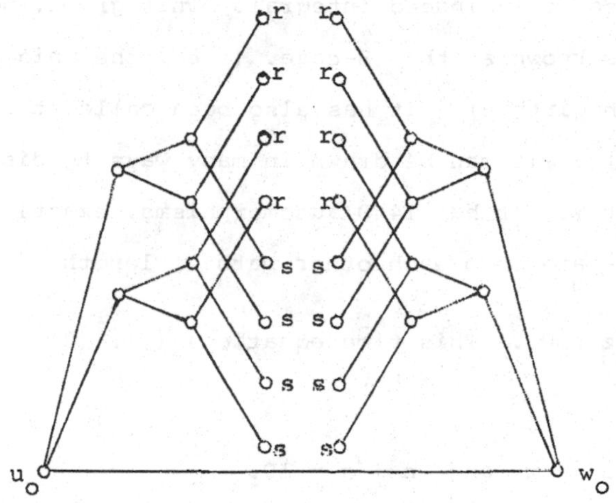

Figure 2. Partial construction of a cubic integral bigraph.

Case 6a. $H = v_0 v_1 \cdots v_{15}$ is a 16-cycle with vertices labeled modulo 16. Each vertex v_i in H is joined by a unique path outside of H of length 2 to one of the vertices v_{i+6}, v_{i+8}, or v_{i+10}. If every v_i is so joined to v_{i+8}, many 6-cycles are formed, so some v_j is joined to either v_{j+6} or v_{j+10}. Without loss of generality, say it is v_{j+6}. But the diameter is 4, and the only way to reach v_{j+11} in at most 4 steps from v_j is to have v_{j+1} joined to v_{j+11} by a 2-path. For the same reason, to get diameter 4 from v_{j+6}, we need v_{j+5} joined v_{j+11}. Thus, v_{j+11} is joined to two different vertices by paths of length 2, contrary to the conditions of the construction.

Case 6b. $H = 2C_8$. Each pair of points joined by a path of length 2 must be from opposite C_8's to avoid forming a cycle of length less than 8. Referring to Figure 2, we conclude that the 8 vertices labeled r comprise one C_8 while the 8 labeled form the other C_8. The former eight are seen to be joined in an essentially unique way in order to avoid forming 6-cycles. Then the latter 8 are forced to be joined in a determined pattern to avoid 6-cycles, producing G_3 of Figure 1. By inspection, it is seen to satisfy equation (8), so it is indeed integral. This graph, our third integral bigraph, is known as the 8-cage. (i.e., the unique smallest cubic graph with girth 8). It has also been called the Levi graph [5, pp. 174-175]. It can be drawn in many ways to display its high degree of symmetry. It has 1440 automorphisms, exactly one mapping any specified 5-path onto each other path of length 5.

Case 7. Only $z = 0$. This time equations (1), (2), and (3) reduce to

(9) $$2n = 8 + 3x \quad \text{and} \quad n + q = 10.$$

Meanwhile, Hoffman's polynomial becomes

(10) $$240\,J = 2n[A^5 - 5A^3 + 4A] + 6n[A^4 - 5A^2 + 4]\ .$$

As in case 6, this requires $n|40$. This restriction applied to equations (9) admit a solution with $x > 0$ and $q \geq 0$ only if we select $n = 10$. We find that B has no quadrilaterals, and (4) requires $h = 20$ hexagons. Equation (10) now reads

$$(11) \quad 12J = [A^5 - 5A^3 + 4A] + 3[A^4 - 5A^2 + 4] .$$

If u and w are selected at distance 2, there must be a unique 2-path, uvw, since B has no quadrilaterals. The (u, w) entry in (11) requires $W_4(u, w) = 9$. Exactly 7 of these 4-walks are obtained by adding two steps to the unique 2-path. The 2 remaining walks cannot contain v, and so each provides a hexagon containing uvw. That is, every 2-path lies in exactly two hexagons!

A direct consequence is that any 3-path can lie in at most two hexagons. We consider two cases.

Case 7a. B contains at least one 3-path $uvwx$ which lies in two hexagons. These two hexagons can only be formed by having three disjoint $u - x$ 3-paths that is uv_iw_ix with $i = 1, 2, 3$ (see Figure 3). Now each w_i must be joined to a distinct new vertex y_i with $d(u, y_i) = 3$ for otherwise some 2-path is in 3 hexagons. Furthermore, each v_i must be joined to a distince new z_i with $d(u, z_i) = 2$. Now, no y_i may be joined to any z_j, without producing either a third hexagon containing uv_iw_i or a forbidden quadrilateral. Consequently, three new vertices s_1, s_2, s_3 are needed, each joined to exactly two z's in order to provide a second hexagon for each v_iuv_j. Finally, each s_i must be joined to a distinct new vertex t_i to avoid quadrilaterals. We have now labeled all 20 vertices, and the final 6 edges can only be added between y's and t's in a unique pattern to produce two hexagons containing each $w_iv_iz_i$. This completes the construction of B. Its adjacency matrix does indeed satsify equation (11), so $B = G_4$ is the fourth cubic integral bigraph. This is the unique graph omitted in [4] but then later discovered by Bussemakter and Cvetković [1]. Their drawing is shown in Figure 1.

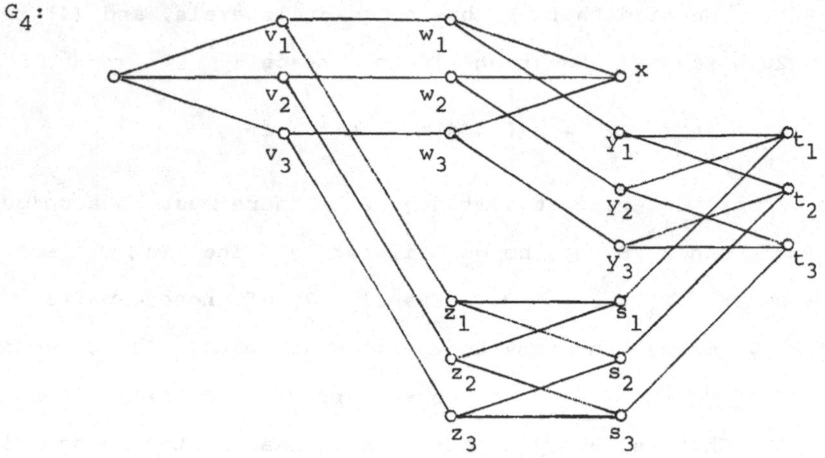

Figure 3. The fourth cubic integral bigraph.

Case 7b. No 3-path $uvwx_1$ lies in two hexagons. We recall that each 2-path lies in two hexagons, so the two possible extensions of uvw, namely $uvwx_1$ and $uvwx_2$ must each lie in exactly one hexagon. In other words, every pair of vertices at distance 3 must be joined by exactly two distinct 3-paths. Let $D_k(u)$ denote the subset of vertices at distance k from u. We now realize that for any vertex u in B, $|D_1(u)| = 3$ and $|D_2(u)| = 6$. There are 12 lines joining $D_2(u)$ to $D_3(u)$, and since each vertex in D_3 is joined by two 3-paths to u, $|D_3(u)| = 6$. Since B is a regular bigraph with 20 vertices, this forces $|D_4(u)| = 3$ and $|D_5(u)| = 1$. It merely remains to specify H, the regular subgraph of degree 2 induced by $D_3(u)$ and $D_4(u)$. (See Figure 4.)

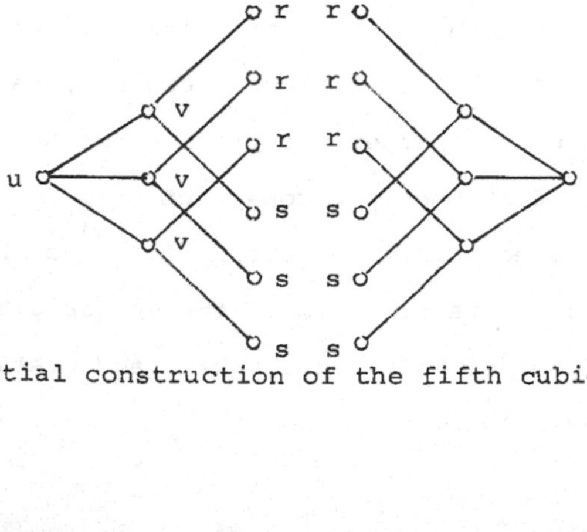

Figure 4. Partial construction of the fifth cubic integral bigraph.

Clearly, H must be either a 12-cycle or two 6-cycles. If $H = w_0 w_1 \cdots w_{11}$ is a 12-cycle, $|D_5(w_0)|$ must be 1 just like $|D_5(u)|$. But only the vertices w_5 and w_7 can be at distance 5. Now w_0 is joined by a 2-path to one of w_4, w_6, or w_8. If it were w_6, $d(w_0, w_5) = d(w_0, w_7) = 3$, a contradiction. Thus, we may assume without loss of generality that $d(w_0, w_4) = 2$. By a similar line of reasoning, $d(w_2, w_6) = 2$. But then w_8 and w_{10} are joined by a 2-path outside of H as well as the path $w_8 w_9 w_{10}$, forming a forbidden quadrilateral. Therefore $H = 2C_6$. To avoid quadrilaterals, the 6 vertices labeled r must form one hexagon while the 6 labeled s form the other. The first hexagon may be filled in arbitrarily without loss of generality, but then the second must match in order that the paths of type rvs may each lie in 2 hexagons. The resulting graph, G_5 of Figure 1, is certainly different from G_4 because it has no pair of vertices joined by three 3-paths. And yet, G_4 and G_5 are cospectral.

Balaban had anticipated the existence of cubic cospectral graphs. He observed that such a pair would invalidate a certain proposed chemical classification scheme.

Case 8. Parameters x, y, and z are all positive. In this case, equations (2) and (3) combine to yield $n = 15 - y - 4q/3 \le 14$. Hoffman's polynomial requires $n|120$. We list the 5 possible sets of parameters and examine each separately.

n	p	x	y	z	q	h
5	10	1	2	1	6	8
6	12	2	1	2	6	2
8	16	3	3	1	3	14
10	20	5	1	3	3	6
12	24	6	3	2	0	12

Case 8a. $n = 5$. The bipartite complement of B is the bigraph $B^* = K_{5,5} - B$. In this case, B^* must be regular of degree 2 . Thus, B^* is either C_{10} or $C_4 \cup C_6$. If $B^* = C_{10}$, then B is the mödius ladder M_{10} . But M_{10} is not integral (see [11]) . Hence $B^* = C_4 \cup C_6$, and $B = G_6$ of Figure 1 is verified to be the sixth cubic integral bigraph.

Case 8b. $n = 6$. Notice that B contains 6 quadrilaterals and 8 hexagons. We first demonstrate that B must not contain the subgraph $K_{2,3}$ formed by two vertices u and w being joined by 3 distinct 2-paths, uv_iw . If it did, the remaining 7 vertices induce a subgraph with 9 edges and at least 3 quadrilaterals in addition to the 3 already contained in $K_{2,3}$. Hence, each v_i must be joined to a unique x_i . Then, to avoid a seventh quadrilateral, the x's must be joined in a hexagon via three new vertices labeled y_i which, in turn, are joined to the last vertex z . But now B contains 10 hexagons, contrary to hypothesis.

An average edge lies in $6 \cdot 4/18 = 4/3$ quadrilaterals. Let uv be chosen to lie in the maximum number. Now uv cannot be in 3 quadrilaterals, for that would force a $K_{2,3}$ subgraph. Therefore, there must be exactly 2 quadrilaterals containing uv which must form the subgraph on the left in Figure 5. The 6 vertices on the right must induce a bigraph with 7 edges. Only two such graphs

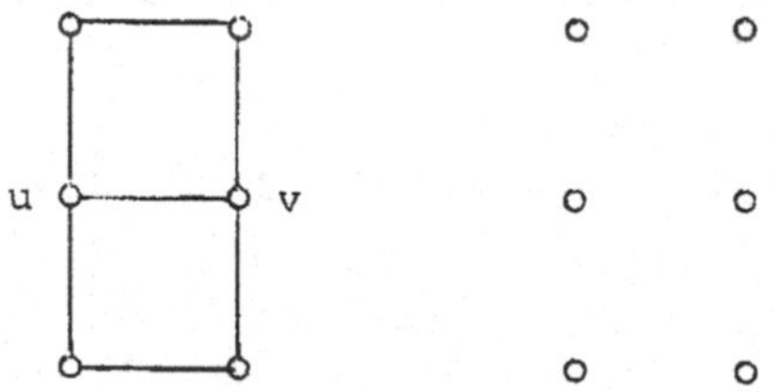

Figure 5. Partial construction of the seventh cubic integral bigraph.

exist, and one of them has the forbidden subgraph $K_{2,3}$. The other can only be joined to the left hand subgraph in two ways to maintain the bipartite structure. One resulting graph, the prism $K_2 \times C_6$, is integral and is G_7 of Figure 1; the other graph has $a = 4 \neq 6$.

Case 8c. The elegant argument in this case is due to Bussemaker and Cvetković [1]. Since B has 16 vertices but only 3 quadrilaterals, we may select v not contained in any quadrilateral. Interpretting the (v , v) entry in Hoffman's polynomial requires:

(12) $$45 = W_6(v, v) - 5W_4(v, v) + 4W_2(v, v)$$

For v not in any quadrilateral, we easily calculate that $W_2 = 3$ and $W_4 = 15$. With a bit more effort, we find

(13) $$W_6(v, v) = 87 + 2a + 2b$$

where a is the number of hexagons containing v and b is the number of quadrilaterals containing a neighbor of v. Substituting these values into (12), we reduce (12) modulo 2 to obtain the contradiction $1 \equiv 0 \bmod 2$.

Case 8d. $n = 10$. Define the weight $w(v)$ to be the sum of the number of hexagons containing v plus 4 times the number of quadrilaterals containing v plus the number of quadrilaterals at distance one from v. Since B contains 3 quadrilaterals and 6 hexagons, we calculate the average weight to be $\overline{w} = \Sigma\, w(v)/20 = (6 \cdot 6 + 3 \cdot 4 \cdot 4 + 3 \cdot 4)/20 = 4.8$. Consequently, B must contain a vertex v_0 with weight at most 4. Clearly v_0 lies in at most one quadrilateral. If v_0 lies in none, we note that $|D_3(v_0)| \leq 7$ (since $n = 10$ and 3 vertices have been used in $D_1(v_0)$). Moreover, there must be 12 edges joining D_3 to D_2. But any arrangement of these 12 edges among the 7 vertices in D_3 contribute at least 5 to the weight $w(v_0)$. Therefore, v_0 must lie in exactly one quadrilateral. But now $|D_2(v_o)| = 5$, $|D_3(v_o)| \le 7$, and there are 9 edges joining D_2 and D_3. Again, these 9 edges contribute 2 to the weight $w(v_o)$ while the quadrilateral it lies in adds 4 more, so $w(v_o) \ge 6$. In other words, every vertex has weight greater than average, an impossibility.

Case 8e. $n = 12$. Hoffman's polynomial becomes

$$(14) \qquad 30\,J = [A^6 - 5A^4 + 4A^2] + 3[A^5 - 5A^3 + 4A.]$$

For any edge uv, this reduces to

$$(15) \qquad 30 = 3[W_5(u,v) - 5W_3(u,v) + 4]\,.$$

Remembering that there are no quadrilaterals, we compute $W_3 = 5$ and $W_5 = 29 + h(uv)$ where $h(uv)$ is the number of hexagons containing edge uv. This is solved to give $h(uv) = 2$ for every edge in B. Consequently, each vertex lies in exactly 3 hexagons. Finally, no path uvw may lie in 2 hexagons, for then the 2 hexagons containing the third edge incident with v gives a total of 4 hexagons containing v. Hence every uvw determines a unique hexagon.

Let u be any fixed vertex. Label its neighbors v_1, v_2, v_3 and their neighbors w_1 through w_6 as shown in Figure 6. To produce a hexagon containing each $v_i u w_j$ we must insert x_1, x_2, x_3 joined as shown to the w's. Any additional vertex joined to two w's would create a fourth hexagon containing u, so we are forced to have 6 vertices y_i each joined to the corresponding w_i. Now if two x's had a common neighbor, some $v_i w_j x_k$ would lie in 2 hexagons, so there must be vertices z_1, z_2, z_3 with each z_i joined to the corresponding x_i. We label the two remaining vertices t_1 and t_2. To avoid forming either a quadrilateral or a second $v_1 w_1 x_1$ hexagon, we are forced to join z_1 to y_3 and y_6. Similarly, z_2 must be joined to y_2 and y_5, and z_3 must be joined to y_1 and y_4 . Finally, for $i = 1, 2, 3$, $w_i v_i w_{i+3}$ already lies in a hexagon, so y_i and y_{i+3} must be joined to opposite t's . Similarly, $w_{2i} x_i w_{2i-1}$ already lies in a hexagon, so y_{2i} and y_{2i-1} must be joined to opposite t's. This requires t_1 to be joined to y_1, y_3, y_5 while t_2 is joined to y_2, y_4, y_6 . The resulting graph is integral and is shown as G_8 in Figure 1 as drawn by Bussemaker and Cvetković [1].

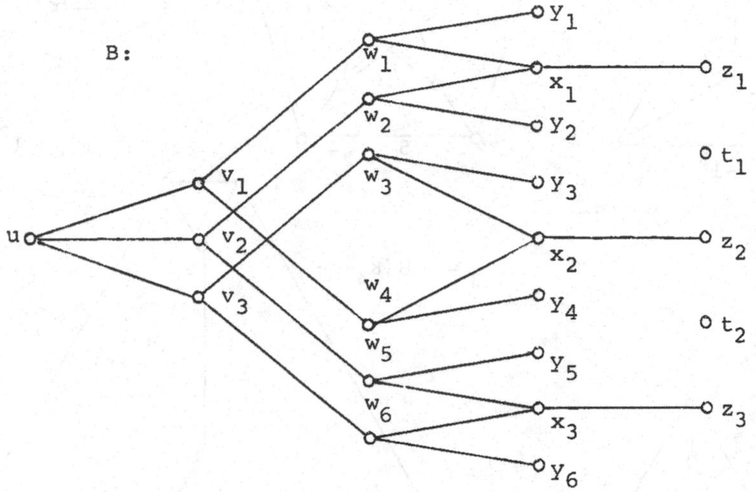

Figure 6. Construction of the last cubic integral bigraph.

Since we have examined every possible case, these are the only connected cubic integral bigraphs.

3. Nonbipartite integral graphs

As we observed in the introduction, any other integral graph G must produce one of the graphs in Figure 1 when we form $G \wedge K_2$. The proof of the following characterization is trivial.

Lemma 4. A bigraph B can be decomposed as a conjunction $G \wedge K_2$ if and only if i) The two vertex sets of B have the same size. ii) These vertices can be labeled $u_1, u_2, \ldots, u_n$ and $v_1, v_2, \ldots, v_n$ so that u_i not adj v_i and u_i adj v_j if and only if u_j adj v_i .

Theorem 2. Exactly five nonbipartite connected cubic graphs have integral spectra. (See Figure 7.)

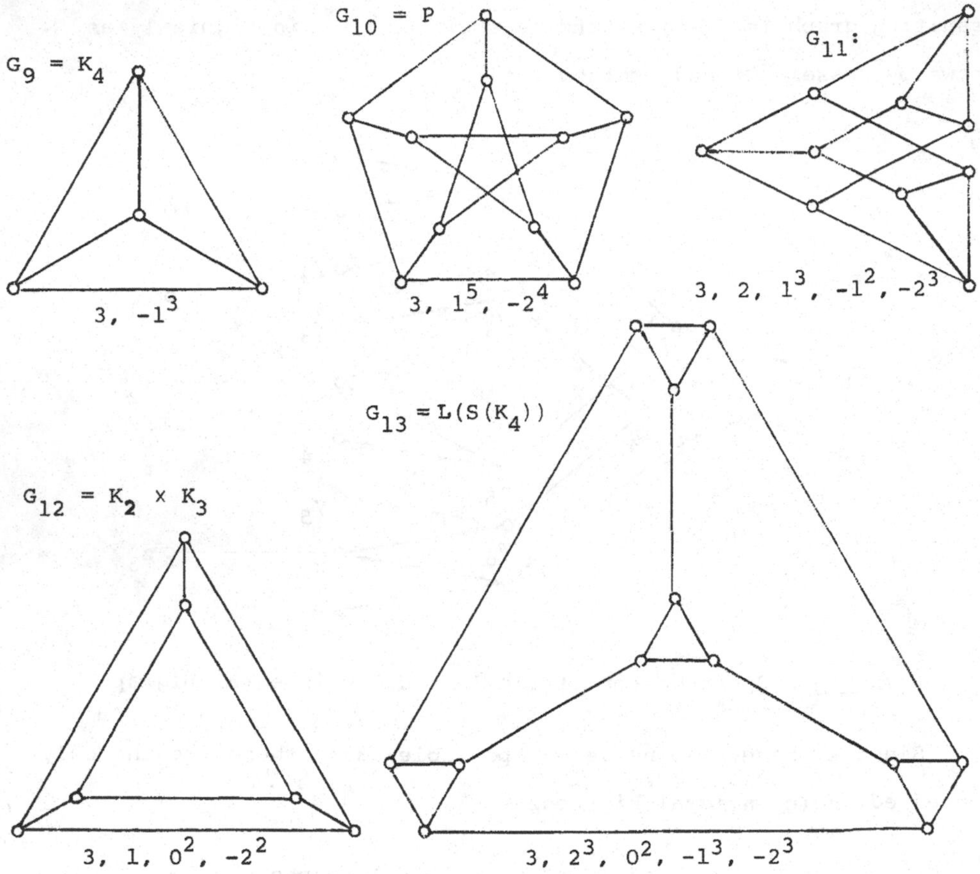

Figure 7. The five nonbipartite connected cubic integral graphs.

Proof. First, we note that G_1, G_3, and G_6 cannot be decomposed as $G \wedge K_2$, for G would then be a cubic graph with an odd number of vertices. Furthermore, we observe that if $u\, v_1 u_2 v_3 u_4 \cdots u_{2n} v$ is a path joining u to v, then Lemma 4 requires that $v\, u_1 v_2 u_3 v_4 \cdots v_{2n} u$ is a second path from u to v. This path is necessarily different, for if they were the same we would be forced to have u_n joined to v_n contrary to Lemma 4. This observation leads to the conclusion that for any i, an even number of shortest paths of odd length $2k+1$ must join u_i and v_i.

Now, considering Lemma 4, we identify each vertex of the cube with the unique vertex at distance 3 to obtain $K_2 \times K_2 \times K_2 = K_4 \wedge K_2$, and so $G_9 = K_4$ is integral.

The graph G_4 of Figure 3 cannot be decomposed because the only vertices at odd distance from u and joined by an even number of shortest path are s_1, s_2, and s_3. But, if we identify u with say s_1, we must successively identify v_1 and z_2, v_2 and z_1, w_1 and s_3, w_2 and s_2, x and z_3, finally v_3 and w_3, contrary to Lemma 4.

In G_5, identifying each vertex with the unique vertex at distance 5 provides $G_5 = G_{10} \wedge K_2$ with $G_{10} = P$, the Petersen graph. Surprisingly, if we identify a vertex v with any other at distance 3, and proceed to identify the remaining vertices as forced, we find a second decomposition $G_5 = G_{11} \wedge K_2$. It comes as a mild shock that G_5 can be decomposed in two different ways, however, after a pause for reflection, there seems to be no reason to expect $G \wedge K_2 = H \wedge K_2$ to imply $G = H$. Evidently, factorization via conjunctions is clearly not unique.

We find $G_7 = (K_2 \times K_3) \wedge K_2$ is the unique decomposition of G_7 which produces $G_{12} = K_2 \times K_3$.

Finally, in Figure 6, we note that u can only be identified with one of the x_i. The remaining identifications are quickly forced to produce $G_8 = G_{13} \wedge K_2$. This, the last connected cubic integral graph, is recognized in [6] as the line graph of the subdivision of K_4.

4. Summary

The purpose of this article has been to identify all graphs with certain highly restrictive properties. We have accomplished this goal without resorting to the ubiquitous computer. Certainly, the cubic integral graphs are no more than a curiosity, but the author believes the techniques which proved useful here might well be applied to related problems. Specifically, the existence of a Moore graph M with 3250 vertices, regular of degree 57, and girth 5 has been an open question for at least sixteen years. (see Hoffman and Singleton [9]). To construct integral graphs, we found it useful to first identify the bigraphs, and then decompose via conjunction. Perhaps we should seek to construct M, or to prove its nonexistence, by searching for the bigraph $M \wedge K_2$. Superficially, this may seem to "double" your problem without doubling your fun, however, the knowledge that $M \wedge K_2$ is bipartite has extremely strong implications which more than compensate for the doubling in size. At least that was our experience with cubic integral graphs.

As a last enticing lead, we point out that the spectrum of the bipartite complement $(M \wedge K_2)^* = K_{3250,3250} - M \wedge K_2$ is simply the spectrum of $M \wedge K_2$ minus the eigenvalues ± 57 and then union the eigenvalues ± 3193. Perhaps $(M \wedge K_2)^*$ will prove easier to identify than M itself.

References

1. F. C. Bussemaker and D. M. Cvetković, There are exactly 13 connected, cubic, integral graphs, Memorandum 1975-15, Eindhoven University of Technology.

2. C. A. Coulson and G. S. Rushbrooke, Note on the method of molecular orbitals, Proc. Cambridge Phil. Soc., 36(1940) 193-199.

3. D. M. Cvetković and R. P. Lučić, A new generalization of the concept of the p-sum of graphs, Univ. Beograd, Publ. Elektrotehn. Fak. Ser. Mat. Fiz., Nos. 302-319 (1970) 67-71.

4. D. M. Cvetković, Cubic integral graphs, Univ. Beograd. Publ. Elektrotehn. Fak. Ser. Mat. Fiz., Nos. 498-541 (1975) 107-113.

5. F. Harary, Graph Theory, Addison-Wesley, Reading, 1969.

6. F. Harary and A. J. Schwenk, Which graphs have integral spectra? Graphs and Combinatorics (R. Bari and F. Harary, eds.) Springer-Verlag, Berlin, 1974, 45-51.

7. F. Harary and A. J. Schwenk, The spectral approach to determining the number of walks in a graph, to appear.

8. A. J. Hoffman, On the polynomial of a graph, Amer. Math. Monthly, 70 (1963) 30-36.

9. A. J. Hoffman and R. R. Singleton, On Moore graphs with diameters 2 and 3, IBM J. Res. Develop., 4(1960) 497-504.

10. D. J. Miller, The categorical product of graphs, Canad. J. Math. 20(1968) 1511-1521.

11. A. J. Schwenk, Computing the characteristic polynomial of a graph, Graphs and Combinatorics, (R. Bari and F. Harary, eds.) Springer-Verlag, Berlin, 1974, 153-172.

A COUNTING THEOREM FOR TOPOLOGICAL GRAPH THEORY

Saul Stahl
Wright State University
Dayton, Ohio 45431

Abstract

Given an embedding of a graph G , the counting theorem provides an expression for the number of distinct regions which contain a specific vertex on their boundary in terms of the orbits of certain permutations. These permutations are derived from Edmonds' rotation systems. This expression leads to new and purely combinatorial proofs of Duke's interpolation theorem and of the Battle, Harary, Kodama, and Youngs Theorem on the additivity of the genus.

For the definition of graph theoretical terms not explained here, the reader is referred to [6]. In this article graphs are finite and admit both loops and multiple edges. An arc is a directed edge, and the inverse of the arc e is denoted by e^{-1}. If u is the initial vertex of the arc e, then e is said to be an arc at u. For any vertex u, the set of all the arcs at u is denoted by D_u. Given a graph G, we denote its vertex and arc sets by $V(G)$ and $D(G)$. The latter will generally be abbreviated to D.

For the definition of group theoretical terminology, the reader is referred to [3]. If the group Γ acts on the set X, then the image of $x \in X$ under the action of $g \in \Gamma$ is written as $(x)g$. We will generally think of the elements of Γ as permutations and make strong use of their decomposition into disjoint cycles. Each cycle in this decomposition of the permutation π is called a cyclic factor of π. The number of distinct cyclic factors of π is denoted by $|\pi|$. Since all of our maps here are permutations, the composition of maps is to be read from left to right.

Given a graph G, a rotation system P of G is a set of cycles P_v, each acting on D_v, where the P_v and D_v are indexed by $V(G)$. Such a rotation system induces the global rotation P_G which is a permutation of $D(G)$ defined in the following manner. If v is the terminal vertex of the arc e, then

$$(e)P_G = (e^{-1})P_v \,. \qquad (1)$$

We now assume that all the orientable closed surfaces have in fact been oriented, and we define an oriented 2-cell embedding of the graph G as a 2-cell embedding of G on some orientable closed surface in which the regions are all clockwise oriented. As the following theorem is merely a restatement of the Edmonds-Youngs Theorem [5,8], no proof is deemed necessary.

1. Theorem (Edmonds, Youngs): The oriented 2-cell embeddings of the graph G are in a one to one correspondence with the rotation systems of G. This cor-

respondence is such that the boundaries of the oriented embedding determined by the rotation P are described by the cyclic factors of P_G .

In view of this correspondence, we identify the regions of the embedding determined by P with the cyclic factors of P_G . Given such a rotation system P on G , $\underline{r(P)}$ denotes the number of regions determined by P ; in other words, $r(P) = |P_G|$. The number of regions which contain arcs at the vertex v is denoted by $\underline{r(P;v)}$; the astute reader will of course observe that when we regard regions as topological subspaces of the ambient surface, then $r(P;v)$ is the number of regions which contain the vertex v in their boundaries. Still working with the same rotation system P , the <u>secondary system</u> of permutations $\underline{S(P)}$, usually abbreviated to S , is induced in the following manner. For every vertex v of G , $\underline{S_v}$ is also a permutation of D_v and if e is an arc at v , then

$$(e)S_v = [(e)P_G^{n(e)}]^{-1}$$

where $\underline{n(e)}$ is the least positive integer such that $(e)P_G^{n(e)+1}$ is also an arc at v . Set $S(P) = \{S_v | v \in V(G)\}$. For example, let P be the rotation system describing the plane embedding given in Figure 1 (the individual rotations of P are to be read in the counterclockwise sense), and let a, b, c, and d be the arcs at v . Then $(a)S_v = d$, $(b)S_v = c$, $(c)S_v = b$, and $(d)S_v = a$. The following counting theorem, to which this article owes its title, is the raison d'être of the secondary system.

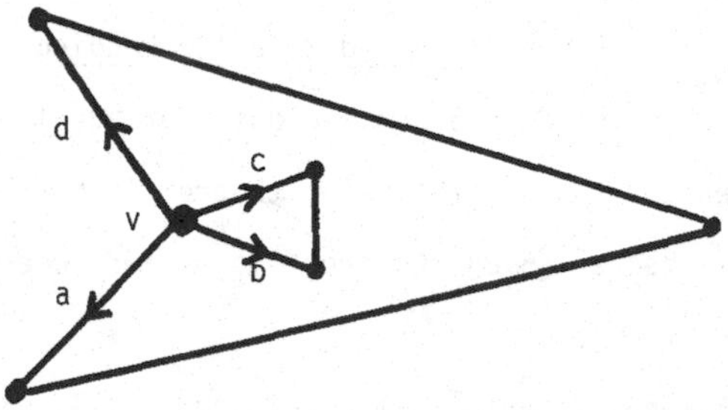

Figure 1.

<u>2. Theorem</u>: Let G be a graph, P a rotation system of G, and let S be the induced secondary system; then $r(P;v) = |S_v \circ P_v|$.

<u>Proof</u>: Suppose e is an arc at v. Then

$$(e)(S_v \circ P_v) = ((e)S_v)P_v = ([(e)P_G^{n(e)}]^{-1})P_v .$$

However, by the definition of $n(e)$, the terminal vertex of $[(e)P_G^{n(e)}]^{-1}$ is v, and hence, by (1),

$$(e)(S_v \circ P_v) = ((e)P_G^{n(e)})P_G = (e)P_G^{n(e)+1} .$$

In other words, $S_v \circ P_v$ is merely the restriction of P_G to D_v and hence the cyclic factors of $S_v \circ P_v$ are obtained from those of P_G by simply deleting all the arcs not at v. In view of the identification of regions with cyclic factors, the proof of this theorem is complete. ∎

This simple theorem has the obvious merit of giving an algebraic expression for the number of regions at a specific vertex. Its usefulness will be demonstrated by showing how it leads to elegant, and purely combinatorial, proofs of two classical theorems of topological graph theory. These are the interpolation theorem of Duke [4] and the Battle, Harary, Kodama, and Youngs Theorem [2]. The topological tools are replaced by the following algebraic lemmas.

<u>3. Lemma</u>: Let ρ and σ be permutations on the set X.

a) If $\sigma = (a_1^1 \ldots a_{i_1}^1)\ldots(a_1^s \ldots a_{i_s}^s)$ is the disjoint cycle decomposition of σ, then $[(a_1^1)\rho \ldots (a_{i_1}^1)\rho]\ldots[(a_1^s)\rho \ldots (a_{i_s}^s)\rho]$ is the disjoint cycle decomposition of $\rho^{-1}\sigma\rho$; consequently, $|\sigma| = |\rho^{-1}\sigma\rho|$.

b) If a and b are in the same orbit of σρ, then $|\rho(ab)\sigma| = |\rho\sigma| + 1$.

c) If a and b are in distinct orbits of $\sigma\rho$, then $|\rho(ab)\sigma| = |\rho\sigma| - 1$.

<u>Proof</u>: a) As this is a well known exercise, its proof is omitted.

b) Evoking part a twice, we see that

$$|\rho(ab)\sigma| = |\sigma\rho(ab)\sigma\sigma^{-1}| = |\sigma\rho(ab)| = |\sigma\rho| + 1 =$$
$$|\sigma^{-1}\sigma\rho\sigma| + 1 = |\rho\sigma| + 1 .$$

c) This part is to be proved in a manner entirely analogous to part b.

∎

The following theorem shows the effect that "small" changes of the rotation system have on the number of regions in the associated embedding.

<u>4. Theorem</u>: Given a graph G and a vertex v, let the rotation system P' be obtained from the given rotation system P by the interchange in P_v of two arcs at v. Then $r(P') = r(P) + \varepsilon$ where $\varepsilon = 2, 0$, or -2.

<u>Proof</u>: Since $P'_u = P_u$ for all vertices u distinct from v, it follows that all the regions of P whose boundaries do not contain arcs at v are also regions of P' and vice versa. Moreover, by Lemma 3a, there are arcs a,b at v such that $P'_v = (ab)P_v(ab)$. Let S_v and S'_v denote the secondary rotations at v. Since in general the secondary rotation at a vertex is independent of the primary rotation at that vertex, it follows that $S_v = S'_v$. Now, evoking lemma 3 twice, we compute:

$$|S'_v \circ P'_v| = |S_v(ab)P_v(ab)| = |S_v(ab)P_v| \pm 1 = |S_v \circ P_v| \pm 1 \pm 1 .$$

In view of Theorem 2, the proof is now complete.

∎

For a given graph G, let <u>$\gamma(G)$</u> and <u>$\gamma_M(G)$</u> denote the minimum and maximum orientable genera, respectively (see [7]). The following is an immediate consequence of the above theorem.

5. Corollary (Duke): The graph G has a 2-cell embedding on the sphere with n handles if and only if $\gamma(G) \leq n \leq \gamma_M(G)$.

Proof: Let P and $\overline{P}$ denote rotation systems which define embeddings of minimum and maximum genus, respectively. Then there exists a sequence of rotation systems $P = P^{(0)}, P^{(1)}, P^{(2)}, \ldots, P^{(s)} = \overline{P}$ and vertices $v_1, v_2, \ldots v_s$ such that for each $i = 1,2,\ldots,s$, $P_v^{(i-1)} = P_v^{(i)}$ for all $v \neq v_i$ and $P_{v_i}^{(i)}$ differs from $P_{v_i}^{(i-1)}$ only by the interchange of two arcs at v_i . An application of Theorem 4 now yields the desired result. ■

Before we proceed to the next graph theoretical theorem, another group theoretical lemma must be proved. If π is a permutation on the set X , then $\langle\pi\rangle = \{x \in X \mid (x)\pi \neq x\}$. If π and σ are both permutations on X then π and σ are said to be disjoint if $\langle\pi\rangle \cap \langle\sigma\rangle = \emptyset$. Now, let $\pi = (a_1 a_2 \cdots a_m)$ be a cyclic permutation and let σ be a permutation on the set $\langle\pi\rangle$ such that $\sigma = \sigma_1 \sigma_2 \cdots \sigma_n$ is a decomposition of σ into disjoint permutations. The cycle π is said to dominate σ if for every i there exist integers j_i, k_i such that $\langle\sigma_i\rangle = \{a_{j_i+1}, a_{j_i+2}, \ldots, a_{k_i}\}$, the indices being added modulo m . Thus, (123456789) dominates (3124)(5)(876)(9) , but it does not dominate (1245)(36789).

6. Theorem: If the cycle π dominates the permutation σ, and $\sigma_1 \sigma_2 \cdots \sigma_n$ is a decomposition of σ into disjoint permutations, then

$$|\sigma\pi| + (n - 1) = \sum_{j=1}^{n} |\sigma_j \pi|$$

Proof: Relabeling the elements of π , if necessary, we may assume that $\pi = (1234\ldots m)$ and that there exist integers $0 = i_0 < i_1 < \ldots < i_n = m$ such that $\langle\sigma_j\rangle = \{i_{j-1}+1, i_{j-1}+2, \ldots, i_j\}$ for each j . Setting $\pi_j = (i_{j-1}+1 \ldots i_j)$ it is easily verified that π can now be factored as:

$$\pi = [\prod_{j=1}^{n} \pi_j][\prod_{j=1}^{n-1} (1\ i_j+1)] .$$

Since disjoint permutations commute and π dominates σ , we have:

$$\sigma\pi = [\prod_{j=1}^{n} \sigma_j\pi_j][\prod_{j=1}^{n-1} (1\ i_j+1)]$$

It is an easy exercise to show that $|\sigma_j\pi| = |\sigma_j\pi_j|$, and hence

$$|\prod_{j=1}^{n} \sigma_j\pi_j| = \sum_{j=1}^{n} |\sigma_j\pi|$$

We note that 1 and i_k+1 are in distinct orbits of $[\prod_{j=1}^{n} \sigma_j\pi_j][\prod_{j=1}^{k-1} (1\ i_j+1)]$ and hence, by induction on k and Lemma 3c ,

$$\begin{aligned} |\sigma\pi| &= |[\prod_{j=1}^{n} \sigma_j\pi_j][\prod_{j=1}^{n-1} (1\ i_j+1)]| = |\prod_{j=1}^{n} \sigma_j\pi_j| - (n-1) \\ &= \sum_{j=1}^{n} |\sigma_j\pi| - (n-1) . \end{aligned}$$

Assume now that π and σ are as above, except that π does not dominate σ . The following lemma enables us to construct a cyclic permutation $\bar{\pi}$ which does dominate σ and such that $|\sigma\bar{\pi}| \geq |\sigma\pi|$. ■

<u>7. Lemma</u>: Let σ be a permutation and π a cyclic permutation on $\{1,2,...,m\}$ Suppose $\sigma(a) = b$ and let π' be the cyclic permutation obtained by removing b from its position in π and placing it immediately to the left of a . Then $|\sigma\pi'| \geq |\sigma\pi|$.

<u>Proof</u>: We may clearly assume that $\pi = (123...m)$. Then $\pi' = \pi(ab)(a\ b+1)$. Now, $(a)[\sigma\pi(ab)] = b+1$, and hence, by Lemma 3b,

$$|\sigma\pi'| = |\sigma\pi(ab)(a\ b+1)| = |\sigma\pi(ab)| + 1 \geq |\sigma\pi| .$$

■

Let $\mathcal{G} = \{G_i\}_{i-1}^{n}$ be a family of disjoint graphs with v_i a vertex of G_i for each i. The graph G obtained by identifying all the v_i is called a <u>one point amalgamation</u> ([1]) of the family $\mathcal{G}$. The following lemma was first proved, by topological means, in [2].

<u>8. Lemma</u>: If G is a one point amalgamation of the family $\{G_i\}_{i=1}^{n}$, then

$$\gamma(G) = \sum_{i=1}^{n} \gamma(G_i) .$$

<u>Proof</u>: For each i let $P^{(i)}$ be a rotation system which determines a minimum orientable genus embedding for G_i. Let v be the identified point of G, and suppose $P_v^{(i)} = (e_1^i \ldots e_{k_i}^i)$. We define a rotation system P for G by setting

$$P_u = P_u^{(i)} \quad \text{if } u \neq v \text{ and } u \in V(G_i)$$

$$P_v = (e_1^1 \ldots e_{k_1}^1 e_1^2 \ldots e_{k_2}^2 \ldots e_1^n \ldots e_{k_n}^n) .$$

Clearly any region of any of the $P^{(i)}$ which does not contain any of the arcs at v is also a region of P, and vice versa. If S and $S^{(i)}$ are the systems secondary to P and $P^{(i)}$, respectively, then it is clear that $S_v = S_v^{(1)} \circ S_v^{(2)} \circ \ldots \circ S_v^{(n)}$ is a decomposition of S_v into disjoint permutations. Thus S_v is dominated by P_v and so, by Theorem 6,

$$r(P;v) = |S_v \circ P_v| = \sum_{i=1}^{n} |S_v^{(i)} \circ P_v| - (n-1) =$$

$$\sum_{i=1}^{n} |S_v^{(i)} \circ P_v^{(i)}| - (n-1) = \sum_{i=1}^{n} r(P^{(i)};v) - (n-1) .$$

Hence, by the Euler-Poincaré formula,

$$\gamma(G) \leq 1 - \tfrac{1}{2}(|V(G)| - |E(G)| + r(P)) =$$

$$1 - \frac{1}{2}[1 + \sum_{i=1}^{n} (|V(G_i)| - 1) - \sum_{i=1}^{n} |E(G_i)| + \sum_{i=1}^{n} r(P^{(i)}) - (n-1)] =$$

$$n - \frac{1}{2}\sum_{i=1}^{n} [|V(G_i)| - |E(G_i)| + r(P^{(i)}] = \sum_{i=1}^{n} \gamma(G_i) . \qquad (2)$$

Conversely, let the rotation system P define a minimum orientable genus embedding for G. It follows from Lemma 7 that there exists a rotation system $\overline{P}$ of G which agrees with P on every vertex but v, and such that $r(\overline{P}) \geq r(P)$ and $\overline{S}_v$ is dominated by $\overline{P}_v$. Since P determines a minimum genus embedding, it follows that $r(\overline{P}) = r(P)$. We now define rotations systems $P^{(i)}$ for each G_i as follows:

$$P_u^{(i)} = P_u \quad \text{if} \quad u \neq v ,$$

and the $P_v^{(i)}$ are defined by specifying that for each i the disjoint cycle decomposition of $P_v^{(i)}$ is obtained from that of $\overline{P}_v$ by deleting all the arcs not in G_i. Once again, the regions of any of the $P^{(i)}$ which do not contain any of the arcs at v are regions of $\overline{P}$ and vice versa. Also, $\overline{S}_v = S_v^{(1)}\circ\ldots\circ S_v^{(n)}$. Thus Lemma 6 applies here as well, and we have:

$$r(\overline{P};v) = |\overline{S}_v \circ \overline{P}_v| = \sum_{i=1}^{n} |S_v^{(i)} \circ P_v^{(i)}| - (n-1) =$$

$$\sum_{i=1}^{n} r(P^{(i)};v) - (n-1) .$$

Recalling that $r(\overline{P}) = r(P)$ we now have

$$r(P) = \sum_{i=1}^{n} r(P^{(i)}) - (n-1) .$$

So, calculations similar to those of (2) above yield:

$$\gamma(G) = 1 - \frac{1}{2}[|V(G)| - |E(G)| + r(P)] =$$

$$\sum_{i=1}^{n} [1 - \frac{1}{2}(|V(G_i)| - |E(G_i)| + r(P^{(i)}))] \geq \sum_{i=1}^{n} \gamma(G_i) .$$

This, together with (2), concludes the proof.

■

The following theorem of [2] now follows as an immediate corollary of Lemma 8.

<u>9. Theorem</u>: If $\{G_1,\ldots,G_n\}$ is the collection of all the blocks of the connected graph G, then

$$\gamma(G) = \sum_{i=1}^{n} \gamma(G_i) .$$

■

<u>Added in proof</u>: Bill Marshall's Master's thesis, written under the direction of Morris Marx at Vanderbilt University (1975), contains an alternate combinatorial proof of Theorem 9.

BIBLIOGRAPHY

1. Alpert, S. R., The Genera of Amalgamations of Graphs. Trans. AMS., Vol. 178 (1973), 1-39.

2. Battle, J., F. Harary, Y. Kodama, and J. W. T. Youngs, Additivity of the Genus of a Graph. Bull. AMS., Vol. 68 (1962), 565-568.

3. Biggs, N., Finite Groups of Automorphisms. Cambridge University Press, New York (1971).

4. Duke, R. A., The Genus, Regional Number, and Betti Number of a Graph. Canad. J. Math., Vol. 18 (1966), 817-822.

5. Edmonds Jr., J., A Combinatorial Representation for Polyhedral Surfaces. AMS Notices, 7 (1960), 646.

6. Harary, F., Graph Theory, Addison-Wesley, Reading, Mass. (1969).

7. White, A. T., Graphs, Groups, and Surfaces, North Holland-American Elsevier, New York (1973).

8. Youngs, J. W. T., Minimal Imbeddings and the Genus of a Graph. J. Math. Mech., Vol. 12, No. 2 (1963), 303-315.

k-ARC HAMILTON GRAPHS

Richard Denman and Dalton Tarwater
Texas Tech University
Lubbock, TX. 79409

Abstract

A simple graph is a k-arc Hamilton graph if every arc of length k is contained in a Hamilton circuit. Various conditions are given for a graph to be k-arc Hamilton. Results of Kronk, Posa and of Chvátal and Erdös are generalized.

k-path Hamilton graphs have been studied for some time, having been examined by Kronk [2] and others. This paper will place Kronk's work in a more general setting and will translate the conditions of Chvátal and Erdös [1] for a graph to be Hamiltonian to conditions for a graph to be k-arc Hamiltonian.

A graph G is a pair (V(G), E(G)) of sets and a 1-1 function assigning each edge $e \in E(G)$ to a pair $\{x_1,x_2\} \in V(G) \times V(G)$, i.e., all graphs are simple. We will denote an edge by $e = (x_1 x_2)$.

An arc $A = (a_0,\ldots,a_j)$ is a subgraph (V(A),E(A)), where $E(A) = \{(a_i a_{i+1}) \mid i=0,\ldots,j-1\}$. Let $P \subset V(G)$ and $v \in V(G) \sim P$. Then a v-P chain $B = (v = b_1,\ldots,b_j = p_i)$ is an arc for which $V(B) \cap P = p_i$. p_i is called a v-P chain termination. The v-P connecting subset, C_{v-P}, of P is the set of all v-P chain terminations.

Proposition 1. Let v and P be as above. For every pair $\{x_1,x_2\} \subset C_{v-P}$, there is an arc $B = (x_1,\ldots,x_2)$ with $V(B) \cap P = \{x_1,x_2\}$.

Proof: Let $\{x_1,x_2\} \subset C_{v-P}$. Then there are arcs $B_1 = (v = b_{11},\ldots, b_{1m} = x_1)$ and $B_2 = (v = b_{21},\ldots,b_{2j} = x_2)$ with $V(B_1) \cap P = \{x_1\}$ and $V(B_2) \cap P = \{x_2\}$. Now since $v \in V(B_1) \cap V(B_2)$ there is a $b_{1i} \in V(B_1) \cap V(B_2)$ such that i is a maximum. Hence the desired arc is (see Figure 1) $B = (x_1 = b_{1m}, b_{1m-i},\ldots,b_{1i} = b_{2e}, b_{2e+1},\ldots,b_{2j} = x_2)$.

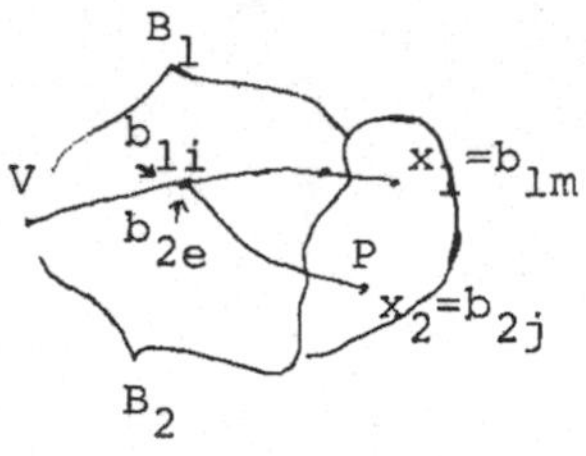

Figure 1

The section graph, or induced graph, $I_G(V)$ on a point set $V \subset V(G)$ is the subgraph $(V, E_{V,G})$ where $E_{V,G} = \{e \in E(G) \mid e = (v_1,v_2)$ for some pair $\{v_1,v_2\} \subset V\}$. A circuit $C = (c_1,\ldots,c_m,c_1)$ is the subgraph (V,E) where $V = V(A)$ for some arc $A = (c_1,\ldots,c_m)$ and $E = E(A) \cup (c_m c_1)$. A Hamilton circuit H in a graph G is a circuit with $V(H) = V(G)$. An arc $B = (b_1,\ldots,b_j)$ is of circuit type if there is a Hamilton circuit in $I_G(V(B))$.

An arc B containing a subgraph $S = A_1 \cup \ldots \cup A_q$ where the $E(A_i)$'s are mutually disjoint is of S-circuit type if there is a Hamilton circuit H on $I_G(V(B))$ and $S \subset H$. The size $|V|$ of an edge or point set is the number of its elements. Let $A = (a_0,\ldots,a_k)$ be an arc contained in a circuit C, and fix the ordering of $C = (c_1,\ldots,c_m,c_1)$ such that $c_1 = a_0$. The successor set S^A_{v-C} with respect to this ordering is the set $\{c_i | c_{i-1} \in C_{v-C} \sim \{a_0,\ldots,a_{k-1}\}\}$.

A graph G is k-arc Hamilton if every arc A in G of length $|E(A)| = k$ is contained in a Hamilton circuit.

Theorem 1. If for every arc $A = (a_0,\ldots,a_k)$ of length k, there is a circuit C containing A and if for every non-Hamilton circuit C containing A, there is a fixed ordering of C and a $v \in V(G) \sim V(C)$ such that at least one of the following conditions holds, then G is k-arc Hamilton:

1) $S^A_{v-C} \cap C_{v-C} \neq \emptyset$
2) there is a pair $\{y_1,y_2\} \subset S^A_{v-C}$ such that at least one of the following conditions holds.
   a) there is an arc $B = (y_1,\ldots,y_2)$ with $V(B) \cap V(C) = \{y_1,y_2\}$.
   b) the arc $(y_1,\ldots,y_2)$ of C not containing A is of (x_2y_2)-circuit type, where $x_2 \in C_{v-C}$.
   c) the arc $(y_2,\ldots,y_1)$ of C containing A is of $[(x_1y_1) \cup A]$-circuit type where $x_1 \in C_{v-C}$.

Proof: It will suffice to show that for every non-Hamilton circuit C containing A, there is a circuit C' containing A, with $V(C) \subset V(C') \sim \{v\}$, for some $v \in V(C') \sim V(C)$. Let C be a non-Hamilton circuit containing A, with the fixed ordering $C = (c_1, \ldots, c_e, c_1)$ and choose $v \in V(G) \sim V(C)$ as given in the hypothesis. Without loss of generality, c_1 can be chosen so that $c_1 = a_0$. If condition 1) holds, then there is a $c_j \in S^A_{v-C} \cap C_{v-C}$ and so $c_{j-1} \in C_{v-C}$. Now, by Proposition 1, there is an arc $B = (c_{j-1}, \ldots, b_{1i}, \ldots, c_j)$ with $V(B) \cap V(C) = \{c_{j-1}, c_j\}$, which gives a circuit $C_1' = (c_{j-1}, \ldots, b_{1i}, \ldots, c_j, c_{j+1}, \ldots, c_e, c_1, \ldots, c_{j-1})$. See Figure 2. Furthermore, $V(C) \subset V(C_1') \sim \{b_{1i}\}$ and $b_{1i} \in V(C') \sim V(C)$.

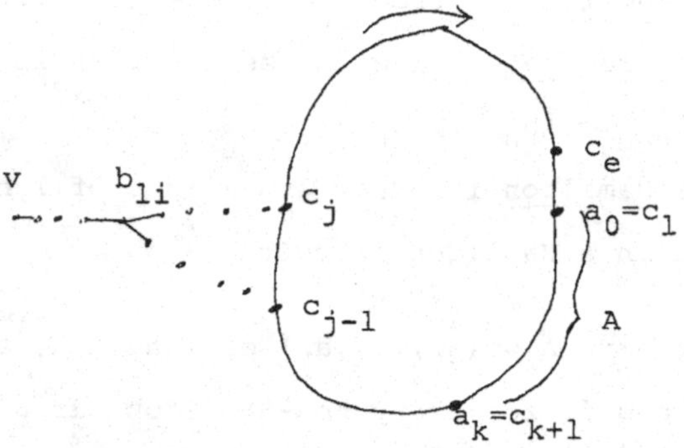

Figure 2

If condition 2a) holds, then the desired circuit is $C_{2a}' = (x_1, \ldots, b_{1i}, \ldots, x_2 = c_m) \cup (c_m, c_{m-1}, \ldots, c_{j+1} = y_1) \cup B \cup (y_2 = c_{m+1}, c_{m+2}, \ldots, c_e, c_1, \ldots, c_j = x_1)$, where x_1 and x_2 are in C_{v-C}. See Figure 3.

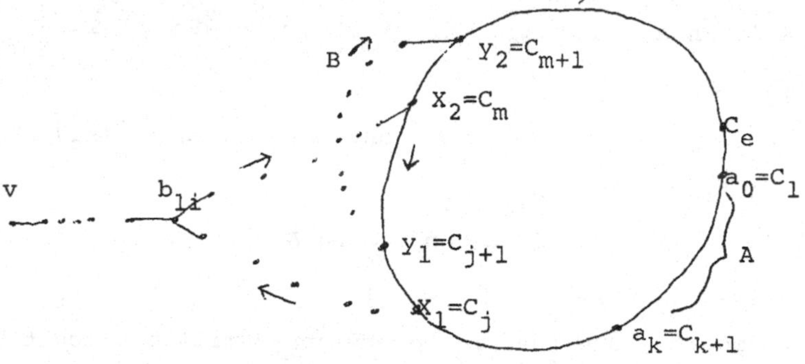

Figure 3

If condition 2b) holds, then the desired circuit is $C'_{2b} = (x_1, \ldots, b_{1i}, \ldots, x_2 = h_1) \cup H \cup (c_{m+1}, c_{m+2}, \ldots, c_e, c_1, \ldots, c_j = x_1)$, where $H = (h_1, \ldots, h_p = c_{m+1})$ is the Hamilton arc on $I_G(V(c_{j+1}, \ldots, c_{m+1}))$ given by 2b). See Figure 4.

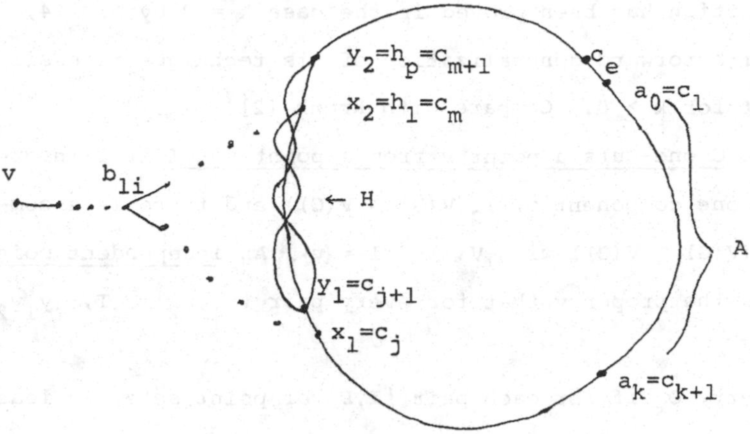

Figure 4

Hence $V(C) \subset V(C'_{2b}) \sim \{b_{1i}\}$ and $b_{1i} \in V(C'_{2b}) \sim V(C)$. If condition 2c) holds, then the desired circuit is $C'_{2c} = (x_1 = c_j, \ldots, b_{1i}, \ldots, x_2 = c_m) \cup (c_m, c_{m-1}, \ldots, c_j = y_1) \cup H$, where $H = (y_1 = h_1, \ldots, a_k, a_{k-1}, \ldots, a_0, \ldots, h_q = x_1)$ is the Hamilton arc on $I_G(V(y_2, \ldots, y_1))$ given by condition 2c. See Figure 5. Hence $V(C) \subset V(C'_{2c}) \sim \{b_{1i}\}$ and $b_{1i} \in V(C'_{2c}) \sim V(C)$. Each C' constructed contains the arc A, thus the theorem is proved.

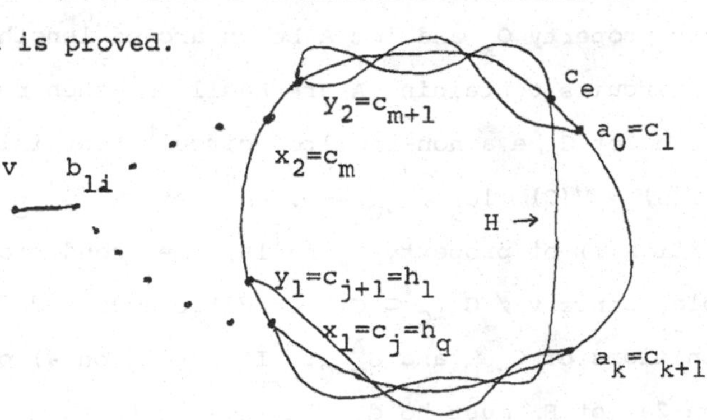

Figure 5

Henceforth, the condition of Theorem 1 will be referred to as property F_k. $\rho_G(v)$ will denote the degree of v in the graph G.

Proposition 2. If $B = (b_1,...,b_j)$ is an arc with $|E(B)| \geq k$ and $\rho_B(b_1) + \rho_B(b_j) \geq |V(B)| + k$, then for every arc A in B of length $|E(A)| \geq k$, B is of $[(b_{j-1}b_j) \cup A]$-circuit type.

This proposition has been proved in the case k = 0 by Ore [4, p. 55]. A straightforward generalization of his technique is easily seen to prove it for k > 0. Compare with Kronk [2].

A point set C one-cuts a point v from a point set I if I intersects more than one component of $I_G(V(G) \sim V(C))$ and there is a component C_1 of $I_G(V(G) \sim V(C))$ with $V(C_1) \cap I = \{v\}$. An independent point set $I \subset V(G)$ has the property that for every pair $\{y_1,y_2\} \subset I$, $(y_1y_2) \not\in E(G)$.

G has property O_k if for each pair {I,D} of point sets, at least one of the following conditions fails:

1) $I \cap D = \emptyset$.
2) $|I| \leq |D| + 1 \leq |I| + k$.
3) D one-cuts a point v from I.
4) I is independent.
5) For every pair $\{y_1,y_2\} \subset I$, $\rho_G(y_1) + \rho_G(y_2) < |V(G)| + k$.

Theorem 2. If G has property O_k, then G has property F_k.

Proof: Suppose G has property O_k and let A be an arc of length $|E(A)| = k$. If all circuits containing A are Hamilton, then property F_k holds vacuously; so let C be a non-Hamilton circuit containing A. Now choose any $v \in V(G) \sim V(C)$, let $C_{v-C} = D$, and let $\{v\} \cup S^A_{v-C} = I$. If, for {I,D}, condition 1) of property O_k fails, then condition 1) of property F_k must hold, since $v \not\in C_{v-C} \subset C$. Conditions 2) and 3) of O_k hold by the definitions of C_{v-C} and S^A_{v-C}. If condition 4) of O_k fails then condition 2a) of F_k must hold. It remains to show that if condition 5) of O_k fails then F_k holds.

Two cases must be considered. First suppose there is a $y \in S^A_{v-C}$ such that $\rho_G(v) + \rho_G(y) \geq |V(G)| + k$. Let $T_1 = \{c_i \in V(C) \sim \{a_1,\ldots,a_k\} | c_{i-1}$ is adjacent to $V\}$, let $T_2 = \{x \in V(C) \sim \{a_1,\ldots,a_k\}|$ x is adjacent to y$\}$, let $T_3 = \{x \in V(G) \sim V(C) | x$ is adjacent to v$\}$, and let $T_4 = \{x \in V(G) \sim V(C)|x$ is adjacent to y$\}$. If $T_1 \cap T_2 \neq \emptyset$, then S^A_{v-C} is not independent so condition 2a) of F_k holds. So let $T_1 \cap T_2 = \emptyset$, then $|T_1| + |T_2| \leq |V(C)| - k$. Hence

$$\begin{aligned} |V(C)| + k + |T_3| + |T_4| &\geq |T_1| + k + |T_2| + k + |T_3| + T_4| \\ &\geq \rho_G(v) + \rho_G(y_i) \\ &\geq V(G) + k \\ &= |V(C)| + |V(G) \sim V(C)| + k. \end{aligned}$$

Therefore $|T_3| + |T_4| \geq |V(G) \sim V(C)|$. But this means that v and y have a common adjacency in $V(G) \sim V(C)$ yielding condition 2a) of property F_k.

Now, in the second case, suppose there is a pair $\{y_1,y_2\} \subset S^A_{v-C}$ with $\rho(y_1) + \rho(y_2) \geq |V(G)| + k$. Let $D_1 = \{x \in V(G) \sim V(C)|x$ is adjacent to y_1 and $y_2\}$, and let $D_2 = \{x \in V(G) \sim V(C)|x$ is adjacent exactly one of $\{y_1,y_2\}\}$. If $D_1 \neq \emptyset$, then there is an arc satisfying condition 2a) of F_k, so let $D_1 = \emptyset$ and note that $|D_2| \leq |V(G)| - |V(C)| - 1$, since neither y_1 nor y_2 is adjacent to v. Hence

$$\begin{aligned} |V(G)| + k &\leq \rho_G(y_1) + \rho_G(y_2) \\ &= |D_2| + \rho_F(y_1) + \rho_F(y_2) \\ &\leq |V(G)| - |V(C)| - 1 + \rho_F(y_1) + \rho_F(y_2), \end{aligned}$$

where $F = I_G(V(C))$. Hence

$$|V(C)| + k + 1 \leq \rho_F(y_1) + \rho_F(y_2).$$

Now let $A_1 = (y_1,\ldots,y_2)$ in C, let $A_2 = (y_2,\ldots,a_0,\ldots,a_k,\ldots,y_1)$ in C, let $F_1 = I_G(V(A_1))$ and let $F_2 = I_G(V(A_2))$. Furthermore, suppose that condition 2b) of F_k fails; for if 2b holds, the proof is complete.

Then $\rho_{F_1}(y_1) + \rho_{F_1}(y_2) < |V(A_1)|$ by Proposition 2.

Now, $$|V(C)| + k + 1 \le \rho_C(y_1) + \rho_C(y_2)$$
$$= \rho_{F_1}(y_1) + \rho_{F_1}(y_2) + \rho_{F_2}(y_1) + \rho_{F_2}(y_2)$$
$$< |V(F_1)| + \rho_{F_2}(y_1) + \rho_{F_2}(y_2)$$
$$= |V(A_1)| + \rho_{F_2}(y_1) + \rho_{F_2}(y_2)$$
$$= |V(C)| - |V(A_2)| + 2 + \rho_{F_2}(y_1) + \rho_{F_2}(y_2),$$
that is,
$$|V(A_2)| + k \le \rho_{F_2}(y_1) + \rho_{F_2}(y_2),$$
which by Proposition 2 gives condition 2c) of F_k. Therefore, Property O_k implies Property F_k.

A graph G is <u>k-spliced</u> if for every disjoint pair {I,D} of point cutsets, if $|I| \le |D| + 1 \le |I| + k$ and if D one-cuts I, then I is dependent. From this definition it is immediate that we have

<u>Theorem 3</u>. If G is k-spliced, then G has property O_k.

A graph G is <u>k-folded</u> if for every pair {I,C} of point sets with $|I| + |I \cap C| - 1 \le |C| < |I| + |I \cap C| + k$, and C one-cuts a point v from I, then I is dependent. Again we have, by immediate application of the definition,

<u>Theorem 4</u>. If G is k-folded, then G is k-spliced.

A graph G is of <u>Ore-type (k_1,k_2)</u> if, in every independent set I of size $|I| \ge k_2 + 1$, there is a pair $\{y_1,y_2\} \subset I$, such that $\rho_G(y_1) + \rho_G(y_2) \ge |V(G)| + k_1$. Note that what has previously been Ore-type k now is defined as Ore-type (k,1).

<u>Theorem 5</u>. If G is of Ore-type (k,1), then G is k-folded.

Proof: Suppose G is of Ore-type (k,1), let {I,C} be a pair of point sets in V(G) and let $v_1 \in I$ be such that C one-cuts v_1 from I with $|I| + |I \cap C| \le |C| + 1 \le |I| + k + |I \cap C|$. Then there is a component C_1 of $I_G(V(G) \sim V(C))$ with $v_1 \in V(C_1)$, and $I \cap V(C_1) = \{v_1\}$.

Now if I is independent, there is a $v_2 \in C_2$, a component of $I_G(V(G) \sim V(C))$. Furthermore, $\rho_G(v_1) \leq |V(C_1)| - 1 + |C| - |I \cap C|$ and $\rho_G(v_2) \leq |V(G)| - |I| - |V(C_1)| + 1$. Therefore

$$\begin{aligned}\rho_G(v_1) + \rho_G(v_2) &\leq |C| - |I \cap C| + |V(G)| - |I|. \\ &< |I| + k + |I \cap C| - |I \cap C| + |V(G)| - |I| \\ &= |V(G)| + k,\end{aligned}$$

which contradicts G being of Ore-type (k,1). Hence I must be dependent, which proves that G is k-folded.

The <u>connectivity number, k(G)</u>, of a graph G is the size of a smallest cutset in G and the <u>independence number, $\beta_0(G)$</u>, is the size of a largest independent point set. Chvátal and Erdös [1] proved that if $k(G) \geq \beta_0(G)$, then G is Hamiltonian. The next theorem shows that if $k(G) \geq \beta_0(G) + k$, then G is k-arc Hamilton.

<u>Theorem 6</u>. If $k(G) \geq \beta_0(G) + k$, then G is k-spliced.

Proof: Let $k(G) \geq \beta_0(G) + k$ and let {I,D} be a disjoint pair of point sets in V(G) such that $|I| \leq |D| + 1 \leq |I| + k$, and D one-cuts v from I. Now, if I is independent, then

$$k(G) + 1 \leq |D| + 1 \leq |I| + k \leq \beta_0(G) + k,$$

contradicting the hypothesis. Therefore I must be dependent, and G is k-spliced.

<u>Theorem 7</u>. If G is of Ore-type (k,k(G)), then G has property O_k.

Proof: Let G be of Ore-type (k,k(G)) and let {I,D} be a pair of point sets. If any of conditions 1) through 4) of property O_k fails, then G has property O_k. If conditions 1) through 4) of O_k hold, then I is independent and $|I \sim v| \geq k(G)$, (where D one-cuts v from I), so there is a pair $\{y_1,y_2\} \subset I$ with $\rho_G(y_1) + \rho_G(y_2) \geq |V(G)| + k$, since G is of Ore-type (k,k(G)). Hence condition 5) of O_k fails, therefore G has property O_k.

For natural numbers j, $S(\leq j)$ is the number of vertices of G with degree less than or equal to j. G has property S_k if the following conditions hold:

$$1) \quad S(\leq j) < j-k, \text{ for } k < j < \frac{|V(G)|+k-1}{2}$$

$$2) \quad S\left(\leq \frac{|V(G)|+k-1}{2}\right) \leq \frac{|V(G)|+k-1}{2} - k.$$

The following theorem extends one of Pósa [3], who proved that if G has property S_0, then G is Hamiltonian.

Theorem 8. If G has property S_k, then G is k-arc Hamilton.

Proof: The following is equivalent to G being k-arc Hamilton: For every arc $A = (a_0,\ldots,a_k)$ of length k, $\{a_1,\ldots,a_{k-1}\}$ does not cut G and there is a longest arc containing A and of A-circuit type.

Let $A = (a_0,\ldots,a_k)$ be an arc of length $|E(A)| = k$, and let $C = (c_0,\ldots,c_p = a_0,\ldots,a_k = c_{p+k},\ldots,c_m)$ be a longest arc containing A. Furthermore choose C such that $\rho_G(c_0) + \rho_G(c_m)$ is a maximum, and let $T = \{c_i \in V(C) | c_{i+1}$ is adjacent to c_0 and $c_{i+1} \in V(C) \sim \{a_1,\ldots,a_k\}\}$. Then for every $c_i \in T$, there is an arc $A_i = (c_i,c_{i-1},\ldots,c_0,c_{i+1},c_{i+2},\ldots,c_m)$ containing A with $|E(A_i)| = |E(C)|$. So by our choice of C, $\rho(c_i) \leq \rho(c_0)$. Now, since $|T| \geq \rho_G(c_0) - k$, $S(\leq \rho_G(c_0)) \geq \rho_G(c_0) - k$ and so

$$\rho_G(c_0) \geq \frac{|V(G)|+k-1}{2}$$

by condition 1) of S_k. Similarly,

$$\rho_G(c_m) \geq \frac{|V(G)|+k-1}{2},$$

and if $\quad \rho_G(c_0) = \rho_G(c_m) = \dfrac{|V(G)|+k-1}{2}, \quad$ then

$S\left(\leq \frac{|V(G)|+k-1}{2}\right) = \rho_G(c_0) - k + 1 = \frac{|V(G)|+k-1}{2} - k + 1 >$ $\frac{|V(G)|+k-1}{2} - k$, contradicting condition 2) of property S_k. Therefore at least one of $\{c_0,c_m\}$ has a strictly larger degree. Hence $\rho_G(c_0) + \rho_G(c_m) \geq |V(G)| + k$, and so, by Proposition 2, C is of A-circuit type. Note that by the choice of C,

$\rho_G(c_m) = \rho_{I_G(V(C))}(c_m)$ and $\rho_G(c_0) = \rho_{I_G(V(C))}(c_0)$; that is, both c_0 and c_m are adjacent only to points in $V(C)$.

In order to show that $\{a_1,\ldots,a_{k-1}\}$ does not cut G, it will suffice to show that $k(G) \geq k + 1$. Let C be a cutset of G, and let C_1 be a component of $I_G(V(G) \sim V(C))$ with a minimum number of points. Then $|V(C_1)| \leq \frac{|V(G)|-|C|}{2}$ and for all $v \in C_1$,

$$\rho_G(v) \leq |V(C_1)| + |C| - 1,$$

$$\begin{aligned} \text{hence } S(\leq|V(C_1)| + |C| - 1) &\geq |V(C_1)| \\ &= (|V(C_1)| + |C| - 1) - (|C| - 1) \\ &\geq (|V(C_1)| + |C| - 1) - k. \end{aligned}$$

Now, by condition 2) of S_k,

$$|V(C_1)| + |C| - 1 \geq \frac{|V(G)|+k-1}{2} \quad ,$$

that is, $|C| \geq k + 1$. Therefore $k(G) \geq k + 1$, and the proof is complete.

The following propositions are easily proved.

<u>Proposition 3</u>. If G is of Ore-type (k,1) then G is of Ore-type (k,k(G)).

<u>Proposition 4</u>. If $k(G) \geq \beta_0(G) + k$ then G is of Ore-type (k,k(G)).

<u>Proposition 5</u>. If G is of Ore-type (0,1), then $k(G) \geq \beta_0$.

<u>Proposition 6</u>. If G is of Ore-type (0,1), then G has property S_0.

The results of this paper show that each of the following graphical properties implies the next: Ore-type (k,1); k-folded; k-spliced; O_k; F_k; k-arc Hamilton. It is not difficult to construct examples to demonstrate that none of the implications are reversible.

This research was partially supported by The Graduate School of Texas Tech University.

REFERENCES

1. Paper
V. Chvátal and P. Erdös, A note on Hamilton Circuits. Discrete Math 2 (1972) 111-113.

2. Paper
H. V. Kronk. A Note on k-Path Hamilton graphs. J. Combinatorial Theory 7 (1969) 104-106.

3. Paper
L. Pósa. A theorem concerning hamilton lines. Magyar Tud. Akad. Mat. Kutato Int. Kozl. 7 (1962), 225-226.

4. Book
O. Ore, Theory of Graphs. Amer. Math. Soc. Colloq. Pub. 38, Providence (1962).

HYPOHAMILTONIAN GRAPHS AND DIGRAPHS

Carsten Thomassen
University of Waterloo
Waterloo, Ontario, Canada

Abstract

Methods for constructing hypohamiltonian graphs and oriented graphs are described. It is shown that every planar hypohamiltonian graph contains a vertex of degree 3 and that for each $n \geq 6$ there exists a planar, hypohamiltonian digraph with n vertices. Finally it is proved that every graph with n vertices contains a set A of at most $\frac{1}{3}n$ vertices such that every longest cycle of the graph intersects A.

1. Introduction

A graph G is hypohamiltonian (resp. hypotraceable) if it has no hamiltonian cycle (resp. path) but every vertex-deleted subgraph $G-v$ has such a cycle (resp. path). Hypohamiltonian digraphs (directed graphs) are defined analogously. Hypohamiltonian graphs are studied in the papers [1], [3], [5], [6], [7], [9], [11], [12], [13], [14], [15], [16].

In this paper we first characterize 3-fragments of hypohamiltonian graphs and generalize the constructions of hypohamiltonian graphs described in [14] and [15]. In section 3 we use this together with a theorem of Tutte [17] on hamiltonian cycles in planar graphs to show that every planar, hypohamiltonian graph contains a vertex of degree 3.

U.S.R. Murty (private communication) raised the question about the existence of a hypohamiltonian oriented graph (i.e., a digraph with no cycle of length 2). We describe in section 4 infinite families of hypohamiltonian oriented graphs and demonstrate in this section the existence of a planar hypohamiltonian digraph with n vertices for each $n \geq 6$.

Zamfirescu [19], [20] and Grünbaum [8] consider more general classes of graphs of which hypohamiltonian and hypotraceable graphs are part. In section 5 we comment briefly upon this, and we prove a conjecture of Zamfirescu [20] by showing that any graph with n vertices contains a set A of at most $\frac{1}{3}n$ vertices such that every longest cycle of G intersects A.

Finally, we exhibit in section 6 a list of problems concerning or related to hypohamiltonian and hypotraceable graphs.

The terminology and notation is that of Harary [10] except that we say vertices and edges instead of points and lines, respectively. The set of vertices (resp. edges) of a graph G is denoted by $V(G)$ (resp. $E(G)$).

2. Hypohamiltonian graphs

If a graph G has connectivity k and is not a complete graph, then G contains two nonempty, proper induced subgraphs G_1, G_2 such that $G = G_1 \cup G_2$ and $V(G_1) \cap V(G_2) = A$, where $|A| = k$. We say that G_1 is a k-fragment of G and that A is the set of vertices of attachment of G_1. It is easy to see that every hypohamiltonian graph is 3-connected. We now characterize 3-fragments of hypohamiltonian graphs.

Lemma 1. Let G_1 be a 3-fragment with at least five vertices of a hypohamiltonian graph, and let $A = \{x,y,z\}$ be the set of vertices of attachment of G_1. Then

(i) G_1 has no hamiltonian path connecting two vertices of A;

(ii) for every vertex u of G_1, $G_1 - u$ has a hamiltonian path connecting two vertices of A.

Remark: The proof of Theorem 1 and Cor. 1 below shows that if G_1 satisfies (i) and (ii) for some set A of three vertices of G_1, then G_1 is a 3-fragment of some hypohamiltonian graph.

Proof of Lemma 1. There exists a hypohamiltonian graph $G = G_1 \cup G_2$ such that $V(G_i) - V(G_{3-i}) \neq \emptyset$ for $i = 1,2$ and $V(G_1) \cap V(G_2) = A$. Suppose first that G_1 has a hamiltonian path P^1 connecting two vertices of A, say x and y. Since G-z has a hamiltonian cycle, $G_2 - z$ has a hamiltonian path P^2 connecting x and y. But now $P^1 \cup P^2$ is a hamiltonian cycle of G. This contradiction proves (i).

Let u be any vertex of G_1. Let S be a hamiltonian cycle of $G - u$. If $u \in \{x,y,z\}$, then $G_1 \cap S$ is a hamiltonian path of $G_1 - u$ connecting the two vertices of $\{x,y,z\} - \{u\}$; so assume $u \notin \{x,y,z\}$. Now x,y,z partition S into three paths P^1, P^2, P^3 such that $S = P^1 \cup P^2 \cup P^3$. Two of these paths, P^1 and P^2 say, are contained in the same graph G_i. If P^1 and P^2 are contained in G_2, then $P^1 \cup P^2$ is a hamiltonian path of G_2 connecting two of the vertices x,y,z. Since G_2 is a 3-fragment of G, we have obtained a contradiction to (i). Hence P^1 and P^2 are contained in G_1, and $P^1 \cup P^2$ is a hamiltonian path of $G_1 - u$ connecting two of the vertices x,y,z. This proves (ii).

Let G be a graph with at least five vertices containing a vertex x_0 of degree 3. Let x_1, x_2, x_3 be the neighbours of x_0. Let H be a 3-fragment with at least 5 vertices of a hypohamiltonian graph. Let y_1, y_2, y_3 be the vertices of attachment of H. Suppose $G \cap H = \emptyset$. Consider the graph G' obtained from $(G - x_0) \cup H$ by identifying x_i and y_i into a vertex z_i for each $i = 1, 2, 3$. We say that G' is obtained from G by <u>replacing</u> x_0 <u>by H</u>. The particular case where H is a vertex-deleted subgraph of the Petersen graph is illustrated in [15, Fig.1].

With the notation above we have:

<u>Lemma 2.</u> (a) G' is hamiltonian if and only if G is hamiltonian;

(b) if G has at least six vertices, then for each $u \in V(G) - \{x_0\}$, $G' - u$ is hamiltonian if and only if $G - u$ is hamiltonian;

(c) if $G - x_i$ is hamiltonian for each $i = 1, 2, 3$, then $G' - u$ is hamiltonian for each $u \in V(H)$.

<u>Proof</u>: Suppose first G' has a hamiltonian cycle S. The vertices z_1, z_2, z_3 partition S into three paths, two of which are contained in either $G - x_0$ or H. Since H has no hamiltonian path joining two of the vertices z_1, z_2, z_3, we conclude that only one of the three paths of S is contained in H. But then $S \cap (G - x_0)$ is a hamiltonian path of $G - x_0$ joining two of the vertices x_1, x_2, x_3, and hence G is hamiltonian.

Suppose next that G has a hamiltonian cycle. Then $G - x_0$ has a hamiltonian path P^1 joining two of the vertices x_1, x_2, x_3, say x_1 and x_2. By Lemma 1, $H - y_3$ has a hamiltonian path P^2 joining y_1 and y_2. Now $P^1 \cup P^2$ is a hamiltonian cycle of G', and (a) is proved.

To prove (b), let u be any vertex of $G - x_0$. If $u \notin \{x_1, x_2, x_3\}$, then $G' - u$ is obtained from $G - u$ by replacing x_0 by H, so in this case (b) follows from (a). So assume $u = x_3$, say. Since $H - y_3$ has a hamiltonian path joining y_1 and y_2, it follows that $G' - u$ is hamiltonian, if and only if $G - \{u, x_0\}$ has a hamiltonian path joining x_1 and x_2. But this is the case if and only if $G - u$ is hamiltonian. This proves (b).

To prove (c), let $u \in V(H)$. By Lemma 1, $H - u$ has a hamiltonian path P^1 joining two of the vertices y_1, y_2, y_3, say y_1 and y_2. By assumption, $G - \{x_0, x_3\}$ has a hamiltonian path P^2 joining x_1 and x_2. Then $P^1 \cup P^2$ is a hamiltonian cycle of $G' - u$, and (c) is proved.

Theorem 1. Let G be a graph with a least five vertices and let $A \subseteq V(G)$. Suppose all vertices of A are pairwise nonadjacent and have degree 3. Suppose furthermore that G is nonhamiltonian and that $G - u$ is hamiltonian for each u in $V(G) - A$. Let G' be any graph obtained from G by replacing each vertex of A by a 3-fragment of a hypohamiltonian graph. Then G' is hypohamiltonian.

Proof. Let $A = \{x_1, x_2, \ldots, x_k\}$. Then there is a sequence of graphs $G^0, G^1, \ldots, G^k$ such that $G^0 = G$, $G^k = G'$, and G^i is obtained from G^{i-1} by replacing x_i by a 3-fragment of a hypohamiltonian graph. By applying Lemma 2, k times, we conclude that G^i is nonhamiltonian and that $G^i - u$ is hamiltonian for each $u \in V(G^i) - \{x_{i+1}, x_{i+2}, \ldots, x_k\}$. In particular, $G^k = G'$ is hypohamiltonian.

Corollary 1. Let G_1, G_2 be 3-fragments of hypohamiltonian graphs with vertices of attachment x_1, y_1, z_1 and x_2, y_2, z_2, respectively. Let G' be the graph obtained from the disjoint union $G_1 \cup G_2$ by identifying x_1, x_2 (resp. y_1, y_2, resp. z_1, z_2) into a new vertex. Then G' is hypohamiltonian.

Proof. If we put $G = K_{2,3}$ and let A denote the set of vertices of degree 3 in

G, then the corollary immediately follows from Theorem 1.

For later purposes we notice that if G_1, G_2 are 3-fragments of planar hypohamiltonian graphs, then also G' is planar.

3. Planar hypohamiltonian graphs.

The existence of planar, hypohamiltonian graphs is demonstrated in [16]. The smallest known planar hypohamiltonian graph has 105 vertices and is shown in [16, Fig.1]. Every known hypohamiltonian graph has a vertex of degree 3. In this section we show that every planar, hypohamiltonian graph contains a vertex of degree 3. We need a theorem of Tutte on hamiltonian cycles in planar graphs.

Theorem 2 (Tutte [17]). Let G be a 2-connected graph drawn in the plane and let e_1, e_2 be two edges of G adjacent to the same region. Then G contains a cycle S such that S contains e_1 and e_2 and such that each component of $G - V(S)$ (if any) is joined by edges to at most three vertices of S.

As a consequence of this result we get the following theorem.

Theorem 3. Let G be a 3-connected, planar graph with at most one separating set of three vertices. Then G is hamiltonian.

Proof. If G has no separating set of three vertices, then we appeal to Tutte's theorem, so assume $\{x, y, z\}$ is a separating set of vertices. Since G contains no subdivision of $K_{3,3}$, $G - \{x, y, z\}$ has precisely two components, say G_1 and G_2.

We shall first consider the case where $G(\{x, y, z\})$ is not a K_3. In this case we can find two edges e_1, e_2 and a vertex in $\{x, y, z\}$ (say x) such that e_i joins x and a vertex in G_i for each $i = 1, 2$, and such that e_1 and e_2 are adjacent to the same region of G. Now by Tutte's theorem, let S be a cycle of G containing e_1, e_2 such that each component of $G - V(S)$ is joined to at most three vertices of S. We shall prove that S is a hamiltonian cycle. For suppose $G - V(S)$ has a component H. Then H is joined to precisely three vertices x', y', z' of S. Since $G - \{x', y', z'\}$ is disconnected and has H as a component we conclude that $\{x', y', z'\} = \{x, y, z\}$ and that $H = G_1$ or $H = G_2$. But S contains vertices of both G_1 and G_2, a contradiction which proves that G is hamiltonian.

Suppose next that $G(\{x, y, z\})$ is a K_3. Put $G_i' = G(V(G_i) \cup \{x, y, z\})$ for $i = 1, 2$. Then $G_1' \cup G_2' = G$. Since G_1' and G_2' have a complete graph in common, the connectivity of G does not exceed the connectivity of G_1' or G_2'. Furthermore, every separating set of vertices of G_i' ($i = 1$ or 2) is also a separating set of G. Since $\{x, y, z\}$ is not a separating set of G_i', we conclude that G_i' is 4-connected for $i = 1, 2$. Hence by Tutte's theorem, G_1' has a hamiltonian cycle including the path xyz, i.e., $G_1 - y$ has a hamiltonian path P^1 joining x and z. Similarly, $G_2 - x$ has a hamiltonian path P^2 joining y and z. Now $P^1 \cup P^2 \cup \{xy\}$ is a hamiltonian cycle of G, and the proof is complete.

Lemma 3. Let G be a 3-connected graph with n vertices ($n \geq 6$). If G contains more than one separating set of three vertices, then G contains a 3-fragment with fewer than $\frac{1}{2}(n+3)$ vertices, or equivalently, G contains a separating set A of three vertices such that $G - A$ has a component with fewer than $\frac{1}{2}(n-3)$ vertices.

Proof. Let $A = \{x,y,z\}$ be a separating set. Then we can write $G = G_1 \cup G_2$, where $V(G_i) - V(G_{3-i}) \neq \emptyset$ for $i = 1, 2$, and $V(G_1) \cap V(G_2) = A$. If $|V(G_i)| < \frac{1}{2}(n+3)$ for $i = 1$ or $i = 2$, we have finished, so assume n is odd and $|V(G_1)| = |V(G_2)| = \frac{1}{2}(n+3)$. Let $A' = \{x', y', z'\}$ be a separating set of vertices distinct from A. If $A' \subseteq V(G_1)$, then G_2 is a proper subgraph of a 3-fragment of G, so G contains in this case a 3-fragment with more than $\frac{1}{2}(n+3)$ vertices, and hence also one with less than $\frac{1}{2}(n+3)$ vertices. Assume therefore without loss of generality that $x' \in V(G_1) - A$, $y' \in V(G_2) - A$, $z' \in V(G_2)$. Similarly, we have finished if one of the sets $\{x', x, y\}$, $\{x', x, z\}$, $\{x', y, z\}$ is a separating set. We therefore assume the opposite, and we shall reach a contradiction. Let v be any vertex of $V(G_1) - \{x, y, z, x'\}$. By the assumption above, each of the graphs $G - \{x', x, y\}$, $G - \{x', x, z\}$, $G - \{x', y, z\}$ contains a path from v to A. Hence the vertices of $\{x, y, z\} - A'$ (there are at least two vertices in this set) belong to the same component, say H, of $G - A'$, and each vertex v of $G_1 - A'$ belongs to H. Since $G - \{y', z'\}$ is connected, also each vertex of $G_2 - A'$ belongs to H. Hence $H = G - A'$, but this contradicts the assumption that A' is a separating set. The proof is complete.

Theorem 4. Every planar, hypohamiltonian graph contains a vertex of degree 3.

Proof. Suppose the theorem is false and let G be a counterexample with as few vertices as possible. Let $n = |V(G)|$. If G has at most one separating set of three vertices, then G is hamiltonian by Theorem 3, so assume G has at least two separating sets with three vertices. Let H be a 3-fragment of G with as few vertices as possible. By Lemma 3, H has at most $\frac{1}{2}(n+2)$ vertices, and by the minimality property of H, each vertex of attachment of H is adjacent to at least two vertices which are not vertices of attachment. Now take two disjoint copies of H and identify their respective vertices of attachment. By Corollary 1, the resulting graph G' is hypohamiltonian. By the remark above, each of the three vertices which the two copies of H have in common has degree at least 4 in G', and each other vertex has the same degree in G' as in G. Clearly G' is planar, and it has fewer vertices than G. Hence we have obtained a contradiction to the minimality property of G, and the theorem is proved.

4. Hypohamiltonian digraphs.

Hypohamiltonian digraphs are easily obtained from hypohamiltonian graphs by replacing each edge by two directed edges going in opposite directions. U.S.R. Murty asked if there exist hypohamiltonian oriented graphs. Corollary 2 below shows that such digraphs can be obtained by forming the cartesian product of cycles. We recall that if D_1, D_2 are digraphs, then the cartesian product $D_1 \times D_2$ of D_1 and D_2 is the digraph with vertex set $V(D_1) \times V(D_2)$ such that the edge from (v_1, v_2) to (u_1, u_2) is present if and only if $v_1 = u_1$ and $\overrightarrow{v_2u_2} \in E(D_2)$, or $v_2 = u_2$ and $\overrightarrow{v_1u_1} \in E(D_1)$. The directed cycle of length k, $2 \leq k$, is denoted $\vec{C}_k$. With this notation we have:

Theorem 5. $\vec{C}_k \times \vec{C}_m$ is nonhamiltonian whenever k and m are relatively prime.

Proof. Let $\vec{C}_k$: $x_1 \to x_2 \to \ldots, \to x_k \to x_1$ and $\vec{C}_m$: $y_1 \to y_2 \to \ldots \to y_m \to y_1$. For $i = 1, 2, \ldots, m$, let S_i denote the following cycle of $\vec{C}_k \times \vec{C}_m$: $(x_1,y_i) \to (x_2,y_i) \to \ldots \to (x_k,y_i) \to (x_1,y_i)$. If S in a cycle of $\vec{C}_k \times \vec{C}_m$ distinct from each S_i $(1 \le i \le m)$, then S intersects each S_i, and moreover, there is an integer $j \ge 1$ such that for each i, there are precisely j edges of S entering (and leaving) S_i. We say that S is a cycle of <u>j'th kind</u>.

Suppose S is a hamiltonian cycle of $\vec{C}_k \times \vec{C}_m$ of j'th kind, and let M be the set of j integers such that the edge from (x_q,y_1) to (x_q,y_2) is in S if and only if $q \in M$. Consider M as a subset of the integers modulo k. Since S contains all vertices of S_2, the edges of S leaving S_2 are precisely the edges from (x_{q-1},y_2) to (x_{q-1},y_3) where $q \in M$. Repeating this, we see that the edges of S leaving S_i are precisely the edges from (x_{q-i+1}, y_i) to (x_{q-i+1}, y_{i+1}), where $q \in M$. In particular, $M = \{q-m \mid q \in M\} = M-m$. Clearly $M-k=M$. So if we define d as the smallest positive integer such that $M-d=M$, then $d > 1$ and d divides k and m. The proof is complete.

<u>Corollary 2</u>. For each $k \ge 3$, $m \ge 1$, $\vec{C}_k \times \vec{C}_{mk-1}$ is a hypohamiltonian oriented graph. Moreover, $\vec{C}_3 \times \vec{C}_{6k+4}$ is hypohamiltonain for each $k \ge 0$.

<u>Proof</u>. By Theorem 5, the digraphs mentioned in the corollary are nonhamiltonian. For each vertex u of $\vec{C}_k \times \vec{C}_{mk-1}$, this digraph has a cycle of first kind missing only u, and for each vertex v of $\vec{C}_3 \times \vec{C}_{6k+4}$, this digraph has a cycle of second kind missing only v. Hence $\vec{C}_k \times \vec{C}_{mk-1}$ and $\vec{C}_3 \times \vec{C}_{6k+4}$ are hypohamiltonian.

The corollary above provides an infinite family of hypohamiltonian oriented graphs, the smallest of which, $\vec{C}_3 \times \vec{C}_4$, has 12 vertices.

We now describe a method for constructing a hypohamiltonian digraph of a given order. Suppose D is a hypohamiltonian digraph containing a vertex x_0 of in-and outdegree 2 such that x_0 is adjacent to three vertices x_1,x_2,x_3, and $\overrightarrow{x_0x_1}$, $\overrightarrow{x_0x_3}$, $\overrightarrow{x_3x_0}$, $\overrightarrow{x_2x_0}$ are in $E(D)$. Suppose further that D satisfies (i) and (ii) below:

(i) $D - \{x_0,x_3\}$ has no hamiltonian path from x_2 to x_1;

(ii)$D - \{x_0,x_2\}$ has no hamiltonian path from x_3 to x_1.

Now construct a digraph D' by deleting x_0 from D and adding two new vertices x_0',x_1' and the edges $\overrightarrow{x_3x_0'}$, $\overrightarrow{x_0'x_1'}$, $\overrightarrow{x_1'x_3}$, $\overrightarrow{x_0'x_2}$, $\overrightarrow{x_2x_0'}$, $\overrightarrow{x_1'x_1}$, $\overrightarrow{x_1x_1'}$ and $\overrightarrow{x_2x_1}$.

See Fig. 1 below:

Fig. 1. Construction of hypohamiltonian digraphs

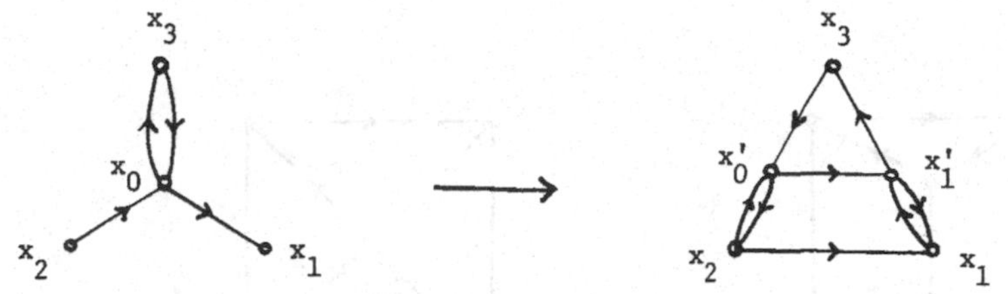

With this notation we have:

Lemma 4. D' is hypohamiltonian and satisfies (i) and (ii) with x_0' instead of x_0.

Proof. We first prove by contradiction that D' has no hamiltonian cycle. For suppose S is a hamiltonian cycle of D'. If S contains the edge $\overrightarrow{x_0'x_1'}$, then an obvious modification of S yields a hamiltonian cycle of D, and if S does not include $\overrightarrow{x_0'x_1'}$, then it contains the path $x_1 \to x_1' \to x_3 \to x_0' \to x_2$ and we get a contradiction to (i).

Let u be any vertex of D'. If $u \in V(D') - \{x_0',x_1'\}$, then a hamiltonian cycle of D - u is easily transformed into a hamiltonian cycle of D' - u. So assume $u \in \{x_0',x_1'\}$.

$D - x_2$ has a hamiltonian cycle, and any such cycle contains the path $x_3 \to x_0 \to x_1$. Replacing this path by $x_3 \to x_0' \to x_2 \to x_1$, we get a hamiltonian cycle of $D' - x_1'$. Similarly we get a hamiltonian cycle of $D' - x_0'$. Hence D' is hypohamiltonian. We leave it to the reader to verify that D' satisfies (i) and (ii) with x_0' instead of x_0.

Theorem 6. There is no hypohamiltonian digraph with fewer than six vertices, and for each $n \geq 6$, there is a planar, hypohamiltonian digraph with n vertices.

Proof. The first part of the statement is left to the reader.

For the second part, consider $\vec{C}_2 \times \vec{C}_m$. This digraph is planar, hypohamiltonian when m is odd. In particular, $\vec{C}_2 \times \vec{C}_3$ is hypohamiltonian, and it is easy to see that it satisfies (i) and (ii) when x_0 is any vertex. By repeated applications of Lemma 4 the theorem follows. Fig. 2 shows some examples of hypohamiltonian digraphs, the 2-cycles being represented by undirected edges.

Fig. 2. Some planar, hypohamiltonian digraphs.

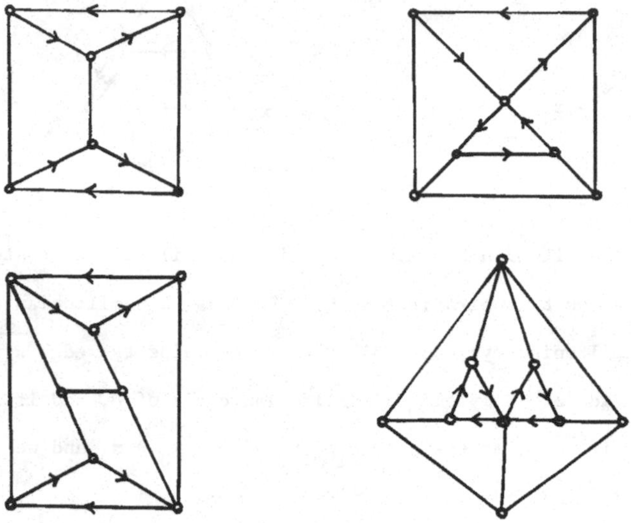

5. Vertex sets meeting all longest cycles.

Zamfirescu [19] defines C_k^j as the smallest integer n such that there exists a k-connected graph G with n vertices and with the property that for every set A of j vertices of G, G contains a longest cycle not intersecting A. If there is no graph G with the property above we put $C_k^j = \infty$. P_k^j is the analogous number when "cycle" is replaced by "path" in the definition above and $\bar{C}_k^j$, $\bar{P}_k^j$ are the analogous numbers for planar graphs. Grünbaum [8] defines $\Gamma(j,m)$, where $m \geq j$, as the family of all graphs G such that a longest cycle of G misses exactly m vertices of G and such that for each set of j vertices, there is a longest cycle missing those j vertices. The family $\Pi(j,n)$ is the family obtained by replacing "cycles" by "paths" in the definition above. Thus $\Gamma(1,1)$ resp. $\Pi(1,1)$ are the families of hypohamiltonian resp. hypotraceable graphs.

By Tutte's theorem [17], $\bar{P}_k^j = \bar{C}_k^j = \infty$ for $j \geq 1$ and $k \geq 4$, and as observed by Zamfirescu [20], $C_1^j \le \bar{C}_1^j \le 3j + 3$. The finiteness of C_k^j, P_k^j, $\bar{C}_k^j$, $\bar{P}_k^j$ has been demonstrated only for finitely many other values of k and j. (For detailed information, see Zamfirescu [20]). It is not known for which values of j,m the classes $\Gamma(j,m)$ and $\Pi(j,m)$ are nonempty.

Zamfirescu [20] conjectures that $C_1^j = \bar{C}_1^j = 3j + 3$. This conjecture follows from Theorem 7 below.

Theorem 7. Let G be any graph with n vertices. Then G contains a set A of at most $[\frac{1}{3}n]$ vertices such that every longest cycle of G intersects A.

Proof. The proof is conducted by induction on n. For $n \le 3$ there is nothing to prove so we proceed to the induction step. If G is disconnected, we write $G = G_1 \cup G_2$ where G_1 and G_2 are nonempty disjoint subgraphs of G. By the induction hypothesis, G_i contains a set A_i of at most $[\frac{1}{3}|V(G_i)|]$ vertices such that every longest cycle of G_i intersects A_i. Now the assertion of the theorem follows with $A = A_1 \cup A_2$.

Suppose next that G is connected but has a cut-vertex x. Then $G = G_1 \cup G_2$ where G_1 and G_2 are induced proper subgraphs of G, and $V(G_1) \cap V(G_2) = \{x\}$. Let $n_i = |V(G_i)|$ for $i = 1,2$. For $i = 1,2$, let A_i' be a set of at most $[\frac{1}{3}n_i]$ vertices of G_i such that every longest cycle of G_i intersects A_i. Put $A' = A_1' \cup A_2'$. If $|A'| \le [\frac{1}{3}n]$ we have finished so assume $|A'| \ge [\frac{n}{3}] + 1$.
Now $[\frac{1}{3}n] + 1 \le |A'| \le |A_1'| + |A_2'| \le [\frac{1}{3}n_1] + [\frac{1}{3}n_2] \le [\frac{1}{3}(n_1+n_2)] = [\frac{1}{3}(n+1)] \le [\frac{1}{3}n] + 1$.
Hence these inequalities are equalities which implies that $n + 1 \equiv n_1 \equiv n_2 \equiv 0 \pmod 3$. By the induction hypothesis, let A_i be a set of at most $[\frac{1}{3}(n_i-1)] = \frac{1}{3}n_i - 1$ vertices such that every longest cycle of $G_i - x$ intersects A_i. Now the theorem follows with $A = A_1 \cup A_2 \cup \{x\}$ which has cardinality
$\frac{1}{3}n_1 + \frac{1}{3}n_2 - 1 = \frac{1}{3}n - \frac{2}{3} = [\frac{1}{3}n]$.

We can therefore assume that G is 2-connected. It is easy to see that any two longest cycles in a 2-connected graph have at least two vertices in common. Let S: $x_1x_2 \ldots x_m x_1$ be a longest cycle of G. If $m \le \frac{1}{3}n$, then the theorem follows with $A = V(S)$; so assume $m > k = [\frac{1}{3}n]$. If $G - \{x_1,x_2,\ldots,x_k\}$ contains no cycle of length m, we have finished (with $A = \{x_1, x_2, \ldots, x_k\}$), so assume it contains a cycle S' of length m. In particular

(i) $n - k \geq m$.

Define $r = \min\{j \mid x_j \in V(S')\}$ and $s = \max\{j \mid x_j \in V(S')\}$. Since S and S' have at least two vertices in common, r and s exist, and $m \geq s > r \geq k+1$. Let H be the subgraph of G consisting of S' and the path P: $x_s x_{s+1} \ldots x_m x_1 x_2 \ldots x_r$. Then P is contained in two cycles of H the longest of which has length at least $\frac{1}{2}m + (m + r - s)$. Hence

(ii) $m \geq \frac{1}{2}m + (m+r-s) \geq \frac{1}{2}m+k+1$,

which implies

(iii) $m \geq 2k+2$.

Now (i) and (ii) imply

(iv) $n \geq 3k+2 = 3[\frac{1}{3}n]+2$.

But the inequality of (iv) cannot be strict, and hence the inequalities of (i) - (iii) are all equalities; in particular, $n = 3k + 2$, $m = 2k + 2$, $s = m$ and $r = k + 1$. It follows that H has the same vertex set as G. Since G contains no cycle of length $> m$, each of the three x_m-x_{k+1} paths of H has length $k+1 = \frac{1}{2}m$, and every edge of G joins vertices belonging to the same of these paths. But then obviously every cycle of length m contains x_{k+1} and x_m and the theorem holds with $A = \{x_m\}$.

Remark. Consider a graph G with $n = 3k$ vertices such that G contains k disjoint 3-cycles and no other cycles. Then every set A of vertices which intersects all longest cycles of G has cardinality at least $k = \frac{1}{3}n$, so in a sense Theorem 7 is best possible. However, if we confine ourselves to 2-connected graphs, the set A can probably be chosen so that its cardinality is considerably less than $[\frac{1}{3}n]$. In particular, it is not known if every 2-connected graph contains a set of three vertices meeting all longest cycles of the graph (see Problem 6 below).

6. Remarks and unsolved problems.

A graph is vertex-transitive if for any two vertices x,y, there is an automorphism of the graph taking x to y.

Problem 1. Do there exist infinitely many connected, vertex-transitive, nonhamiltonian graphs? If so, are infinitely many of these hypohamiltonian?

The Petersen graph and the Coxeter graph (see [18]) are the only known vertex-transitive, hypohamiltonian graphs. The $\frac{3}{2}$-tough nonhamiltonian graph described by Chvátal [4] is an example of a vertex-transitive graph which is neither hamiltonian nor hypohamiltonian.

Note that every hypohamiltonian oriented graph described in this paper is vertex-transitive. U.S.R. Murty conjectures (private communication) that every planar, connected, vertex-transitive graph is hamiltonian.

Problem 2. Does there exist a connected vertex-transitive graph with no hamiltonian path? If so, can such a graph be hypotraceable?

L. Lovasz conjectures (see [2]) that every connected, vertex-transitive graph has a hamiltonian path.

The following problem is due to Grünbaum [8].

Problem 3. Are the classes $\Gamma(2,2)$, $\Pi(2,2)$ empty?

A graph in $\Gamma(2,2)$ (resp. $\Pi(2,2)$) has the property that every vertex-deleted subgraph is hypohamiltonian (resp. hypotraceable). We agree with Grünbaum's conjecture [8] that $\Gamma(2,2)$ and $\Pi(2,2)$ are empty.

Problem 4. For which integers k does there exist a hypohamiltonian graph of connectivity (or minimum degree) k?

Every known hypohamiltonian graph has connectivity and minimum degree 3, so the case $k = 4$ is particularly interesting.

Grünbaum [8] mentions a related problem:

Problem 5. Are C_k^1 or P_k^1 finite for some $k \geq 4$?

Problem 6. Are C_2^j and P_1^j finite for some $j \geq 3$?

Several interesting problems concerning the numbers C_k^j, P_k^j and the classes $\Gamma(j,m)$, $\Pi(j,m)$ are mentioned in Grünbaum [8] and Zamfirescu [20].

Problem 7. Does there exist hypohamiltonian graphs of girth > 5?

Every known hypohamiltonian graph has girth 3, 4 or 5 (see [15]). Corollary 2 shows that hypohamiltonian digraphs may have arbitrary high (directed) girth.

Problem 8. (Chvátal [3]). Is every graph an induced subgraph of some hypohamiltonian graph?

An important partial answer is provided by Collier and Schmeichel [5] who prove that every bipartite graph is an induced subgraph of some hypohamiltonian graph. The author [15] and Collier and Schmeichel [5] prove that K_3 is a subgraph of some hypohamiltonian graph. It is not known if the same holds for K_4.

We conclude with two problems on hypohamiltonian oriented graphs.

Problem 9. Does there exist a planar hypohamiltonian oriented graph?

Problem 10. Does there exist a hypohamiltonian oriented graph whose underlying undirected graph is also hypohamiltonian?

REFERENCES

[1] J.A. Bondy, Variations on the hamiltonian theme Can. Math. Bull. 15 (1972) 57 - 62.

[2] J.A. Bondy and U.S.R. Murty, Graph Theory with Applications. MacMillan, London 1975.

[3] V. Chvátal, Flip-flops in hypohamiltonian graphs. Can. Math. Bull.16 (1973) 33-41.

[4] V. Chvátal, New directions in hamiltonian graph theory, in: F. Harary, Ed., New Directions in Graph Theory. Academic Press, New York (1973).

[5] J.B. Collier and E.F. Schmeichel, New flip-flop constructions for hypohamiltonian graphs, to appear.

[6] J.B. Collier and E.F. Schmeichel, Systematic searches for hypohamiltonian graphs, to appear.

[7] J. Doyen and V. Van Diest, New families of hypohamiltonian graphs. Discrete Math. 13 (1975) 225-236.

[8] B. Grünbaum, Vertices missed by longest paths or circuits. J. Combinatorial Theory 17 (1974) 31-38.

[9] F. Harary and C. Thomassen, Anticritical graphs, Math.Proc.Camb.Phil.Soc. 79 (1976) 11-18.

[10] F. Harary, Graph Theory. Addison-Wesley, Reading, Mass. (1969).

[11] J.C. Herz, J.J. Duby and F. Vigué, Rechereche systématique des graphes hypohamiltonien, in: P. Rosenstiehl, Ed., Theorie des Graphes. Dunod, Paris (1967) 153-160.

[12] J.C. Herz, T. Gaudin and P. Rossi, Solution du problème No. 29. Rev. Franc. Rech. Opérat. 8 (1964) 214-218.

[13] W.F. Lindgren, An infinite class of hypohamiltonian graphs. Am.Math.Monthly 74 (1967) 1087-1089.

[14] C. Thomassen, Hypohamiltonian and Hypotraceable graphs. Discrete Math.9 (1974) 91-96.

[15] C. Thomassen, On hypohamiltonian graphs. Discrete Math.10 (1974) 383-390.

[16] C. Thomassen, Planar and infinite hypohamiltonian and hypotraceable graphs. Discrete Math. 14 (1976) 377-389.

[17] W.T. Tutte, A theorem on planar graphs. Trans. Am. Math. Soc. 82 (1956) 99-116.

[18] W.T. Tutte, A non-Hamiltonian graph. Can. Math. Bull. 3 (1960) 1-5.

[19] T. Zamfirescu, A two-connected planar graph without concurrent longest paths, J. Combinatorial Theory, Ser. B. 13 (1972) 116-121.

[20] T. Zamfirescu, On longest paths and circuits in graphs. Math. Scand., to appear

ORDER PRESERVING EMBEDDINGS OF AOGRAPHS

William T. Trotter, Jr.
University of South Carolina
Columbia, SC 29208

Abstract

We call an oriented graph which does not have any directed cycles an aograph. In this paper we discuss the problem of embedding an aograph on a surface in an order preserving fashion. The general problem is motivated by recent research involving partially ordered sets with planar Hasse diagrams.

We call an oriented graph G which does not have any directed cycles an aograph (short for acyclic oriented graph). With an aograph G, we associate a partial order P(G) on the vertex set of G defined by a < b in P(G) iff there is a directed path from b to a in G. It is convenient to draw a graph diagram of an aograph so that b is higher in the plane than a whenever a < b in P(G). In such a diagram, which we will call an order diagram, it is not necessary to include the orientation on the edges. Figure 1b is an order diagram for the aograph in Figure 1a.

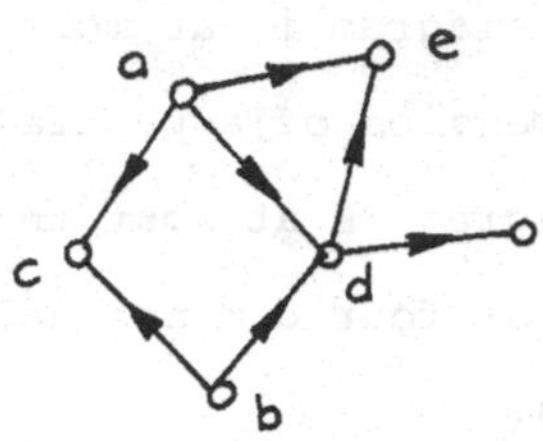

Figure 1a

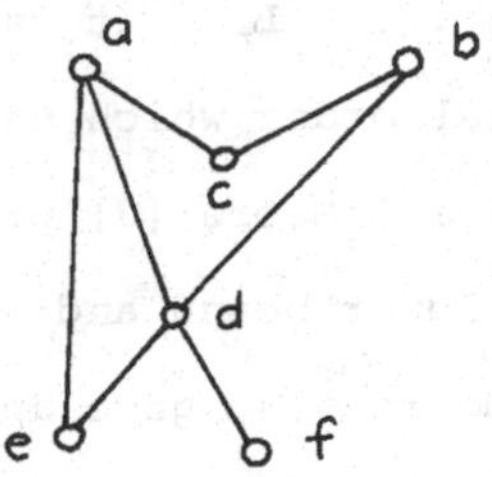

Figure 1b

As is the case with ordinary diagrams for graphs, incidental edge crossings are permitted in order diagrams. We say an aograph is planar when it is possible to draw, in the plane, an order diagram for the aograph with no incidental edge crossings. For example, the aograph in Figure 1 is planar.

Problem 1: Find the minimum list L_1 of aographs so that an aograph is planar if and only if it does not contain a subgraph isomorphic to a graph from L_1.

The aographs in Figure 3 belong to L_1; however, we comment that Problem 1 is most likely a very difficult problem.

A partial order P is said to have an upper bound when there exists a point x so that $y \leq x$ for all points y. Lower bounds are defined analogously. A partial order is said to be bounded when it has both an upper bound and a lower bound. We will use the symbol 0 to denote a lower bound and 1 to denote an upper bound. We will say that an aograph G is bounded when P(G) is bounded.

Any Hasse diagram of a partial order is also an order diagram of an aograph. Conversely, a Hasse diagram for P(G) can be obtained from an order diagram of G by removing (if necessary) some of the edges in the diagram.

Dushnik and Miller [1] defined the dimension of a partial order P on a set X, denoted Dim(P), as the smallest positive integer n for which there exist n linear extensions $L_1, L_2, \ldots, L_n$ of P so that $P = L_1 \cap L_2 \cap \ldots \cap L_n$. It is well known that the dimension of a bounded partial order which has a planar Hasse diagram is at most two. Trotter and Moore [3] proved that the dimension of a partial order with a lower bound and a planar Hasse diagram is at most three. Trotter and Moore also gave an infinite family of four dimensional partial orders which have planar Hasse diagrams.

Problem 2: Determine whether planar posets with dimension larger than four exist.

In [3] Trotter and Moore proved that if G is an aograph formed by orienting an ordinary tree (G is also called a tree), then the aograph H obtained from G by adding a point 0 with a directed edge from x to 0 for each $x \in G$ is also planar. We say such an aograph is outerplanar.

Problem 3: Determine the minimum list L_2 of aographs so that an aograph is outerplanar if and only if it does not contain a subgraph isomorphic to a graph from L_2.

Some of the aographs in L_2 are shown in Figure 2.

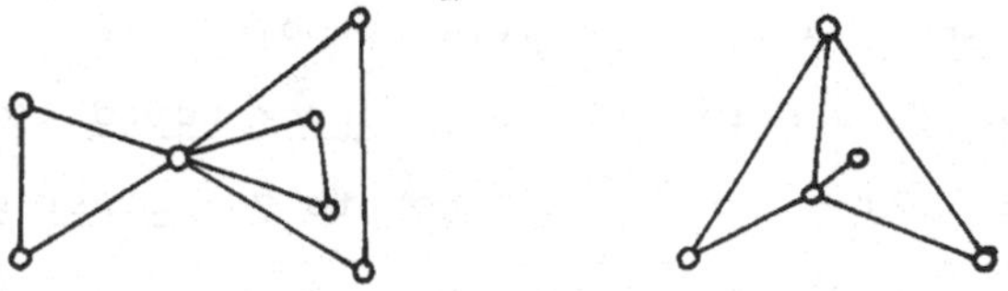

Figure 2

We now discuss the problem of embedding aographs in order preserving fashion on a sphere with n-handles. We consider such spheres in ordinary 3-space using the z-axis to determine downwardness. With the usual notions of piecewise linearity implied but not stated, we define the <u>order preserving genus</u> of an aograph G, denoted $\gamma_d(G)$ to be the smallest positive integer n for which there exists an embedding of G on a sphere with n-handles so that edges do not cross, and whenever there is a directed edge from b to a in G, the edge from b to a in the embedding flows downward.

We note that embedding an aograph on a sphere and on a plane are related but not equivalent problems. For example, the aographs in Figure 3 are non-planar but each has order preserving genus zero.

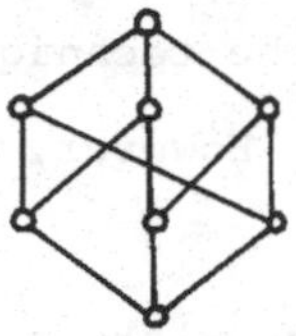

Figure 3a

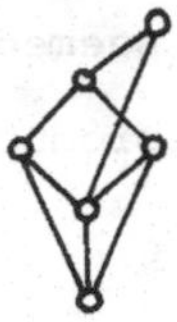

Figure 3b

<u>Problem 4</u>: Find the minimum list L_3 of aographs so that an aograph has order preserving genus zero if and only if it does not contain a subgraph isomorphic to a graph in L_3.

In [2] the author defined for $n \geq 3$ and $k \geq 0$ the crown S_n^k as the poset of height one with maximal elements $\{a_i : 1 \leq i \leq n + k\}$, minimal elements $\{b_i : 1 \leq i \leq n + k\}$, and partial ordering defined cyclically as follows: Each b_i is incomparable with $a_i, a_{i+1}, \ldots, a_{i+k}$ and is less than the remaining $n - 1$ maximal elements. The author proved that the crown S_n^k was a poset of dimension $\{2(n + k)/(k + 2)\}$.

The crown S_n^0 is isomorphic to the 2n-element poset formed by the n - 1-element and 1-element subsets of an n-element set ordered by inclusion. In dimension theory, the poset S_n^0 plays an analogous role to the complete graph in chromatic number theory for graphs. e.g.

S_n^0 is the standard example of an n-dimensional poset. The Hasse diagram for S_n^0 has, as its underlying ordinary graph, the complete bipartite graph $K_{n,n}$ minus a 1-factor. It follows that the order preserving genus of S_n^0 is at least as large as the ordinary genus of $K_{n,n}$ - 1-factor. By elementary reasoning, it follows that $\gamma(K_{n,n}$ - 1-factor) $\geq \{(n - 1)(n - 4)/4\}$. A. T. White and M. Jungerman [6] have made substantial progress towards determining that equality actually holds. It is reasonable to conjecture that $\gamma_d(S_n^0)$ is also $\{(n - 1)(n - 4)/4\}$.

In view of the results involving the embedding of posets with bounds on the plane, the author conjectured that there existed a function f(n) so that if G is a bounded aograph with $\gamma_d(G) = n$, then Dim P(G) $\leq$ f(n). It seemed plausible that the techniques of [3] could be modified to produce such a result. However, we will now prove:

<u>Theorem 1</u>: For every $n \geq 3$ and $k \geq 0$, there exists a bounded aograph G with $\gamma_d(G) = 0$ so that the crown S_n^k is a subposet of P(G).

<u>Proof</u>: Given integers $n \geq 3$ and $k \geq 0$, consider an aograph H whose order diagram has a grid-like pattern of the following type.

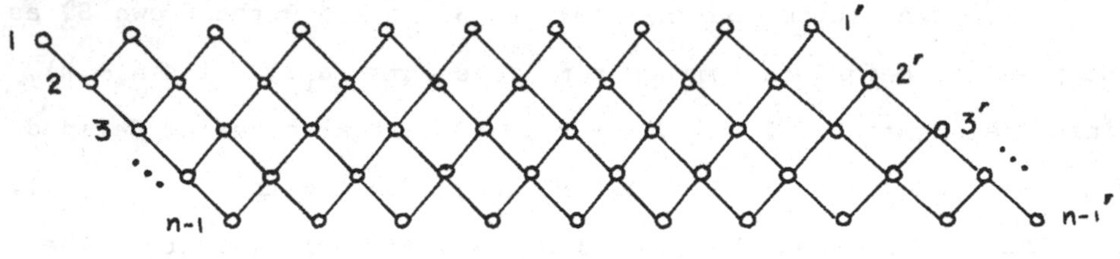

Figure 4

We choose the size of the grid so that P(H) has n + k + 1 maximal elements and the length of the longest chain in P(H) is n - 1.

Now form an aograph G by identifying the points marked i and i' for i = 1, 2,...,n - 1. $\tilde{G}$ is then formed from G by attaching a point

1 directed to each maximal element of P(G) and a point 0 with an edge from each minimal element of P(G) to 0.

The diagram in Figure 3a is an order diagram for $\tilde{G}$ when n = 3 and k = 0. The subposet of $P(\tilde{G})$ determined by the maximal elements and the minimal elements of P(G) is S_n^k. Finally, we note that $\gamma_d(\tilde{G}) = 0$ since it is easy to embed $\tilde{G}$ on a sphere by wrapping G around the equator with 1 and 0 located at the North and South poles respectively.

As a consequence of this theorem, we see that bounded aographs with order preserving genus zero and arbitrarily large dimension exist. (In fact the aograph $\tilde{G}$ has the same dimension as S_n^k).

It is easy to see that the aograph G is planar only when n is 3 or 4. An embedding of G when n = 4 and k = 5 is shown in Figure 5.

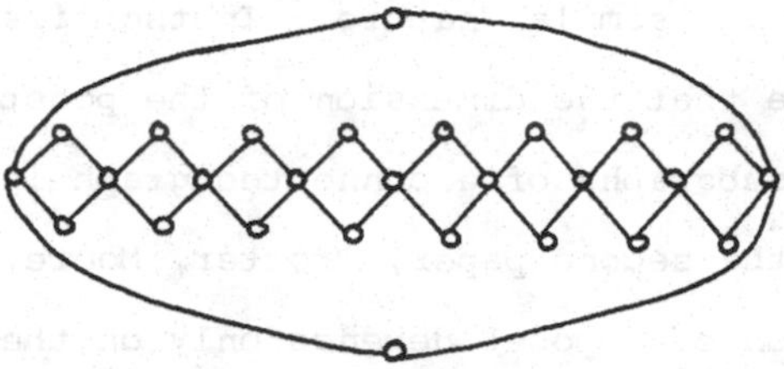

Figure 5

However, the dimension of this aograph is four when k = 0 or 1 and is three when k ≥ 2. In view of the fate of the author's conjecture concerning the dimension of aographs with order preserving genus t, we are reluctant to count this as evidence in support of four as an upper bound on the dimension of planar posets.

In retrospect, the existence of posets with order preserving genus zero and arbitrarily large dimension as established in Theorem 1 is not overly surprising in view of the characterization of dimension in terms of TM-cycles presented in [3]. However, this does suggest another area for investigation, specifically the embedding of aographs in surfaces which do not have an inherent circular nature.

Consider a number of half-planes in 3-space with each half plane containing the points on and to one side of the z-axis. These half-planes form a surface like the pages of a book. It is easy to see that any aograph can be embedded on this surface provided the book has sufficiently many pages; e.g., the aographs in Figure 3 require 3 pages.

Problem 5: For $n \geq 3$, do there exist (bounded) aographs with arbitrarily large dimension which can be embedded in a book with n pages?

In this brief paper we have given some recent results further detailing the interplay between the dimension theory of partial orders and graph theory. We refer the reader to [4] and [5] for other research papers of a similar nature. In the first paper, Trotter and Moore prove that the dimension of the poset consisting of all connected induced subgraphs of a connected graph is the number of non-cut vertices. In the second paper, Trotter, Moore, and Sumner prove that the dimension of a poset depends only on the underlying comparability graph.

REFERENCES

1. B. Dushnik and E. Miller, "Partially Ordered Sets", Amer. J. Math 63 (1941) 600-610.

2. W. T. Trotter, "Dimension of the Crown S_n^k", Discrete Math. 8 (1974) 85-103.

3. W. T. Trotter and J. I. Moore, "The Dimension of Planar Posets", J. Comb. Theory B, To Appear.

4. W. T. Trotter and J. I. Moore, "Some Theorems on Graphs and Posets", Discrete Math., To Appear.

5. W. T. Trotter, J. I. Moore, and D. P. Sumner, "The Dimension of a Comparability Graph", Proceedings AMS, To Appear.

6. A. T. White and M. Jungerman, Personal Communication.

CIRCULAR ARC GRAPHS: NEW USES AND A NEW ALGORITHM

Alan Tucker
State University of New York at Stony Brook
Stony Brook, N. Y. 11794

Abstract

An undirected graph is called a circular-arc graph if there exists a family F of arcs on a circle and a 1-1 correspondence between vertices and arcs such that two distinct vertices are adjacent if and only if the corresponding arcs overlap. Circular-arc graphs have recently been applied to problems in multi-dimensional scaling, computer compiler design, and the characterization of a certain class of lattices. An efficient algorithm for recognizing a circular-arc graph is outlined.

1. Interval graphs.

An undirected graph G is called a circular-arc graph if there exists a family F of arcs on a circle and a 1-1 correspondence between vertices and arcs such that two distinct vertices are adjacent if and only if the corresponding arcs overlap. The family F is called a circular-arc model for G. Circular-arc graphs were first considered as a natural mathematical generalization of interval graphs, graphs modeled by a family of overlapping intervals on a line. While the problem of characterizing interval graphs had been posed by Hajos [7] in 1957, the real interest in these graphs grew out of the implicit use make of them by the Nobel laureate biologiest S. Benzer. Part of the work that was to earn him a Nobel prize in biology involved verifying the hypothesis that the local (intra-gene) structure of a DNA molecule was linear. Benzer [1] isolated a collection of connected segments ("blemishes") within a certain gene of a virus and by recombination techniques, determined which segments overlapped. Although he worked with (0,1)-matrices (the adjacency matrix of the overlap graph), his question really was, is the overlap graph generated by this data an interval graph. It turned out that by ad hoc procedures that Benzer could easily demonstrate the existence of an interval model.

Within six years after Benzer's paper, three different characterizations of interval graphs had been found [4, 5, 10] and several other applications (surveyed in [8]). Berge [2] showed that interval graphs are perfect. Fulkerson and Gross [3] gave a characterization based on the following fact. In an interval graph, a maximal clique (i.e., a complete subgraph not properly contained in another complete subgraph) corresponds in an interval model to a point on the line where a maximal overlap of intervals occurs. Thus, if all maximal cliques are properly indexed, each vertex (interval) will be in a consecutive set of cliques. Let M(G), the <u>maximal clique incidence matrix</u>, be defined to have entry (i,j) = 1 if the i-th vertex is in the j-th maximal clique, and = 0 otherwise. If G is an interval graph, then M(G) has <u>the consecutive 1's property for columns</u>, that is, the rows can be arranged so that the 1's in each column are consecutive. This property can readily be shown to be sufficient as well as necessary. Note that Lueker [12] has an algorithm to find M(G) in linear time. So the Fulkerson-Gross approach reduced the interval graph problem to a problem of checking for the consecutive 1's property for columns, a property that arises naturally in a variety of clustering problems (see [8].

2. Circular-arc applications.

Circular-arc models for overlap and other relations among data are of some potential interest to molecular biologists (see Stahl [15]) and are relevant to other applications where they generalize the role of an interval model. For example, in measurement theory one can build a linear model to explain why a person is indifferent between various pairs of items by using a model of overlapping intervals (see Roberts [13]). One can also examine circular indifference situations which arise with a color wheel or musical tones (see Hubert [8] and Luce [11]). However, circular-arc graphs and the circular 1's property for columns, in which the 1's in each column are cyclicly consecutive (with the first row following the last row), have been used as an important tool in their own right by researchers in many different fields. In lattice theory, Trotter and Moore [17] have reduced the problem of characterizing posets of interval dimension at most two to the problem of characterizing circular-arc graphs that decompose into two cliques. In designing an optimal sequence of green signals for traffic signals at a complex intersection, Stoffers [15] was naturally led to the use of circular-arc models. An arc represents the period that a certain signal is green. Arcs may overlap if and only if the respective traffic streams they control do not cross. See Roberts [14, p. 129] for a good discussion of this application.

Now we present, in more detail, two recent applications, one of circular-arc graphs and the other of the circular 1's property. When designing a compiler or any other basic software system for a computer, it is very important to have efficient inner loops in subroutines. This means that the words of information involved in the loops should not, if possible, be stored in the regular memory, but should be placed in registers (and possibly even accumulators) in the central processing unit. Since the number of registers is limited, efficient use of the registers is most important. During every subsequence of steps in the loop when a word is not being used, that word must be stored (possibly just left in an accumulator that will not be used). There may be several such subsequences in the loop for a single word. The word would not have to be stored in the same place each time. However, the storage operations will be the same on each successive iteration of the loop. Suppose we have a 20 step loop and will store words W_1, W_2, W_3, W_4 as follows: W_1 is stored from step 2 to step 8 and from step 15 to step 19; W_2 is stored from step 5 to step 13; W_3 is stored from step 12 to to step 16; and W_4 is stored from step 18 to step 3 (of the next iteration of the loop). Although no more than two words are stored at any one time, the reader will quickly see that three registers are required for this example. We can consider each subsequence during which a particular word is stored as a circular arc (not an interval, because of the cyclic nature of the loop). Now our register problem is to find a minimum number of labels so that overlapping arcs get different labels. This is just a "coloring" problem - in the associated graph, overlapping arcs correspond to adjacent vertices. This writer [20] has investigated problems related to coloring arcs. It was conjectured that the chromatic number of a family of arcs is bounded by three halves the size of the largest clique. This bound was verified for two classes of families of arcs. A procedure to convert arc coloring to an integer-valued multi-commodity flow was obtained. This procedure is very useful for small and moderate sized problems but it is not, in general, efficient. Indeed, no efficient algorithm for obtaining a minimal coloring of arcs is known. (However, Gavril [5] has developed an efficient algorithm to obtain a minimal clique covering of a family of arcs.) Finally, we note that this author was able to prove that the strong perfect graph conjecture is true for circular-arc graphs.

The next example involves the following problem in data analysis. Our discussion is based on the work of Hubert [8]. We have a symmetric matrix of "closeness" measures, typically correlations, of different test scores, and we wish to index the items (rows and columns) so that each item is immediately surrounded in the indexing by all the items which are "close" to it.

TABLE 1

Test	1	2	3	4	5	6
1	1.000	0.260	0.222	0.126	0.144	0.358
2	0.260	1.000	0.254	0.115	0.073	0.115
3	0.222	0.254	1.000	0.248	0.090	0.010
4	0.126	0.115	0.248	1.000	0.307	0.081
5	0.144	0.073	0.090	0.307	1.000	0.243
6	0.358	0.115	0.010	0.081	0.243	1.000

Table 1 represents sample correlations for six tests (in order): rhyming words, digit span, arithmetic, geometrical forms, identical pictures, and picture naming. This data was first analysed by Guttman in 1964. There are a variety of techniques for indexing the rows to cluster the high values in each row (and column) together (around the main diagonal). If we dichotomize at the value of 0.2 (below 0.2 becomes 0 and above 0.2 becomes 1), we get the (0,1)-matrix in Table 2. We want it to have consecutive 1's

TABLE 2

Test	1	2	3	4	5	6
1	1	1	1	0	0	1
2	1	1	1	0	0	0
3	1	1	1	1	0	0
4	0	0	1	1	1	0
5	0	0	0	1	1	1
6	1	0	0	0	1	1

in each row (and column). The best possible result is given in Table 2. Actually, the standard procedures work with the original correlation matrix and while some methods give the indexing shown here: 1, 2, 3, 4, 5, 6, some techniques give other indexings. The problem,Hubert recognized, is that this matrix represents a circular relationship, as is clear in Table 2. That is, a geometric model should associate the items with points on a circle, not points on a line. Hubert showed that the effect of the standard indexing techniques' attempt to get a linear model from this data is either to project the circle onto a line or to break the circle at some place and open it out into a line. Thus when linear models are inconsistent, a circular structure should be sought (before one turns to multi-dimensional scaling). The only current procedure for detecting an underlying circular structure is by dichotomizing, as in Table 2, and testing for the circular 1's property (such a test, based on the consecutive 1's test, is given by Tucker [18]). Hubert notes that other well known hard-to-index correlation matrices can also be understood with a circular model.

3. Circular-arc algorithm.

Unlike the case with interval graphs, there are no good characterizations known of circular-arc graphs. This writer [18] has published a characterization of circular-arc graphs in terms of a weakened type of circular 1's property in the adjacency matrix. This characterization gives little useful information about circular-arc graphs and does not produce an efficient algorithm. While Lekkerkerker and Boland [10] obtained a structure theorem (a list of types of forbidden subgraphs) for interval graphs with five types of critical graphs, it is not hard to produce a vast number of different types of critical non-circular-arc graphs. Similarly, although the complement of an interval graph satisfies the transitive relation "is to the right (left) of", and from this fact Gilmore and Hoffman [6] developed a characterization of interval graphs in terms of comparability graphs, the complement of a circular-arc graph has no such simple structure. (However, it is worth noting that working in the complement turns out, for technical reasons, to be the appropriate way to prove a structure theorem for proper circular-arc graphs, graphs with a circular-arc model in which no arc properly contains another arc. Proper circular-arc graphs and the closely related unit (unit-length) circular-arc graphs are the only classes of circular-arc graphs for which structure theorems are known (see [19]). These graphs generalize the proper and unit interval graphs which Roberts [13] related to problems in psychophysics.)

The third characterization of interval graphs, mentioned above, in terms of the consecutive 1's property for columns in the maximal clique incidence matrix does not have a natural generalization to circular-arc graphs, as we shall see shortly. However, it does provide a natural starting point for explaining this author's new algorithm for recognizing circular-arc graphs. This new algorithm depends on an efficient test for the consecutive 1's property for columns. Fulkerson and Gross [4] provided the first such test, and recently Booth [3] and Lueker [12] developed an elegant linear-time algorithm (thus yielding a linear-time test for interval graphs). Before discussing the basis of the new algorithm, we should mention that Gavril [5] has succeeded in characterizing certain subclasses of circular-arc graphs and obtaining efficient algorithms for recognizing them. He also has produced efficient algorithms for determining parameters of circular-arc graphs, such as the size of a largest clique.

As mentioned above, maximal cliques correspond in an interval model to a point on the line of maximal overlap. Then when the rows are indexed by the order of these associated points, the maximal clique incidence matrix has consecutive 1's in each column. However, in a circular-arc model one could have three arcs, say, each covering 40% of the circle, such that each pair of two arcs overlap but all three have no common point. That is, maximal cliques in a circular-arc graph need not correspond to a common overlap point on the circle. If somehow one could determine a "local" maximal clique incidence matrix whose rows represented all maximal cliques of vertices which must overlap at some point (provided a circular-arc existed), then this matrix would have the circular 1's property for columns if and only if the graph were a circular-arc graph. However, there is a huge amount of indeterminacy in the arrangement of arcs in a circular-arc graph. Consider the circular-arc model in Figure 1. Observe that arcs A_o, A_1, A_4 and arcs A_o, A_2, A_3 both represent cliques but only A_o, A_1, A_4 have a point of common overlap in the model in Figure 1. Suppose now that A_1 and A_2 interchange positions and that A_3 and A_4 interchange positions. Now it is A_o, A_2, A_3 which have the point of common overlap. This is a simple example of the indeterminacies that have made it so difficult for researchers to test for the existence of a circular-arc model for a graph. The proposed algorithm tries to minimize these difficulties by reducing the question of whether a certain graph has a circular-arc model to the solvable question of whether a related graph has an interval model.

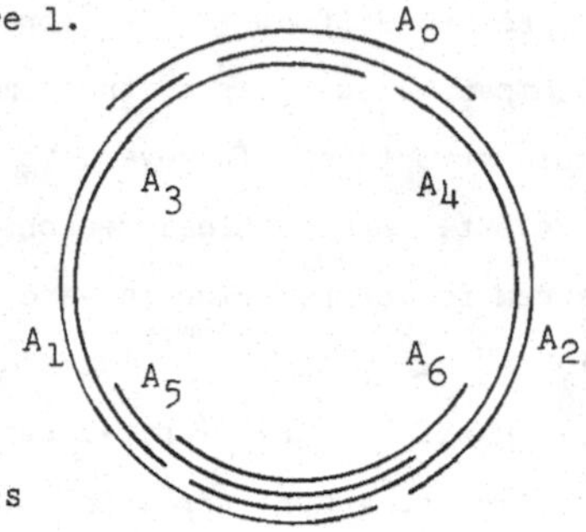

Figure 1

We motivate this reduction with the simple technique (Tucker [18]) for reducing the circular 1's property for columns to the consecutive 1's property for columns. Suppose the rows of a (0,1)-matrix are arranged so that the 1's in each column are circular (cyclicly consecutive). Observe then that in such an arrangement, the 0's in each column would also be circular. Thus if any subset of columns were complemented (1's and 0's interchanged), the resulting matrix M^1 would have the circular 1's property for columns if and only if the original matrix M did. Suppose that the subset of columns that were complemented were precisely the columns with 1 in the last row, so that M^1 has all 0's in the last row R^*. Suppose now that the rows of M^1 can be rearranged to yield circular 1's. Then a cyclic permutation (if necessary) will move R^* down to the last row (this cyclic permutation does not affect circular 1's), and now M^1 has consecutive 1's in each column (the last row of 0's blocks cyclic 1's). So M has the circular 1's property for columns if and only if M^1 has the consecutive 1's property for columns.

The circular-arc version of this circular 1's reduction is easy to describe but very difficult to do. First one finds a set of vertices V^* that must correspond to the set S^* of all arcs that contain a certain point in some circular-arc model (assuming circular-arc model(s) exist). Second, one complements each arc in S^* (replaces it by the complementary section of the circle) to obtain an interval model; actually one determines the "total" maximal clique incidence matrix (mentioned above) which must result from complementing the arcs of S^*. The details of picking S^* and determining the "local" maximal cliques are quite complex. We will sketch the method used to pick S^*. Let $N(x)$ and $\overline{N(x)}$ represent the vertices adjacent, and not adjacent, to x, respectively. Pick a vertex x_o (in the algorithm, $|N(x_o)|$ should be maximal). Then in a circular-arc model, $\overline{N(x_o)}$ must correspond to an interval model in the part of the circle that does not contain A_o the arc associated with x_o. In Figure 1, A_5, A_6 are the arcs corresponding to $\overline{N(x_o)}$. A given maximal clique C in $\overline{N(x_o)}$ will then correspond to a point in the interval model - actually it is a small interval segment, call it T, which is contained in all the arcs corresponding to the maximal clique. In Figure 1, $T = A_5 \cap A_6$. Now we seek to determine a set of vertices S^* that could correspond to the set of arcs containing p, the counterclockwise endpoint of T. Clearly $C \subseteq S^*$. The question is which $x \in N(x_o)$ could be in S^* in some circular-arc model. In Figure 1, the vertices are x_2, x_4 (corresponding to A_2, A_4). As noted above by interchanging A_1, A_2 and A_3, A_4, we obtain an equivalent model, or we could interchange just A_1, A_2 or just A_3, A_4. So S^* could be x_2, x_4, x_5, x_6 or x_1, x_3, x_5, x_6 or x_1, x_4, x_5, x_6 or x_2, x_3, x_5, x_6. However, because $x_1, x_2 \in N(x_5) \cap N(x_6)$ but x_1, x_2 are non-adjacent, only one of x_1, x_2 can be in S^*. The same applies for x_3, x_4. On the other hand, placing x_1 in S^* implies neither that x_3 is or is not in S^*. There are many more complex ways in which placing x_i in S^* may imply that x_j is or is not in S^*. The algorithm checks 15 different possibilities to build up a bipartite relation indicating which pairs of vertices may not, or must, be in S^* together. For example, if for $x_i, x_j \in N(x_o) x_i, x_j \in N(x_5) \cap N(x_6)$, x_i, x_j are adjacent, $N(x_o) \not\subseteq N(x_i)$, $N(x_o) \not\subseteq N(x_j)$ but $N(x_o) \subseteq N(x_i) \cup N(x_j)$, then this means corresponding arcs A_i, A_j overlap A_o on different sides and so A_i, A_j overlap T on different sides. Thus x_i, x_j cannot both be in S^* (actually an additional condition is required or else A_i or A_j might contain all of T and both x_i, x_j should be in S^*).

Different, but equally unpleasant, difficulties arise in the second part of the algorithm. If n is the number of vertices, the algorithm requires $O(n^4)$ operations. The full details of the algorithm will appear in [21].

REFERENCES

1. S. Benzer, On the topology of the genetic fine structures. Proc. Nat. Acad. Sci. 45 (1959) 1607-1620.

2. C. Berge, Some classes of perfect graphs. Graph Theory and Theoretical Physics. Academic Press, New York (1967) 155-165.

3. K. S. Booth, PQ - like algorithms, Ph.D. thesis. University of California, Berkeley (November, 1975).

4. D. R. Fulkerson and O. Gross, Incidence matrices and interval graphs. Pac. J. Math. 15 (1965) 835-855.

5. F. Gavril, Algorithms on circular-arc graphs. Networks 4 (1974) 357-369.

6. P. C. Gilmore and A. J. Hoffman, A characterization of comparability graphs and of interval graphs. Canad. J. Math. 16 (1964) 539-548.

7. G. Hajos, Uber eine Art von Graphen. Internat. Math. Nachrichten 11 (1957) Sondechummer 65.

8. L. Hubert, Some applications of graph theory and related non-metric techniques to problems of approximate seriation: the case of symmetry proximity measures. Brit. J. Math. Stat. Psych. 27 (1974) 133-153.

9. V. Klee, What are the intersection graphs of arcs in a circle. Math. Monthly 76 (1969) 810-813.

10. C. B. Lekkerkerker and J. C. Boland, Representation of a finite graph by a set of intervals on the real line. Fund. Math. 51 (1962) 45-64.

11. R. D. Luce, Periodic extensive measurement. Compositio Math. 23 (1971) 189-198.

12. G. S. Lueker, Interval graph algorithms, Ph.D. thesis. Princeton University (August, 1975).

13. F. S. Roberts, Indifference graphs. Proof Techniques in Graph Theory. Academic Press, New York (1969) 139-146.

14. F. S. Roberts, Discrete Mathematical Models. Prentice-Hall, Englewood Cliffs, N. J. (1976).

15. F. W. Stahl, Circular genetic maps. J. Cell Physiol.70 (Sup. 1) (1967) 1-12.

16. K. E. Stoffers, Scheduling of traffic lights - a new approach. Transport. Research 2 (1968) 199-234

17. W. T. Trotter and J. I. Moore, Characterization problems for graph partially ordered sets, lattices and families of sets. To appear.

18. A. C. Tucker, Matrix characterizations of circular-arc graphs. Pac. J. Math. 39 (1971) 535-545.

19. A. C. Tucker, Structure theorems for some circular-arc graphs. Disc. Math. 7 (1974) 167-195.

20. A. C. Tucker, Coloring a family of circular-arcs. SIAM J. Appl. Math. 29 (1975) 493-502.

21. A. C. Tucker, An efficient recognition algorithm for circular-arc graphs. To be submitted to SIAM J. Comp.

CHROMATIC SUMS

W.T. Tutte

Department of Combinatorics and Optimization

University of Waterloo

Waterloo, Ontario, Canada

SUMMARY

This is an expository article without proofs. The detailed theory of chromatic sums is explained in five papers in the Canadian Journal of Mathematics ([6]). Here we abbreviate the title of each of these as "Chromatic Sums", with the appropriate Roman numeral. After summarizing these papers we give some account of more recent work.

1. Triangulations of the sphere.

We define a triangulation of the sphere as a map M on the sphere whose faces are all triangles. In the graph of a triangulation we allow no loops, but we do permit double joins. A double join is defined by two distinct edges with the same pair of ends. We may say equally well that two such edges define a 2-circuit or digon. Of course both residual domains of the digon must be subdivided into triangular faces.

We root a triangulation by specifying one vertex as the root-vertex, one edge incident with the root-vertex as the root-edge, and one face incident with the root-edge as the root-face. The number of combinatorially distinct rooted triangulations with 2m vertices is known to be

(1) $$\frac{2^m.(3m)!}{(m+1)!(2m+1)!}.$$

This result is established in [2], albeit in dual form. It is derived also in [1] and in Chromatic Sums, IV.

2. Near-triangulations of the sphere.

We define a near-triangulation of the sphere as a map M on the sphere in which at most one face is non-triangular. As in Section 1 loops are not permitted, but double joins are. A non-triangular face of M could thus be a digon. We recognize a degenerate near-triangulation having a single edge, its two incident vertices and a single face. Apart from this degenerate case the non-triangular face, if any, must be bounded by a simple closed curve. This requirement ensures that the graphs of the near-triangulationsare non-separable.

We root a near-triangulation just as we root a triangulation, but we require that the non-triangular face, it there is one, must be the root-face.

Let M be a rooted near-triangulation. If M is non-degenerate we write $m(M)$ for the number of edges incident with the root-face. In the degenerate case we write $m(M) = 2$, remarking that the root-face is then doubly incident with the single edges of M. We note that M is a triangulation if and only if $m(M) = 3$.

We also write $n(M)$ for the number of edges of M incident with the root-vertex. Thus $n(M) = 1$ if M is degenerate, and $n(M) \geq 2$ in all other cases. We denote the number of non-root faces of M by $t(M)$.

3. Chromatic polynomials.

Let λ be a non-negative integer, and let X be a set of λ objects called colours. We may think of the colours as being the integers from 1 to λ, if λ is non-zero. We define a λ-colouring of a near-triangulation M as a mapping of the vertex-set $V(M)$ of M into X, with the property that the two ends of any edge of M are mapped onto distinct elements of X. We denote the number of λ-colourings of M by $P(M, \lambda)$.

We observe that $P(M, 0)$ and $P(M, 1)$ are both zero. The number $P(M, 2)$ is non-zero for the degenerate near-triangulation, but it is zero in all other cases.

It is known that $P(M, \lambda)$ can be expressed as a polynomial in λ whose degree is the number of vertices of M and whose coefficients depend only on the structure of M. These coefficients are integers. This is the chromatic polynomial of M. Once the coefficients are known $P(M, \lambda)$ takes a definite value for any real or complex number λ, but it is to be interpreted as a number of colourings only when λ is a non-negative integer.

In the theory of chromatic polynomials the number

(2) $$B_n = 2 + 2\cos(2\pi/n),$$

where n is a positive integer, is called the nth Beraha number. There is some reason to believe that the Beraha numbers, considered as possible values of λ, are of special importance in the theory. (See [5]). In particular there are some interesting theorems that hold for $= B_5 = \tau + 1$, but not for general λ. ([3], [4])

4. Generating functions.

In Chromatic Sums, I we define a formal power series g in four variables x, y, z and λ, as follows.

(3) $$g = g(x, y, z, \lambda) = \sum_M x^{m(M)} y^{n(M)} z^{t(M)} P(M, \lambda).$$

Here the sum is over all the combinatorially distinct rooted near-triangulations M. The power series g is well-defined in the sense that each product $x^p y^q r^z \lambda^s$ has a definite integer as its coefficient in g. This is because only a finite number of rooted near-triangulations can have a given value of $t(M)$. We can take this a little further; when g is regarded as a power series in z the coefficient of each power of z is a polynomial in the other variables. We state this property by saying that the power series g is z-restricted.

We go on to define some auxiliary power series. One such is

(4) $$q = q(x, z, \lambda) = g(x, 1, z, \lambda).$$

A second is $\ell = \ell(y, z, \lambda)$, defined as the coefficient of x^2 in g. A third is $h = h(z, \lambda) = \ell(1, z, \lambda)$. This can equally well be defined as the coefficient of x^2 in q. We refer to the generating series, g, q, ℓ and h as chromatic sums.

If we could somehow determine the function g we could substitute some positive integer for λ and then read off the coefficient of $x^p y^q z^r$. This number would be the number of λ-coloured rooted near-triangulations M satisying $m(M) = p$, $n(M) = q$ and $t(M) = r$. Similarly if we knew h we could determine the sum of the chromatic polynomials of all the rooted near-triangulations with t non-root faces and with a digon for root-face. When t is non-zero it is easily verified that this is the sum of the chromatic polynomials of the rooted triangulations with t faces in all. In summary we can hope to obtain enumerative results about coloured triangulations and near-triangulations by identifying g, q, ℓ or h as analytic functions.

Chromatic Sums, I makes the first step towards this goal by deriving a functional equation for g. It runs as follows.

$$xg = x^3y\lambda(\lambda - 1) + \lambda^{-1}yzgq + yz(g - x^2\ell) - x^2y^2z\Delta g. \tag{5}$$

Here the symbol Δ operates on an arbitrary function $F(y)$ of y, with the following effect.

$$\Delta F = \frac{F(y) - F(1)}{y - 1}. \tag{6}$$

It is found that (5) can be solved term by term for successive powers of z. This "solution" is not satisfactory for enumerative purposes, but its possibility does show that (5) has a unique solution for the formal power series g.

When $\lambda = 0$, 1 or 2 the solution of (5) for g is entirely trivial. Incidentally these values of λ are the 2nd, 3rd and 4th Beraha numbers. It is possible, but not trivial, to determine the values of $dg/d\lambda$ at $\lambda = 1$ and $\lambda = 2$. The second half of Chromatic Sums, I, is concerned with this determination.

We may try to solve (5) by finding a way to eliminate g, so as to get a functional equation in q, ℓ and h, or perhaps only in two of these. Such an equation is obtained in Chromatic Sums, II, for the case $\lambda = B_5$. It involves only the power series ℓ and h, and is as follows.

$$\tau^3(1 + \tau y)(\ell - y\tau^3) = y^2z^2\ell^2 + \tau^2y^3z^2\Delta\ell. \tag{7}$$

This equation is obtained in Chromatic Sums, II not as a consequence of (5) but by an independent graph-theoretical argument using a theorem on chromatic polynomials that is peculiar to B_5. The author knows no way of deducing (7) from (5) that does not depend on the rather deep theory of "invariants". (See below).

Equation (7) is simple enough to solve. In Chromatic Sums, II the power series ℓ is identified as an analytic function, and the coefficients in h are explicitly determined.

Chromatic Sums, III deals analogously with the case $\lambda = B_6 = 3$. We know that a triangulation T is 3-colourable if and only if it is Eulerian. Moreover if T is Eulerian, then, allowing for permutations of the three colours, we have $P(T, 3) = 6$. By exploiting the properties of Eulerian triangulations we obtained, independently of (5), the functional equation

$$(8) \qquad 6(1 + y)(\ell - 6y) = (1 + y)yz^2\ell^2 - 6y^2z^2\ell + 6y^3z^2\Delta\ell.$$

This equation is solved, and the coefficients in h are determined. The coefficient of z^{2r}, where r is positive, is six times the number of rooted Eulerian triangulations with $2r$ faces. (Only even powers of z occur in h, for any value of λ.)

Chromatic Sums, IV is really concerned with non-chromatic enumeration. We ask how many rooted near-triangulations there are with given restrictions on $m(M)$, $n(M)$ and $t(M)$. We find that this enumerational problem can be included in the theory of chromatic sums as the limiting case $\lambda \to \infty$. Formula (1) is verified. Moreover an explicit expression is obtained for the number of rooted triangulations T with a given value of $n(T)$ and $t(T)$.

In Chromatic Sums, V we begin with an attempt to eliminate g from Equation (5). The attempt is successful, at the cost of introducing other forms of complication. The variable y has to be replaced by two new quantities y_1 and y_2. These are to be regarded as power series in z, λ and s, where $x = zs$. They are the two solutions of a quadratic equation whose coefficients depend on q. As the result of the elimination we obtain two simultaneous functional equations for ℓ, in terms of z, λ, y_1 and y_2. They are as follows.

$$(9) \qquad \lambda^{-1}(1 - y_1)y_1y_2^2\ell(y_1, z, \lambda) = (1 - y_1)(1 - y_2)(y_1 - \mu y_2 + \mu y_1y_2) + y_1^2y_2^2z^2,$$

$$(10) \qquad \lambda^{-1}(1 - y_2)y_2y_1^2\ell(y_2, z, \lambda) = (1 - y_2)(1 - y_1)(y_2 - \mu y_1 + \mu y_2y_1) + y_2^2y_1^2z^2.$$

Here $\mu = \lambda - 2$. We observe that each equation is converted into the other by an interchange of y_1 and y_2.

The constant terms in the power series y_1 and y_2 can be taken as 0 and 1 respectively. If we prefer to avoid the use of the variable s we can treat y_1 as a new independent variable and regard y_2 as an unknown function of y_1, z and λ. It is then found that Equations (9) and (10) can be solved term by term, for successive powers of z, to give the unknown functions y_2 and ℓ. We infer that Equations (9) and (10) contain all the information necessary to determine the power series ℓ, it being unnecessary to know the form of y_1 and y_2 as power series in s, z and λ.

Chromatic Sums, V goes on to the study of "invariants". An invariant is defined as a function of y, z and λ such that the substitution of either y_1 or y_2 for y yields the same function of z, s and λ. Thus a constant, or a power series in z and λ only, is trivially an invariant. The first non-trivial invariant to be noted in the paper is the following function J.

$$J = \lambda^{-1}z^2\ell(y, z, \lambda) - \frac{yz^2}{1 - y} + y^{-2} - y^{-1}. \tag{11}$$

Let J_1 and J_2 denote the results of substituting y_1 and y_2 respectively for y in J. Then we can use (9) and (10) to show that $J_1 = J_2$. Hence J is an invariant.

Sums and products of invariants are invariants. Hence any polynomial in J, with coefficients depending only on z and λ, is an invariant. In later work it was often found convenient to replace J by the invariant

$$K = J + (2 - \mu)^{-1}, \text{ or} \tag{12}$$

$$k = K/(4 - \mu^2). \tag{13}$$

It should be noted that we are now dealing, not just with formal power series, but with quotients of such power series. However, the denominators involve only powers of y and (1 - y). Chromatic Sums, V now makes a study of those invariants that can be expressed as quotients of this kind. The investigation concludes with the following Theorem.

THEOREM. If p(y, z, λ) is a z-restricted power series in y, z and λ, if b and c *are non-negative integers, and if*

$$Y = y^{-b}(1 - y)^{-c}z^{2c}p(y, z, \lambda)$$

is an invariant, then Y can be expressed as a polynomial in J (or K) of degree not exceeding c, the coefficients in this polynomial being z-restricted power series in z and λ, (or simply power series in z if λ is a constant).

In order to make use of this theorem we must find some new invariants. Moreover the fact that these are polynomials in J must not be an obvious consequence of their method of construction. *Chromatic Sums, V* constructs such invariants as follows.

First we write

(14) $$v_1 = y_1^{-1}, \quad v_2 = y_2^{-1}.$$

Then we define v_3, v_4 and so on by the recursion formula

(15) $$v_{n+2} = \mu v_{n+1} - v_n + 1 - \mu.$$

It can be shown that when $\lambda = B_m$ the sequence $(v_1, v_2, v_3, \ldots)$ is periodic, with period m, and that polynomials in the v_j with suitable symmetry are invariants. The theory in the paper leads to an invariant $I(m)$ that can be identified as follows.

(16) $$I(m) = (-1)^{[(m-1)/2]} \prod_{j=1}^{m} (1 - v_j).$$

Let us write $m = 2M$ or $m = 2M - 1$, according as m is even or odd. It is found that $I(m)$ is an invariant of the kind considered in our Theorem, with $c = M - 2$. So we can write

(17) $$I(m) = \sum_{i=0}^{M-2} f_i J^i,$$

where the f_i are power series in z.

5. *Recent work.*

In *Chromatic Sums, V* the explicit formula for $I(m)$ is a product of somewhat formidable appearance. Since the publication of the paper the author has noticed that a more satisfactory formulation is possible.

First we transform from v to a new variable V related to it as follows.

$$v = V + 1 - \frac{1}{2 - \mu}. \tag{18}$$

The new expressions for $I(m)$ are as follows.

$$I(2M) = 2k^{M}\left\{\cos\left(2M \sin^{-1} \frac{V}{2\sqrt{k}}\right) - \cos\left(2M \sin^{-1} \frac{1}{2(2 - \mu)\sqrt{k}}\right)\right\} \tag{19}$$

$$I(2M - 1) = 2k^{M-\frac{1}{2}}\left\{\sin\left((2M - 1)\sin^{-1} \frac{1}{2(2 - \mu)\sqrt{k}}\right) - \sin\left((2M - 1)\sin^{-1} \frac{V}{2\sqrt{k}}\right)\right\}. \tag{20}$$

The trigonometrical expressions in these formulae can be written as finite sums Thus instead of (19) we can write

$$I(2M) = 2M \sum_{r=1}^{M} \frac{(-1)^r (M + r - 1)!}{(2r)!(M - r)!} k^{M-r} \left\{V^{2r} - \frac{1}{(2 - \mu)^{2r}}\right\}, \tag{21}$$

and instead of (2)) we can have

$$I(2M - 1) = (2M - 1) \sum_{r=0}^{M} \frac{(-1)^r (M + r)!}{(2r + 1)!(M - r)!} k^{M-r} \left\{\frac{1}{(2 - \mu)^{2r+1}} - V^{2r+1}\right\}. \tag{22}$$

The function f_{M-2} in (17) can be evaluated as a multiple of z^2 by comparing coefficients of $(1 - y)^{2-M}$. The true number of unknown functions f_j in (17) is thus $M - 2$. Hence when $m = 5$ or 6 the equation can be regarded as a functional equation for the power series ℓ, involving one unknown function of z. It can then be verified that the equation is equivalent to (7) or (8) respectively. (The expression $\Delta\ell$ in (7) and (8) involves an unknown function h).

For larger values of m we still have a functional equation for ℓ, but there are more unknown functions. It is shown in <u>Chromatic Sums, V</u> that the equation still has a unique solution, but as yet no explicit form of this solution has been obtained.

REFERENCES

1. R.C. Mullin, On counting rooted triangular maps, Can. J. Math., 17 (1965), 373-382.

2. W.T. Tutte, A census of Hamiltonian polygons, Can. J. Math. 14, (1962), 402-417.

#. _____, On chromatic polynomials and the golden ratio, J. Comb. Theory, 9, (1970), 289-296.

4. _____, The golden ratio in the theory of chromatic polynomials, Annals of New York Academy of Sciences, 175, Article 1, 391-402 (1970).

5. _____, Chromials. In Studies in Graph Theory, ed. D.R. Fulkerson, (MAA Studies in Mathematics, Vol. 12), (1975).

6. _____, Chromatic Sums for rooted planar triangulations, I-V, Can J. Math. 25 (1973), 426-447, 657-671, 780-790, 929-940; 26 (1974), 893-907.

INFINITE CAYLEY GRAPHS OF CONNECTIVITY ONE

Mark E. Watkins
Syracuse University
Syracuse, New York 13210

Abstract

A characterization of infinite Cayley graphs with vertex-connectivity 1 is given in terms of the multiplicities of isomorphic lobe subgraphs incident with each vertex.

1. INTRODUCTION. The purpose of this note [+] to characterize completely the Cayley graphs of vertex-connectivity 1. Cayley graphs are vertex-transitive, and the only finite, vertex-transitive Cayley graph of vertex-connectivity 1 is the graph K_2 which is the unique Cayley graph of the 2-element group. Hence we are concerned only with infinite Cayley graphs. For any collection of isomorphism classes of biconnected graphs or K_2, there exists an infinite Cayley graph whose lobe subgraphs yield precisely the given collection of isomorphism classes. The essential matter is the multiplicities of representatives of each isomorphism class of lobe graphs incident with each vertex.

In [1], H. A. Jung and I characterized graphs of vertex-connectivity 1 whose automorphism groups act transitively (respectively: primitively, regularly) on their vertex sets. Several arguments in this paper will rely either upon results directly from [1] or upon results whose proofs differ from proofs in [1] in only minor or formal ways. In either case the proof will be omitted.

2. TERMINOLOGY AND NOTATION. Throughout this note a graph may be either finite or infinite but will be understood to have no loops or multiple edges. Capital Greek letters without superscripts, especially the symbol Γ, will always denote a graph. The vertex set of Γ will be denoted by $V(\Gamma)$, its edge set by $E(\Gamma)$, and its automorphism group by $A(\Gamma)$.

The terms lobe and lobe subgraph are used as defined in O. Ore [2, p.86]. We let $L(\Gamma)$, or more briefly L, denote the set of lobe subgraphs of Γ. Certainly any two elements of L have at most one vertex in common.

Let G be an abstract group, let e denote the identity of G, and let H be a subset of G such that $e \notin H$. The Cayley graph $\Gamma_{G,H}$ of G with respect to H in the graph with vertex set $V(\Gamma_{G,H}) = G$ and edge set

$$E(\Gamma_{G,H}) = \{[g,gh] : g \in G ; h \in H\}.$$

It is immediate that $\Gamma_{G,H}$ is connected if and only if H generates G. For each $g \in G$, define the function $\lambda_g : G \to G$ by $\lambda_g(x) = gx$ for all $x \in G$. Clearly $\lambda_g \in A(\Gamma_{G,H})$ for any H. Moreover, the set $*G = \{\lambda_g : g \in G\}$ is a subgroup of $A(\Gamma_{G,H})$. It is a regular permutation group on $V(\Gamma_{G,H})$ and is (abstractly) isomorphic to G. By "regular" we mean that $*G$ is transitive and semi-regular. By "semi-regular" we mean that the stabilizer of any single vertex is trivial. If M is a subgroup of G, then $*M = \{\lambda_g : g \in M\}$ is a subgroup of $*G$ which acts semi-regularly on $V(G)$.

3. THE MAIN RESULT. Some further notation is required in order that we may state the main result. We assume that Γ is infinite and of vertex-connectivity 1.

Let I be any set and let $\{L_i : i \in I\}$ be an indexed partition of $L = L(\Gamma)$ such that if two lobe subgraphs belong to the same cell L_i then they are isomorphic (but not necessarily conversely). For each $i \in I$, select $\Lambda_i \in L_i$, and let B_i be a subgroup of $A(\Lambda_i)$ whose action on $V(\Lambda_i)$ is semi-regular. Let $\{\Lambda_i^{(j)} : j \in J_i\}$ be an indexed partition of $V(\Lambda_i)$ into orbits with respect to B_i. Clearly this indexing of B_i-orbits may be assumed for all $\Lambda \in L_i$. More precisely, if $\Lambda \in L_i$, let $\phi : \Lambda_i \to \Lambda$ be an isomorphism, and define $\Lambda^{(j)} = \phi[\Lambda_i^{(j)}]$ for all $j \in J_i$.

For each $x \in V(\Gamma)$, $i \in I$, and $j \in J_i$, let $L_i^{(j)}(x) = \{\Lambda \in L_i : x \in \Lambda^{(j)}\}$. Finally, let $b_i^{(j)}(x) = |L_i^{(j)}(x)|$.

Theorem. Let Γ be an infinite graph of vertex-connectivity 1. A necessary and sufficient condition for Γ to be a Cayley graph is that there exist an indexed partition $\{L_i : i \in I\}$ and semi-regular subgroups $\{B_i\}_{i\in I}$ as described above such that for all $i \in I$, the orbits $\Lambda^{(j)}$ for all $j \in J_i$ and all $\Lambda \in L_i$ may be assigned so that the functions $b_i^{(j)}$ are identically 1.

Before proceeding with the proof, we include the following notation.

If $\Lambda_o \in L$ and $n \geq -1$ is an integer, we define $\Gamma_n(\Lambda_o)$ recursively as follows:

$\Gamma_{-1}(\Lambda_o)$ is the empty graph,

$\Gamma_0(\Lambda_o) = \Lambda_o$,

$\Gamma_{n+1}(\Lambda_o) = \bigcup\{\Lambda \in L : V(\Lambda) \cap V(\Gamma_n(\Lambda_o)) \neq \emptyset\}$.

Clearly if Γ is connected, then $\Gamma = \bigcup_{n=0}^{\infty} \Gamma_n(\Lambda)$ for any $\Lambda \in L$.

Given the indexed partition $\{L_i : i \in I\}$ of L we define the subgraph Δ_i of Γ for each $i \in I$ by

$$V(\Delta_i) = V(\Gamma) \ ; \ E(\Delta_i) = \bigcup_{\Lambda \in L_i} E(\Lambda).$$

4. PROOF OF NECESSITY. Suppose that Γ is the Cayley graph $\Gamma_{G,H}$ where H generates G. and assume that some appropriate identification of $V(\Gamma)$ with G has been given.

Let Λ be any lobe subgraph incident with the vertex labeled e. The set $V(\Lambda)$ generates a subgroup of G, which in turn induces a connected subgraph Θ of Γ. Note that if $\{x,xh\} \in E(\Lambda)$ for some $h \in H$, then $h \in V(\Theta)$. It follows immediately that Θ is a union of lobe subgraphs of Γ, all of which are isomorphic to Λ. Since left cosets with respect to any subgroup of G induce pairwise-disjoint isomorphic subgraphs of $\Gamma_{G,H}$, we consider the spanning subgraph $\Delta = \bigcup_{g \in G} g\Theta$, where $g\Theta$ indicates the subgraph induced by the left coset $gV(\Theta)$. Note that Θ is uniquely determined by Λ, and that any other lobe subgraph Λ' incident with e would determine in this manner a subgraph Δ', where either $\Delta' = \Delta$ or Δ' is edge-disjoint from Δ. In this manner we construct a family $\{\Delta_i\}_{i \in I}$ of spanning subgraphs of Γ which are pairwise edge-disjoint and whose union covers Γ. For each $i \in I$, let L_i be the set of lobe subgraphs of Δ_i and let Θ_i be the component of Δ_i containing e.

We now fix $i \in I$, and consider a lobe $V(\Lambda_i)$ containing e which generates the subgroup $V(\Theta_i)$. Note that the restriction of $*[V(\Theta_i)]$ to Θ_i is a subgroup of $A(\Theta_i)$. Let B_i denote the set of restrictions to $V(\Lambda_i)$ of those elements of $*[V(\Theta)]$ which fix $V(\Lambda_i)$ setwise. Since $*[V(\Theta)]$ is semi-regular, B_i is a faithful group of permutations and acts semi-regularly on $V(\Lambda_i)$.

Let $\{\Lambda_i^{(j)} : j \in J_i\}$ be the set of B_i-orbits of $V(\Lambda_i)$. We use the elements of $*G$ to define similarly labeled orbit decompositions of each lobe of Δ_i. (By the definition of B_i, one easily verifies that this labeling is well-defined.) The functions $b_i^{(j)}$ for $j \in J_i$ may now be defined on $V(\Gamma)$ as in §3.

Fix $j \in J_i$. Because Γ is vertex-transitive, it suffices to prove $b_i^{(j)}(e) = 1$. To show that $L_i^{(j)}(e) \neq \emptyset$, let $y \in \Lambda_i^{(j)}$. The automorphism $\lambda_y^{-1} \in *[V(\Theta_i)]$ maps Λ_i onto a lobe $\Lambda \in L_i$ so that $\lambda_y^{-1}[\Lambda_i^{(j)}] = \Lambda^{(j)}$. In particular, $e = \lambda_y^{-1}(y) \in \Lambda^{(j)}$.

On the other hand, suppose $\Lambda_i, \Lambda \in L_i^{(j)}(e)$ and that Λ_i and Λ are distinct. For some $\tau \in *[V(\Theta_i)]$, we have $\tau[\Lambda_i^{(j)}] = \Lambda^{(j)}$. There exists $\beta' \in B_i$ such that $\beta'(e) = \tau^{-1}(e)$. If β' is the restriction of $\beta \in *[V(\Theta_i)]$, then $\tau\beta(e) = e$. But $\tau\beta \neq \lambda_e$, contrary to the regularity of $*G$.

The proof of necessity is now complete. We remark that each subgraph Θ_i is a Cayley graph of the subgroup $V(\Theta_i)$ and that [1: Lemma 5.2] G is isomorphic to the free product of the family of groups $\{V(\Theta_i)\}_{i\in I}$.

5. PROOF OF SUFFICIENCY.

Let us assume that the indexed partition $\{L_i : i \in I\}$ of L and the subgroups $\{B_i\}_{i\in I}$ as defined in §3 have been given and that the functions $b_i^{(j)}$ have been defined on $V(\Gamma)$.

<u>Lemma</u>. Suppose that for each $i \in I$, $j \in J_i$, the function $b_i^{(j)}$ is constant on $V(\Gamma)$. Let $\Lambda, \Lambda' \in L_i$ and let n be a non-negative integer. Let $\sigma_n : \Gamma_n(\Lambda) \twoheadrightarrow \Gamma_n(\Lambda')$ be an isomorphism which maps $\Gamma_n(\Lambda) \cap \Delta_i$ onto $\Gamma_n(\Lambda') \cap \Delta_i$ for all $i \in I$. Then σ_n admits an extension $\sigma \in A(\Gamma)$ such that for all $i \in I$, the restriction of σ to Δ_i belongs to $A(\Delta_i)$.

We omit the proof of this lemma since it varies only slightly from the proof of [1: Lemma 3.1].

Let us now suppose that $b_i^{(j)}(x) = 1$ for all $i \in I$, $j \in J_i$, and $x \in V(\Gamma)$. We use the Lemma to deduce that Γ as well as each subgraph Δ_i is vertex-transitive. Let $i \in I$, $j \in J_i$, and $x_1, x_2 \in V(\Gamma)$. There exist (unique) lobe subgraphs $\Lambda_1, \Lambda_2 \in L_i$ such that $x_1 \in \Lambda_1^{(j)}$ and $x_2 \in \Lambda_2^{(j)}$. There exists an isomorphism $\sigma_0 : \Lambda_1 \twoheadrightarrow \Lambda_2$ such that $\sigma_0(x_1) = x_2$, and hence there exists an extension $\sigma \in A(\Gamma)$ of σ_0 such that $\sigma_{|\Delta_i} \in A(\Delta_i)$.

It follows that each component of Δ_i is also vertex-transitive and that any two components of Δ_i are isomorphic. Once it can be shown that for each $i \in I$ a component of Δ_i is a Cayley graph (say, of the group M_i), then the hypothesis of [1: Lemma 5.2] will be satisfied, and the sufficiency of the condition will follow. In particular, Γ will be a Cayley graph of the free product of the family $\{M_i\}_{i \in I}$. We may therefore assume that I consists of a single element, and we may without ambiguity continue all the above notation but with the subscript i suppressed. Thus $\Gamma = \Delta$, and the problem consists of identifying $V(\Gamma)$ with an appropriate group G.

Let $\Lambda_o \in L$. The action of B on $V(\Lambda_o)$ induces orbits on $V(\Lambda_o) \times V(\Lambda_o)$. Since the elements of B are graph-automorphisms, each of these orbits consists either wholly of oriented edges of Λ_o or wholly of "non-edges". Ignoring the latter type, we let H and H^{-1} be sets of orbits of oriented edges subject only to the following two conditions: (1) For any edge

$\{x,y\} \in E(\Lambda_o)$, (x,y) belongs to an orbit in H if and only if (y,x) belongs to an orbit in H^{-1}. If the orbit containing (x,y) is denoted by h we let h^{-1} denote the orbit containing (y,x). Thus $(h^{-1})^{-1} = h$. (2) An orbit h belongs to $H \cap H^{-1}$ if and only if $h = h^{-1}$. The existence of sets H and H^{-1} satisfying (1) and (2) is an immediate consequence of the semi-regularity of B. The group G with which $V(\Gamma)$ will be identified is generated by the set H.

To begin the identification, let some vertex of Λ_o be labeled e. Let $\{e,x\} \in E(\Lambda_o)$. Then $(e,x) \in h$ for some $h \in H \cup H^{-1}$. If $h \in H$, let x be relabeled h; otherwise identify x with h^{-1}. (Note that one cannot have (e,x_1), $(e,x_2) \in h$ for $x_1 \neq x_2$ due to the semi-regularity of B. Our identification thus far is well-defined.)

To continue the identification, consider an edge $\{h,y\} \in E(\Lambda_o)$, where $(h,y) \in h'$ for some $h' \in H \cup H^{-1}$. We may then identify y with the two-letter word hh'. We continue in this manner until all of $V(\Lambda_o)$ has been identified with words in $H \cup H^{-1}$. Note that the same vertex may be identified with many different words. These identifications, in fact all directed circuits in Λ_o, generate the defining relations for G.

It may be enlightening to interrupt our proof at this point with an example. Suppose that Λ_o is represented by Figure 1 and that B is the 2-element group generated by the reflection across the dashed line as shown, inducing three vertex-orbits indicated by the different shapes around the vertices. A possible resulting set $H = \{f, g, h, k, m\}$ of orbits of oriented edges is indicated by Figure 2, and a consequent labeling of $V(\Lambda_o)$ is shown in Figure 3. The relations are generated by hmf^{-1}, g^2, k^2, and $kmgm^{-1}$

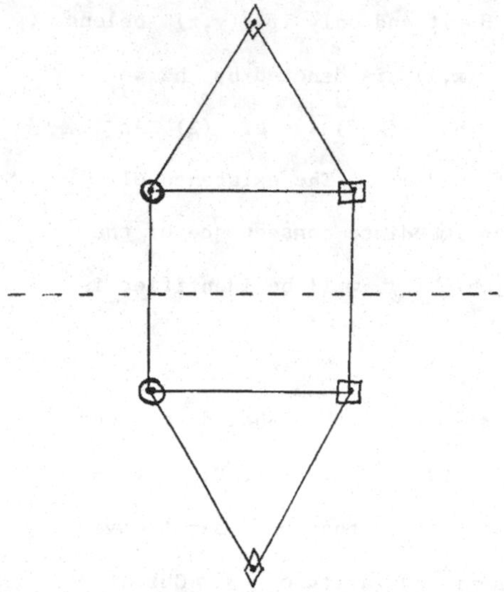

Figure 1

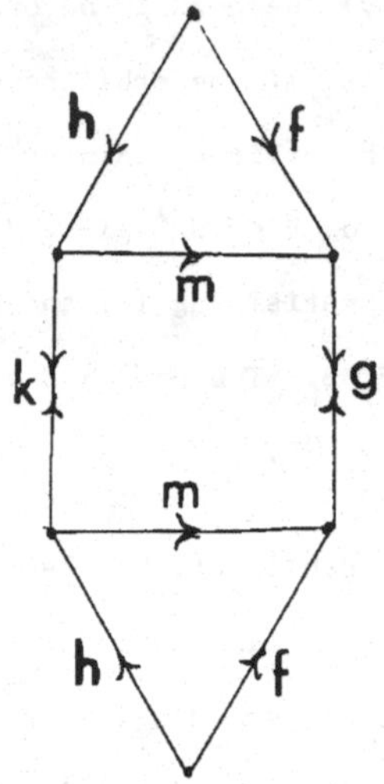

Figure 2

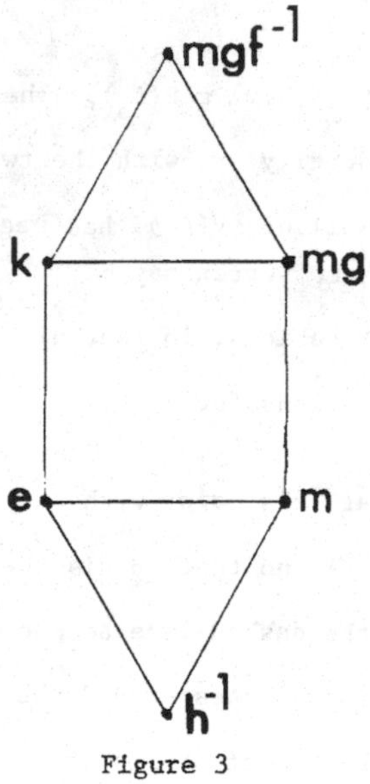

Figure 3

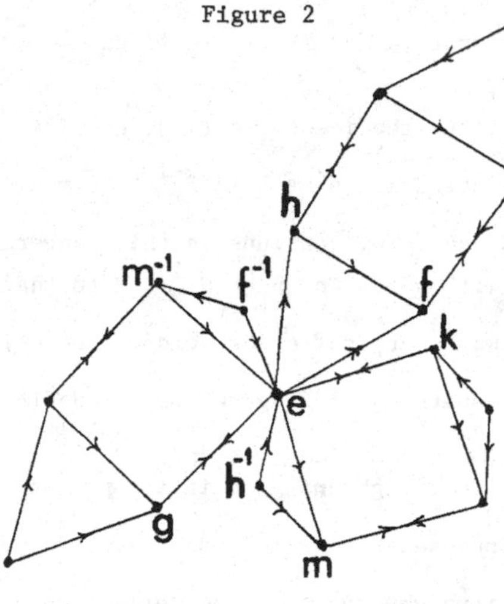

Figure 4

Returning to the proof, we next extend the labeling of Λ_o to the other lobes of Γ inductively as follows. Suppose that for some $n \geq 0$, the vertices of $\Gamma_n(\Lambda_o)$ have been consistently labeled, and let x be a vertex of $\Gamma_n(\Lambda_o) - \Gamma_{n-1}(\Lambda_o)$. Then x belongs to exactly one lobe Λ_n of $\Gamma_n(\Lambda_o)$.

Let Λ be any lobe subgraph other than Λ_n which contains x. There exists an isomorphism $\phi : \Lambda_o \longrightarrow \Lambda$ such that $\phi [\Lambda_o^{(j)}] = \Lambda^{(j)}$ for all $j \in J$. The labels for $V(\Lambda)$ are then determined by the formula

$$\phi(u) = x(\phi^{-1}(x))^{-1} u, \qquad u \in V(\Lambda_o).$$

In this manner a labeling is obtained for $\Gamma_{n+1}(\Lambda_o)$, and the proof is complete.

In Figure 4, the three lobes incident with e and the set $H \cup H^{-1}$ are indicated when Λ_o and B are as in the previous example.

REFERENCES

1. H. A. Jung and M. E. Watkins, On the structure of infinite vertex-transitive graphs. (Submitted for publication.)

2. O. Ore, The Theory of Graphs, Amer. Math. Soc., Colloq. Publ. 38, Providence, 1962.

EDGE-COLORINGS OF GRAPHS - A SURVEY

Robin J. Wilson
The Open University

1. Introduction

In the past hundred years, problems involving the coloring of the vertices or the faces of a graph have received much attention in the mathematical literature. The reason for this interest is undoubtedly due mainly to one problem - the four-color problem - and in view of this, it is somewhat surprising that various closely-related problems involving the coloring of the <u>edges</u> of a graph seem to have received little attention until comparatively recently. In this survey we shall outline the progress that has been made in connection with edge-coloring problems in the last few years. The reader who is interested in seeing some of the applications of this theory to such fields as the designs of experiments, scheduling and tensor calculus, is referred to [10].

Historically, the first papers to appear on edge-colorings were two brief abstracts by P. G. Tait in the Proceedings of the Royal Society of Edinburgh in 1880 [19]. In these notes, Tait asserted that the four-color conjecture is true if and only if the edges of every trivalent planar graph can be properly colored (in a sense to be explained) using only three colors. Related results were obtained in the 1890s by J. Petersen and P. J. Heawood, and in 1916 D. König [15] proved that, if G is a bipartite graph with maximum valency ρ, then the edges of G can be properly colored using exactly ρ colors. After this, little was done until 1949, when C. E. Shannon [18] proved that, if G is a graph with maximum valency ρ, then the edges of G can be properly colored using at most $[\frac{3}{2}\rho]$ colors. But the great breakthrough occurred in 1964, when V. G. Vizing [20,21] proved that if G is a simple graph with maximum valency ρ, then the number of colors needed to color the edges of G is either ρ or $\rho + 1$ - a surprisingly strong result. Much of this survey will be concerned with the 'classification problem' - the problem of determining which graphs can be colored with ρ colors, and which need $\rho + 1$ colors.

Just as in vertex-coloring problems, it turns out that arbitrary graphs are often too clumsy to deal with in full generality, and we sometimes find it convenient to restrict our attention to graphs which are 'critical' in some sense . Critical graphs have received a certain amount of attention recently, and in the latter half of this survey, we outline some of the progress made in this direction.

Unless otherwise stated, we assume throughout that all graphs G are finite, connected and simple. The valency of a vertex v is denoted by $\rho(v)$, and the largest valency in G is denoted by $\rho(G)$, or simply by ρ . The number of vertices of G is denoted by n , and the number of edges by m .

2. The Classification Problem

If G is a graph, we define the chromatic index of G (written $\chi'(G)$) to be the least number of colors needed to color the edges of G in such a way that adjacent edges are assigned different colors. For example, if C_n is the circuit graph consisting of a single n-gon $(n \geq 4)$, then $\chi'(C_n) = 2$ if n is even, and $\chi'(C_n) = 3$ if n is odd. Note that if G contains a vertex of valency k , then $\chi'(G) \geq k$; it follows that the maximum valency $\rho(G)$ is a lower bound for the chromatic index of G .

As mentioned in the introduction, Vizing gave very sharp bounds for the chromatic index of a simple graph. We state his theorem formally as follows:

Vizings's Theorem. If G is a simple graph with maximum valenct ρ then

$$\rho \leq \chi'(G) \leq \rho + 1 .$$

Vizing's theorem immediately gives us a simple way of classifying graphs into two classes, according to their chromatic index. We shall say that a graph G is of class one if $\chi'(G) = \rho$, and that G is of class two if $\chi'(G) = \rho + 1$. The Classification Problem is the problem of deciding which graphs are of class one, and which are of class two. The importance and difficulty of this problem become apparent when we realise that its solution would immediately settle the four-color conjecture; this follows from Tait's result mentioned earlier, that the four-color conjecture is true if and only if every trivalent bridgeless planar graph is of class one.

For reference, we include in the following table a list of some of the types of graph whose chromatic class is known; several of these entries will be discussed later in the survey.

CLASS 1	CLASS 2
C_n, K_n (if n is even)	C_n, K_n (if n is odd)
The Platonic graphs	The Petersen graph
Bipartite graphs (König, 1916)	Regular graphs (if n is odd)
Planar graphs with $\rho \geq 8$ (see later)	Regular graphs with a cut-vertex

Although the problem of deciding which graphs belong to which class is unsolved, it is certainly the case that graphs of class two are relatively scarce. For example, of the 143 connected graphs with at most six vertices, only eight are of class two (see Figure 1). A more general result is due to P. Erdős and R. J. Wilson [5] who have proved that 'almost all' graphs are of class one, in the sense that if P(n) is the probability that a random graph is of class one, then $P(n) \to 1$ as $n \to \infty$.

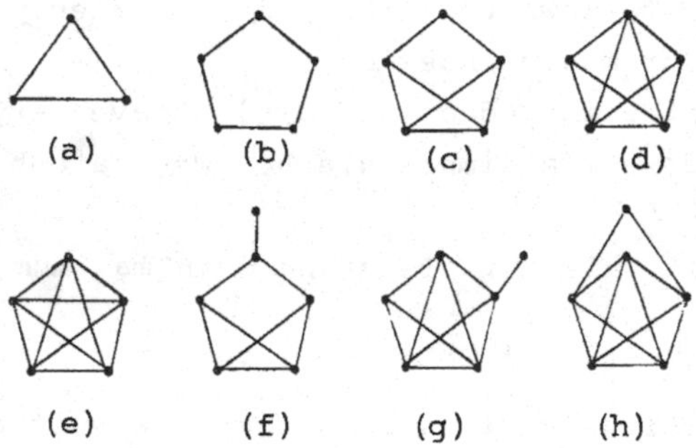

Figure 1.

It seems natural to expect that the more edges a graph has, the more likely it is to be of class two. This idea is made precise in the following result which gives a sufficient, but not necessary, condition for a graph to be of class two. This elementary result is implicit in the work of Vizing, but was first formulated explicitly by L. W. Beineke and R. J. Wilson [2].

<u>Theorem</u>. If G has n vertices, m edges and maximum valency ρ, and if $m > \rho[\frac{1}{2}n]$, then G is of class two.

<u>Proof</u>. If G is of class one, then any ρ-coloring of the edges of G partitions these edges into at most ρ independent sets. But the number of edges in each independent set cannot exceed $[\frac{1}{2}n]$, since otherwise two of these edges would be adjacent. It follows that $m \leq \rho[\frac{1}{2}n]$, giving the required contradiction. ∎

Using this theorem, we can obtain several important constructions for graphs of class two. In each case, the proof consists of calculating m in terms of n and ρ, and checking that $m > \rho\cdot[\frac{1}{2}n]$.

<u>Corollary</u> 1. If G is a regular graph with an odd number of vertices, then G is of class two.

<u>Corollary</u> 2. If H is a regular graph with an odd number of vertices, and if G is any graph obtained from H by deleting not more than $\frac{1}{2}\rho - 1$ edges, then G is of class two.

<u>Corollary</u> 3. If H is a regular graph with an even number of vertices, and if G is any graph obtained from H by inserting a new vertex into one edge of H, then G is of class two.

Examples of these constructions are the graphs (e), (d) and (c) of Figure 1. There are, of course, many other consturctions for graphs of class two. Many of these are quite complicated, and there is no room to discuss them here. The reader is referred to [2], [9], and [13] for details of these constructions.

3. <u>Generalizations of the Petersen Graph</u>.

Before continuing with our main theme, we shall digress to present some recent work on the chromatic index of graphs which resemble, in one way or another, the Petersen graph. Although the Petersen graph itself is of class two, the variations we shall describe fall into both classes.

(a) <u>The Generalized Petersen Graphs P(n, k)</u>.

The Petersen graph (see Figure 2(i)) consists of an outer 5-circuit, five spokes incident with the vertices of this 5-circuit, and an inner 5-circuit attached by joining its vertices to every <u>second</u> spoke.

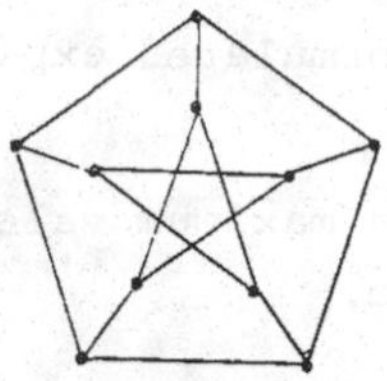

Figure 2(i).

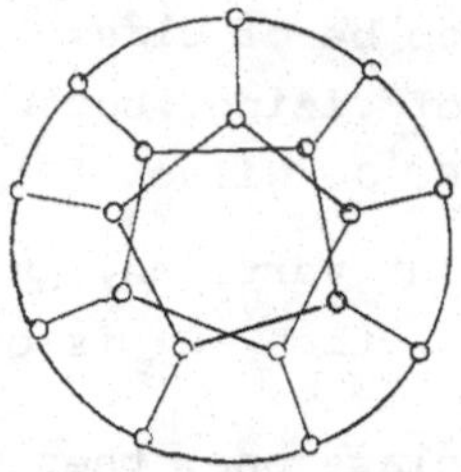

Figure 2(ii)

M. E. Watkins [23] has defined the <u>generalized</u> <u>Petersen</u> <u>graph</u> $P(n, k)$, which consists of an outer n-circuit, n spokes incident with the vertices of this n-circuit, and an inner n-circuit attached by joining its vertices to every <u>kth</u> spoke. Thus the Petersen graph is the graph $P(5, 2)$, and $P(9, 2)$ is the graph shown in Figure 2(ii). Note that $P(n, k)$ is isomorphic to $P(n, n-k)$. Watkins conjectured that, with the single exception of the Petersen graph itself, each of the graphs $P(n, k)$ is of class one. This conjecture was eventurally proved in 1972 by F. Castagna and G. Prins [4].

<u>The</u> <u>Odd</u> <u>Graphs</u> O_k.

N. L. Biggs [3] has defined the <u>odd</u> <u>graph</u> O_k to be the graph obtained by taking as vertices each of the $(k-1)$-subsets of the set $\{1, 2, \ldots, 2k-1\}$, and joining two of these vertices by an edge whenever the corresponding subsets are disjoint. It is easy to verify that O_2 is simply a 3-gon, and that O_3 is the Petersen graph. More generally, the graph O_k is always a regular graph of valency k, and is of odd order if and only if k is a power of 2. It has been proved [17] that O_k is of class two if $k = 3$ or 4, and that O_k is of class one if $k = 5$ or 6, and it is clear from Corollary 1 above that O_k is of class two if k is a power of 2. Biggs has conjectured that for all other values of k, O_k is a graph of class one, and this conjecture remains open.

<u>Meredith's</u> <u>Graphs</u>.

Let $n \geq 2$, and let m be the integer nearest to $\frac{n}{3}$ (so that $n = 3m$ or $3m \pm 1$). G. H. J. Meredith [16] has defined a graph G_n by taking the Petersen graph, and replacing each vertex by a copy of the complete bipartite graph $K_{n,n-1}$, joined up as shown in Figure 3, which represents G_4. Meredith

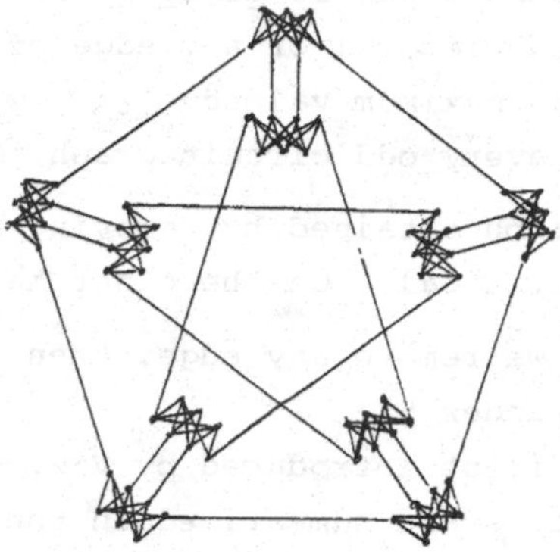

Figure 3.

proved that the graph G_n is regular of valency n, and is always n-connected and non-Hamiltonian. He also investigated the chromatic index of these graphs, and proved that G_n is of class one if m is even, and of class two if m is odd.

(d) Isaacs' Trivalent Graphs.
Regular trivalent graphs of class two are hard to find. In fact, until 1975, only four of these graphs were known - namely, the Petersen graph, a graph with 18 vertices due to Blanuša, a graph with 50 vertices due to Szekeres, and a graph with 210 vertices due to Tutte. Since then, R. Isaacs has described two infinite sequences of such graphs; the first of these sequences contains the four graphs already mentioned. There is no space to describe Isaacs' constructions here, and the interested reader is referred to his paper in the American Math. Monthly [12].

4. Critical Graphs.

In the study of vertex-colorings of graphs, those graphs which are critical in some sense have played an important role. The reason for this is not hard to find - since every graph contains a critical graph, and since critical graphs generally have more structure than arbitrary graphs, we certainly lose nothing, and frequently gain a lot, by restricting our attention to critical graphs.

A similar situation holds in the theory of edge-colorings. In this case, we define a graph G to be <u>critical</u> if G is connected and of class two, and if the removal of any edge of G lowers the chromatic index; if G has maximum valency ρ, we say that G is <u>ρ-critical</u>. For example, every odd circuit graph C_{2k+1} is a 2-critical graph, and the graph obtained by removing any edge from the complete graph K_5 is 4-critical. On the other hand, K_5 itself is not critical, since if we remove any edge, then the remaining graph still has chromatic index 5.

Critical graphs were first introduced by Vizing (see [21]). Some of their main properties are summarized in the following theorem; proofs of these results may be found in [21] and [13].

<u>Theorem</u>. (i) A critical graph has no cut-vertex (so that every critical graph is 2-connected)

(ii) If G is ρ-critical, and if v and w are adjacent vertices of G, then $\rho(v) + \rho(w) \geq \rho + 2$.

(iii) If I is an independent set of edges in a critical graph G, then the edges in I can all be given the same color; in other words, $\chi'(G - I) = \chi'(G) - 1$.

(iv) If G is a graph of class two with maximum valency ρ, then G contains a k-critical subgraph for all values of k satisfying $2 \leq k \leq \rho$.

(v) If G is a ρ-critical graph, where $\rho \geq 3$, then G cannot be regular; in other words, the only regular critical graphs are the odd circuit graphs.

A result which is considerably deeper than those just mentioned is also due to Vizing. It has two forms, both of which we present here:

<u>Vizing's 'Adjacency Lemma'</u> (First form [21]): If G is a ρ-critical graph, then every vertex of G is adjacent to at least two vertices of valency ρ; in particular, G has at least three vertices of valency ρ.

<u>Vizing's 'Adjacency Lemma'</u> (Second form [22]): If G is a ρ-critical graph, and if v and w are adjacent vertices with $\rho(v) = k$, then w is adjacent to at least $\rho - k + 1$ vertices of valency ρ.

The Adjacency Lemma is probably the most useful and important of all the results on critical graphs, and is used in the proofs of most of the results which follow. To give some idea of its use, we shall prove a simple lower bound (due to Vizing) for the number of edges in a critical graph:

Theorem. If G is a ρ-critical graph, then G has at least $\frac{1}{8}(3\rho^2 + 6\rho - 1)$ edges.

Proof. If σ is the smallest valency occurring in G, then by Vizing's Adjacency Lemma, G has at least $\rho - \sigma + 2$ vertices of valency ρ. Since the total number of vertices is at least $\rho + 1$, it follows that the number of edges of G is at least

$$\frac{1}{2}\{\rho(\rho - \sigma + 2) + \sigma(\sigma - 1)\} .$$

It is easy to check that this expression is smallest when $\sigma = [\frac{1}{2}(\rho + 1)]$, and the result follows. ∎

5. Further Results.

In the previous section we outlined some of the most important properties of critical graphs. We shall now give a brief survey of some recent work on critical graphs, and of the role they play in edge-coloring problems.

(a) Jakobsen's Results on 3-critical Graphs.

If G is a 3-critical graph, then every vertex of G has valency 2 or 3. But by Vizing's Adjacency Lemma, every vertex of valency 2 is adjacent to two vertices of valency 3, and every vertex of valency 3 is adjacent to at most one vertex of valency 2. It follows by a simple calculation that, if G has n vertices, then the number of edges is at least $\frac{4}{3}n$; also, since G cannot be regular, the number of edges is at most $\frac{1}{2}(3n - 1)$.

These results were first proved by I. T. Jakobsen [14], who also showed that a 3-critical graph cannot contain exactly two vertices of valency 2. It follows easily from this that there can be no 3-critical graphs with 4, 6, 8 or 10 vertices. These results prompted Jakobsen to make the following conjecture, also formulated independently by Beineke and Wilson [2].

Critical Graph Conjecture: Every critical graph has an odd number of vertices.

In support of this conjecture, L. W. Beineke and S. Fiorini [1] have proved that there are no critical graphs with 4, 6, 8 or 10 vertices, and no 3-critical graphs with 12 vertices. Between them, Jakobsen, Beineke and Fiorini have found all the critical graphs with 3, 5 and 7 vertices, and all the 3-critical graphs with 9 vertices.

(b) Bounds on the Number of Edges.
We have already seen that every ρ-critical graph has at least $\frac{1}{8}(3\rho^2 + 6\rho - 1)$ edges, and that if $\rho = 3$, then the number of edges lies between $\frac{4}{3}n$ and $\frac{1}{2}(3n - 1)$, where n is the number of vertices. The first of these bounds is unsatisfactory if n is large compared with ρ, and the latter inequalities are valid only if $\rho = 3$. It is therefore of interest to see whether we can determine upper and lower bounds which relate to a general value of ρ, and which involve both ρ and n explicitly.

The best upper bound of this kind is due to S. Fiorini, who modified Jakobsen's methods, and deduced that if n is odd, then every ρ-critical graph has at most $\frac{1}{2}(n - 1)\rho + 1$ edges, and if n is even, then it has at most $\frac{1}{2}(n - 2)\rho + (\sigma - 1)$ edges, where σ is the minimum valency of the graph.

As far as lower bounds are concerned, it is easy to prove, using Vizing's Adjacency Lemma and a simple counting argument, that every ρ-critical graph has at least $\frac{2n}{\rho}(\rho - 1)$ edges. However, the true result is likely to be much higher than this; indeed, Vizing has conjectured that every ρ-critical graph has at least $\frac{1}{2}\{n(\rho - 1) + 3\}$ edges, and this conjecture remains unproved. The best lower bound known to date is $\frac{1}{4}n(\rho + 1)$, due to Fiorini [8].

(c) Topological Results.
In [21], Vizing proved that if G is a planar graph whose maximum valency is at least 10, then G must be of class one; the proof of this uses Vizing's Adjacency Lemma, together with the fact that every planar graph contains a vertex whose valency is at most five. Later, in [22], he strengthened this result by proving that if G is a planar graph whose maximum valency is at least 8, then G must be of class one. Since planar graphs with maximum valency 2, 3, 4, or 5 can be of class one or class two, the question arises as to what happens when $\rho = 6$ or 7. It has been conjectured that every planar graph with $\rho = 6$ or 7 is of

class one, but this has never been proved. It is interesting to note that this conjecture with $\rho = 7$ follows easily from Vizing's conjecture on the number of edges of a ρ-critical graph.

Results similar to those described in the previous paragraph have been obtained for graphs on higher surfaces, and for graphs with a larger girth, but there is no space to describe them here. In addition, Fiorini [7] has proved that if G is an outer-planar graph, other than an odd circuit graph C_{2k+1}, then G is necessarily of class one.

(d) Uniquely Edge-colorable Graphs.

A graph G is said to be uniquely edge-colorable if there is only one way of partitioning its edges into color-classes; if the chromatic index of G is k, then we say that G is uniquely k-colorable. An example of a uniquely 3-colorable graph is the graph K_4, and we can get further examples of such graphs by successively replacing each vertex by a triangle. It has been conjectured that every planar uniquely 3-colorable graph can be obtained in this way; this is easily seen to be equivalent to the conjecture that apart from $K_{1,3}$, every planar uniquely 3-colorable graph contains a triangle. (Note that if we remove the word 'planar', the conjecture is certainly false; for example, the generalized Petersen graph $P(9,2)$ (Figure 3) is a non-planar uniquely 3-colorable graph which contains no triangle.)

In 1973, Greenwell and Kronk [11], and (independently) Fiorini [6], proved that apart from K_3, all uniquely colorable graphs are of class one. Somewhat surprisingly, no-one has ever been able to give any examples of uniquely k-colorable graphs with $k \geq 4$, apart than the star graph $K_{1,k}$ which clearly has this property ; it has been conjectured that such graphs do not exist.

6. Summary of Conjectures.

In this survey we have mentioned six conjectures, which still remain unresolved. For convenience, we repeat these conjectures here; in [24], I offered a prize of $10 for the first correct proof of, or counter-example to, any of these conjectures.

(a) The odd graph O_k is of class two only if $k = 3$ or k is a power of 2 (Biggs).

(b) The critical graph conjecture: Every critical graph has an odd number of vertices (Jakobsen; Beineke and Wilson).

(c) Every ρ-critical graph has at least $\frac{1}{2}[n(\rho - 1) + 3]$ edges (Vizing).

(d) Every planar graph with maximum valency 6 or 7 is of class one (Vizing).

(e) Apart from $K_{1,3}$, every planar uniquely 3-colorable graph contains a triangle (Fiorini).

(f) Apart from $K_{1,k}$, there is no uniquely k-colorable graph with $k \geq 4$ (Fiorini).

REFERENCES

1. L. W. Beineke and S. Fiorini, On small graphs critical with respect to edge colourings, Discrete Math. (to appear).

2. L. W. Beineke and R. J. Wilson, On the edge-chromatic number of a graph, Discrete Math. 5 (1973), 15-20.

3. N. Biggs, An edge-colouring problem, Amer. Math. Monthly 79 (1972), 1018-1020.

4. F. Castagna and G. Prins, Every generalized Petersen graph has a Tait coloring, Pacific J. Math. 40 (1972), 53-58.

5. P. Erdős and R. J. Wilson, On the chromatic index of almost all graphs, J. Combinatorial Theory (to appear).

6. S. Fiorini. On the chromatic index of a graph, III: Uniquely edge-colourable graphs, Quart. J. Math. (Oxford) (3) 26 (1975), 129-140.

7. S. Fiorini, On the chromatic index of outerplanar graphs, J. Combinatorial Theory 18 (1975), 35-38.

8. S. Fiorini, Some remarks on a paper by Vizing on critical graphs, Math. Proc. Camb. Phil. Soc. 77 (1975), 475-483.

9. S. Fiorini and R. J. Wilson, On the chromatic index of a graph, II, in Combinatorics (Proceedings of the British Combinatorial Conference, Aberystwyth, 1973), London Math. Soc. Lecture Notes No. 13, pp. 37-51.

10. S. Fiorini and R. J. Wilson, Edge-colourings of graphs - some applications, Proceedings of the Fifth British Combinatorial Conference, Aberdeen, 1975, Utilitas Mathematica (Winnipeg), pp. 193-202.

11. D. Greenwell and H. Kronk, Uniquely line-colorable graphs, Canad. Math. Bull. 16 (1973), 525-529.

12. R. Isaacs, Infinite families of nontrivial trivalent graphs which are not Tait colorable, Amer. Math. Monthly 82 (1975), 221-239.

13. I. T. Jakobsen, Some remarks on the chromatic index of a graph, Arch. Math. (Basel) 24 (1973), 440-448.

14. I. T. Jakobsen, On critical graphs with chromatic index 4, Discrete Math. 9 (1974), 265-276.

15. D. König, Über Graphen und ihre Anwendung auf Determinantentheorie und Mengenlehre, Math. Ann. 77 (1916), 453-465.

16. G. H. J. Meredith, Regular n-valent, n-connected, non-Hamiltonian, non-n-edge-colourable graphs, J. Combinatorial Theory 14 (1973), 55-60.

17. G. H. J. Meredith and E. K. Lloyd, The footballers of Croam, J. Combinatorial Theory 15 (1973), 161-166.

18. C. E. Shannon, A theorem on colouring the lines of a network, J. Math. Phys. 28 (1949), 148-151.

19. P. G. Tait, On the colouring of maps, Proc. Roy. Soc. Edinburgh 10 (1878-80), 501-503, 729.

20. V. G. Vizing, On an estimate of the chromatic class of a p-graph, Diskret. Analiz 3 (1964), 25-30.

21. V. G. Vizing, The chromatic class of a multigraph, Cybernetics 1 No. 3 (1965), 32-41.

22. V. G. Vizing, Critical graphs with a given chromatic class, Diskret. Analiz 5 (1965), 9-17.

23. M. E. Watkins, A theorem on Tait colorings with an application to the generalized Petersen graphs, in Proof Techniques in Graph Theory (ed. F. Harary), Academic Press 1969, pp. 171-177.

24. R. J. Wilson, Some conjectures on edge-colourings of graphs, in Recent Advances in Graph Theory, Academia Prague 1975, pp. 525-528.

MENGER AND KÖNIG SYSTEMS

D. R. Woodall[+]
University of Nottingham
Nottingham, England NG7 2RD

ABSTRACT

A system of sets is called a Menger system or a König system if it has properties that were originally proved to hold for certain special systems by Menger and by König, respectively, in the theorems that customarily bear their names. In this survey paper, these properties are specified, and the relationship between them is described. Some known theorems about them are stated. Some systems of sets of edges of graphs with these properties are exhibited. The main (and very substantial) omission from this survey is any mention of rational analogues of these properties and the connection with polyhedra and linear programming, for which see, for example, [2-5, 10, 16, 17] and references cited therein.

1. The Menger and König properties. Let $(S, \mathfrak{F})$, or just $\mathfrak{F}$, be a finite set-system, i.e., a collection $\mathfrak{F}$ of distinct subsets of a finite set S [elsewhere called a configuration, design, hypergraph, etc.]. The elements of S will be called points and the sets of $\mathfrak{F}$ blocks. Suppose for the moment that $\mathfrak{F}$ contains at least one non-empty block, and that the blocks are mutually incomparable as sets [$\mathfrak{F}$ is a Sperner family or clutter]. Clearly then $\emptyset \notin \mathfrak{F}$.

+ This paper was written while the author was visiting the University of Calgary.

The Menger dual [or blocking clutter, or blocker] $(S, \mathfrak{F}_{\geq 1})$ of $\mathfrak{F}$ consists of all the minimal subsets of S that have at least one element in common with each block of $\mathfrak{F}$.

PROPOSITION 1. $(\mathfrak{F}_{\geq 1})_{\geq 1} = \mathfrak{F}$. ∎

If $\mathfrak{F}$ contains k disjoint blocks, then clearly every block in $\mathfrak{F}_{\geq 1}$ has cardinality at least k . $\mathfrak{F}$ is a Menger system, or has the Menger property [or packs, or has the strong (or Boolean) max-flow min-cut property] if the maximum number $\nu(\mathfrak{F})$ of disjoint blocks in $\mathfrak{F}$ is equal to the cardinality $\tau(\mathfrak{F})$ of the smallest block in $\mathfrak{F}_{\geq 1}$.

Examples of Menger systems:

1a. Edges of a bipartitie graph G , regarded as pairs of vertices (S = V(G)) .

1b. Minimal sets of vertices of a bipartitie graph that are incident with all edges.

2a. AB-paths in a directed or an undirected graph G , the paths being regarded either as sets of vertices or as sets of edges. (These are Menger's theorem and its edge-separation analogue, A and B being disjoint sets of vertices of G.)

2b. AB-separating sets of vertices or of edges of G (clearly).

Note that 1a and 1b are Menger duals of each other, as are 2a and 2b. Note also that 1a and 1b are special cases of 2a and 2b. Further examples are given in §3.

In a similar way, the König dual [or antiblocking clutter or antiblocker] $(S, \mathfrak{F}_{\leq 1})$ of $\mathfrak{F}$ consists of all the maximal subsets of S that have at most one element in common with each block of $\mathfrak{F}$. However, $(\mathfrak{F}_{\leq 1})_{\leq 1} \neq \mathfrak{F}$ in general. $\mathfrak{F}$ is clique-complete [or conformal (Berge [1])] if it has the following property, possessed by the maximal cliques of a graph: if $S' \subseteq S$, and every pair of elements of S' is contained in some block of $\mathfrak{F}$, then S' is contained in some block of $\mathfrak{F}$. (Equivalently, if $B_1, B_2, B_3 \in \mathfrak{F}$, then there is a block of $\mathfrak{F}$ that contains all points that are in at least two of B_1, B_2, B_3.) The <u>clique-completion</u> $\mathfrak{F}_{cc}$ of $\mathfrak{F}$ consists of all the maximal subsets of S whose pairs of elements are all contained in blocks of $\mathfrak{F}$.

<u>PROPOSITION 2</u>. $(\mathfrak{F}_{\leq 1})_{\leq 1} = \mathfrak{F}$ <u>if and only if</u> $\mathfrak{F}$ <u>is clique-complete</u>. <u>In general</u>, $(\mathfrak{F}_{\leq 1})_{\leq 1} = \mathfrak{F}_{cc}$. <u>But</u> $((\mathfrak{F}_{\leq 1})_{\leq 1})_{\leq 1} = \mathfrak{F}_{\leq 1}$ <u>always</u> (<u>since</u> $\mathfrak{F}_{\leq 1}$ <u>is clique-complete</u>). ∎

Suppose that S is the union of all the blocks of $\mathfrak{F}$. If $\mathfrak{F}_{\leq 1}$ contains a block of cardinality k, then clearly S cannot be expressed as the union of fewer than k blocks of $\mathfrak{F}$. $\mathfrak{F}$ is a <u>König system</u>, or has the <u>König property</u>, if the smallest number $\rho(\mathfrak{F})$ of blocks of $\mathfrak{F}$ whose union is S is equal to the cardinality $\alpha(\mathfrak{F})$ of the largest block in $\mathfrak{F}_{\leq 1}$. $\mathfrak{F}$ has the <u>dual König</u> property if the smallest number $\gamma(\mathfrak{F}) = \rho(\mathfrak{F}_{\leq 1})$ of blocks of $\mathfrak{F}_{\leq 1}$ whose union is S is equal to the cardinality $\mu(\mathfrak{F})$ of the largest block in $\mathfrak{F}$. If $\mathfrak{F}$ is clique-complete, this is the same as saying that $\mathfrak{F}_{\leq 1}$ has the König property.

<u>Examples of König systems</u>:

1a. Vertex coboundaries in a bipartite graph. (This is König's theorem (see §3 for terminology). Equivalently: the sets of non-zero elements of a matrix that lie in the same row or the same column.)

1b. Maximal sets of non-adjacent edges in a bipartite graph [1, Theorem 2 on p. 250].

2a. Maximal chains in a partially ordered set (Dilworth's theorem).

2b. Maximal antichains in a partially ordered set (clearly).

Note that 1a and 1b are König duals of each other, as are 2a and 2b.

Further examples are given in §3.

If $\mathfrak{F}$ is any finite set-system (not necessarily a clutter), and if $\mathfrak{F}_{min}$ denotes the set of minimal blocks of $\mathfrak{F}$, then we define $\mathfrak{F}_{\geq 1} := (\mathfrak{F}_{min})_{\geq 1}$ and say that $\mathfrak{F}$ has the Menger property if $\mathfrak{F}_{min}$ does. Similarly, if $\mathfrak{F}_{max}$ denotes the set of maximal blocks of $\mathfrak{F}$, then we define $\mathfrak{F}_{\leq 1} := (\mathfrak{F}_{max})_{\leq 1}$ and say that $\mathfrak{F}$ has the König property or dual König property if $\mathfrak{F}_{max}$ does. We adopt the convention that $\{\emptyset\}$ and $\emptyset$ are Meneger duals of each other that both have the Menger properties, and that $(S, \mathfrak{F})$ has the König property if S is not the union of all the blocks of $\mathfrak{F}$.

The two examples 1a described above for the Menger and König properties are easily seen to be equivalent to each other, and it is well known that Menger's theorem and Dilworth's theorem are related. This suggests that there may be a connection between the Menger and König properties. To exhibit this connection, let $\mathfrak{F}_x$ be the collection of blocks of $\mathfrak{F}$ containing x, for each $x \in S$, and let $\mathfrak{F}^T$, the _transpose_ [or _design dual_, or _hypergraph dual_] of $\mathfrak{F}$, be the collection $(\mathfrak{F}_x : x \in S)$ of subsets of $\mathfrak{F}$. The _incidence matrix_ of $\mathfrak{F}$ has a row for each block B of $\mathfrak{F}$ and a column for each point $x \in S$, with a 1 in the (B, x)-position if $x \in B$ and a 0 otherwise: so the incidence matrix of $\mathfrak{F}^T$ is just the transpose of that of $\mathfrak{F}$. Clearly $(\mathfrak{F}^T)^T = \mathfrak{F}$.

PROPOSITION 3. $\mathfrak{F}$ _has the Menger property if and only if_ $\mathfrak{F}^T$ _has the König property, and vice versa._

Proof. If $X \subseteq S$, $X \in \mathfrak{F}_{\geq 1}$ means the same as $\bigcup_{x \in X} \mathfrak{F}_x = \mathfrak{F}$; similarly, the largest number of disjoint blocks in $\mathfrak{F}$ is equal to the cardinality of the largest block in $(\mathfrak{F}^T)_{\leq 1}$. Thus $\mathfrak{F}$ has the Menger property if and only if $\mathfrak{F}^T$ has the König property. The "vice versa" follows on taking transposes. (What we have shown is that $\tau(\mathcal{F}) = \rho(\mathcal{F}^T)$ and $\nu(\mathcal{F}) = \alpha(\mathcal{F}^T)$; hence $\tau(\mathcal{F}^T) = \rho(\mathcal{F})$ and $\nu(\mathcal{F}^T) = \alpha(\mathcal{F})$.) ∎

2. Restrictions and minors. In this section we attempt to explain the occurrence of Menger and König systems in dual pairs in §1.

If $(S, \mathcal{F})$ is a finite set-system and $S' \subseteq S$, we define the point-restriction $\mathcal{F}\setminus S'$ of $\mathcal{F}$ to $S\setminus S'$ [or subhypergraph (Berge [1]) or contraction $\mathcal{F}/S'$ (Seymour [14])] to be the collection of sets $(B\setminus S' : B \in \mathcal{F})$, and the strong point-restriction $\mathcal{F}/S'$ [or section hypergraph (Lovasz, [9]) or deletion $\mathcal{F}\setminus S'$ (Seymour [14])] to be $(B \in \mathcal{F} : B \cap S' = \emptyset)$. So a point-restriction of $\mathcal{F}$ is obtained by deleting a set of columns of the incidence matrix, i.e., deleting a set of points from S and from any block containing them, while a strong point-restriction is obtained by deleting a set of points from S and deleting all blocks containing any of them. A block-restriction [or partial hypergraph (Berge [1])] is obtained by deleting a set of blocks from $\mathcal{F}$, i.e., deleting a set of rows of the incidence matrix. (Strong block-restriction would be defined in the obvious way, but is not needed.) A restriction is anything obtained by a sequence of point-restrictions and block-restrictions, and a minor is anything obtained by a sequence of point-restrictions and strong point-restrictions; note that all of these operations commute. It is not difficult to see that $(\mathcal{F}\setminus S')_{\leq 1} = (\mathcal{F}_{\geq 1}\setminus S')_{max}$, $(\mathcal{F}\setminus S')_{\geq 1} = \mathcal{F}_{\geq 1}/S'$, and $(\mathcal{F}/S')_{\geq 1} = (\mathcal{F}_{\geq 1}\setminus S')_{min}$. It is easy to see that in the examples of König systems quoted in §1, not only the systems described but all point-restrictions of them have the König property (since the point-restrictions are themselves of the same form as the original systems). Similarly, not only the Menger systems quoted, but all minors of them, have the Menger property (although different constructions are needed to show this for the four cases in 2a: directed or undirected, vertex-sets or edge-sets).

A set-system is called balanced if it has no "proper" odd circuit, i.e., if its incidence matrix has no square submatrix of odd order of the form $\begin{bmatrix} 0 & 1 & 1 \\ 1 & 1 & 0 \\ 1 & 0 & 1 \end{bmatrix}$, $\begin{bmatrix} 0 & 0 & 0 & 1 & 1 \\ 0 & 0 & 1 & 1 & 0 \\ 0 & 1 & 1 & 0 & 0 \\ 1 & 1 & 0 & 0 & 0 \\ 1 & 0 & 0 & 0 & 1 \end{bmatrix}$, etc.

All parts of the following theorem can be found in Berge's book [1, Chapter 20, theorems quoted] except for part (h), which follows from the fact that every restriction of $\mathfrak{F}$ is clique-complete if and only if $\mathfrak{F}$ has no "proper" triangle; I am indebted to D. C. Sweetman for this observation, and for drawing these results to my attention.

THEOREM 4. The following statements about a set-system $(S, \mathfrak{F})$ are all equivalent.

(a) $\mathfrak{F}$ (and hence every restriction of $\mathfrak{F}$) is balanced.

(b) $\mathfrak{F}^T$ (and hence every restriction of $\mathfrak{F}^T$) is balanced.

(c) (Theorem 2) Every restriction of $\mathfrak{F}$ has property B (if one-element blocks are ignored), i.e., is (weakly) 2-colourable.

(d) (Theorem 5) Every restriction of $\mathfrak{F}$ has the Menger property.

(e) (Theorem 6) Every restriction of $\mathfrak{F}$ has the König property.

(f) (Theorem 3) For every restriction $\mathfrak{F}'$ of $\mathfrak{F}$, $\mathfrak{F}'_{\geq 1}$ has the Menger property.

(g) (Theorem 4) Every restriction of $\mathfrak{F}$ has the dual König property.

(g) $\mathfrak{F}$ has no "proper" triangle (i.e., no submatrix $\begin{bmatrix} 0 & 1 & 1 \\ 1 & 1 & 0 \\ 1 & 0 & 1 \end{bmatrix}$, and, for every restriction $\mathfrak{F}'$ of $\mathfrak{F}$, $\mathfrak{F}'_{<1}$ has the König property.

■

However, this theorem does not explain the occurrence of Menger and König systems in dual pairs in §1, since the classes of such systems exhibited there are not closed under block-restrictions. For the König property, the appropriate result was proved (in transpose form) by Fulkerson and Lovasz [5; 8; and 1, Theorem 7 of Chapter 20].

THEOREM 5. *Every point-restriction of* $\mathfrak{F}$ *has the König property if and only if every point-restriction of* $\mathfrak{F}$ *has the dual König property. This in turn implies (and, if* $\mathfrak{F}$ *is clique-complete, is implied by) the fact that every point-restriction of* $\mathfrak{F}_{\leq 1}$ *has the König property.* ∎

$\mathfrak{F}^T$ is then called *normal*. ($\mathfrak{F}$ is normal if every block-restriction of it has the Menger property, and *semi-normal* (Lovasz, [9]) if every strong point-restriction of it has the Menger property). Theorem 5 shows why the König systems in §1 occur in dual pairs. (They are clearly all clique-complete.) In particular, it provides an alternative proof of Dilworth's chain-decomposition theorem. For, as Fulkerson pointed out in [5], it is clear that the maximal antichains of a poset have the König property, and hence all point-restrictions of them do (since these are just the maximal antichains of another poset); thus Dilworth's theorem follows immediately from Theorem 5. The other main consequence of Theorem 5 is the perfect-graph theorem, that the maximal independent sets of vertices of a graph G and all its induced subgraphs have the König property if and only if the same is true of the maximal cliques of G and all its induced subgraphs; this follows immediately from Theorem 5.

In view of Theorem 5, one might be tempted to try to explain the occurrence of Menger systems in pairs by conjecturing that if every minor of $\mathfrak{F}$ has the Menger property, then $\mathfrak{F}_{\geq 1}$ has the Menger property. I am indebted to P. D. Seymour for pointing out that this is false. The system of seven sets (abc, cde, bdf, aef, cf, be, ad), with Menger dual (abc, cde, bdf, aef), has the Menger property, and all minors of it do, but its Menger dual does not. (The same example is cited by Lovasz in [9].) However, Seymour conjectures that there is a sense in which this example is contained in every other counter-example. If this could be proved, it might provide the required explanation of the occurrence of Menger systems in dual pairs.

3. Further examples of Menger and König systems, and unsolved problems.

First we need some terminology. If (A, B) is a partition of the vertices of a graph G into two disjoint sets, then $\langle A, B\rangle$ denotes the set of edges joining a vertex in A to a vertex in B. If G is undirected, then $\langle A, B\rangle = \langle B, A\rangle$ and is called a coboundary; if A consists of just one vertex, then $\langle A, B\rangle$ is a vertex coboundary. If G is directed, then $\langle A, B\rangle$ and $\langle B, A\rangle$ are disjoint sets, and are called semidirected coboundaries; if $\langle B, A\rangle = \emptyset$, then $\langle A, B\rangle$ is a directed coboundary. Similarly, if C is a circuit (elementary) in a directed graph G, then the sets C_- and C_+ of "clockwise" (or "negatively") and "anticlockwise" (or "positively") oriented edges round C are called semidirected circuits (the choice of positive direction being arbitrary); if $C_- = \emptyset$, then C_+ is a directed circuit.

Almost all the König systems that I know of arise from perfect graphs or from Dilworth's theorem (or its dual). A graph G is perfect if, for each induced subgraph H of G, the maximal cliques of H have the König property (then, by Theorem 5 above, the maximal independent sets of vertices also have the König property) Lovasz [9] lists many examples of perfect graphs. There are other examples of König systems. If a and b are vertices in a directed graph G without directed circuits, then the ab-paths (regarded as sets of vertices or as sets of edges) have the König property (but this is just Dilworth's theorem). In the same graph, all the directed co-boundaries [18, and the ab-separating directed co-boundaries, are König systems (but these are both the dual of Dilworth's theorem). The semidirected coboundaries of a bipartite directed graph G have the König property, since E(G) is the union of at most two of them. But K_4 with every edge replaced by two oppositely-oriented directed edges (either in series or in parallel) gives a counterexample to most more general conjectures involving circuits and coboundaries. The main outstanding problem on the König property is probably the strong version of the perfect-graph conjecture (Berge [1, p. 361]): a graph is perfect if and only if neither it nor its complement contains a chordless odd circuit. One can also make mischievous conjectures, such as: in a bridgeless cubic planar graph, the maximal independent sets of edges have the König property and the minimal covering sets of edges have the Menger property.

The situation with regard to the Menger property seems more interesting. Since bipartite graphs are perfect and maximal cliques in a bipartite graph are just edges, Proposition 3 shows that the vertex coboundaries of a bipartite graph have the Menger property. The Menger dual of this sytem, the system of minimal covering sets of edges of a bipartite graph, also has the Menger property (Gupta [6; 1, p. 455]). All minors of these systems are also Menger systems (the incidence matrix of the first is totally unimodular).

Edmonds [19; 9, p. 113; 11] has proved that the spanning out-trees with fixed root in a directed graph G, when regarded as sets of edges, have the Menger property. All minors of this system also have the Menger property (by applying the theorem to the results of delecting and replicating edges of G). The Menger dual of this example has the Menger property (clearly: consider the sets of edges entering the vertices other than the root), and so do all point-restrictions of it (which correspond to deleting edges from G), and Fulkerson [20] has shown that the strong point-restrictions do too.

Rothschild and Whinston [10] have proved the following Menger-type theorem: if a, a', b, b' are four distinct vertices of an (undirected) Eulerian graph, then the set of all aa'- and bb'-paths (regarded as sets of edges) has the Menger property. All point-restrictions of this example have the Menger property (by contracting edges), but not all strong point-restrictions (or the theorem would hold in a non-Eulerian graph, which it needn't). The Menger dual of this system need not have the Menger property (take a, b, a', b' in order round a square of the octahedron).

Lucchesi and Younger ([12]; see also [11]) have proved that the directed coboundaries of a directed graph have the Menger property. All minors of them also have the Menger property (by contraction of edges, and replication of edges in series). The situation for the Menger dual of this system seems completely open: I strongly suspect that it does have the Menger property, but this seems quite a hard unsolved problem.

By dualizing (graphically) the result of Lucchesi and Younger, as they themselves pointed out, we see that the directed circuits in a directed planar graph have the Menger property: but not in an arbitrary graph (take $K_{3,3}$ with its three "vertical" edges directed upwards and its six "sloping" edges directed downwards).

The semidirected coboundaries in a Hamiltonian directed graph have the Menger property (clearly: the Hamiltonian circuit of p edges intersects every semidirected coboundary, and the p sets of edges entering the p vertices are disjoint semidirected coboundaries), but the same does not necessarily hold for their point-restrictions, strong point-restrictions or Menger dual. The result does not hold in an arbitrary graph: consider the Petersen graph with every edge replaced by two oppositely-oriented directed edges in parallel. The semidirected circuits of an arbitrary graph also do not have the Menger property: consider $K_{3,3}$ with every edge replaced by two oppositely-oriented edges in series. However, we have the following theorem.

THEOREM 6. *Let G be a directed graph.*

(a) *If G contains no directed circuit (i.e., every edge is in a directed coboundary), let* $\mathfrak{F} = \mathfrak{F}(G)$ *consist of the semidirected coboundaries of G.*

(b) *If G contains no directed coboundary (i.e., every edge is in a directed circuit), let* $\mathfrak{F} = \mathfrak{F}(G)$ *consist of the semidirected circuits of G.*

Then $\mathfrak{F}$ *and* $\mathfrak{F}_{\geq 1}$, *and all minors of them, have the Menger property.*

Proof. We shall prove in each case that the minimal blocks of $\mathfrak{F}$ are pairwise disjoint. It immediately follows that $\mathfrak{F}$ and all strong point-restrictions of it have the Menger property, and so do $\mathfrak{F}_{\geq 1}$ and all point-restrictions of it. The rest of the theorem follows from the fact that, for each point-restriction $\mathfrak{F}'$ of $\mathfrak{F}$, $\mathfrak{F}' = \mathfrak{F}(G')$ for some graph G' obtained from G by deleting edges in cases (a) and contracting edges in case (b), using the observation made in §2 that $(\mathfrak{F}\backslash S')_{\geq 1} = \mathfrak{F}_{\geq 1}/S'$ for each $S' \subseteq S$.

(a) If $a, b \in V(G)$, let $E(a, b)$ be the set of all edges joining a to b (at most one, unless G has multiple edges). If there is no other directed path in G from a to b, then $E(a, b) = \langle A, B \rangle$, where A contains a and all vertices of G reachable from a by directed paths, and $B := V(G) \backslash A$; so $E(a, b) \in \mathfrak{F}$; and $E(a, b)$ is clearly a *minimal* block of $\mathfrak{F}$. If, on the other hand, there is another directed path in A from a to b, choose a longest such path, with vertices

$$a = a_0 , a_1 , \ldots , a_n = b$$

in order along it; then each of $E(a_0 , a_1) , \ldots , E(a_{n-1} , a_n)$ is a minimal block of $\mathfrak{F}$ (by the previous sentence), and every semidirected couboundary containing any edge of $E(a , b)$ must contain one of these sets; thus no minimal semidirected coboundary contains any edges of $E(a , b)$. It follows that the minimal semidirected coboundaryies are pairwise disjoint, as required.

(b) Let C_1 and C_2 be circuits in G such that C_{1+} and C_{2+} are distinct minimal semidirected circuits that are not disjoint, say $a \in C_{1+} \cap C_{2+}$, and let C be a directed circuit in G containing a . Suppose first that $C_{1+} \subseteq C$ and $C_{2+} \subseteq C$. Since $C_{1+} \neq C_{2+}$, there exist edges b and c in C such that $b \in C_{1+} \cap C_{2+}$, $c \in C_{1+} \setminus C_{2+}$ (without loss of generality), and c is the first edge after b round C that is in either C_{1+} or C_{2+}. Now construct a pseudocircuit P (i.e., possibly with repeated edges and vertices) as follows: start along b, then follow C_2 (in the positive sense) until the first time it hits C outside the segment strictly between b and c, go backwards round C to the terminal vertex (head) of c, and then follow C_1 back to b . Since $C_{2+} \subseteq C$, the only edges of P that are used in their correct sense (forwards) are edges of $C_{1+} \setminus \{c\}$, including b; all other edges are used backwards. Thus P contains a circuit Q such that $b \in Q_+ \subsetneqq C_{1+}$, which contradicts the supposition that C_{1+} was a minimal semidirected circuit. So suppose instead that (without loss of generality) $C_{1+} \nsubseteq C$, say $b \in C_{1+} \setminus C$. Consider the following set H of edges: all edges of $C \setminus \{a\}$ with their correct orientations, all edges of $C_{1+} \setminus \{a\}$ with orientation reversed (thus "doubling up" any edges in $(C \cap C_{1+}) \setminus \{a\})$, and all edges of C_{1-} with their correct orientations (possibly doubling up more edges of C, but in the same sense this time). Every component of H is

Eulerian, and so b is contained in a directed circuit P in H which does not contain a . Restoring to all edges of C_{1+} their correct orientations, we convert P into a circuit Q such that $b \in Q_+ \subsetneqq C_{1+}$, contradictiong the minimality of C_{1+} as before. This completes the proof. ∎

Finally we consider a matroid $\underset{\sim}{M}$ with sets $\underset{\sim}{C}$ of circuits and $\underset{\sim}{D}$ of cocircuits. Gallai [13] proved that if $\underset{\sim}{M}$ is representable over every field (regular) then, for each $x \in S$, the set-systems $\underset{\sim}{C}_x := (C \setminus \{x\} : x \in C \in \underset{\sim}{C})$ and $\underset{\sim}{D}_x := (D \setminus \{x\} : x \in D \in \underset{\sim}{D})$, which are Menger duals, both have the Menger property, and so do all minors of them. Seymour [14] extended this by proving that $\underset{\sim}{C}_x$ and all its minors have the Menger property if and only if M does not have a minor U_4^2 or F* containing x , and $\underset{\sim}{D}_x$ and all its minors have the Menger property if and only if M does not have a minor U_4^2 or F containing x . I do not know whether any of the other examples above have been extended to matroids: most of them involve directed graphs, and so would presumably need some sort of concept of a directed matroid, such as that introduced by Las Vergnas [7]. Perhaps Lucchesi and Younger's axiomatic approach [12], combined with [7], would give a results on matroids.

4. <u>Alternative characterizations</u>. Edmonds and Fulkerson [2 , 3] give an alternative characterization of the Menger dual of a finite set-system $(S, \mathfrak{F})$: it is the uniuqe clutter $\mathfrak{F}_{\geq 1}$ such that, for every function f from S to the real numbers,

$$\min_{B \in \mathfrak{F}} \max_{x \in B} f(x) = \max_{B \in \mathfrak{F}} \min_{x \in B} f(x) .$$

They also show that it is the unique clutter $\mathfrak{F}_{\geq 1}$ such that, for each partition of S into two disjoint sets S_1 and S_2, there is either a block of $\mathfrak{F}$ contained in S_1 or a block of $\mathfrak{F}_{\geq 1}$ contained in S_2, but not both-which is the same as the condition of the previous sentence with the image of f restricted to the set $\{0,1\}$. I know of no analogous characterizations of the König dual.

Lovasz [9, Theorem 2] proves (still in transpose form) that $\mathfrak{F}^T$ is normal, i.e., every point-restriction of $\mathfrak{F}$ has the König property, if and only if every point-restriction $\mathfrak{F}\setminus S'$ of $\mathfrak{F}$ satisfies the inequality $\alpha(\mathfrak{F}\setminus S')\,\mu(\mathfrak{F}\setminus S') \geq |S\setminus S'|$, i.e. $\mu(\mathfrak{F}\setminus S')\mu(\mathfrak{F}_{\leq 1}\setminus S') \geq |S\setminus S'|$. The inequality $\mu(\mathfrak{F})\mu(\mathfrak{F}_{\leq 1}) \geq |S|$ is a very special case of the "max-max" inequality of [5] for antiblocking pairs of clutters, and is an obvious consequence of $\mathfrak{F}$ having the König property. The analogous inequality for blocking pairs of clutters is $\lambda(\mathfrak{F})\ \lambda(\mathfrak{F}_{\geq 1}) < |S|$, where $\lambda(\mathfrak{F})$ denotes the cardinality of the smallest block in $\mathfrak{F}$. This is a very special case of the min-min or length-width inequality of [2, 3, 4], and is an obvious consequence of $\mathfrak{F}$ having the Menger property; but its precise connection with the Menger property is not clear (to me).

It is a consequence of [15] that every point-restriction of $\mathfrak{F}$ has the König property if and only if the function $f\colon \mathcal{P}(S) \to \mathbb{Z}$, defined by

$$f(x) := \rho(\mathfrak{F}\setminus(S\setminus X)) \quad \text{for each } X \subseteq S$$

(so that $f(X)$ is the samllest number of blocks of $\mathfrak{F}$ that cover X), is <u>strongly subadditive</u>, in the sense that $f(X) - \frac{1}{k}\sum_{i\in I} f(X_i)$ for each positive integer k, $X \subseteq S$, and family $(X_i : i \in I)$ of not-necessarily-distinct subsets of X such that $|\{i ; x \in X_i\}| = k$ for each $x \in X$. For every point-restriction of $\mathfrak{F}$ has the König property if and only if f is the rank function of $\mathfrak{F}_{\leq 1}$, and in this case it is certainly strongly subadditive; and if it is strongly subadditive, then [15, Theorem 1] shows that it is the rank function of a subclusive set-system, which it is easy to see must be $\mathfrak{F}_{\leq 1}$.

In a similar way, every strong point-restriction of $\mathfrak{F}$ has the Menger property if and only if the function $g: \mathcal{P}(S) \to \mathbb{Z}$, defined by

$$g(X) := \nu(\mathfrak{F}'(S \setminus X)) \quad \text{for each} \quad X \subseteq S$$

(so that $g(X)$ is the largest number of disjoint blocks of $\mathfrak{F}$ that are contained in X), is strongly subadditive. For every strong point-restriction of $\mathfrak{F}$ has the Menger property if and only if g is the rank function of $\mathfrak{F}_{\geq 1}$.

References.

1. C. Berge, Graphs and Hypergraphs. North-Holland, Amsterdam (1973).

2. J. Edmonds and D. R. Fulkerson, Bottleneck extrema. J. Combinatorial Theory 8 (1970), 299-306.

3. D. R. Fulkerson, Networks, frames, blocking systems. Mathematics of the Decision Sciences Part I (G. B. Dantzig and A. F. Veinott, eds). American Mathematics Society Lectures in Applied Mathematics 11 (1968), 303-334.

4. D. R. Fulkerson, Blocking polyhedra. Graph Theory and its Applications (B. Harris, ed.). Academic Press, New York (1970), 93-112.

5. D. R. Fulkerson, Antiblocking polyhedra. J. Combinatorial Theory Ser. B 12 (1972), 50-71.

6. R. P. Gupta, A decomposition theorem for bipartite graphs. Theory of Graphs. (P. Rosenstiehl, ed.). Gordon and Breach, New York and Dunod, Paris (1967), 135-138.

7. L. Las Vergnas, Matroides orientables. Duplicated typescript. 1974.

8. L. Lovasz, Normal hypergraphs and the perfect graph conjecture. Discrete Math. 2 (1972), 253-267.

9. L. Lovasz, Minimax theorems for hypergraphs. Hypergraph Seminar (A. Dold and B. Eckmann, eds.) Springer Lecture Notes in Mathematics 211 (1974), 111-126.

10. B. Rathschild and A. Whinston, On two commodity network flows. Operations Reserach 14 (1966), 377-387.

11. L. Lovasz, Two minimax theorems in graph theory. J. Combinatorial Theory Ser. B, to appear.

12. C. L. Lucchesi and D. H. Younger, A minimax theorem for directed graphs. To appear.

13. T. Gallai, Über reguläre Kettengruppen. Acta Math. Acad. Sci. Hungar 10 (1959), 227-240.

14. P. D. Seymour, The matroids with the max-flow min-cut property. J. Combinatorial Theory Ser. B, to appear.

15. D. R. Woodall, A note on rank functions and integer programming. Discrete Math., submitted (1974).

16. D. R. Woodall, Applications of polymatroids and linear programming to transversals and graphs. Combinatorics (T. P. McDonough and V. C. Mavron, eds). London Mathematicsl Society Lecture Note Series 13, Cambridge University Press (1974), 195-200.

17. D. R. Woodall, Applications of polymatroids and linear programming to transversals and graphs. Unpublished preprint (contains the proofs of results in [15]).

18. K. Vidyasankar and D. H. Younger, A minimax equality related to the longest directed path in an acyclic graph. Canad. J. Math. 27 (1975), 348-351.

19. J. Edmonds, Edgepdisjoint branchings. Combinatorial Plgonthoms (R. Rustin, ed.). Algorithmic Press, New York (1973), 91-96.

20. D. R. Fulkerson, Packing weighted cuts in rooted directed graphs, Math. Programming 6 (1974), 1-13.

Vol. 489: J. Bair and R. Fourneau, Etude Géométrique des Espaces Vectoriels. Une Introduction. VII, 185 pages. 1975.

Vol. 490: The Geometry of Metric and Linear Spaces. Proceedings 1974. Edited by L. M. Kelly. X, 244 pages. 1975.

Vol. 491: K. A. Broughan, Invariants for Real-Generated Uniform Topological and Algebraic Categories. X, 197 pages. 1975.

Vol. 492: Infinitary Logic: In Memoriam Carol Karp. Edited by D. W. Kueker. VI, 206 pages. 1975.

Vol. 493: F. W. Kamber and P. Tondeur, Foliated Bundles and Characteristic Classes. XIII, 208 pages. 1975.

Vol. 494: A Cornea and G. Licea. Order and Potential Resolvent Families of Kernels. IV, 154 pages. 1975.

Vol. 495: A. Kerber, Representations of Permutation Groups II. V, 175 pages. 1975.

Vol. 496: L. H. Hodgkin and V. P. Snaith, Topics in K-Theory. Two Independent Contributions. III, 294 pages. 1975.

Vol. 497: Analyse Harmonique sur les Groupes de Lie. Proceedings 1973–75. Edité par P. Eymard et al. VI, 710 pages. 1975.

Vol. 498: Model Theory and Algebra. A Memorial Tribute to Abraham Robinson. Edited by D. H. Saracino and V. B. Weispfenning. X, 463 pages. 1975.

Vol. 499: Logic Conference, Kiel 1974. Proceedings. Edited by G. H. Müller, A. Oberschelp, and K. Potthoff. V, 651 pages 1975.

Vol. 500: Proof Theory Symposion, Kiel 1974. Proceedings. Edited by J. Diller and G. H. Müller. VIII, 383 pages. 1975.

Vol. 501: Spline Functions, Karlsruhe 1975. Proceedings. Edited by K. Böhmer, G. Meinardus, and W. Schempp. VI, 421 pages. 1976.

Vol. 502: János Galambos, Representations of Real Numbers by Infinite Series. VI, 146 pages. 1976.

Vol. 503: Applications of Methods of Functional Analysis to Problems in Mechanics. Proceedings 1975. Edited by P. Germain and B. Nayroles. XIX, 531 pages. 1976.

Vol. 504: S. Lang and H. F. Trotter, Frobenius Distributions in GL_2-Extensions. III, 274 pages. 1976.

Vol. 505: Advances in Complex Function Theory. Proceedings 1973/74. Edited by W. E. Kirwan and L. Zalcman. VIII, 203 pages. 1976.

Vol. 506: Numerical Analysis, Dundee 1975. Proceedings. Edited by G. A. Watson. X, 201 pages. 1976.

Vol. 507: M. C. Reed, Abstract Non-Linear Wave Equations. VI, 128 pages. 1976.

Vol. 508: E. Seneta, Regularly Varying Functions. V, 112 pages. 1976.

Vol. 509: D. E. Blair, Contact Manifolds in Riemannian Geometry. VI, 146 pages. 1976.

Vol. 510: V. Poènaru, Singularités C^{∞} en Présence de Symétrie. V, 174 pages. 1976.

Vol. 511: Séminaire de Probabilités X. Proceedings 1974/75. Edité par P. A. Meyer. VI, 593 pages. 1976.

Vol. 512: Spaces of Analytic Functions, Kristiansand, Norway 1975. Proceedings. Edited by O. B. Bekken, B. K. Øksendal, and A. Stray. VIII, 204 pages. 1976.

Vol. 513: R. B. Warfield, Jr. Nilpotent Groups. VIII, 115 pages. 1976.

Vol. 514: Séminaire Bourbaki vol. 1974/75. Exposés 453 – 470. IV, 276 pages. 1976.

Vol. 515: Bäcklund Transformations. Nashville, Tennessee 1974. Proceedings. Edited by R. M. Miura. VIII, 295 pages. 1976.

Vol. 516: M. L. Silverstein, Boundary Theory for Symmetric Markov Processes. XVI, 314 pages. 1976.

Vol. 517: S. Glasner, Proximal Flows. VIII, 153 pages. 1976.

Vol. 518: Séminaire de Théorie du Potentiel, Proceedings Paris 1972–1974. Edité par F. Hirsch et G. Mokobodzki. VI, 275 pages. 1976.

Vol. 519: J. Schmets, Espaces de Fonctions Continues. XII, 150 pages. 1976.

Vol. 520: R. H. Farrell, Techniques of Multivariate Calculation. X, 337 pages. 1976.

Vol. 521: G. Cherlin, Model Theoretic Algebra – Selected Topics. IV, 234 pages. 1976.

Vol. 522: C. O. Bloom and N. D. Kazarinoff, Short Wave Radiation Problems in Inhomogeneous Media: Asymptotic Solutions. V. 104 pages. 1976.

Vol. 523: S. A. Albeverio and R. J. Høegh-Krohn, Mathematical Theory of Feynman Path Integrals. IV, 139 pages. 1976.

Vol. 524: Séminaire Pierre Lelong (Analyse) Année 1974/75. Edité par P. Lelong. V, 222 pages. 1976.

Vol. 525: Structural Stability, the Theory of Catastrophes, and Applications in the Sciences. Proceedings 1975. Edited by P. Hilton. VI, 408 pages. 1976.

Vol. 526: Probability in Banach Spaces. Proceedings 1975. Edited by A. Beck. VI, 290 pages. 1976.

Vol. 527: M. Denker, Ch. Grillenberger, and K. Sigmund, Ergodic Theory on Compact Spaces. IV, 360 pages. 1976.

Vol. 528: J. E. Humphreys, Ordinary and Modular Representations of Chevalley Groups. III, 127 pages. 1976.

Vol. 529: J. Grandell, Doubly Stochastic Poisson Processes. X, 234 pages. 1976.

Vol. 530: S. S. Gelbart, Weil's Representation and the Spectrum of the Metaplectic Group. VII, 140 pages. 1976.

Vol. 531: Y.-C. Wong, The Topology of Uniform Convergence on Order-Bounded Sets. VI, 163 pages. 1976.

Vol. 532: Théorie Ergodique. Proceedings 1973/1974. Edité par J.-P. Conze and M. S. Keane. VIII, 227 pages. 1976.

Vol. 533: F. R. Cohen, T. J. Lada, and J. P. May, The Homology of Iterated Loop Spaces. IX, 490 pages. 1976.

Vol. 534: C. Preston, Random Fields. V, 200 pages. 1976.

Vol. 535: Singularités d'Applications Differentiables. Plans-sur-Bex. 1975. Edité par O. Burlet et F. Ronga. V, 253 pages. 1976.

Vol. 536: W. M. Schmidt, Equations over Finite Fields. An Elementary Approach. IX, 267 pages. 1976.

Vol. 537: Set Theory and Hierarchy Theory. Bierutowice, Poland 1975. A Memorial Tribute to Andrzej Mostowski. Edited by W. Marek, M. Srebrny and A. Zarach. XIII, 345 pages. 1976.

Vol. 538: G. Fischer, Complex Analytic Geometry. VII, 201 pages. 1976.

Vol. 539: A. Badrikian, J. F. C. Kingman et J. Kuelbs, Ecole d'Eté de Probabilités de Saint Flour V-1975. Edité par P.-L. Hennequin. IX, 314 pages. 1976.

Vol. 540: Categorical Topology, Proceedings 1975. Edited by E. Binz and H. Herrlich. XV, 719 pages. 1976.

Vol. 541: Measure Theory, Oberwolfach 1975. Proceedings. Edited by A. Bellow and D. Kölzow. XIV, 430 pages. 1976.

Vol. 542: D. A. Edwards and H. M. Hastings, Čech and Steenrod Homotopy Theories with Applications to Geometric Topology. VII, 296 pages. 1976.

Vol. 543: Nonlinear Operators and the Calculus of Variations, Bruxelles 1975. Edited by J. P. Gossez, E. J. Lami Dozo, J. Mawhin, and L. Waelbroeck, VII, 237 pages. 1976.

Vol. 544: Robert P. Langlands, On the Functional Equations Satisfied by Eisenstein Series. VII, 337 pages. 1976.

Vol. 545: Noncommutative Ring Theory. Kent State 1975. Edited by J. H. Cozzens and F. L. Sandomierski. V, 212 pages. 1976.

Vol. 546: K. Mahler, Lectures on Transcendental Numbers. Edited and Completed by B. Diviš and W. J. Le Veque. XXI, 254 pages. 1976.

Vol. 547: A. Mukherjea and N. A. Tserpes, Measures on Topological Semigroups: Convolution Products and Random Walks. V, 197 pages. 1976.

Vol. 548: D. A. Hejhal, The Selberg Trace Formula for PSL (2, ℝ). Volume I. VI, 516 pages. 1976.

Vol. 549: Brauer Groups, Evanston 1975. Proceedings. Edited by D. Zelinsky. V, 187 pages. 1976.

Vol. 550: Proceedings of the Third Japan – USSR Symposium on Probability Theory. Edited by G. Maruyama and J. V. Prokhorov. VI, 722 pages. 1976.

Vol. 551: Algebraic K-Theory, Evanston 1976. Proceedings. Edited by M. R. Stein. XI, 409 pages. 1976.

Vol. 552: C. G. Gibson, K. Wirthmüller, A. A. du Plessis and E. J. N. Looijenga. Topological Stability of Smooth Mappings. V, 155 pages. 1976.

Vol. 553: M. Petrich, Categories of Algebraic Systems. Vector and Projective Spaces, Semigroups, Rings and Lattices. VIII, 217 pages. 1976.

Vol. 554: J. D. H. Smith, Mal'cev Varieties. VIII, 158 pages. 1976.

Vol. 555: M. Ishida, The Genus Fields of Algebraic Number Fields. VII, 116 pages. 1976.

Vol. 556: Approximation Theory. Bonn 1976. Proceedings. Edited by R. Schaback and K. Scherer. VII, 466 pages. 1976.

Vol. 557: W. Iberkleid and T. Petrie, Smooth S^1 Manifolds. III, 163 pages. 1976.

Vol. 558: B. Weisfeiler, On Construction and Identification of Graphs. XIV, 237 pages. 1976.

Vol. 559: J.-P. Caubet, Le Mouvement Brownien Relativiste. IX, 212 pages. 1976.

Vol. 560: Combinatorial Mathematics, IV, Proceedings 1975. Edited by L. R. A. Casse and W. D. Wallis. VII, 249 pages. 1976.

Vol. 561: Function Theoretic Methods for Partial Differential Equations. Darmstadt 1976. Proceedings. Edited by V. E. Meister, N. Weck and W. L. Wendland. XVIII, 520 pages. 1976.

Vol. 562: R. W. Goodman, Nilpotent Lie Groups: Structure and Applications to Analysis. X, 210 pages. 1976.

Vol. 563: Séminaire de Théorie du Potentiel. Paris, No. 2. Proceedings 1975–1976. Edited by F. Hirsch and G. Mokobodzki. VI, 292 pages. 1976.

Vol. 564: Ordinary and Partial Differential Equations, Dundee 1976. Proceedings. Edited by W. N. Everitt and B. D. Sleeman. XVIII, 551 pages. 1976.

Vol. 565: Turbulence and Navier Stokes Equations. Proceedings 1975. Edited by R. Temam. IX, 194 pages. 1976.

Vol. 566: Empirical Distributions and Processes. Oberwolfach 1976. Proceedings. Edited by P. Gaenssler and P. Révész. VII, 146 pages. 1976.

Vol. 567: Séminaire Bourbaki vol. 1975/76. Exposés 471–488. IV, 303 pages. 1977.

Vol. 568: R. E. Gaines and J. L. Mawhin, Coincidence Degree, and Nonlinear Differential Equations. V, 262 pages. 1977.

Vol. 569: Cohomologie Etale SGA 4½. Séminaire de Géométrie Algébrique du Bois-Marie. Edité par P. Deligne. V, 312 pages. 1977.

Vol. 570: Differential Geometrical Methods in Mathematical Physics, Bonn 1975. Proceedings. Edited by K. Bleuler and A. Reetz. VIII, 576 pages. 1977.

Vol. 571: Constructive Theory of Functions of Several Variables, Oberwolfach 1976. Proceedings. Edited by W. Schempp and K. Zeller. VI, 290 pages. 1977

Vol. 572: Sparse Matrix Techniques, Copenhagen 1976. Edited by V. A. Barker. V, 184 pages. 1977.

Vol. 573: Group Theory, Canberra 1975. Proceedings. Edited by R. A. Bryce, J. Cossey and M. F. Newman. VII, 146 pages. 1977.

Vol. 574: J. Moldestad, Computations in Higher Types. IV, 203 pages. 1977.

Vol. 575: K-Theory and Operator Algebras, Athens, Georgia 1975. Edited by B. B. Morrel and I. M. Singer. VI, 191 pages. 1977.

Vol. 576: V. S. Varadarajan, Harmonic Analysis on Real Reductive Groups. VI, 521 pages. 1977.

Vol. 577: J. P. May, E_∞ Ring Spaces and E_∞ Ring Spectra. IV, 268 pages. 1977.

Vol. 578: Séminaire Pierre Lelong (Analyse) Année 1975/76. Edité par P. Lelong. VI, 327 pages. 1977.

Vol. 579: Combinatoire et Représentation du Groupe Symétrique, Strasbourg 1976. Proceedings 1976. Edité par D. Foata. IV, 339 pages. 1977.

Vol. 580: C. Castaing and M. Valadier, Convex Analysis and Measurable Multifunctions. VIII, 278 pages. 1977.

Vol. 581: Séminaire de Probabilités XI, Université de Strasbourg. Proceedings 1975/1976. Edité par C. Dellacherie, P. A. Meyer et M. Weil. VI, 574 pages. 1977.

Vol. 582: J. M. G. Fell, Induced Representations and Banach *-Algebraic Bundles. IV, 349 pages. 1977.

Vol. 583: W. Hirsch, C. C. Pugh and M. Shub, Invariant Manifolds. IV, 149 pages. 1977.

Vol. 584: C. Brezinski, Accélération de la Convergence en Analyse Numérique. IV, 313 pages. 1977.

Vol. 585: T. A. Springer, Invariant Theory. VI, 112 pages. 1977.

Vol. 586: Séminaire d'Algèbre Paul Dubreil, Paris 1975–1976 (29ème Année). Edited by M. P. Malliavin. VI, 188 pages. 1977.

Vol. 587: Non-Commutative Harmonic Analysis. Proceedings 1976. Edited by J. Carmona and M. Vergne. IV, 240 pages. 1977.

Vol. 588: P. Molino, Théorie des G-Structures: Le Problème d'Equivalence. VI, 163 pages. 1977.

Vol. 589: Cohomologie l-adique et Fonctions L. Séminaire de Géométrie Algébrique du Bois-Marie 1965–66, SGA 5. Edité par L. Illusie. XII, 484 pages. 1977.

Vol. 590: H. Matsumoto, Analyse Harmonique dans les Systèmes de Tits Bornologiques de Type Affine. IV, 219 pages. 1977.

Vol. 591: G. A. Anderson, Surgery with Coefficients. VIII, 157 pages. 1977.

Vol. 592: D. Voigt, Induzierte Darstellungen in der Theorie der endlichen, algebraischen Gruppen. V, 413 Seiten. 1977.

Vol. 593: K. Barbey and H. König, Abstract Analytic Function Theory and Hardy Algebras. VIII, 260 pages. 1977.

Vol. 594: Singular Perturbations and Boundary Layer Theory, Lyon 1976. Edited by C. M. Brauner, B. Gay, and J. Mathieu. VIII, 539 pages. 1977.

Vol. 595: W. Hazod, Stetige Faltungshalbgruppen von Wahrscheinlichkeitsmaßen und erzeugende Distributionen. XIII, 157 Seiten. 1977.

Vol. 596: K. Deimling, Ordinary Differential Equations in Banach Spaces. VI, 137 pages. 1977.

Vol. 597: Geometry and Topology, Rio de Janeiro, July 1976. Proceedings. Edited by J. Palis and M. do Carmo. VI, 866 pages. 1977.

Vol. 598: J. Hoffmann-Jørgensen, T. M. Liggett et J. Neveu, Ecole d'Eté de Probabilités de Saint-Flour VI – 1976. Edité par P.-L. Hennequin. XII, 447 pages. 1977.

Vol. 599: Complex Analysis, Kentucky 1976. Proceedings. Edited by J. D. Buckholtz and T. J. Suffridge. X, 159 pages. 1977.

Vol. 600: W. Stoll, Value Distribution on Parabolic Spaces. VIII, 216 pages. 1977.

Vol. 601: Modular Functions of one Variable V, Bonn 1976. Proceedings. Edited by J.-P. Serre and D. B. Zagier. VI, 294 pages. 1977.

Vol. 602: J. P. Brezin, Harmonic Analysis on Compact Solvmanifolds. VIII, 179 pages. 1977.

Vol. 603: B. Moishezon, Complex Surfaces and Connected Sums of Complex Projective Planes. IV, 234 pages. 1977.

Vol. 604: Banach Spaces of Analytic Functions, Kent, Ohio 1976. Proceedings. Edited by J. Baker, C. Cleaver and Joseph Diestel. VI, 141 pages. 1977.

Vol. 605: Sario et al., Classification Theory of Riemannian Manifolds. XX, 498 pages. 1977.

Vol. 606: Mathematical Aspects of Finite Element Methods. Proceedings 1975. Edited by I. Galligani and E. Magenes. VI, 362 pages. 1977.

Vol. 607: M. Métivier, Reelle und Vektorwertige Quasimartingale und die Theorie der Stochastischen Integration. X, 310 Seiten. 1977.

Vol. 608: Bigard et al., Groupes et Anneaux Réticulés. XIV, 334 pages. 1977.